Urban Deformation Monitoring using Persistent Scatterer Interferometry and SAR tomography

Urban Deformation Monitoring using Persistent Scatterer Interferometry and SAR tomography

Special Issue Editors

Michele Crosetto
Oriol Monserrat
Alessandra Budillon

MDPI • Basel • Beijing • Wuhan • Barcelona • Belgrade

MDPI

Special Issue Editors
Michele Crosetto
Centre Tecnològic de Telecomunicacions
de Catalunya (CTTC)
Spain

Oriol Monserrat
Centre Tecnològic de Telecomunicacions
de Catalunya (CTTC)
Spain

Alessandra Budillon
Universita' degli studi di Napoli Parthenope
Ialy

Editorial Office
MDPI
St. Alban-Anlage 66
4052 Basel, Switzerland

This is a reprint of articles from the Special Issue published online in the open access journal *Remote Sensing* (ISSN 2072-4292) from 2018 to 2019 (available at: https://www.mdpi.com/journal/remotesensing/special_issues/PSI_tomoSAR)

For citation purposes, cite each article independently as indicated on the article page online and as indicated below:

LastName, A.A.; LastName, B.B.; LastName, C.C. Article Title. *Journal Name* **Year**, *Article Number*, Page Range.

ISBN 978-3-03921-126-5 (Pbk)
ISBN 978-3-03921-127-2 (PDF)

Contents

About the Special Issue Editors

Michele Crosetto holds a civil engineering degree from the Politecnico di Torino (1993) and a doctorate in Topographic and Geodesic Sciences from the Politecnico di Milano (1998). He specialized in Geodesy, Photogrammetry and GIS in Lausanne (EPFL) and Zurich (ETHZ) from 1993 to 1995. He has worked in the Joint Research Centre of the European Commission in Ispra, Italy (January 1999–July 2000) and as a researcher at the Cartographic Institute of Catalonia. He has been a member of the Institute of Geomatics since 2002. Since January 2014 he has worked with CTTC, where he is now Head of the Geomatics Division. His main research activities are related to the analysis of spaceborne, airborne and ground-based remote sensing data and the development of scientific and technical applications using active sensor types, such as Synthetic Aperture Radar (SAR), Real Aperture Radar (RAR) and laser scanners. In recent years he has been involved in a number of projects of the Fifth, Sixth and Seventh and H2020 Framework Programmes of the EU. In addition, he has been involved in different projects funded by the European Space Agency.

Oriol Monserrat holds a PhD in aerospace science and technology from the Polytechnic University of Catalonia (2012) and a degree in mathematics from the University of Barcelona (2004). In 2003 he started working as a researcher in the Active Remote Sensing Unit of the Geomatics Institute. Since January 2014 he has worked as Head of the Remote Sensing Department of the Division of Geomatics at the Technological Centre of Telecommunications of Catalunya. His research activities are related to the analysis of satellite, airborne and terrestrial remote sensing data and the development of scientific and technical applications using mainly active sensors, such as Synthetic Aperture Radar (SAR), Real Aperture Radar (RAR) and laser scanners. From the point of view of applications, Dr. Monserrat is specialized in the measurement and monitoring of deformations using SAR interferometry techniques (InSAR). In his research career he has participated in different projects (most of them related to geohazards) of the Sixth and Seventh Framework Programmes of the EU (Galahad, SubCoast, PanGeo and Aphorism) as well as H2020 (HEIMDALL, GIMS). He has also participated in outstanding projects funded by the European Space Agency. He has been the coordinator of the SAFETY and U-Geohaz projects.

Alessandra Budillon received her "Laurea" degree (cum laude) in Electronic Engineering in 1996, and earned a PhD in Electronic Engineering and Computer Science in 1999 at the Università degli Studi di Napoli Federico II, Naples, Italy. From January to July 1998 she carried out, within her PhD course of study, research activity at the Brain and Cognitive Sciences Department, MIT, Boston, USA. In February 2001 she became assistant professor of Telecommunication at the Department of Information Engineering at the Seconda Università degli Studi di Napoli, Aversa, Italy. In November 2004 she moved to the Department of Engineering at the Università degli Studi di Napoli Parthenope. Her main scientific interests have been focused on Statistical Signal Processing, with applications in data and signals compression and coding, and on remote sensing, with applications in Synthetic Aperture Radar (SAR) processing, SAR interferometry and tomography. She has papers published in international journals, she has attended several national and international conferences and she acts as a referee for several international journals.

remote sensing

MDPI

Editorial

Editorial for the Special Issue "Urban Deformation Monitoring using Persistent Scatterer Interferometry and SAR Tomography"

Alessandra Budillon [1],*, **Michele Crosetto [2] and Oriol Monserrat [2]**

[1] Engineering Department, Universita' degli studi di Napoli Parthenope, Centro Direzionale, Isola C4, 80143 Napoli, Italy
[2] Centre Tecnològic de Telecomunicacions de Catalunya (CTTC), Remote Sensing Department, Division of Geomatics, Av. Gauss, 7 E-08860 Castelldefels, Spain; mcrosetto@cttc.cat (M.C.); omonserrat@cttc.cat (O.M.)
*   Correspondence: alessandra.budillon@uniparthenope.it; Tel.: +39-081-54-76-725

Received: 29 May 2019; Accepted: 31 May 2019; Published: 31 May 2019

Abstract: This Special Issue hosts papers related to deformation monitoring in urban areas based on two main techniques: Persistent Scatterer Interferometry (PSI) and Synthetic Aperture Radar (SAR) Tomography (TomoSAR). Several contributions highlight the capabilities of Interferometric SAR (InSAR) and PSI techniques for urban deformation monitoring. In this Special Issue, a wide range of InSAR and PSI applications are addressed. Some contributions show the advantages of TomoSAR in un-mixing multiple scatterers for urban mapping and monitoring. This issue includes a contribution that compares PSI and TomoSAR and another one that uses polarimetric data for TomoSAR.

Keywords: synthetic aperture radar; persistent scatterers; tomography; differential interferometry; polarimetry; radar detection; urban areas; deformation

---

Our capability to monitor deformation using satellite-based Synthetic Aperture Radar (SAR) sensors has increased substantially in recent years, thanks to the availability of multiple SAR sensors and the development of several data processing and analysis procedures. Differential interferometric SAR (DInSAR) [1] and Persistent Scatterer Interferometry (PSI) [2] involve the exploitation of at least a pair of complex SAR images to measure surface deformation. Both the DInSAR and PSI techniques exploit the phase of the SAR images. Most of the InSAR and PSI techniques assume the presence of only one dominant scatterer per resolution cell [3,4]. This assumption cannot be valid when observing ground scenes with a pronounced extension in the elevation direction for which more than one scatterer can fall in the same range-azimuth resolution cell. This potential limitation can be overcome by using SAR tomography (TomoSAR) techniques [5]. In fact, in such techniques, the use of a stack of complex-valued interferometric images makes it possible to separate the scatterers interfering within the same range-azimuth resolution cell [6,7]. This Special Issue is focused on deformation monitoring in urban areas based on PSI and TomoSAR. It collects the latest innovative research results related to these two techniques. These published papers show the capability of both techniques in mapping and monitoring urban areas.

The papers related to PSI describe methodological and application-oriented research work. In reference [8], the authors assess the deformations associated with the construction of a new metro tunnel. In reference [9], PSI results are used as a key input for geological and geomorphological analyses in urban areas. In reference [10], the subsidence phenomena over an entire metropolitan area (Rome) are studied using Sentinel-1 data and open source tools. In reference [11], the applicability for urban monitoring of pursuit monostatic data from the very high-resolution TanDEM-X mission is addressed. A new PSI procedure is described in reference [12], which is used to monitor the land deformation in an urban area induced by aquifer dewatering. The most original part of this work includes the estimation of the atmospheric phase component using stable areas located in the vicinity of the monitoring area. In reference [13], the observations coming from PSI are used to contribute to the assessment of the health state of two bridges. The use of PSI to study the long-term land deformation patterns in earthquake-prone areas is addressed in reference [14]. A methodology to exploit PSI time series from Sentinel-1 data for the detection and characterization of uplift phenomena in urban areas is described in reference [15]. In reference [16], PSI is used to identify and measure ground deformations in urban areas to determine the vulnerable parts of the cities that are prone to geohazards. In reference [17], the authors address the use of PSI data to study the pattern of temporal evolution in reclamation settlements. Finally, in reference [18], the authors study the wide-area surface subsidence characteristics of a large metropolitan area (Wuhan) using Sentinel-1 data.

In an urban environment, one of the most important tasks is to resolve layover, which causes multiple coherent scatterers to be mapped in the same range-azimuth image cell. In references [19–22] the use of tomographic techniques that synthesize apertures along the elevation direction exploiting a stack of SAR images, allows the separation of the scatterers interfering within the same range-azimuth cell. In particular, in reference [19], the detection strategy for multiple scatters is reported in the context of "tomography as an add-on to PSI", i.e., tomographic analysis is subsequent to a prior PSI processing. The paper also highlights that while the instabilities in phase are typically modeled as additive noise, their impact on tomography is multiplicative in nature. In reference [20], a Generalized Likelihood Ratio Test (GLRT) with the use of multi-look is proposed to separate multiple scatterers and shows tangible improvements in the detection of single and double interfering persistent scatterers at the expense of a minor spatial resolution loss. In reference [21], an inter-comparison of the results from PSI and TomoSAR is carried out on Sentinel-1 data. The analysis of the parameters estimated by the two techniques allows us to achieve a level of precision comparable to other studies. The paper also addresses the complementarity of the two techniques, and in particular, it assesses the increase of measurement density that can be achieved by adding the double scatterers from SAR tomography to the Persistent Scatterer Interferometry measurements. Finally, in reference [22], the use of polarimetric channels in TomoSAR is explored. This paper shows that using a GLRT approach and dual pol data is possible to reduce the number of baselines required to achieve a given scatterer detection performance.

Author Contributions: The authors contributed equally to all aspects of this editorial.

Acknowledgments: The authors would like to thank the authors who contributed to this Special Issue and to the reviewers who dedicated their time to providing the authors with valuable and constructive recommendations.

Conflicts of Interest: "The authors declare no conflict of interest."

References

1. Gabriel, A.K.; Goldstein, R.M.; Zebker, H.A. Mapping small elevation changes over large areas: Differential radar interferometry. *J. Geophys. Res.* **1989**, *94*, 9183–9191. [CrossRef]
2. Ferretti, A.; Prati, C.; Rocca, F. Nonlinear subsidence rate estimation using permanent scatterers in differential SAR interferometry. *IEEE Trans. Geosci. Remote Sens.* **2000**, *38*, 2202–2212. [CrossRef]
3. Gernhardt, S.; Adam, N.; Eineder, M.; Bamler, R. Potential of very high resolution SAR for persistent scatterer interferometry in urban areas. *Ann. GIS* **2010**, *16*, 103–111. [CrossRef]
4. Crosetto, M.; Monserrat, O.; Iglesias, R.; Crippa, B. Persistent scatterer interferometry: Potential, limits and initial C- and X-band comparison. *Photogramm. Eng. Remote Sens.* **2010**, *76*, 1061–1069. [CrossRef]
5. Reigber, A.; Moreira, A. First Demonstration of Airborne SAR Tomography Using Multibaseline L-band Data. *IEEE Trans. Geosci. Remote Sens.* **2000**, *38*, 2142–2152. [CrossRef]
6. Budillon, A.; Johnsy, A.; Schirinzi, G. Extension of a fast GLRT algorithm to 5D SAR tomography of Urban areas. *Remote Sens.* **2017**, *9*, 844. [CrossRef]
7. Budillon, A.; Ferraioli, G.; Schirinzi, G. Localization Performance of Multiple Scatterers in Compressive Sampling SAR Tomography: Results on COSMO-SkyMed Data. *IEEE J. Sel. Top. App. Earth Obs. Remote Sens.* **2014**, *7*, 2902–2910. [CrossRef]

8. Khorrami, M.; Alizadeh, B.; Ghasemi Tousi, E.; Shakerian, M.; Maghsoudi, Y.; Rahgozar, P. How Groundwater Level Fluctuations and Geotechnical Properties Lead to Asymmetric Subsidence: A PSInSAR Analysis of Land Deformation over a Transit Corridor in the Los Angeles Metropolitan Area. *Remote Sens.* **2019**, *11*, 377. [CrossRef]

9. Floris, M.; Fontana, A.; Tessari, G.; Mulè, M. Subsidence Zonation Through Satellite Interferometry in Coastal Plain Environments of NE Italy: A Possible Tool for Geological and Geomorphological Mapping in Urban Areas. *Remote Sens.* **2019**, *11*, 165. [CrossRef]

10. Delgado Blasco, J.M.; Foumelis, M.; Stewart, C.; Hooper, A. Measuring Urban Subsidence in the Rome Metropolitan Area (Italy) with Sentinel-1 SNAP-StaMPS Persistent Scatterer Interferometry. *Remote Sens.* **2019**, *11*, 129. [CrossRef]

11. Wang, Z.; Balz, T.; Zhang, L.; Perissin, D.; Liao, M. Using TSX/TDX Pursuit Monostatic SAR Stacks for PS-InSAR Analysis in Urban Areas. *Remote Sens.* **2019**, *11*, 26. [CrossRef]

12. Crosetto, M.; Devanthéry, N.; Monserrat, O.; Barra, A.; Cuevas-González, M.; Mróz, M.; Botey-Bassols, J.; Vázquez-Suñé, E.; Crippa, B. A Persistent Scatterer Interferometry Procedure Based on Stable Areas to Filter the Atmospheric Component. *Remote Sens.* **2018**, *10*, 1780. [CrossRef]

13. Huang, Q.; Monserrat, O.; Crosetto, M.; Crippa, B.; Wang, Y.; Jiang, J.; Ding, Y. Displacement Monitoring and Health Evaluation of Two Bridges Using Sentinel-1 SAR Images. *Remote Sens.* **2018**, *10*, 1714. [CrossRef]

14. Aimaiti, Y.; Yamazaki, F.; Liu, W. Multi-Sensor InSAR Analysis of Progressive Land Subsidence over the Coastal City of Urayasu, Japan. *Remote Sens.* **2018**, *10*, 1304. [CrossRef]

15. Bonì, R.; Bosino, A.; Meisina, C.; Novellino, A.; Bateson, L.; McCormack, H. A Methodology to Detect and Characterize Uplift Phenomena in Urban Areas Using Sentinel-1 Data. *Remote Sens.* **2018**, *10*, 607. [CrossRef]

16. Aslan, G.; Cakır, Z.; Ergintav, S.; Lasserre, C.; Renard, F. Analysis of Secular Ground Motions in Istanbul from a Long-Term InSAR Time-Series (1992–2017). *Remote Sens.* **2018**, *10*, 408. [CrossRef]

17. Yang, M.; Yang, T.; Zhang, L.; Lin, J.; Qin, X.; Liao, M. Spatio-Temporal Characterization of a Reclamation Settlement in the Shanghai Coastal Area with Time Series Analyses of X-, C-, and L-Band SAR Datasets. *Remote Sens.* **2018**, *10*, 329. [CrossRef]

18. Zhou, L.; Guo, J.; Hu, J.; Li, J.; Xu, Y.; Pan, Y.; Shi, M. Wuhan Surface Subsidence Analysis in 2015–2016 Based on Sentinel-1A Data by SBAS-InSAR. *Remote Sens.* **2017**, *9*, 982. [CrossRef]

19. Siddique, M.A.; Wegmüller, U.; Hajnsek, I.; Frey, O. SAR Tomography as an Add-On to PSI: Detection of Coherent Scatterers in the Presence of Phase Instabilities. *Remote Sens.* **2018**, *10*, 1014. [CrossRef]

20. Dănișor, C.; Fornaro, G.; Pauciullo, A.; Reale, D.; Datcu, M. Super-Resolution Multi-Look Detection in SAR Tomography. *Remote Sens.* **2018**, *10*, 1894. [CrossRef]

21. Budillon, A.; Crosetto, M.; Johnsy, A.C.; Monserrat, O.; Krishnakumar, V.; Schirinzi, G. Comparison of Persistent Scatterer Interferometry and SAR Tomography Using Sentinel-1 in Urban Environment. *Remote Sens.* **2018**, *10*, 1986. [CrossRef]

22. Budillon, A.; Johnsy, A.C.; Schirinzi, G. Urban Tomographic Imaging Using Polarimetric SAR Data. *Remote Sens.* **2019**, *11*, 132. [CrossRef]

remote sensing

MDPI

Article

How Groundwater Level Fluctuations and Geotechnical Properties Lead to Asymmetric Subsidence: A PSInSAR Analysis of Land Deformation over a Transit Corridor in the Los Angeles Metropolitan Area

Mohammad Khorrami [1,*], Babak Alizadeh [2], Erfan Ghasemi Tousi [3], Mahyar Shakerian [1], Yasser Maghsoudi [4] and Peyman Rahgozar [5]

[1] Department of Civil Engineering, Faculty of Engineering, Ferdowsi University of Mashhad, Mashhad 91779, Iran; mahyar.shakeryan@gmail.com
[2] Department of Civil Engineering, University of Texas at Arlington, Arlington, TX 76019, USA; babak.alizadeh@mavs.uta.edu
[3] Department of Civil and Architectural Engineering and Mechanics, University of Arizona, Tucson, AZ 85721, USA; erfang@email.arizona.edu
[4] Department of Photogrammetry and Remote Sensing, Faculty of Geodesy and Geomatics Engineering, K. N. Toosi University of Technology, Tehran 19967, Iran; ymaghsoudi@kntu.ac.ir
[5] M. E. Rinker, Sr. School of Construction Management, University of Florida, P.O. Box 115703, Gainesville, FL 32611, USA; peymanrahgozar@ufl.edu
* Correspondence: mohammad.khorrami@mail.um.ac.ir

Received: 31 December 2018; Accepted: 9 February 2019; Published: 12 February 2019

Abstract: Los Angeles has experienced ground deformations during the past decades. These ground displacements can be destructive for infrastructure and can reduce the land capacity for groundwater storage. Therefore, this paper seeks to evaluate the existing ground displacement patterns along a new metro tunnel in Los Angeles, known as the Sepulveda Transit Corridor. The goal is to find the most crucial areas suffering from subsidence or uplift and to enhance the previous reports in this metropolitan area. For this purpose, we applied a Persistent Scatterer Interferometric Synthetic Aperture Radar using 29 Sentinel-1A acquisitions from June 2017 to May 2018 to estimate the deformation rate. The assessment procedure demonstrated a high rate of subsidence in the Inglewood field that is near the study area of the Sepulveda Transit Corridor with a maximum deformation rate of 30 mm/yr. Finally, data derived from in situ instruments as groundwater level variations, GPS observations, and soil properties were collected and analyzed to interpret the results. Investigation of geotechnical boreholes indicates layers of fine-grained soils in some parts of the area and this observation confirms the necessity of more detailed geotechnical investigations for future constructions in the region. Results of investigating line-of-sight displacement rates showed asymmetric subsidence along the corridor and hence we proposed a new framework to evaluate the asymmetric subsidence index that can help the designers and decision makers of the project to consider solutions to control the current subsidence.

Keywords: subsidence monitoring; persistent scatterer interferometry; asymmetric subsidence; groundwater level variation; Sepulveda Transit Corridor; Los Angeles

---

1. Introduction

Ground subsidence is mainly due to fluid overexploitation and expanding construction [1–4]. There are several cities and regions suffering from land subsidence, such as Mexico City [5,6],

Shanghai, China [7–9], Lhokseumawe, Medan, Jakarta, Bandung, Blanakan, Pekalongan, Bungbulang, and Semarang, Indonesia [10–13], Ravenna, Prato, Bologna, Italy [14–18], Tehran, Rafsanjan, Neyshabour, Mashhad, Iran [19–25], Los Angeles, United States [26–32], and many more places around the world. In the present study, we studied land deformation in Los Angeles metropolitan area, Southern California, with a focus on the study area of a new transit corridor, known as Sepulveda Transit Corridor. This investigation is crucial because land displacement will affect the design and depth of a tunnel [33–36] and should be assessed based on soil properties. Also, all the information about the location, soil and groundwater needs to be carefully managed, analyzed and investigated in planning and design phase of the road construction to ensure the reliability of the subgrade [37–39]. Based on the previous researches in Los Angeles [26–31,40], the ground displacements in this area are mainly due to the groundwater level variations and oil extraction [26].

Advances in technology and science have made accurate measurement of ground deformation simple. Interferometric Synthetic Aperture Radar (InSAR) technique is a geodetic tool to image ground displacement in centimeter-scale and can be a very helpful technique in understanding the earthquakes, volcanos and glaciers [41]. InSAR can also benefit geomorphologists and hydrologist by providing an accurate measurement of slope motion, sediment erosion and deposition, water level fluctuation and soil moisture content [42–46]. InSAR has been considered as a powerful method to monitor ground surface deformations [47] and is an alternative technique to measure surface displacement. InSAR can measure small surface deformations in different situations and projects such as ground settlement and excavations [48]. Using the high spatial and temporal resolution of radar images, the InSAR technique can provide reliable results in the application of subsidence monitoring of such infrastructures as roads [49], subways, rails, and tunnels. Tunnels are visible because of localized subsidence of the above ground surface along their tunnel path. It means that it is possible to determine the effect of tunnel excavation on the ground surface. Highways, standing over the ground surface, in most cases show reliable stability compared to the surrounding areas [50].

A number of studies have used geodetic and InSAR techniques to evaluate the ground deformation in Los Angeles Basin. For example, the radar data acquired by the European Remote Sensing Satellites (ERS-1 and ERS-2) from 1992 to 1999 were analyzed [51] using InSAR to study the ground deformations along the southern San Andreas fault system. In addition, the interseismic crustal movement was measured [52] near Los Angeles, along the San Andreas Fault (SAF), by a new technique for integrating InSAR analysis on ERS descending and ALOS ascending radar images, and GPS data. The outputs display the vertical velocity of land deformation between −2 to +2 mm/yr, and shows uplift on the SAF in the Los Angeles area. Several researchers investigated the ground displacements related to groundwater level changes and fluid extraction in the Los Angeles Basin. For instance, radar images of ERS-1/2 satellite and GPS data were deployed [29] to infer the seasonal land deformations related to groundwater extraction in the Los Angeles basin. Also, a study on metropolitan Los Angeles [40] evaluated seasonal oscillations of the Santa Ana aquifer (uplift and subsidence), located in Los Angeles Basin, using InSAR technique from 1998 to 1999. The analysis provided estimates of ground displacement in the Line of Sight (LOS) of the European Remote Sensing (ERS) satellite in the time between satellite passes. The InSAR outputs showed uplift and subsidence in metropolitan Los Angeles to in response to extraction of fluid resources.

The subsidence associated with groundwater pumping and faulting in Santa Ana basin, CA was measured using InSAR technique from 1997 to 1999 and GPS data from 1999 to 2000 [53]. The results showed subsidence as high as 12 mm/yr is happening by groundwater withdrawal and re-injection in metropolitan Los Angeles. A time series analysis of ground deformation by InSAR based on small baseline subset (SBAS) algorithm was carried out [28] for Santa Ana basin in Los Angeles metropolitan area. ERS satellite data from 1995 to 2002 were used and it was found that ground deformations time series from InSAR significantly agree with GPS time series from Southern California Integrated GPS Network (SCIGN). A temporarily coherent point InSAR method [30] was applied on the Los Angeles Basin, using 32 ERS-1/2 images acquired during 1995 to 2000 to detect land subsidence. InSAR and GPS measurements were used [26] for detecting ground deformations caused by injection of groundwater and oil in Los Angeles from 2003 to 2007. A dataset of 64 TerraSAR-X images has been processed [27] in Los Angeles in the period 2010–2014 and showed a cumulative displacement of −50 mm in oil extraction fields. In 2018, a research [54] conducted to quantify ground deformation in the Los Angeles Basin due to groundwater withdrawal and showed −20 to +10 mm/yr LOS displacement rate.

A number of studies have been carried out to measure surface deformation along the transit corridors and their near infrastructures such as aqueducts and levees in California [55,56] and Rome (Italy) [57]. For instance, land subsidence rate of Hampton Roads in Virginia, USA, was estimated [58] using GPS observation and InSAR applied to ALOS-1 radar data. The outputs showed decent agreement between GPS data and InSAR-generated subsidence rate map. In a study in Shanghai, China [50], the X-band sensor Cosmo-SkyMed was used to monitor the subway tunnels and highways by Persistent Scatterer Interferometric Synthetic Aperture Radar (PSInSAR) analysis. In order to detect and monitor ground subsidence caused by tunneling, InSAR time series analysis was applied [59] on RADARSAT-1 and RADARSAT-2 radar data in the urban area of Vancouver, Canada. InSAR technique was also used to monitor landslide displacements induced by excavations related to tunneling in the Northern Apennines, Italy [48]. The tunnel was part of a larger project that contains the improvement of a highway that connects Bologna and Florence. The InSAR outputs showed high agreement with inclinometer and GPS as ground-based monitoring data.

Land surface deformation depends on many factors such as the depth of sediments and the amount of fluid extraction. Therefore, each area may behave differently at different places and different periods. In geotechnical engineering, land subsidence is estimated by considering the following parameters: deformable soil thickness, effective stress variation, and modulus relating the two previous parameters. The changes in the stress state are due to variations in the groundwater level. As the piezometric levels were measured frequently during a period, they are used to determine the groundwater table depth and pore water pressure changes are assumed equal to changes of ground water table [24,60]. Drainage of groundwater in soil deposits can induce huge ground subsidence. Thus, it is imperative to investigate the soil properties of deep geotechnical wells to detect thick compressible sediments particularly in the areas suffering from groundwater extraction.

In this research, we focused on the study area of the Sepulveda Transit Corridor which is planned to improve transportation means between the Los Angeles International Airport and the San Fernando Valley. The previous studies considered the displacements of constructed or under-construction infrastructures such as ground deformations caused by tunnel excavations. The main goal of conducting the present study is to obtain the current ground deformation pattern of a new transit corridor, which can affect its designing criteria and help the designers and decision makers of future constructions. In addition, it is necessary to investigate the subsidence rates in recent years to modify and update the past reports. This paper is organized as follows. First, the study area and the Sepulveda Transit Corridor project is introduced. Second, a brief description of the basic concepts of PSInSAR and the dataset is given. In this study, we used Sentinel-1A SAR images, provided by the European Space Agency (ESA) [61], acquired over the study area from June 2017 to May 2018. Third, the subsidence map derived from PSInSAR analysis is presented. Fourth, piezometric data, GPS observations, and geotechnical properties are provided to assess the outputs. Finally, a framework for evaluation of asymmetric subsidence is proposed. The research objectives of this research are:

- To assess and complement the previous studies on subsidence monitoring in Los Angeles using more recent data.
- To evaluate the PSInSAR results considering soil properties, and hydrological data and GPS information in the area.
- To identify deformation patterns over the study area of the corridor to inform and warn the managers, designers and other stakeholders about the future hazardous consequences.

- To show the variation in displacement rates along the alignment of corridor to help the designers and decision makers of the project to detect the places that require considering immediate solutions to control the current displacements.

2. Study Area: Sepulveda Transit Corridor, Los Angeles, California

The main aim of the Sepulveda Transit Corridor is to enhance transportation between the Los Angeles International Airport (LAX) and the San Fernando Valley. In the current situation, the I-405 highway in this area bear more than 400,000 travel every day and known as one of the most traveled urban freeways in the US [62]. As such, the Los Angeles County Metropolitan Transportation Authority (known as Metro), the agency that controls public transportation for the County of Los Angeles, is conducting a study to assess a range of high-capacity rail transit alternatives between the San Fernando Valley and LAX. The study conducted by Metro is expected to take approximately 20 months, from December 2017 (study kickoff) to Summer/Fall 2019 (study completion). It should be noted that due to the importance of the Sepulveda project, it is funded by the Measure M expenditure plan, with around $5.7 billion for construction of new transportation service to connect the San Fernando Valley and the Westside, and around $3.8 billion for extending that transit service between the Westside and LAX [62]. Figure 1 shows the study area of the Sepulveda Transit Corridor covering an area of about 229 km^2.

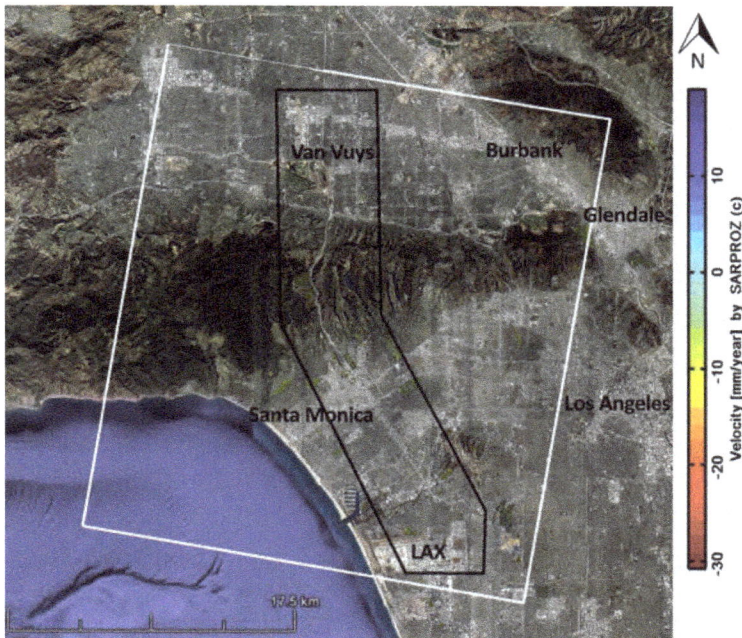

Figure 1. The study area for PSInSAR analysis, including the Sepulveda Transit Corridor.

3. Methodology

3.1. PSInSAR Time Series Analysis

The PSInSAR technique [63] was used in this research to monitor ground deformation through the study are. This technique is one of the powerful SAR time series applications which can analyze land displacements, particularly in urban areas [64]. PSInSAR looks for Permanent Scatterers [65] with stable scattering properties and also relatively good coherence, over long period intervals in multi-temporal data [66]. For mapping ground deformation, a stack of SAR images of the same area is selected. Afterwards, one single master acquisition is chosen from the stack based on the measured baselines in time and space to achieve an appropriate coherence in interferograms. A reference point is chosen, among the selected Persistent Scatterer Candidates (PSCs), which is relatively unaffected by ground surface displacement. Then, a stack of co-registered Single Look Complex (SLC) images is created using this single master configuration. Phases of each pixel are acquired when the topography and earth curvature influence is removed from the phase. There are a number of factors influenced the acquired phases, such as external DEM inaccuracy, Atmospheric Phase Screen (APS), linear phase ramp, the scatterer movement, and decorrelation and speckle noise. The following equation [67] shows the main factors in the phase calculation.

$$\phi^k = \frac{4\pi}{\lambda}\left(\frac{B_{\perp}^k}{R\sin\theta}\right)h + \frac{4\pi}{\lambda}T^k v + \phi_{atm}^k + \phi_{orb}^k + \phi_{noise}^k \qquad (1)$$

where the first term is related to the DEM error (h) because of the external DEM inaccuracy, the second term is related to the linear deformation velocity (v) during the acquisition period. In this equation, ϕ_{atm}^k, ϕ_{orb}^k and ϕ_{noise}^k denote the atmospheric phase delay, the residual orbital error phase, and the temporal and geometrical decorrelation noise, respectively. In this study, we implemented PSInSAR analysis in SARPROZ [4] and the applied processing steps are as the following:

First, each pixel could be a PS candidate if it satisfies the amplitude stability index for the pixel have a value of at least 0.85. The amplitude stability index can be calculated as follow:

$$D_{stab} = 1 - \frac{\sigma_a}{\bar{a}} \qquad (2)$$

where D_{stab}, σ_a and $\bar{a}$ are the amplitude stability index, the standard deviation and the mean of amplitude values, respectively. This condition resulted in 57,667 points in the present study.

Second, the unknown parameters of DEM error and the velocity are estimated. For this purpose, the spatial graph of connections between points is considered and the initial parameters are estimated along the connections. Then, the absolute values are achieved by numerical integration considering a reference point as a starting point for the integration. Careful selection of the reference point is a key factor in the accuracy of outputs, as careless reference selection will result in biased parameters for all points.

Finally, a wider set of points are selected considering a spatial coherence of 0.80 and temporal coherence of 0.85 conditions. At this stage, a second approximation of the parameters were applied on the new dataset. Then, all PS points above the temporal coherence threshold were selected for the final estimation. The DEM error, the linear deformation rate along the Line of Sight and the subsidence time series are approximately calculated for the selected PS points.

It should be noted that differentiating between the contributions made to the phase by deformation and atmosphere would be difficult, if we only had two SAR images. As we are using a time-series of SAR images, we can take advantage of this fact that often the atmospheric perturbations exhibit typically high spatial correlation but low temporal correlation [66]. Therefore, we can estimate the atmospheric signal by applying a high-pass filtering in time and a low-pass filtering in space [63]. This is how the atmospheric phase signal was computed and removed from the total phase. Furthermore, the displacement measured by InSAR can be decomposed into two main components: a periodical component and a linear component. The periodical signal is a seasonal deformation phenomenon which is occurred due to the thermal expansion and contraction particularly evident on skyscrapers, bridges, etc. which is not the case in our study area. Therefore, in our work, we only considered the linear trend signal and did not take the seasonality signal into account. Here, we used descending images which resulted in LOS displacement. So, in order to compare the PSInSAR and GPS data, we used the following equation to obtain GPS measurements in LOS direction [66]:

$$
\begin{aligned}
GPS_{LOS} = {}& GPS_{up} \times \cos(\theta_{inc}) \\
& - GPS_{north} \times \cos(\theta_{azi} - 3\pi/2) \times \sin(\theta_{inc}) \\
& - GPS_{east} \times \sin(\theta_{azi} - 3\pi/2) \times \sin(\theta_{inc})
\end{aligned}
\tag{3}
$$

where GPS_{LOS} is the converted value of GPS data in LOS direction. GPS_{up}, GPS_{north}, and GPS_{east} are the values of GPS observation vector in the up, north, and east directions. θ_{inc} represents incidence angle. The radar images were taken from different incidence angles and the average incident angel is about 43.97° in this study. θ_{azi} represents the heading angle of the satellite from the North (azimuth angle) and is about −9.66° in this study.

3.2. Data Collection

Land deformation measurement by PSInSAR needs sufficient number of SAR images. From literature [67,68], the PS analysis requires at least 20 to 25 SAR images to achieve reliable outputs. Considering this important condition on number of images, we collected 29 descending Sentinel-1A SAR images acquired over the study area during June 2017 and May 2018. After collecting the raw data, we defined the study area with an area of 1019 km² to cover the corridor and its neighborhoods. Figure 1 displays the study area. The white line indicates the master area and the black line shows the boundary of the study area of Sepulveda Transit Corridor.

Figure 2 displays the SLC data used in this study and the spatiotemporal baseline configuration of interferometric pairs. To form the interferograms, all images were connected with the master image (5 December 2017). The master image is chosen at the barycenter of the temporal baseline, x-axis, and normal baseline, y-axis, distributions. The dots and lines represent the images and the interferograms, respectively.

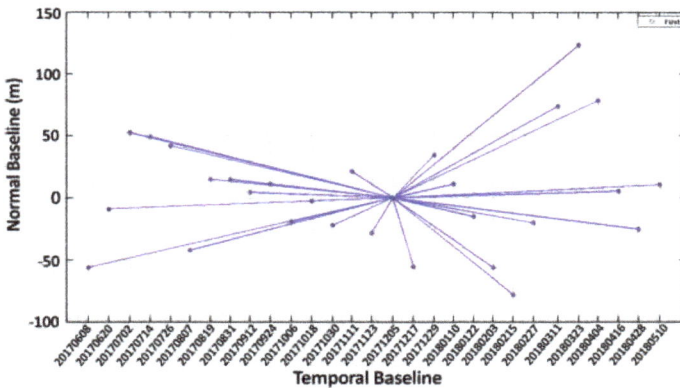

Figure 2. The spatiotemporal baseline configuration of interferometric pairs showing the SLC data in this study (29 images): Sentinel-1A, descending mode (track 71), and polarization VV.

4. Results and Discussion

4.1. Ground Deformation

Applying PSInSAR on a dataset of 29 descending Sentinel-1A radar images resulted in mean velocity map of land deformation in the interest area covering a period between June 2017 and May 2018. It should be noted that based on the spatial coherence of the PSs calculated in the area, most of the region is covered by a coherence around 0.85 or higher and it can prove the reliability of the monitoring process (Figure 3).

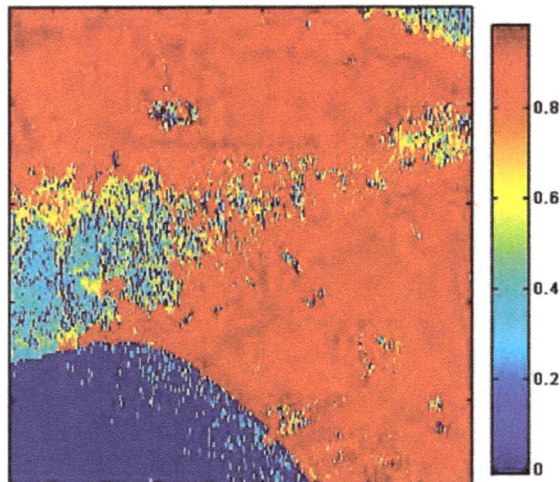

Figure 3. The coherence map obtained in PSInSAR analysis in the study area.

Figure 4 shows the deformation map along the study area of Sepulveda Transit Corridor and its vicinity. In an overall view, we can categorize the corridor into three zones based on the trend of displacement rates: (a) from 0 to 12 km; (b) from 12 to 24 km; and (c) from 24 to 34 km. The red spots in Figure 4 indicates the southeast of the corridor, located over oil extraction sites. In Particular, Figure 5 shows the deformation map in the Inglewood oil field with the maximum subsidence rate about 30 mm/yr. Therefore, it is essential to investigate such engineering solutions as ground stabilization in this site during the study phase and construction phase of Sepulveda corridor. The deformation pattern, also, displays low amounts of uplift (blue features) in south and east of the region meaning that water or gas probably pumped underground to stabilize the subsidence, or it may be as a result of an increase in groundwater level which will be discussed in Section 4.3 of this paper. For instance, the Los Angeles International Airport (LAX) is located in the regions suffering from low amounts of uplift. Green and yellow features through the corridor demonstrate subsidence rates between −15 and 0 mm/yr. Vegetation areas include less coherent PS points; so, there are some regions without sufficient outputs in the extracted maps. It was one of the main reasons to select a large study area to provide more PS points and obtain the deformation trend.

Figure 6 shows the variations in displacement rates (average) along the corridor from south (0 m) to north (34,000 m). In order to estimate the displacement rates in an arbitrary point through the corridor, we proposed a function as Equation (4) derived from the available deformation rates in the location of PS points. Such categorizations can help the designers and decision makers of the project to detect the places, which require solutions to control the probable asymmetric subsidence along the corridor. The asymmetric subsidence is fully explained in Section 4.5.

$$DR\left(\frac{mm}{year}\right) = \begin{cases} 0.016x + 0.51, & 0 < x < 12\ km \\ -1.028x + 11.44, & 12 < x < 24\ km \\ 0.314x - 16.11, & 24 < x < 34\ km \end{cases} \tag{4}$$

where DR is the displacement rate in each point through the alignment of Sepulveda Transit Corridor, and x (km) is the distance from the start point (LAX).

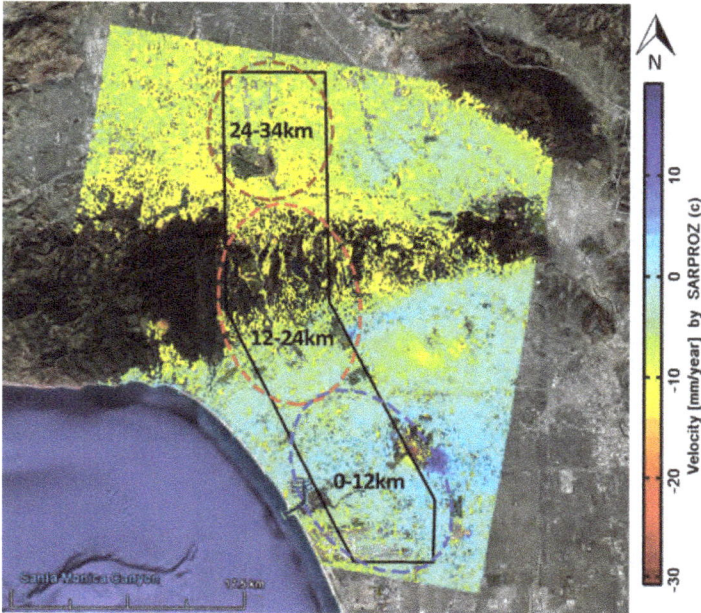

Figure 4. Mean velocity map of land deformation (mm/yr) in the region covering a period between June 2017 and May 2018 overlapped onto Google Earth high-resolution imagery. The black line shows the boundary of Sepulveda Transit Corridor study area. The corridor categorized into three zones based on the trend of displacement rates: (a) from 0 to 12 km; (b) from 12 to 24 km; and (c) from 24 to 34 km.

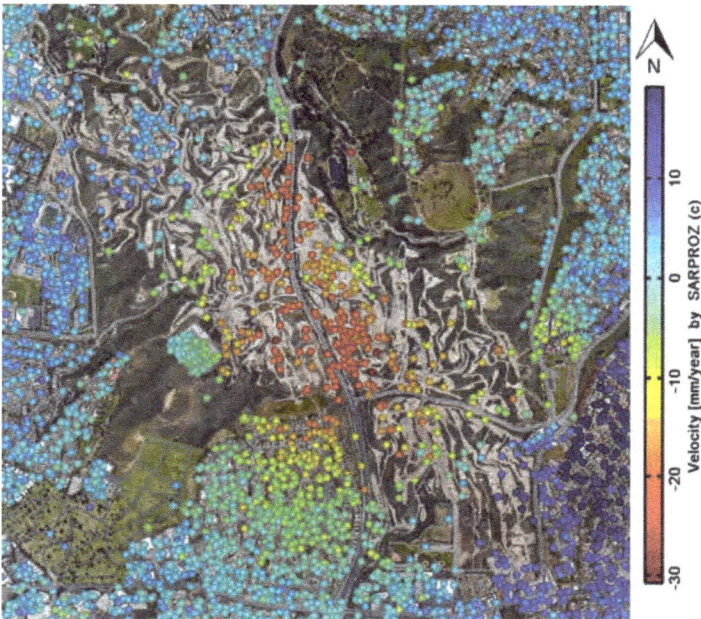

Figure 5. Deformation map in the Inglewood area.

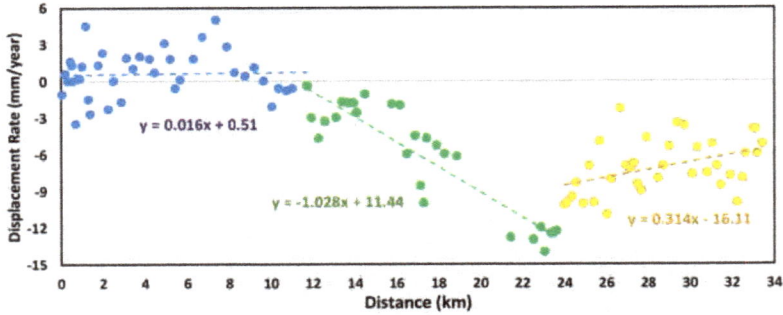

Figure 6. Average rate of displacement along the Sepulveda Transit Corridor from south to north.

4.2. GPS Monitoring

In order to assess the results of PSInSAR analysis, the GPS observations and piezometric data are collected in the present study. Figure 7 shows the location of GPS stations and piezometric wells (P1 to P6). The characteristics and information of the well points are fully explained in Section 4.3.

GPS data has high temporal resolution because of continuous measurements while the PSInSAR method provides high spatial resolution and lower temporal resolution compared to data from GPS stations. Thus, the integration of GPS and PSInSAR measurements can be used to interpret the land displacements. In order to evaluate the PSInSAR results in the previous section, GPS data [69] were collected and introduced in Table 1. Eight stations are represented which two of them (DSHS and FXHS) are inactive since 2011. So, we considered six active stations to compare their results with the PSInSAR outputs in their locations and then, two stations (BRAN and NOPK) with more noises and insufficient observations were removed. Figure 8 shows the comparison between PSInSAR-derived time series deformation and the corresponding GPS observations. For this comparison, the RMSE was computed between each PSInSAR output and GPS measurement and demonstrated relatively good agreement between them. Lacking sufficient number of GPS stations is a significant weakness of GPS stations in monitoring the land displacements compared to SAR Interferometry. Also, the fluctuations in GPS results referred to seasonal effects and the instrument inherit errors [69]. These are the main disadvantageous or weaknesses of GPS observations compared to the SAR analysis performed in the present study.

Table 1. GPS Stations in the Study Area.

GPS Station	Start Date	Location		Current Situation
		Long.	Lat.	
DSHS	1999	−118.3485°	34.0239°	Inactive (since 2011)
FXHS	1999	−118.3595°	34.0806°	Inactive (since 2011)
BRAN	1994	−118.2771°	34.1849°	Active
NOPK	1999	−118.3480°	33.9797°	Active
LAPC	1999	−118.5747°	34.1819°	Active
LFRS	1999	−118.4128°	34.0951°	Active
CSN1	1999	−118.5238°	34.2536°	Active
WRHS	1999	−118.4276°	33.9582°	Active

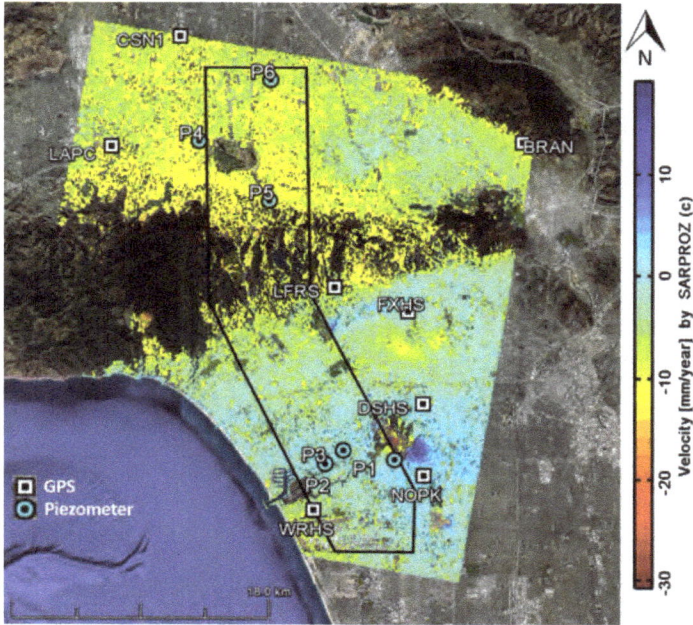

Figure 7. The location of GPS stations (squares) and piezometers (circles) in the study area.

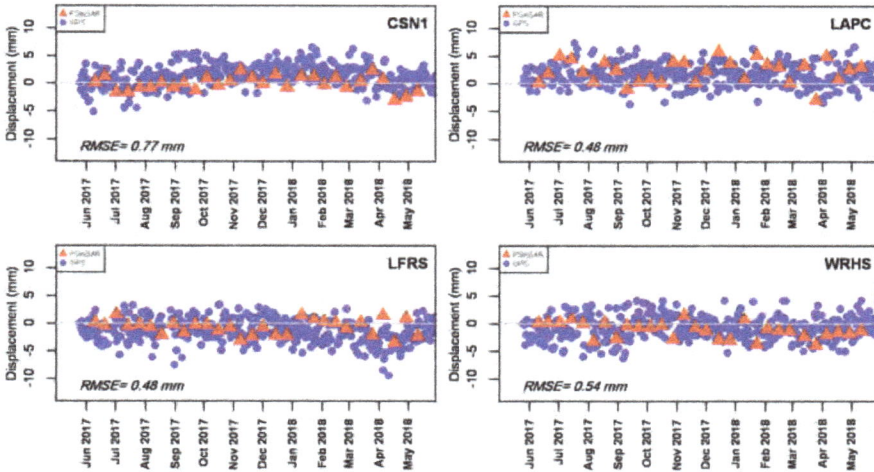

Figure 8. Comparison between PSInSAR-derived time series deformation (red triangles) and GPS observations (blue dots), Line-of-Sight direction, from June 2017 to May 2018.

Table 2 shows the comparison between GPS and PSInSAR deformation rates in long-term and short period. Overall, from Table 2 it can be found that the Standard Deviation of GPS data is in average (2.04) bigger than the Standard Deviation of PSInSAR data (1.56) which is because there are more noises in GPS measurements.

Table 2. The comparison between GPS and PSInSAR outputs.

GPS Station	GPS Observation		PSInSAR Deformation Rate from 2017 to 2018 (mm/yr)	Standard Deviation of GPS Data (mm)	Standard Deviation of PSInSAR Data (mm)	RMSE (mm)
	Long-Term Deformation Rate from 1995 to 2018 (mm/yr)	Deformation Rate from 2017 to 2018 (mm/yr)				
WRHS	+0.73	−2.61	−2.23	2.16	1.41	0.54
LAPC	+0.84	+0.49	+0.31	1.94	2.13	0.48
LFRS	−0.19	−0.72	−0.49	2.09	1.37	0.48
CSN1	+0.20	+0.08	+0.06	1.95	1.34	0.77

4.3. Monitoring of Groundwater Level Variations

According to the literature, one of the main reasons for ground displacements in the study area is water withdrawal or increase in groundwater level, except the red spots in the deformation map which suffer from oil extraction in the region. Based on the project's official report [70], groundwater is highly variable along the extent of the project corridor. Unfortunately, the groundwater depths and elevations are not well-documented throughout the Santa Monica Mountains. The historical groundwater level data of the Inglewood quadrangle [71] shows the groundwater depths for the southern end of the Sepulveda corridor and indicates that groundwater level in the southerly part of the project alignment is about 12 m below grade and deepens to 15 m as the corridor extends northward through Inglewood city. In addition, data from another project in the region, called the Crenshaw/LAX Transit Corridor Project, show the areas along the southern part of the project corridor have measured depths of groundwater ranging between 12 and 27.5 m below grade. As the corridor bends northwest, the groundwater moves closer to the ground surface, with an approximate depth of 3 m or less [70].

Much of the I-405 highway in Sepulveda Canyon along the Santa Monica Mountains is not known to encounter shallow groundwater [72]. Based to groundwater monitoring data of the widening project of I-405 corridor from 2008 to 2009, groundwater was reached at depths greater than 21 m below the corridor surface. However, higher groundwater levels were observed during drilling between 1958 and 2007 for the purpose of as-built data at bridge locations through the existing Sepulveda Pass. This data contains groundwater depths between 0.6 and 24 m below existing grade [70]. The historical groundwater level data of the Van Nuys quadrangles [72] and the San Fernando [73] displays groundwater to be progressively shallower northward from the base of the Santa Monica Mountains where the groundwater depth is 12 m below grade and rises to 0 m below grade where the transit corridor intersects the 101 freeway. From the 101 freeway north along the corridor, the groundwater ascends progressively northward along alignment up to approximately 67 m below grade, where it reaches an abrupt groundwater barrier at the location of the Mission Hills fault. At this area, where the I-405 meets SR-118, the groundwater jumps to 12 m below grade. This site is where the San Fernando fault exists and groundwater data is probably not sufficient enough to show accurate contours due to the extensive faulting and deformation within the area [70].

We monitored the variations in groundwater level in the study area of Sepulveda Transit Corridor. Figure 9 shows temporal evaluation of groundwater level changes for the piezometers (the locations of piezometers are shown in Figure 7). Table 3 shows the overall trend of groundwater level changes at the studied piezometric wells and their corresponding PSInSAR deformation rate. The groundwater level in the location of P5 experienced several fluctuations and dramatically decreased since 2008. Surprisingly, this point shows the maximum subsidence rate among the piezometers with 11 mm/yr. On the other hand, the water level remained stable during the period at P1 and P2. Both piezometers have negligible displacements at their locations based on PSInSAR outputs. The rising trend of groundwater in piezometer P3 confirms the PSInSAR analysis which shows uplift of almost 3 mm/yr in P3 location. It should be noted that PSInSAR computes the total displacement rate and there may be some other factors as parts of ground movements. In order to investigate the relation between land deformation rate and water level variation it is imperative to know soil properties that are thoroughly explained in Section 4.4.

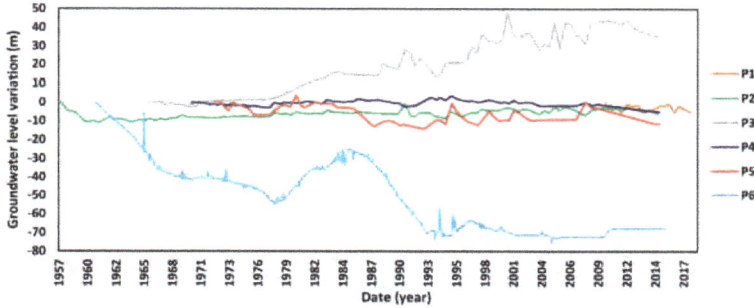

Figure 9. Temporal evaluation of groundwater level variations for six piezometers (P1 to P6) located in the study area of Sepulveda Transit Corridor.

Table 3. Overall characteristics of the studied piezometric wells.

Piezometric Well	Deformation Rate by PSInSAR (mm/yr)	Δh (m)	Δt (year)	$\Delta h/\Delta t$ (mm/yr)	Overall Trend of Ground Water Level Variations
P1	0	5.08	8	0.64	Remained stable.
P2	0	6.5	59	0.11	Remained stable.
P3	+3	−34.9	49	−0.71	Increased during the whole period and slightly decreased after 2009.
P4	−6.7	5	45	0.11	Relatively stable up to 1995 and then decreased slightly up to now.
P5	−11	11.4	43	0.27	Experienced several fluctuations, but decreased after 2008 up to now.
P6	−5	67.9	55	1.23	Decreased about 40 m between 1985 and 1995 and relatively stable up to now.

4.4. Geological Characteristics of the Sepulveda Project and Hydrogeology of Basins

The Los Angeles area consists of several basins containing groundwater systems. The Sepulveda project extends through numerous geologic characteristics of Los Angeles County within the Santa Monica (SM) and San Fernando (SF) Groundwater Basins. Table 4 shows the overall properties of SM and SF basins. The recharge of SF is by natural streamflow from the surrounding mountains, precipitation falling on impervious areas, reclaimed wastewater, and industrial discharges [74]. The replenishment of SM is mainly by percolation of precipitation and surface runoff onto the sub-basin from the SM Mountains [75].

The SF Valley Basin is bounded by the Santa Susana Mountains on the north and northwest, the San Gabriel Mountains on the north and northeast, the San Rafael Hills on the east, the Santa Monica Mountains and Chalk Hills on the south, and the Simi Hills on the west. The groundwater in this basin is mainly unconfined with some confinement. Also, several structures disturb the flow of groundwater through this basin such as faults and subsurface dams [74]. The groundwater in the SM Basin is mainly confined and this basin underlies the northwestern part of the Coastal Plain of Los Angeles Basin. SM bounded by impermeable rocks of the SM Mountains on the north and by the Ballona escarpment on the south [75].

The main water-producing units of SM include the relatively coarse-grained sediments of the Recent Alluvium, Lakewood Formation, and San Pedro Formation [76]. The Recent Alluvium reaches a maximum thickness of around 27 m and comprises the clays of the Bellflower aquiclude and the underlying Ballona aquifer, depositing gravels resulting in the present Ballona Gap structure. These gravels are dominant at an approximate depth of 15 m. The Ballona aquifer is generally separated from the underlying San Pedro Formation by the confining layer [77]. The Lakewood Formation seems to be present only in the northern half of the SM Basin. The most significant water-bearing units are the sands and gravels within the San Pedro Formation. The Silverado aquifer of the San Pedro Formation has the greatest lateral extent and saturated thickness, and is considered as the main source of groundwater. The average thickness of San Pedro Formation is about 60 m in the SM Basin. Beneath the Silverado aquifer are relatively low-permeability sediments of the lower San Pedro and upper Pico formations [77].

The Sepulveda project cuts through San Fernando Valley in the north and extends through the Santa Monica Mountains in the south. The corridor is underlain by a layer of horizontal Quaternary sediment and also Tertiary-age sediments and sedimentary rocks which faced deformation into folds and offset by faults. Sedimentary and metamorphic bedrock are exposed with colluvial and alluvial soil at the surface at high elevations such as Santa Monica Mountains. In the north and south of Santa Monica Mountains, there is a thick layer of alluvial sediments. Also, the portion of the corridor located above San Fernando Valley is underlain by up to 600 m of alluvial deposits and a layer of Cretaceous-aged crystalline bedrock which exists below the alluvium [78]. The southern part of the project corridor, located in the Los Angeles Basin, is underlain by unconsolidated Quaternary-aged sandy deposits. These deposits can be subdivided into a loose unconsolidated Holocene-age layer and late-Pleistocene sediments. Also, hard rocks only exist in the mountainous portion of the basin at depth of 1500 m to 9000 m.

Figure 10 shows surface soil map of the study area including various soil types (the map is created based on raw soil data provided by the Los Angeles County Department of Public Works). In order to investigate the subsidence and uplift in the region, it is needed to study the soil properties in depth. Figure 11 displays the location of nine geotechnical boreholes in the region. The raw data of boreholes are collected from geotechnical report of the corridor and a number of geotechnical reports in the area [79–84].

Groundwater pumping has the potential to cause subsidence which can induce structural impacts. Induced subsidence is caused by the lowering of groundwater levels causing compaction of the aquifer materials to a point that the ground surface changes elevation. As water is withdrawn and groundwater levels declines, the effective pressure in the drained sediments increases. Compressible layers then compact under the over-pressure burden that is no longer compensated by hydrostatic pressure. The subsequent subsidence, includes both a component of elastic (recoverable) and inelastic (unrecoverable) subsidence, and is most pronounced in poorly compacted sediments. As a historical subsidence example, there is evidence for subsidence near Redondo Beach, in south of SM, that is attributed to oil and gas extraction [85]. From literature, a review of the geotechnical logs for wells completed in the SM Basin does not show considerable evidence of a thick compressible layer. Groundwater levels have also experienced significant drawdown in the past prior to the importation of water into the area. So, inelastic subsidence, which is of most concern, by nature can only occur once; consequently, any potential subsidence would have already occurred. Land subsidence in the study area does not appear to be a significant concern [76]. It should be added that as shown in Figure 12, investigation of the boreholes indicates some layers of fine-grained materials in some parts of the study area, which are susceptible to variations in groundwater level, an indication of the necessity of more detailed geotechnical investigations for the future constructions in the region.

Table 4. Overall characteristics of San Fernando (SF) and Santa Monica (SM) basins [74,75].

Basin	Confined/Unconfined	Recharge	Groundwater Level Trend
SF	Mainly unconfined with some confinement	Natural streamflow from the surrounding mountains, precipitation falling on impervious areas, reclaimed wastewater, and industrial discharges.	fairly stable over about the past 20 years
SM	Confined	Mainly by percolation of precipitation and surface runoff onto the sub-basin from the SM Mountains.	fairly stable over about the past 20 years

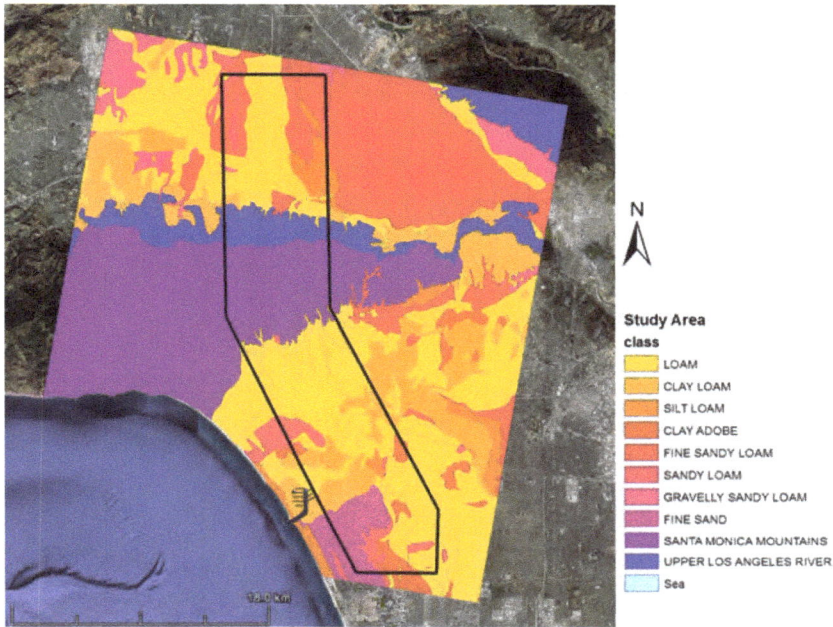

Figure 10. Surface soil map of the study area. The map is created based on raw soil data provided by the Los Angeles County Department of Public Works, Water Resources Division.

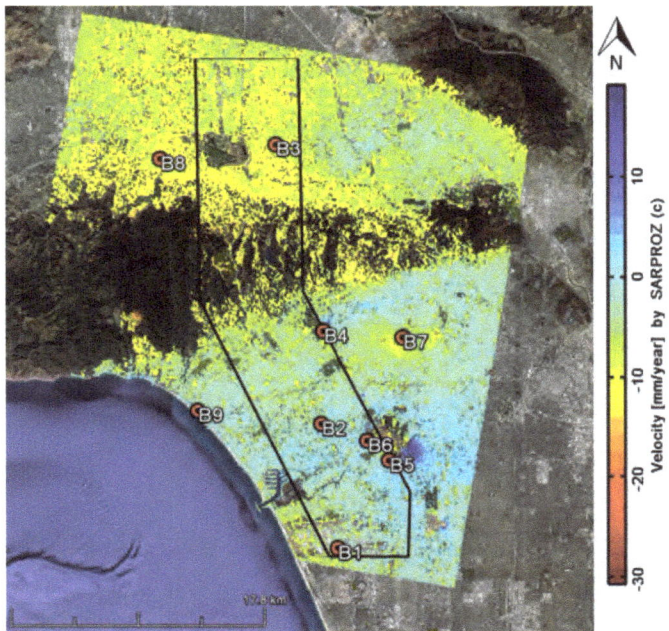

Figure 11. The location of Lithological logs.

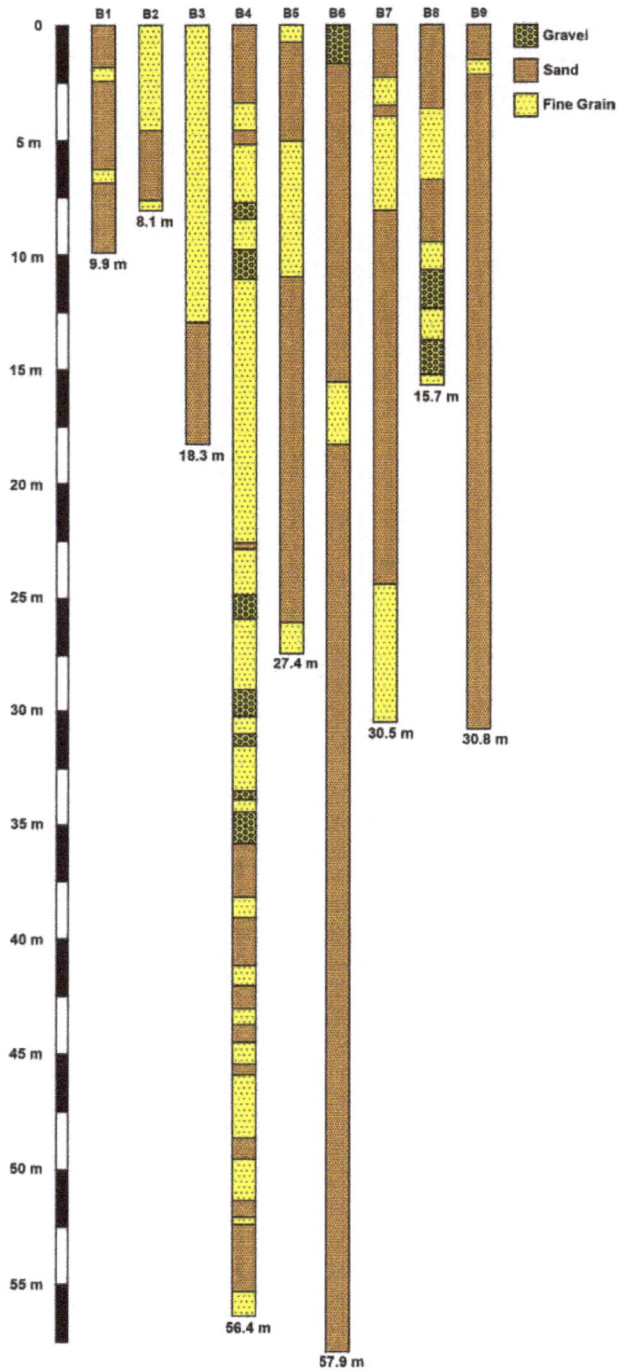

Figure 12. Lithological logs in the interest area.

4.5. Asymmetrical Subsidence

In most cases, the subsidence profile during the design phase of construction is considered symmetrical due to the assumptions inherent in the analysis and oversimplifications of the ground behavior. However, the ground behavior is not simple but instead very complex with different types of materials and different stress-strain responses. Such complexities can lead to the ground surface displacement to occur in such a way that it is not symmetrical (asymmetrical). In other words, asymmetrical subsidence means the difference in the amount of ground deformation between two near points which could be devastating especially to the available infrastructure and the infrastructure under construction such as the Sepulveda Transit Corridor. Asymmetry in subsidence can be observed in such industries [86–88] as mining, tunneling, groundwater withdrawal, oil and gas extraction, and geothermal fluid withdrawal.

Asymmetrical ground subsidence can be economically devastating to structures at surface. The heterogeneity of the ground layers (soil or rock) contribute to difficult estimation of asymmetrical subsidence [86,89]. As discussed in the Section 4.4, the Sepulveda project extends through numerous geologic characteristics and the region suffers from ground deformations. Accordingly, it is necessary to provide a certain procedure for the evaluation of asymmetric subsidence. Therefore, to detect the areas suffering from asymmetric subsidence, we propose a simplified version of strain rate based on the PSInSAR outputs to calculate Asymmetric Subsidence Index (*ASI*) as the following steps:

In step 1, consider two close PS points through the corridor length (L_L) and two PS points on/near both sides of the corridor width (L_W).

In step 2, determine the displacement rate (*DR*) of the selected points in step 1 based on PSInSAR analysis.

In step 3, calculate the ASI along length (ASI_L) by the ratio between the displacement rates (step 2) and length (L_L), Equation (5). Calculate the ASI along width (ASI_W) by the ratio between the displacement rates and length (L_W), Equation (6).

$$ASI_L = \frac{d_L}{L_L} = \left| \frac{DR_2 - DR_1}{L_L} \right| \tag{5}$$

where DR_1 and DR_2 are the displacement rate of the PS points in length. For instance, the value of DR_1 and DR_2 of the Sepulveda Transit Corridor can be estimated by Equation (4).

$$ASI_W = \frac{d_W}{L_W} = \left| \frac{DR_{Right} - DR_{Left}}{L_W} \right| \tag{6}$$

where DR_{Right} and DR_{Left} are the displacement rate of the PS points in width.

In step 4, the final Asymmetric Subsidence Index of the interest area is defined as the maximum of ASI_L and ASI_W, Equation (7).

$$ASI = \max\{ASI_L, ASI_W\} \tag{7}$$

It should be noted that for practical purposes, it is more accurate and better to use vertical displacements, to provide much more meaningful result, instead of line-of-sight deformations in engineering problems. The higher the value of *ASI*, the higher the asymmetry of the deformation and the value of allowable *ASI* depends on the sensitivity of each especial structure. Therefore, we suggest computing *ASI* for new constructions for considering possible evaluations and solutions. For practical calculation, we suggest considering the average of *ASI* values for a several points. Clearly, the amount of d_L and d_W must be less than allowable displacement which depends on the sensitivity of each particular project. Figure 13 displays a simple example in the study area to show how to calculate the *ASI*. The required calculations for this example are shown in Table 5 and the computed *ASI* in this example is negligible; so, it can be assumed symmetrical. The proposed framework can be easily used in engineering applications compared to the more common strain rate analysis.

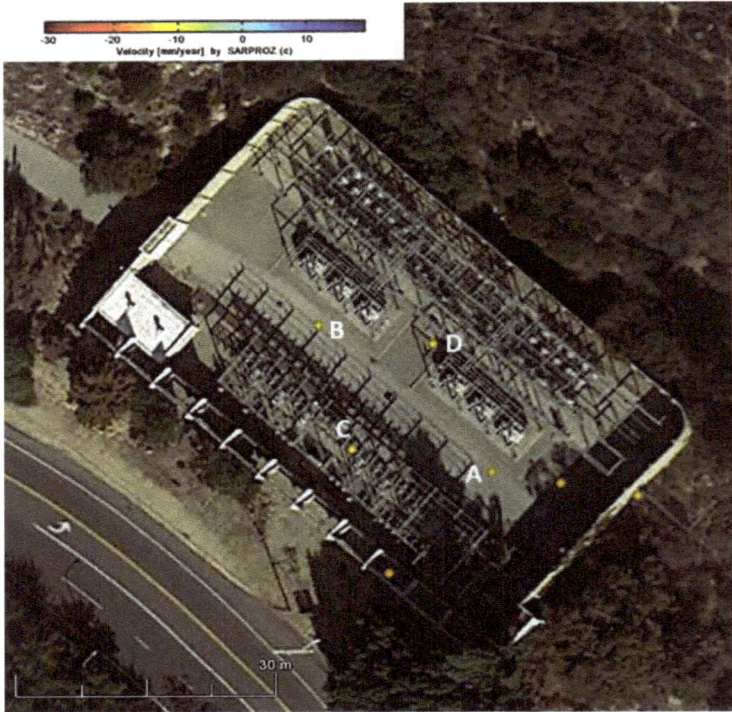

Figure 13. An example for ASI calculation for an infrastructure in the study area.

Table 5. The ASI calculation for the example.

Point	Temporal Coherence	DR (mm/yr)	Length (m)	ASI	Max ASI
A	0.99	−14.6	$L_{AB} = 53.4$	$ASI_L = \left\| \frac{-14.6-(-14.1)}{53.4 \times 1000} \right\| = -9 \times 10^{-6}$	
B	0.99	−14.1			3×10^{-5}
C	0.99	−13.1	$L_{CD} = 31.2$	$ASI_W = \left\| \frac{-13.1-(-14.1)}{31.2 \times 1000} \right\| = 3 \times 10^{-5}$	
D	0.99	−14.1			

5. Conclusions

The main aim of this research was to obtain the land displacements along a new metro tunnel under preliminary study in Los Angeles, CA called Sepulveda Transit Corridor; to detect the most crucial areas suffering from subsidence or uplift; and to complement the previous reports in Los Angeles. For this purpose, we applied Persistent Scatterer Interferometric Synthetic Aperture Radar using 29 Sentinel-1A radar images from 2017 to 2018. The outputs demonstrated a high-rate of subsidence in the Inglewood field that is near the south portion of the Sepulveda Transit Corridor. Finally, we used the PSInSAR outputs to calculate Asymmetric Subsidence Index (ASI). The main conclusions of the present study can be drawn as the following:

- The results of this paper showed that the ground subsidence in northern portion of the Sepulveda Transit Corridor is continuous with subsidence rates between 1 and 14 mm/yr and a high-rate of subsidence (30 mm/yr) occurs in the Inglewood field near the south portion of the corridor, which may cause irreversible consequences in the future.

- Based on the variation in displacement rates along the corridor, we categorized the corridor into three zones to help the designers and decision makers of the project to detect the places which require considering solutions to control the probable asymmetric subsidence along the corridor.
- The ground water extraction rate and geotechnical properties in the area both strongly influence the rate and the distribution of subsidence.
- Collecting deep geotechnical boreholes indicated fine-grained layers in the region. This observation confirmed the necessity of more detailed geotechnical investigations in the interest area.
- There are not a sufficient number of piezometers to detect the groundwater level and accurate in-situ instruments such as GPS stations and extensometers to monitor the land displacements in this area. Therefore, for future researches, we recommend adding more piezometers and instruments particularly in the places suffering from continuous subsidence or uplift.
- Asymmetrical subsidence can be devastating to structures. Because of the heterogeneity of the ground layers, it is difficult to estimate asymmetrical subsidence. So, a simplified framework was proposed based on PSInSAR outputs to evaluate asymmetric subsidence.

Author Contributions: Conceptualization, M.K.; Data curation, M.K. and B.A.; Formal analysis, M.K. and B.A.; Investigation, M.K., B.A., E.G.T., M.S. and P.R.; Methodology, M.K., B.A., Y.M., P.R., E.G.T. and M.S.; Software, M.K. and B.A.; Supervision, Y.M.; Validation, M.K., B.A., E.G.T., P.R. and M.S.; Writing—original draft, M.K. and B.A.; Writing—review & editing, M.K., B.A., Y.M., P.R., E.G.T., and M.S.

Funding: This research received no external funding.

Acknowledgments: Authors are sincerely grateful to the European Space Agency (ESA) for providing Sentinel-1A data. We convey our gratitude to the United States Geological Survey (USGS) for GPS data, Los Angeles County Department of Public Works (LACDPW) for providing groundwater data, the Los Angeles County Metropolitan Transportation Authority for the required data of Sepulveda Pass Corridor, Daniele Perissin for providing SARPROZ, Aron Meltzner and Fatemeh Foroughnia for helpful discussions.

Conflicts of Interest: The authors declare no conflict of interest.

References

1. Bull, W.B.; Poland, J.F. *Land Subsidence Due to Ground-Water Withdrawal in the Los Banos-Kettleman City Area, California: Part 3. Interrelations of Water-Level Change, Change in Aquifer-System Thickness, and Subsidence*; US Government Printing Office: Washington, DC, USA, 1975; Volume 3.
2. Galloway, D.L.; Hudnut, K.W.; Ingebritsen, S.; Phillips, S.P.; Peltzer, G.; Rogez, F.; Rosen, P. Detection of aquifer system compaction and land subsidence using interferometric synthetic aperture radar, Antelope Valley, Mojave Desert, California. *Water Resour. Res.* **1998**, *34*, 2573–2585. [CrossRef]
3. Poland, J.; Davis, G. Subsidence of the land surface in the Tulare—Wasco (Delano) and Los Banos—Kettleman City area, San Joaquin Valley, California. *EOS Trans. Am. Geophys. Union* **1956**, *37*, 287–296. [CrossRef]
4. Perissin, D.; Wang, Z.; Wang, T. The SARPROZ InSAR tool for urban subsidence/manmade structure stability monitoring in China. In Proceedings of the 34th International Symposium on Remote Sensing of Environment, Sydney, Australia, 10–15 April 2011.
5. Osmanoğlu, B.; Dixon, T.H.; Wdowinski, S.; Cabral-Cano, E.; Jiang, Y. Mexico City subsidence observed with persistent scatterer InSAR. *Int. J. Appl. Earth Obs.* **2011**, *13*, 1–12. [CrossRef]
6. López-Quiroz, P.; Doin, M.-P.; Tupin, F.; Briole, P.; Nicolas, J.-M. Time series analysis of Mexico City subsidence constrained by radar interferometry. *J. Appl. Geophys.* **2009**, *69*, 1–15. [CrossRef]
7. Chai, J.C.; Shen, S.L.; Zhu, H.H.; Zhang, X.L. Land subsidence due to groundwater drawdown in Shanghai. *Geotechnique* **2004**, *54*, 143–147. [CrossRef]
8. Shen, S.L.; Xu, Y.S. Numerical evaluation of land subsidence induced by groundwater pumping in Shanghai. *Can. Geotech. J.* **2011**, *48*, 1378–1392. [CrossRef]
9. Shen, S.L.; Ma, L.; Xu, Y.S.; Yin, Z.Y. Interpretation of increased deformation rate in aquifer IV due to groundwater pumping in Shanghai. *Can. Geotech. J.* **2013**, *50*, 1129–1142. [CrossRef]
10. Maghsoudi, Y.; van der Meer, F.; Hecker, C.; Perissin, D.; Saepuloh, A. Using PS-InSAR to detect surface deformation in geothermal areas of West Java in Indonesia. *Int. J. Appl. Earth Obs.* **2017**. [CrossRef]

11. Abidin, H.Z.; Andreas, H.; Gumilar, I.; Fukuda, Y.; Pohan, Y.E.; Deguchi, T. Land subsidence of Jakarta (Indonesia) and its relation with urban development. *Nat. Hazards* **2011**, *59*, 1753. [CrossRef]

12. Chaussard, E.; Amelung, F.; Abidin, H.; Hong, S.-H. Sinking cities in Indonesia: ALOS PALSAR detects rapid subsidence due to groundwater and gas extraction. *Remote Sens. Environ.* **2013**, *128*, 150–161. [CrossRef]

13. Marfai, M.A.; King, L. Monitoring land subsidence in Semarang, Indonesia. *Environ. Geol.* **2007**, *53*, 651–659. [CrossRef]

14. Teatini, P.; Ferronato, M.; Gambolati, G.; Bertoni, W.; Gonella, M. A century of land subsidence in Ravenna, Italy. *Environ. Geol.* **2005**, *47*, 831–846. [CrossRef]

15. Stramondo, S.; Saroli, M.; Tolomei, C.; Moro, M.; Doumaz, F.; Pesci, A.; Loddo, F.; Baldi, P.; Boschi, E. Surface movements in Bologna (Po plain—Italy) detected by multitemporal DInSAR. *Remote Sens. Environ.* **2007**, *110*, 304–316. [CrossRef]

16. Raucoules, D.; Le Mouélic, S.; Carnec, C.; Maisons, C.; King, C. Urban subsidence in the city of Prato (Italy) monitored by satellite radar interferometry. *Int. J. Remote Sens.* **2003**, *24*, 891–897. [CrossRef]

17. Rosi, A.; Tofani, V.; Agostini, A.; Tanteri, L.; Stefanelli, C.T.; Catani, F.; Casagli, N. Subsidence mapping at regional scale using persistent scatters interferometry (PSI): The case of Tuscany region (Italy). *Int. J. Appl. Earth Obs.* **2016**, *52*, 328–337. [CrossRef]

18. Del Soldato, M.; Farolfi, G.; Rosi, A.; Raspini, F.; Casagli, N. Subsidence evolution of the Firenze–Prato–Pistoia plain (Central Italy) combining PSI and GNSS data. *Remote Sens.* **2018**, *10*, 1146. [CrossRef]

19. Motagh, M.; Djamour, Y.; Walter, T.R.; Wetzel, H.U.; Zschau, J.; Arabi, S. Land subsidence in Mashhad Valley, northeast Iran: results from InSAR, levelling and GPS. *Geophys. J. Int.* **2007**, *168*, 518–526. [CrossRef]

20. Akbari, V.; Motagh, M. Improved ground subsidence monitoring using small baseline SAR interferograms and a weighted least squares inversion algorithm. *IEEE Geosci. Remote Sens.* **2012**, *9*, 437–441. [CrossRef]

21. Dehghani, M.; Valadan Zoej, M.J.; Entezam, I.; Mansourian, A.; Saatchi, S. InSAR monitoring of progressive land subsidence in Neyshabour, northeast Iran. *Geophys. J. Int.* **2009**, *178*, 47–56. [CrossRef]

22. Motagh, M.; Shamshiri, R.; Haghighi, M.H.; Wetzel, H.-U.; Akbari, B.; Nahavandchi, H.; Roessner, S.; Arabi, S. Quantifying groundwater exploitation induced subsidence in the Rafsanjan plain, southeastern Iran, using InSAR time-series and in situ measurements. *Eng. Geol.* **2017**, *218*, 134–151. [CrossRef]

23. Sadeghi, Z.; Zoej, M.J.V.; Dehghani, M.; Chang, N.B. Enhanced algorithm based on persistent scatterer interferometry for the estimation of high-rate land subsidence. *J. Appl. Remote Sens.* **2012**, *6*. [CrossRef]

24. Khorrami, M.; Abrishami, S.; Maghsoudi, Y. Mashhad Subsidence Monitoring by Interferometric Synthetic Aperture Radar Technique. *AUT J. Civ. Eng.* **2018**. [CrossRef]

25. Foroughnia, F.; Nemati, S.; Maghsoudi, Y.; Perissin, D. An iterative PS-InSAR method for the analysis of large spatio-temporal baseline data stacks for land subsidence estimation. *Int. J. Appl. Earth Obs.* **2019**, *74*, 248–258. [CrossRef]

26. Hu, J.; Ding, X.L.; Li, Z.W.; Zhang, L.; Zhu, J.J.; Sun, Q.; Gao, G.J. Vertical and horizontal displacements of Los Angeles from InSAR and GPS time series analysis: Resolving tectonic and anthropogenic motions. *J. Geodyn.* **2016**, *99*, 27–38. [CrossRef]

27. Perissin, D. Interferometric SAR Multitemporal Processing: Techniques and Applications. In *Multitemporal Remote Sensing*; Springer: New York, NY, USA, 2016; pp. 145–176. [CrossRef]

28. Lanari, R.; Lundgren, P.; Manzo, M.; Casu, F. Satellite radar interferometry time series analysis of surface deformation for Los Angeles, California. *Geophys. Res. Lett.* **2004**, *31*. [CrossRef]

29. Watson, K.M.; Bock, Y.; Sandwell, D.T. Satellite interferometric observations of displacements associated with seasonal groundwater in the Los Angeles basin. *J. Geophys. Res.-Sol. Earth* **2002**, *107*. [CrossRef]

30. Zhang, L.; Lu, Z.; Ding, X.; Jung, H.S.; Feng, G.; Lee, C.W. Mapping ground surface deformation using temporarily coherent point SAR interferometry: Application to Los Angeles Basin. *Remote Sens. Environ.* **2012**, *117*, 429–439. [CrossRef]

31. Borchers, J.W.; Carpenter, M.; Grabert, V.; Dalgish, B.; Cannon, D. Land subsidence from groundwater use in California. In *California Water Foundation Full Report of Findings*; California Water Foundation: Sacramento, CA, USA, 2014; p. 151.

32. Khorrami, M.; Hatami, M.; Alizadeh, B.; Khorrami, H.; Rahgozar, P.; Flood, I. Impact of Ground Subsidence on Groundwater Quality: A Case Study in Los Angeles, California. In Proceedings of the 2019 ASCE International Conference on Computing in Civil Engineering, Atlanta, GA, USA, 17–19 June 2019. (In Press)

33. Nikvar Hassani, A.; Katibeh, H.; Farhadian, H. Numerical analysis of steady-state groundwater inflow into Tabriz line 2 metro tunnel, northwestern Iran, with special consideration of model dimensions. *Bull. Eng. Geol. Environ.* **2016**, *75*, 1617–1627. [CrossRef]

34. Rowe, R.K.; Lee, K.M. Subsidence owing to tunnelling. II. Evaluation of a prediction technique: Reply. *Can. Geotech. J.* **1994**, *31*, 467–469. [CrossRef]

35. Sharma, J.R.; Najafi, M.; Marshall, D.; Kaushal, V.; Hatami, M. Development of A Model For Estimation of Buried Large Diameter Thin-walled Steel Pipe Deflection Due To External Loads "In Press". *J. Pipeline Syst. Eng.* **2019**. [CrossRef]

36. Hatami, M.; Ameri Siahooei, E. Examines criteria applicable in the optimal location new cities, with approach for sustainable urban development. *Middle-East J. Sci. Res.* **2013**, *14*, 734–743.

37. Riding, K.A.; Peterman, R.J.; Guthrie, S.; Brueseke, M.; Mosavi, H.; Daily, K.; Risovi-Hendrickson, W. Environmental and Track Factors That Contribute to Abrasion Damage. In Proceedings of the 2018 Joint Rail Conference, Pittsburgh, PA, USA, 18–20 April 2018.

38. Shi, Y.J.; Li, M.G.; Chen, J.J.; Wang, J.H. Long-Term Settlement Behavior of a Highway in Land Subsidence Area. *J. Perform. Constr. Fac.* **2018**, *32*. [CrossRef]

39. Lahijanian, B.; Zarandi, M.F.; Farahani, F.V. Double coverage ambulance location modeling using fuzzy traveling time. In Proceedings of the Fuzzy Information Processing Society (NAFIPS), 2016 Annual Conference of the North American, El Paso, TX, USA, 31 October–4 November 2016; pp. 1–6.

40. Argus, D.F.; Heflin, M.B.; Peltzer, G.; Crampé, F.; Webb, F.H. Interseismic strain accumulation and anthropogenic motion in metropolitan Los Angeles. *J. Geophys. Res.-Sol. Earth* **2005**, *110*. [CrossRef]

41. Gens, R.; Van Genderen, J.L. Review Article SAR interferometry—Issues, techniques, applications. *Int. J. Remote Sens.* **1996**, *17*, 1803–1835. [CrossRef]

42. Klees, R.; Massonnet, D. Deformation measurements using SAR interferometry: potential and limitations. *Geologie en Mijnbouw* **1998**, *77*, 161–176. [CrossRef]

43. Smith, L.C. Emerging applications of interferometric synthetic aperture radar (InSAR) in geomorphology and hydrology. *Ann. Assoc. Am. Geogr.* **2002**, *92*, 385–398. [CrossRef]

44. Alsdorf, D.E.; Melack, J.M.; Dunne, T.; Mertes, L.A.; Hess, L.L.; Smith, L.C. Interferometric radar measurements of water level changes on the Amazon flood plain. *Nature* **2000**, *404*, 174. [CrossRef]

45. Peduto, D.; Huber, M.; Speranza, G.; van Ruijven, J.; Cascini, L. DInSAR data assimilation for settlement prediction: case study of a railway embankment in the Netherlands. *Can. Geotech. J.* **2016**, *54*, 502–517. [CrossRef]

46. Daghighi, A.; Nahvi, A.; Kim, U. Optimal Cultivation Pattern to Increase Revenue and Reduce Water Use: Application of Linear Programming to Arjan Plain in Fars Province. *Agriculture* **2017**, *7*, 73. [CrossRef]

47. Hooper, A.; Bekaert, D.; Spaans, K.; Arıkan, M. Recent advances in SAR interferometry time series analysis for measuring crustal deformation. *Tectonophysics* **2012**, *514*, 1–13. [CrossRef]

48. Bayer, B.; Simoni, A.; Schmidt, D.; Bertello, L. Using advanced InSAR techniques to monitor landslide deformations induced by tunneling in the Northern Apennines, Italy. *Eng. Geol.* **2017**, *226*, 20–32. [CrossRef]

49. Shafieardekani, M.; Hatami, M. Forecasting Land Use Change in suburb by using Time series and Spatial Approach; Evidence from Intermediate Cities of Iran. *Eur. J. Sci. Res.* **2013**, *116*, 199–208.

50. Perissin, D.; Wang, Z.; Lin, H. Shanghai subway tunnels and highways monitoring through Cosmo-SkyMed Persistent Scatterers. *ISPRS J. Photogr.* **2012**, *73*, 58–67. [CrossRef]

51. Lundgren, P.; Hetland, E.A.; Liu, Z.; Fielding, E.J. Southern San Andreas—San Jacinto fault system slip rates estimated from earthquake cycle models constrained by GPS and interferometric synthetic aperture radar observations. *J. Geophys. Res.-Sol. Earth* **2009**, *114*. [CrossRef]

52. Wei, M.; Sandwell, D.; Smith-Konter, B. Optimal combination of InSAR and GPS for measuring interseismic crustal deformation. *Adv. Space Res.* **2010**, *46*, 236–249. [CrossRef]

53. Bawden, G.W.; Thatcher, W.; Stein, R.S.; Hudnut, K.W.; Peltzer, G. Tectonic contraction across Los Angeles after removal of groundwater pumping effects. *Nature* **2001**, *412*, 812. [CrossRef] [PubMed]

54. Riel, B.; Simons, M.; Ponti, D.; Agram, P.; Jolivet, R. Quantifying ground deformation in the Los Angeles and Santa Ana Coastal Basins due to groundwater withdrawal. *Water Resour. Res.* **2018**, *54*, 3557–3582. [CrossRef]

55. Sharma, P.; Jones, C.E.; Dudas, J.; Bawden, G.W.; Deverel, S. Monitoring of subsidence with UAVSAR on Sherman Island in California's Sacramento–San Joaquin Delta. *Remote Sens. Environ.* **2016**, *181*, 218–236. [CrossRef]

56. Jeanne, P.; Farr, T.G.; Rutqvist, J.; Vasco, D.W. Role of agricultural activity on land subsidence in the San Joaquin Valley, California. *J. Hydrol.* **2019**, *569*, 462–469. [CrossRef]

57. Tapete, D.; Morelli, S.; Fanti, R.; Casagli, N. Localising deformation along the elevation of linear structures: An experiment with space-borne InSAR and RTK GPS on the Roman Aqueducts in Rome, Italy. *Appl. Geogr.* **2015**, *58*, 65–83. [CrossRef]

58. Bekaert, D.; Hamlington, B.; Buzzanga, B.; Jones, C. Spaceborne Synthetic Aperture Radar Survey of Subsidence in Hampton Roads, Virginia (USA). *Sci. Rep.-UK* **2017**, *7*, 14752. [CrossRef] [PubMed]

59. Rabus, B.; Eppler, J.; Sharma, J.; Busler, J. Tunnel monitoring with an advanced InSAR technique. In Proceedings of the Radar Sensor Technology XVI, Baltimore, MD, USA, 23–25 April 2012; p. 83611F.

60. Nahvi, A.; Daghighi, A.; Nazif, S. The environmental impact assessment of drainage systems: a case study of the Karun river sugarcane development project. *Arch. Agron. Soil Sci.* **2018**, *64*, 185–195. [CrossRef]

61. Davidson, M.; Attema, E.; Rommen, B.; Floury, N.; Partricio, L.; Levrini, G. ESA sentinel-1 SAR mission concept. In Proceedings of the EUSAR, Dresden, Germany, 16–18 May 2006.

62. Metro. Webpage of Sepulveda Transit Corridor Project. Available online: https://www.metro.net/projects/sepulvedacorridor/ (accessed on 5 August 2018).

63. Ferretti, A.; Prati, C.; Rocca, F. Permanent scatterers in SAR interferometry. *IEEE Trans. Geosci. Remote* **2001**, *39*, 8–20. [CrossRef]

64. Canaslan Comut, F.; Ustun, A.; Lazecky, M.; Perissin, D. Capability of Detecting Rapid Subsidence with COSMO SKYMED and Sentinel-1 Dataset over Konya City. In Proceedings of the Living Planet Symposium, Prague, Czech Republic, 9–13 May 2016; p. 295.

65. Perrone, G.; Morelli, M.; Piana, F.; Fioraso, G.; Nicolò, G.; Mallen, L.; Cadoppi, P.; Balestro, G.; Tallone, S. Current tectonic activity and differential uplift along the Cottian Alps/Po Plain boundary (NW Italy) as derived by PS-InSAR data. *J. Geodyn.* **2013**, *66*, 65–78. [CrossRef]

66. Hanssen, R.F. *Radar Interferometry: Data Interpretation and Error Analysis*; Springer Science & Business Media: New York, NY, USA, 2001; Volume 2.

67. Ferretti, A.; Prati, C.; Rocca, F. Nonlinear subsidence rate estimation using permanent scatterers in differential SAR interferometry. *IEEE Trans. Geosci. Remote* **2000**, *38*, 2202–2212. [CrossRef]

68. Colesanti, C.; Ferretti, A.; Novali, F.; Prati, C.; Rocca, F. SAR monitoring of progressive and seasonal ground deformation using the permanent scatterers technique. *IEEE Trans. Geosci. Remote* **2003**, *41*, 1685–1701. [CrossRef]

69. Murray, J.R.; Svarc, J. Global Positioning System data collection, processing, and analysis conducted by the US Geological Survey Earthquake Hazards Program. *Seismol. Res. Lett.* **2017**, *88*, 916–925. [CrossRef]

70. Metro. *Final Compendium Report of Sepulveda Transit Corridor: Appendix B, Geotechnical Evaluation Memorandum*; The Los Angeles County Metropolitan Transportation Authority: Los Angeles, CA, USA, 2012.

71. CGS. *California Geological Survey: Seismic Hazard Zones, Beverly Hills Quadrangle, Official Map*; CGS: Sacramento, CA, USA, 1999.

72. CGS. *California Geological Survey: Seismic Hazard Zone Report for the Van Nuys 7.5-Minute Quadrangle*; CGS: Sacramento, CA, USA, 1997.

73. CGS. *California Geological Survey: Seismic Hazard Zone Report for the Venice 7.5-Minute Quadrangle*; CGS: Sacramento, CA, USA, 1998.

74. California's Groundwater. *Groundwater Basin Number: 4-12, San Fernando Valley Groundwater Basin, Bulletin 118*; California's Groundwater, 2004. Available online: https://www.bhusd.org/pdf/seismic_reports//___36_Leighton%20El%20Rodeo%20Geohazards%20Report_3-2-15.pdf (accessed on 8 September 2018).

75. California's Groundwater. *Groundwater Basin Number: 4-11.01, Coastal Plain of Los Angeles Groundwater Basin, Santa Monica Subbasin, Bulletin 118*; California's Groundwater, 2004. Available online: http://www.water.ca.gov/groundwater/bulletin118/basindescriptions/4-11.01.pdf (accessed on 12 September 2018).

76. LADWP. *Feasibility Report for Development of Groundwater Resources in the Santa Monica and Hollywood Basins*; LADWP: Los Angeles, CA, USA, 2011.

77. Poland, J.F.; Garrett, A.A.; Sinnott, A. *Geology, Hydrology, and Chemical Character of Ground Waters in the Torrance-Santa Monica Area, California*; US Government Printing Office: Washington, DC, USA, 1959.

78. Norris, R.M.; Webb, R.W. *Geology of California*; John Wiley & Sons Inc.: Hoboken, NJ, USA, 1990.

79. Ninyo & Moore Geotechnical & Environmental Sciences Consultants. Geotechnical Evaluation Sepulveda Feeder Interconnection Project Culver City, California. 2009. Available online: https://dpw.lacounty.

gov/wwd/web/Documents/Reports/Sepulveda%20Feeder%20Service%20Connection.pdf (accessed on 7 August 2018).

80. Shannon & Wilson, Inc. Geology and Soil Discipline Report: the Academy Museum of Motion Picture, Los Angeles. 2014. Available online: https://planning.lacity.org/eir/AcademyMuse_MotionPictures/DEIR/DEIR/Technical_Appendices/Appendix_H-1_Methane_Report.pdf (accessed on 14 June 2018).

81. Geotechnologies, Inc. Geotechnical Report: Geotechnical Engineering Investigation Proposed Mixed-Use Development 6001–6059 Van Nuys Boulevard, Van Nuys, California. 2015. Available online: https://planning.lacity.org/eir/nops/6001VanNuys/appendixa.pdf (accessed on May 2018).

82. Leighton Consulting, Inc. Geohazard Report: El Rodeo K-8 School 605 Whittier Drive, Beverly Hills, Los Angeles. 2015. Available online: https://www.bhusd.org/pdf/seismic_reports//____36_Leighton%20El%20Rodeo%20Geohazards%20Report_3-2-15.pdf (accessed on 16 June 2018).

83. Ninyo & Moore Geotechnical & Environmental Sciences Consultants. Runway 25R Reconstruction Project: Runway 25R-7L Improvements, Los Angeles International Airport (LAX), California. 2017. Available online: http://www.labavn.org/contracts/documents/81/30368/Runway%2025R%20Reconstruction%20RFB.pdf (accessed on 19 April 2018).

84. SCS Engineers, Inc. Groundwater Monitoring Program, Inglewood Field. 2018. Available online: https://inglewoodoilfield.com/csd-related-plans (accessed on 2 April 2018).

85. Hodgkinson, K.M.; Stein, R.S.; Hudnut, K.W.; Satalich, J.; Richards, J.H. *Damage and Restoration of Geodetic Infrastructure Caused by the 1994 Northridge, California, Earthquake*; US Geological Survey: Reston, VA, USA, 1996.

86. Martz, P. Asymmetrical Subsidence Resulting from Material and Fluid Extraction. 2009. Available online: https://open.library.ubc.ca/media/download/pdf/52966/1.0053573/1 (accessed on 23 May 2018).

87. Jozaghi, A.; Alizadeh, B.; Hatami, M.; Flood, I.; Khorrami, M.; Khodaei, N.; Ghasemi Tousi, E. A Comparative Study of the AHP and TOPSIS Techniques for Dam Site Selection Using GIS: A Case Study of Sistan and Baluchestan Province, Iran. *Geosciences* **2018**, *8*, 494. [CrossRef]

88. Hatami, M.; Shafieardekani, M. The Effect of Industrialization on Land Use Changes; Evidence from Intermediate Cities of Iran. *Int. J. Curr. Life Sci.* **2014**, *4*, 11899–11902.

89. Desir, G.; Gutiérrez, F.; Merino, J.; Carbonel, D.; Benito-Calvo, A.; Guerrero, J.; Fabregat, I. Rapid subsidence in damaging sinkholes: Measurement by high-precision leveling and the role of salt dissolution. *Geomorphology* **2018**, *303*, 393–409. [CrossRef]

remote sensing

MDPI

Article

Subsidence Zonation Through Satellite Interferometry in Coastal Plain Environments of NE Italy: A Possible Tool for Geological and Geomorphological Mapping in Urban Areas

Mario Floris [1],*, Alessandro Fontana [1], Giulia Tessari [2] and Mariachiara Mulè [1]

[1] Department of Geosciences, University of Padua, 35131 Padua, Italy; alessandro.fontana@unipd.it (A.F.); mulemariachiara28@gmail.com (M.M.)

[2] sarmap SA, Cascine di Barico, 6989 Purasca, Switzerland; giulia.tessari@sarmap.ch

* Correspondence: mario.floris@unipd.it; Tel.: +39-049-827-9121

Received: 29 November 2018; Accepted: 11 January 2019; Published: 16 January 2019

Abstract: The main aim of this paper is to test the use of multi-temporal differential interferometric synthetic aperture radar (DInSAR) techniques as a tool for geological and geomorphological surveys in urban areas, where anthropogenic features often completely obliterate landforms and surficial deposits. In the last two decades, multi-temporal DInSAR techniques have been extensively applied to many topics of Geosciences, especially in geohazard analysis and risks assessment, but few attempts have been made in using differential subsidence for geological and geomorphological mapping. With this aim, interferometric data of an urbanized sector of the Venetian-Friulian Plain were considered. The data derive by permanent scatterers InSAR processing of synthetic aperture radar (SAR) images acquired by ERS 1/2, ENVISAT, COSMO SKY-Med and Sentinel-1 missions from 1992 to 2017. The obtained velocity maps identify, with high accuracy, the border of a fluvial incised valley formed after the last glacial maximum (LGM) and filled by unconsolidated Holocene deposits. These consist of lagoon and fluvial sediments that are affected by a much higher subsidence than the surrounding LGM deposits forming the external plain. Displacement time-series of localized sectors inside the post-LGM incision allowed the causes of vertical movements to be explored, which consist of the consolidation of recent deposits, due to the loading of new structures and infrastructures, and the exploitation of the shallow phreatic aquifer.

Keywords: geological and geomorphological mapping; Late-Quaternary deposits; differential compaction; multi-temporal DInSAR; Venetian-Friulian Plain

1. Introduction

In coastal areas and urbanized zones, the recent sedimentation or shallow deposits, even anthropogenic, generally bury the previous deposits that can often be rather different from surface formations. This setting frequently hampers the correct assessment of the subsoil, even in the first 5–30 m. This paper analyzes the possible relationship existing between the geological and geomorphological features of an urbanized sector of the coastal plain located in north eastern Italy, and its rate of subsidence measured by multi-temporal differential synthetic aperture radar interferometry (DInSAR) techniques. Here, we test the potential of this method on reconstructing the shallow stratigraphic sequence in areas where traditional in situ and remote sensing surveys, such as geological and geomorphological field work and air-photo interpretation, are difficult or impossible because of the presence of anthropogenic structures.

Land subsidence commonly affects urban areas as a consequence of intensive groundwater exploitation, which reduces the pore water pressure and activates soil consolidation processes. Several cases are deeply studied all around the world as Kolkata [1], Bucharest [2], most of the big cities in Central Mexico area [3]. A well-known case study corresponds to the area of Venice and its mainland, which is rather close to the study area and where several pioneering researches were carried out [4–6] (and reference therein). Another frequent cause of subsidence is the realization of new buildings and infrastructure that are underground excavations and tunneling, which alter the subsoil stress conditions and trigger compaction effects [7–9].

Monitoring of land subsidence could be performed through conventional techniques, which include repeated leveling or global positioning system (GPS) surveys [10–12]. Despite the relatively high horizontal and vertical accuracy, the main limitations of these monitoring strategies are the punctual nature and low resolution of the measurements. Alternatively, remote sensing techniques, like unmanned aerial vehicle (UAV) [13], airborne laser scanning [14], or airborne surveys in general, lead to distributed information over the area of interest. Unfortunately, a dense temporal resolution of these measurements is costly and time-consuming, limiting their availability to very few areas. Therefore, in the last two decades, DInSAR techniques have been extensively applied to estimate displacements caused by subsidence [15–19]. DInSAR techniques provide a good compromise between the temporal and spatial resolution of these measurements, which could be effective for the analysis of surface deformation over extended areas.

When applied to urbanized areas, DInSAR techniques are generally used to detect and assess surface deformations and damage induced on buildings, or other anthropogenic structures by natural or human-induced processes. An increasing number of studies have been performed on the effects of recent urbanization and subsidence effects, exploiting space-borne satellite data and trying to find a connection between interferometric remote sensing techniques, civil engineering, and urban developing planning. Some recent applications focused on cross-rail, being in London [20], on the effect of differential subsidence affecting buildings in some Rome neighborhoods [21], or on bridge-monitoring [22].

Due to the numerous advantages of DInSAR techniques and the growing availability of synthetic aperture radar (SAR) satellite data, amplitude and phase information from SAR images have also been used by applying different techniques in order to investigate many topics of geosciences, including geology [23,24]. Most of the geological applications are related to earthquakes [25–27], volcanic eruptions [28–30], tectonics [31–34], and landslides [35–38], while few research has been focused on the potential of multi-temporal DInSAR as a tool for geological and geomorphological mapping [39]. This represents the main aim of our research.

To test our hypothesis, we considered an area near the city of Portogruaro, in the eastern part of the Venetian Plain (Figure 1) at the passage from the alluvial to the coastal plain, where fluvial and lagoon/coastal deposits are present. In this area, the on-going subsidence was investigated at a regional scale. The combined use of DInSAR and DGPS measurements highlighted the occurrence of a zone where the subsidence value reaches up to 2–7 mm/year, while in the surrounding zones the average values are between 0 and 1 mm/year [6,12,40,41]. This down-lifting area is elongated in N-S direction, has an average width between 1 and 2 km [41] and, when compared to geological maps [42], seems to coincide with a major incised filled fluvial valley existing in the area. The sedimentary deposits filling this valley are very different from the ones forming the external alluvial plain and they are characterized by a larger compressibility. These characteristics led to mapping the differential subsidence currently affecting the area by multi-temporal DInSAR techniques, and check whether the pattern of down-lift matches with the planform of the buried valley so that it can eventually improve the detection of its boundaries. The comparison between remote-sensed data and geological ground truth is supported by the availability of recent geological maps, and a huge database of stratigraphic cores and geotechnical tests [42,43].

In the next sections, the main geological and geomorphological features of the study area are reported (Section 2); after the description of SAR data and processing and post-processing methods used to test the contribution of interferometric data in geological and geomorphological mapping

(Section 3), the obtained results are presented (Section 4). Finally, the results are discussed in Section 5 where, due to the large amount of gathered information, we will preliminary explore the possible causes of the subsidence.

2. Study area

This research analyzes the distal sector of the alluvial megafan of Tagliamento River (Figure 1), that is fed by the Carnic and Julian Alps and is one of the major streams of the Venetian-Friulian Plain [44]. The plain corresponds to the foreland basin of the south-eastern Alps and is formed by Plio-Quaternary deposits, which along the coastal sector, between Tagliamento and Livenza rivers, have a thickness from 500 to 800 meters [45]. Active tectonic structures are not present in the study area, but the distal plain is affected by a long-term subsidence related to crustal flexuring and the compaction of Quaternary deposits, with an average vertical rate of −0.4 mm/year in the last 125 kyr [40,46].

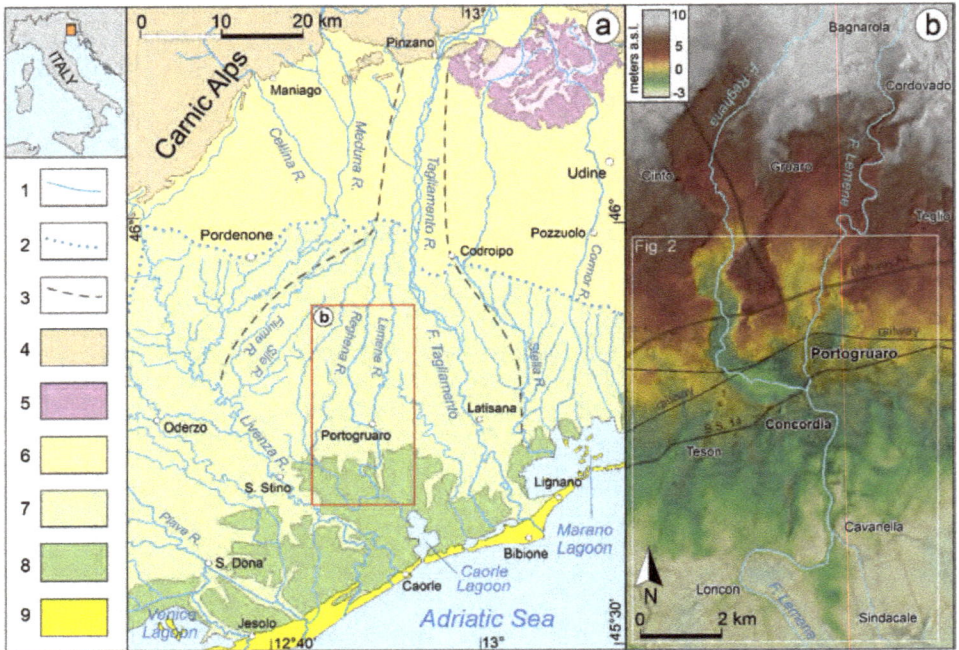

Figure 1. (a) Simplified geomorphological sketch of north-eastern Italy, with an indication of the study area (red square). Legend: (1) rivers; (2) upstream limit of the spring line; (3) boundary of the Tagliamento alluvial megafan; (4) Alps; (5) morainic amphitheater; (6) gravelly plain; (7) fine-dominated distal plain; (8) reclaimed areas currently under sea level; 9) coastal sand ridges and beaches. (b) Digital elevation model of the study area (modified from [47]).

In the study area the first subsoil consists of Late-Quaternary alluvial sediments, alternated with coastal deposits. A major phase of deposition occurred during the Last Glacial Maximum (LGM, 29–19 kyr BP [48]), when the Tagliamento alluvial megafan was formed and 15-30 m of alluvial sediments aggraded over the whole Venetian-Friulian Plain. During that period, the mountain catchment of Tagliamento hosted a major Alpine glacier, which reached the plain with its front (#5 in Figure 1a [49]). The Tagliamento River was one of the main glacial outwashes but, at that time, it was characterized by an unconfined channel, which transported the gravel only up to 15-25 km from the glacial front, while sands, silts, and clays reached the distal sector of the plain [50]. Thus, the distal portion of the LGM megafan of Tagliamento is dominated by fine sediments and along the boundary, between coarse (permeable) and fine sediments (impermeable), a belt of springs feed a dense network of minor streams (Figure 1a). These are groundwater-fed rivers, which are characterized by a rather steady water discharge along the year and almost no sedimentary load, as they originate in the middle of the plain [51] (and reference therein).

Since 19.5 kyr BP, the front of Tagliamento glacier withdrawn from the plain and, consequently, the fluvial system experienced a severe starvation in the sediment supply that led the river to entrench along few narrow incised valleys [50]. This process induced the river to abandon, almost completely, the alluvial megafan, leading the LGM surface to be exposed over large sectors of the plain up to the present (Figure 2a). Where the LGM surface is still cropping out, is marked by a rather well-developed soil, which is over consolidated and characterized by the occurrence of calcic horizon [51] (and reference therein) [52].

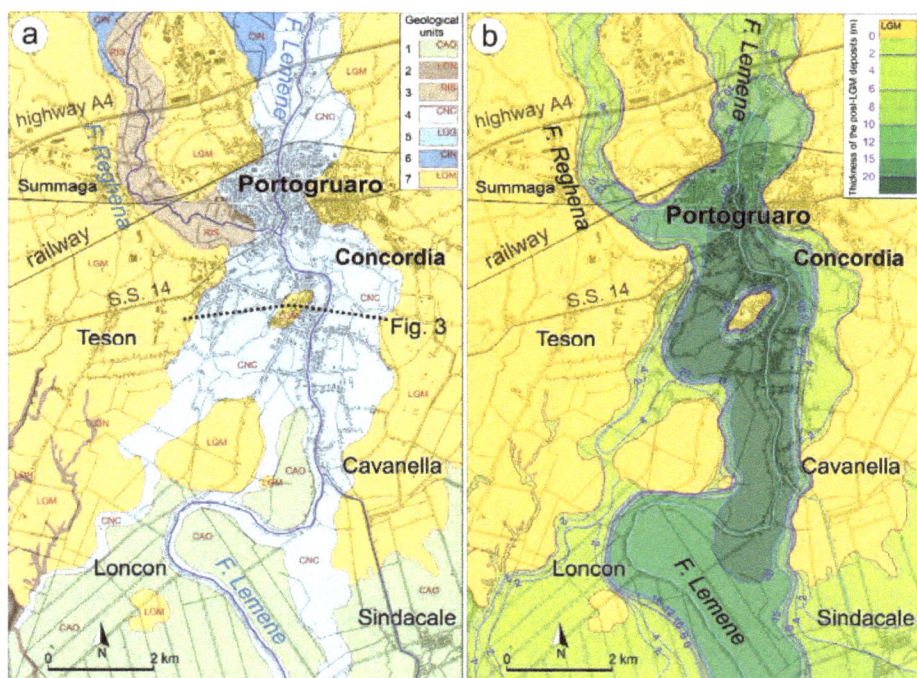

Figure 2. (**a**) Map of the geological units (after [43]). Legend: (1) lagoon deposits of late Holocene; (2) swamp organic deposits; (3) organic deposits at the bottom of the valley of Reghena River; (4) alluvial deposits of Early Middle Age; (5) alluvial deposits of Roman age; (6) alluvial deposits of early Holocene; (7) Last Glacial Maximum (LGM) alluvial deposits. (**b**) Map of the thickness of the post-LGM deposits (modified from [42]).

Two of the major fluvial valleys incised by the ancient Tagliamento, in the post-LGM, have been occupied by Lemene and Reghena River, which are important groundwater-fed streams (Figure 2b). The incised landforms can be recognized in the landscape up to Portogruaro, where the rivers join. The geomorphological evolution occurred along the Holocene brought to the abandonment of the incised valleys and their progressive infill, leading to the obliteration of their topographic evidence in the coastal plain. The combined analyses of detailed digital elevation models (DEMs) and stratigraphic cores that can recognize and characterize the fluvial incision between Portogruaro and the Lagoon of

Caorle (Figure 2b). This buried sector of the incised valley has been also named the valley of Concordia, after the name of the Roman city of *Julia Concordia*, that was built over a remnant terrace of LGM plain isolated inside the valley [42,47,50].

In the distal plain, the fluvial incisions were active between Late Glacial and Early Holocene (i.e., 19–8 kyr BP) and these landforms were between 500 and 2000 m wide and reached a maximum depth of 20 m to the top of the LGM (Figures 2b and 3 [42,51]). Because of the funneling of the river flux, at the bottom of the incised valleys the gravels could be transported far more downstream than during the LGM and reached the present coastal plain. Avulsion processes occurred upstream of the study area between 9.6 and 8.4 kyr BP and caused the eastern shifting of the Tagliamento River, leading to the abandonment of the incision. The valley of Concordia was rapidly waterlogged and occupied by swampy environments that favored the accumulation of up to 1.5 m of peat and organic sediments (#5 in Figure 3).

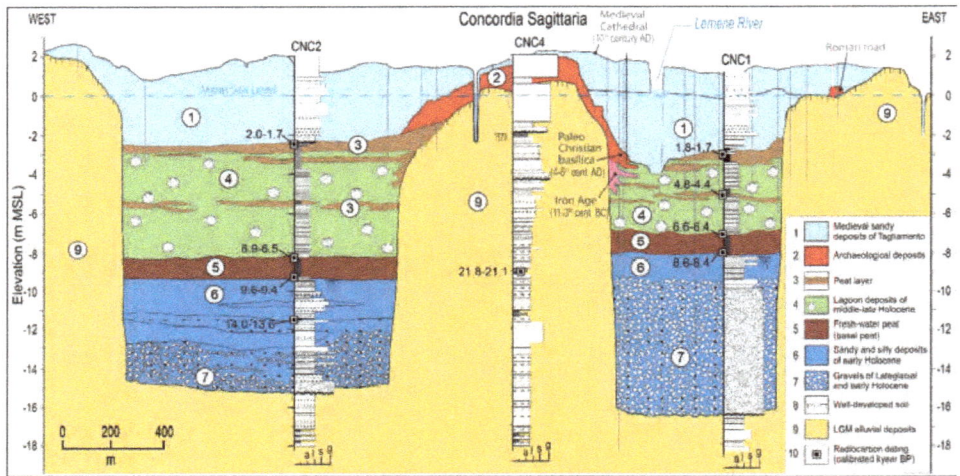

Figure 3. Reference cross section of the stratigraphic setting near Concordia Sagittaria (modified after [47]). The location of the section is reported in Figure 2.

Between 8.5 and 7.5 kyr BP, the post-LGM marine transgression reached the present coast [53] and led the lagoon waters to expand along the pre-existing depressed areas, as the abandoned fluvial incisions. Thus, the brackish environment occupied the bottom of the valley up to the center of Portogruaro, and deposited within the incised valley a greenish gray muddy unit characterized by the common occurrence of lagoon fossils and some lenses of peat. This brackish and swampy setting characterized the valley of Concordia until the early Medieval, when an important avulsion phase led the Tagliamento to temporarily activate a branch along the present Lemene River [42,47]. Between the 6th and 8th century AD, the river floods deposited a huge quantity of sediment that completely buried the valley downstream of Portogruaro and sealed large sectors of the ancient city of *Julia Concordia* [47] (and references therein). This phase formed a remarkable fluvial ridge, which is visible from the highway A4 almost to the present lagoon (Figure 1b). The Lemene River is currently flowing along the residual channel of Tagliamento, that was maintained open and prone to the activity of groundwater after a sudden avulsion, which moved the Alpine river to its present direction near Latisana (Figure 1a [47]).

The last important phase in shaping the present landscape occurred in the first part of the 20th century, when large sectors of the Caorle Lagoon had been reclaimed for agricultural purposes. Nowadays, between Tagliamento and Livenza rivers, about 100 km^2 are lower than sea and are drained thanks to the lagoon dykes and a complex network of ditches, canals, and pumping stations (#8 in Figure 1b).

3. Materials and Methods

The evolution and rate of surface deformations have been obtained through the processing of several space borne synthetic aperture radar (SAR) datasets, acquired by different national and international missions, and characterized by various ground resolutions, satellite revisiting time, and acquisition geometries. As listed in Table 1, ERS-1/2, ENVISAT, COSMO-SkyMed, and Sentinel-1 datasets have been considered. The main specifics of each dataset influence the expected results. One of the parameters, which could condition the multi-temporal DInSAR results is the wavelength, which determines the data sensitivity to surface variation and vegetation changes. Moreover, the satellite revisiting time acts on the data temporal decorrelation, therefore the multi-temporal coherence and the persistent scatterers density tends to increase as the time span between subsequent images decreases. All the technical details of the considered SAR data are reported in Table 1. The availability of this archive data has allowed the reconstruction of almost 26 year deformations, from 1992 up to 2017, with some limited temporal gaps. For all the datasets, descending acquisition geometry has been considered because of the larger amount of available scenes, in particular for the ERS, ENVISAT and COSMO-SkyMed (CSK) datasets. The unique acquisition geometry allows a consistent comparison of the results along the line-of-sight (LOS), despite some differences in the incidence angles. Only the ERS data, both ascending and descending datasets, have been considered to verify whether the expected vertical direction of deformation, common for the subsidence phenomenon, could be confirmed.

Table 1. Main characteristics of synthetic aperture radar (SAR) data considered in this study.

Satellite Mission	Orbit	Period	N. of Images	Revisiting Time (Days)	Band/ Wavelength (cm)	Resol. az./Range (m)	Line-of-Sight (LOS) Incidence Angle, θ	LOS Azimut, α
ERS-1/2	Desc. Asc.	06/14/1992– 12/13/2000 08/01/1995– 08/30/2000	63 37	36	C/5.6	6/24	~23°	~274° ~85°
ENVISAT	Desc.	04/02/2003– 07/14/2010	71	36	C/5.6	6/24	~23°	~274°
COSMO– SkyMED	Desc.	02/18/2012– 01/12/2016	66	12	X/3.1	2.5/2.5	~33°	~277°
Sentinel-1	Desc.	12/23/2014– 07/22/2017	91	6/12	C/5.6	5/20	~37°	~277°

The multi-temporal DInSAR techniques extend InSAR analyses to retrieve the spatio-temporal evolution of deformations over large areas, considering a stack of data. In this context, the numerous approaches, developed in the last two decades, can be classified into two main categories, the persistent scatterers interferometry (PSI) [54,55] and the small baseline subset (SBAS) [56]. Generally, the PSI approach generates all the interferograms referred to as a common master image, detecting point targets characterized by a stable back-scattered signal over time, and a high coherence between different acquisitions. The SBAS algorithm maximizes the spatio-temporal coherence by relying on interferograms characterized by small perpendicular baseline values. Therefore, PSI is generally applied to analyze deformation affecting urban areas while SBAS is more adequate on distributed scattering conditions.

Here, data processing was performed through the PSI technique as the study area is densely urbanized. This remote sensing technique can measure Earth surface displacement from space, with millimetric sensitivity. This method exploits multiple SAR scenes acquired over the same area and, through the algorithm proposed by [54,55], is able to separate the displacement component of the phase from the back-scattered signal. Identifying the persistent scatterers (PS) candidates depends on the dispersion of the amplitude of the backscatter signal in time, but additionally a multi-temporal coherence threshold could be defined to filter the most reliable points. Thus, the density of the output results is strongly dependent on the land use and, in our case, on the urbanization density.

The main output of the PSI processing is a mean deformation velocity map along the satellite line of sight (LOS) showing a velocity value for each of the selected PS. Furthermore, the time-series of displacements is obtained, providing not only the mean displacement rate but also the evolution in time of the deformation trends. This is essential information to support the results interpretation and its connection with a specific triggering factor. The precision of the deformation velocity depends on several factors, as the number of scenes and their temporal distribution, the PS density, their coherence, the characteristics of the deformation evolution in time, the reference point quality, the distance from the reference point [57]. According to [24], numerous PSI validations have been carried out in the last 15 years, providing an inter-comparison of common PSI results from different groups. In the framework of Terrafirma project [58], the standard deviation of the deformation velocity differences ranges between 0.4 and 0.5 mm/yr, while it reaches 1.1 to 4 mm for deformation time-series. These values refer to ERS and Envisat datasets, considering urban areas with zero or moderate deformation velocities.

ERS-1/2 and ENVISAT interferometric data, derived by PSI processing performed by TRE srl, were provided by the Italian Ministry of the Environment and for Protection of the Land and Sea in the framework of the "Not Ordinary Plan of Remote Sensing" project (http://www.pcn.minambiente.it/mattm/en/). COSMO SKY-Med and Sentinel-1 interferometric data were derived by the PSI processing of SAR images through the SARscape software developed by sarmap SA.

To support one of the main aims of this work, that is, the comparison of PSI results with ground information and specifically with thickness of post-LGM sediments, sparsely distributed PS velocities have been interpolated. The interpolation has been performed on ERS ascending and descending PSI results through the inverse distance weighted (IDW) method. To avoid the underestimation of the ground displacements and better investigate the possible causes of subsidence, interpolated maps have been combined to assess the vertical and horizontal (E-W) deformation components using the following equations [59–62]:

$$V_{horizontal} = \frac{\left(V_{descending}/h_{descending}\right) - \left(V_{ascending}/h_{descending}\right)}{\left(e_{descending}/h_{descending}\right) - \left(e_{ascending}/h_{ascending}\right)} \tag{1}$$

$$V_{vertical} = \frac{\left(V_{descending}/e_{descending}\right) - \left(V_{ascending}/e_{ascending}\right)}{\left(h_{descending}/e_{descending}\right) - \left(h_{ascending}/e_{ascending}\right)} \tag{2}$$

where h and e are the LOS directional cosines. If there are no horizontal components, as in the case hypothesized in this study, the vertical displacement rate can be easily derived by taking into account the LOS incidence angle (θ):

$$V_{vertical} = \frac{V_{ascending,descending}}{\cos \theta} \tag{3}$$

4. Results

Figure 4 reports the PS velocity maps derived by the PSI processing of the different SAR datasets with superimposed the isopach of the post-LGM sediments. Negative values (red) indicate an increase in the distance between satellites and PS measured in the LOS, a green color indicates a decrease in the LOS, and yellow colored PS indicate points supposed stables having a velocity between −1.5 and 1.5 mm/year. As mentioned in Section 3, SAR data, acquired in descending orbit, have been considered, but we have also reported the results from the processing of ERS 1/2 data acquired in ascending mode to detect whether horizontal components of the displacement are present. As the distribution and the values of velocity are very similar in the two maps, it can be inferred that the displacement is mainly vertical and indicates the subsidence of the area. This evidence was confirmed by the estimation of E-W and the vertical components, combining ERS ascending and descending PS LOS velocities (Figure 4a,b) through Equations (1) and (2) (Figure 5). A slightly horizontal component is present in few isolated pixels (Figure 5a), probably as a consequence of the deformation being related to buildings that can react in different ways to vertical displacements. Figure 5b shows that the rate of displacement is mainly vertical, with an increment of about 8.7%, compared to those measured along the LOS, taking into account the incidence angle of ERS acquisitions (Table 1) (see Equation (3)). In the case of ENVISAT, COSMO-SKyMed and Sentinel-1 acquisitions, the expected increment is about 8.7%, 19% and 25%, respectively.

PS density, as expected, increases from the oldest to the more recent interferometric data and with the resolution of SAR images: 61 PS/km^2(ERS), 102 PS/km^2(ENVISAT), 4765 PS/km^2(COSMO-SkyMED) and 1266 PS/km^2(Sentinel). This could be explained by considering the higher resolution of COSMO-SKyMed data with respect to all other datasets, the shorter revisiting time, and the small orbital tube of Sentinel-1 acquisitions, which reduced decorrelation effects. To get an acceptable density of PSI points in all four datasets, in the case of ERS and ENVISAT, the final results were filtered using a PS coherence >=0.6, while in the other two cases a coherence >=0.75.

As it can be seen in Figure 4, most of the negative values in the velocity fell inside the post-LGM incision in the entire observation period. The measured maximum velocities reach −10 mm/year, few and isolated PS showed positive values, which considered errors in the processing and were excluded from the results. The theoretical precisions of the measured deformation velocities are equal to 0.35 mm/year and 0.42 mm/year for ERS ascending and descending datasets, and 0.38 for the Envisat descending datasets. The velocity precisions are even better for Sentinel-1 and COSMO-SkyMed datasets, 0.28 mm/year and 0.25 mm/year respectively.

The relationship between the different displacement rates and the spatial distribution of surficial deposits is clearly evidenced in Figure 6, where a velocity map of the study area was created interpolating both ascending and descending ERS interferometric data by using the IDW method. Inside the post-LGM incision LOS velocities vary from −1 up to −10 mm/year, while outside the area can be considered stable with velocities ranging from −0.9 to +1 mm/year. Velocity cross sections through the post-LGM incision (Figure 6b–g) show the different displacement rates between Holocene deposits filling the incision and LGM deposits forming the external plain in different sectors. The sections clearly identify the border of the incision marked by a sudden increase in the velocity, but the rate of displacement can change both into the same section and in the different sectors. These variations can be related to the edification of new buildings during the observation period, which in most of cases represents the triggering factor of the ground displacement. Thus, the different velocities can be explained taking into account the time when soils have been overloaded, as will be better discussed in the next section. Velocity sections D-D′ and F-F′ (Figure 6e,g) are of particular interest because they show the effectiveness of PSI technique in geological and geomorphological mapping. Section D-D′ shows that, by assessing the subsidence rate, it is possible to identify the limits of the post-LGM incision and also the presence of the remnants of the alluvial terrace existing in correspondence of the ancient city of Concordia Sagittaria, which was built on the fully consolidated sediments of the LGM alluvial plain. In fact, the velocities range from −1 to 0.0 mm/year at the margin of the cross section and in correspondence of this isolated terrace. Also in section F-F′, it is possible to identify the margins of the post-LGM incision, which are marked by the sudden variation in the subsidence rate. In this case, the decrease of ground displacement rate recorded almost at the center of the profile (around 1500 m) is not related to the presence of a buried remnant of the LGM plain, but to the artificial embankment existing along the Lemene River. These anthropogenic structures were built during the 19th century and, probably, this rather long period led to an almost complete consolidation of the subsoil, limiting the on-going subsidence.

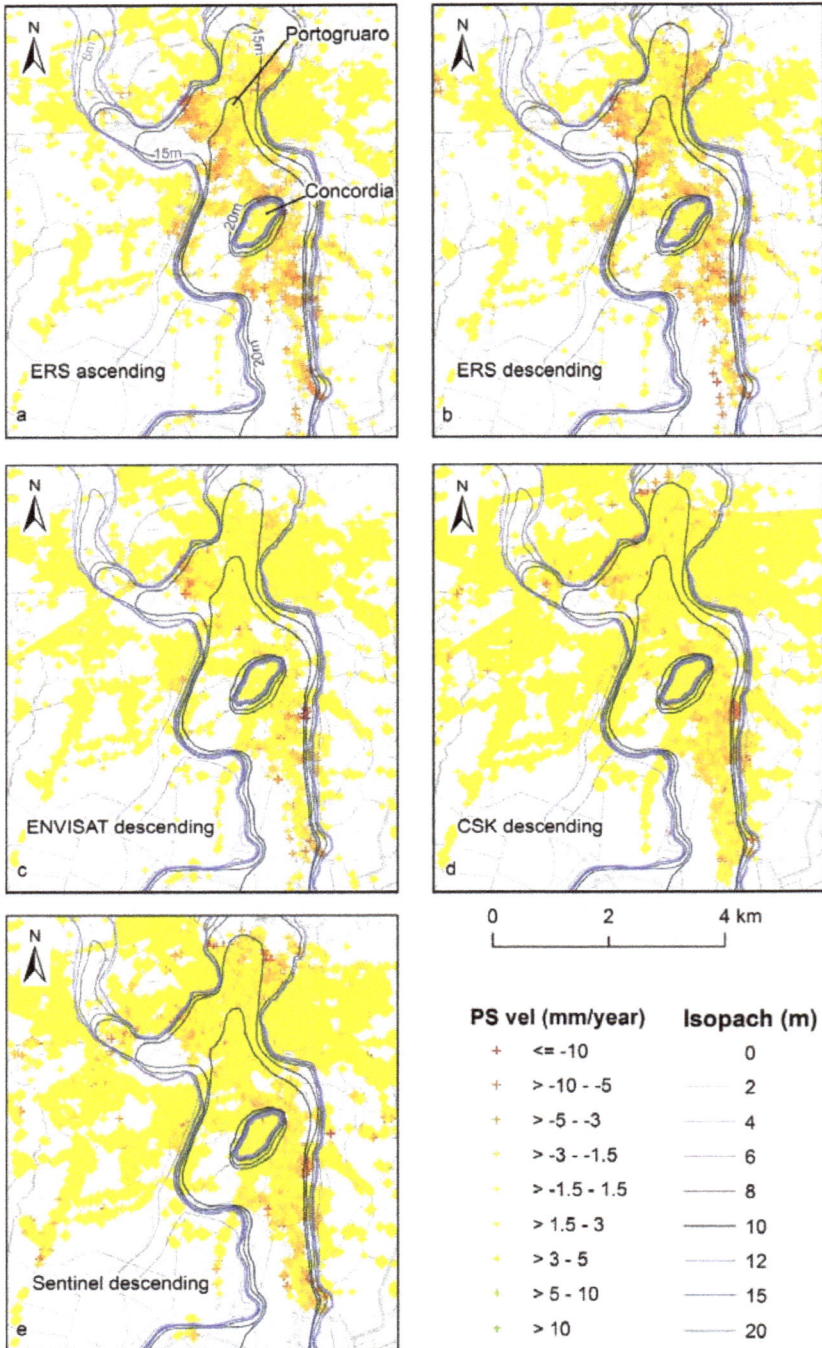

Figure 4. Persistent Scatterers (PS) velocity maps derived from PSI processing of ERS (1992-2000) (**a**,**b**), ENVISAT (2003–2010) (**c**), CSK (2012–2016) (**d**), and Sentinel-1 (2014–2017) (**e**) synthetic aperture radar (SAR) data. Graduated blue lines (Isopach) show the thickness of post-LGM sediments.

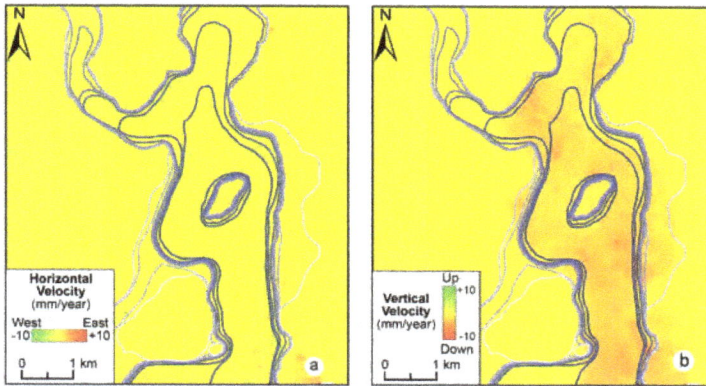

Figure 5. East-West (**a**) and vertical (**b**) components of displacement rate estimated by combining ERS ascending and descending PSI results. The green color (positive values) indicates displacements toward East (**a**) and uplift (**b**), red color (negative values) indicates movements to West (**a**) and down-lift (**b**). Graduated blue lines (Isopach) show the thickness of post-LGM sediments (same classification of Figure 4).

Comparing the PS velocity maps of Figure 4, it is possible to note that the subsidence rate of the sectors inside the post-LGM incision decreases over time. This evidence can be better observed considering the area included in the red square of Figure 7a, located slightly west of the historical center of Portogruaro town. From 1989 to 2012 this sector has undergone intense urbanization with the construction of new residential buildings (mainly mono-familiar houses with two floors and detached houses). PS velocity maps, superimposed on the land cover changes in the observation period, show that all the houses built inside the incision are affected by subsidence, while those located outside it are stable (Figure 7c–f). The sharp difference in the subsidence rate showed by all interferometric datasets allows high accuracy detection of the border post-LGM incision.

By observing a single new building, it can be noted that the rate of subsidence decreases over time. This is the case of the house indicated with the purple circle #1 in in Figure 7b. In the aerial view taken in May 1989, it is not present, but it was built in the time span between May 1989 (Figure 7c) and June 1992 (date of the first ERS SAR acquisition in descending mode). In fact, it is present in the aerial view from August 1998, but it has been identified as a PS, which means that it is included in all the SAR images acquired by ERS satellites, otherwise it could not be recognized as a PS. Immediately after construction, the house was affected by a vertical deformation rate of −7.3 mm/year and a total vertical displacement of about −75 mm measured by PSI processing of ERS SAR images in the period 1992–2000 (Figure 8a); then, from 2003 to 2010, the velocity decreased to −2.8 mm/year as measured by ENVISAT SAR data processing (Figures 7d and 8a). Finally, from 2012 the subsidence is still ongoing with a rate between −3.0 and −3.4 mm/year, which was measured by processing COSMO and Sentinel SAR images (Figure 7e–f and Figure 8a). A similar behavior can be observed in the case of the house indicated by the purple circle #2 in Figure 7b, which was built between May 1989 and January 1995 (date of the first ERS SAR acquisition in ascending mode). After the construction it was affected by a vertical displacement with a rate of −8.0 mm/year, measured through the processing of ERS images (Figures 7c and 8b), while during the ENVISAT period the rate decreased to −3.2 mm/year (Figures 7d and 8b) and, finally, from February 2012 to the present, the velocity measured by COSMO and Sentinel interferometric data is −2.8 mm/year (Figures 7e–f and 8b). Hence, also in this case, subsidence is still ongoing. The last case relating to the house is indicated by the purple circle #3 in Figure 7b, which was built between August 1998 and February 2003 (first ENVISAT acquisition in descending mode). In fact, it is not present in the 1998 aerial view (Figure 7c), but it was detected by ENVISAT SAR sensor. In this case, the monitoring starts by processing ENVISAT acquisitions, which shows a vertical displacement rate of −9.6 mm/year (Figure 8c), then the rate decreased to −4.5 and −4.1 mm/year, measured by the processing of COSMO, and Sentinel SAR data, respectively (Figure 8c).

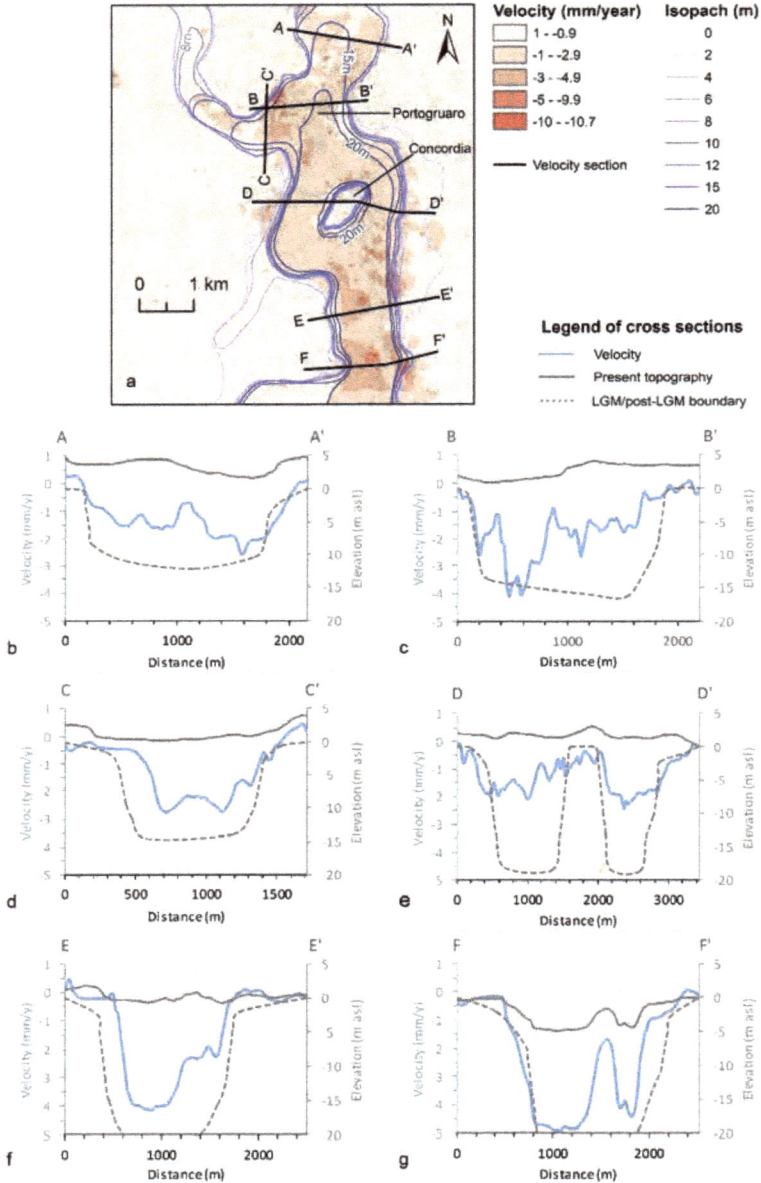

Figure 6. Velocity map (**a**) and cross sections (**b–g**) showing the variation in the displacement rate due to the presence of post-LGM sediments. Gray continuous lines indicate the variation in the ground surface elevation. Gray dashed lines indicate the boundary of LGM and post-LGM deposits.

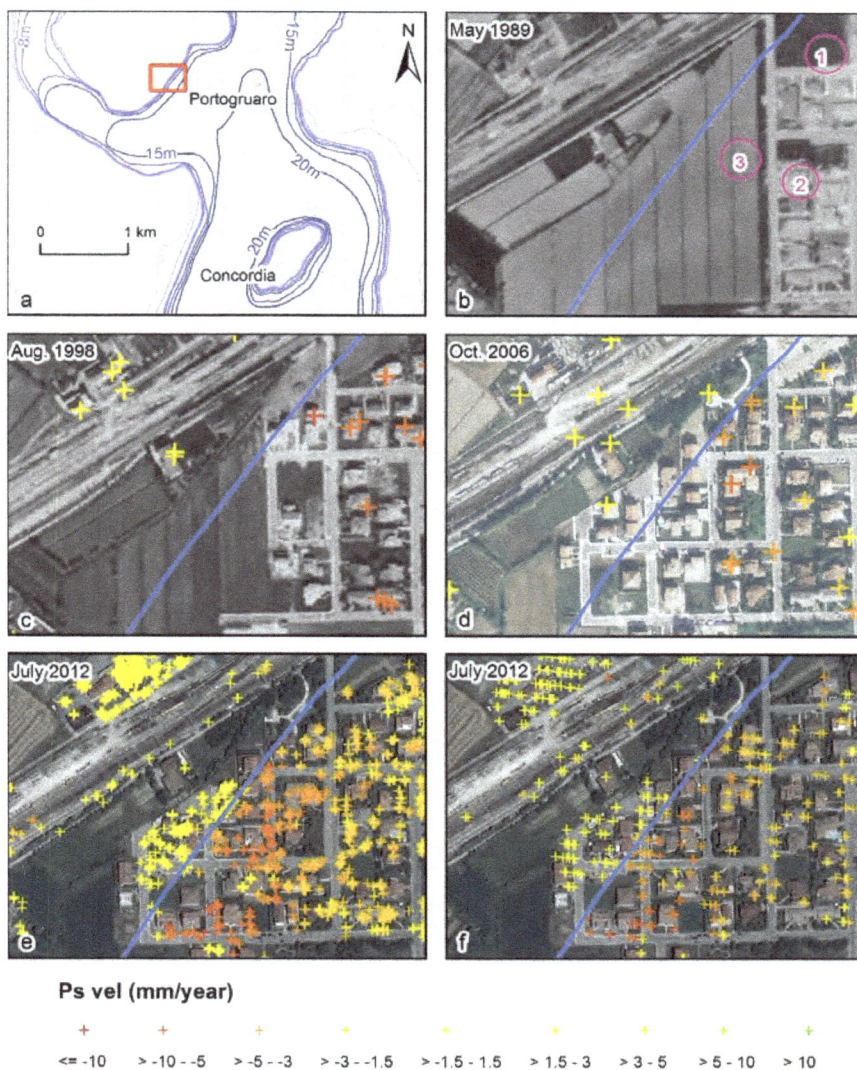

Ps vel (mm/year)

+	+	+	+	+	+	+	+	+
<= -10	> -10 - -5	> -5 - -3	> -3 - -1.5	> -1.5 - 1.5	> 1.5 - 3	> 3 - 5	> 5 - 10	> 10

Figure 7. Land cover change from 1989 to 2012 in the area indicated by the red rectangle in (**a**) and PS velocities calculated through PSI processing of ERS (**c**), ENVISAT (**d**), COSMO SkyMED (**e**) and Sentinel-1 (**f**) SAR data. Purple circles in (**b**) indicate the sectors where the time series of displacement have been plotted in Figure 8. The blue line (**b–f**) indicates the border of the post-LGM incision.

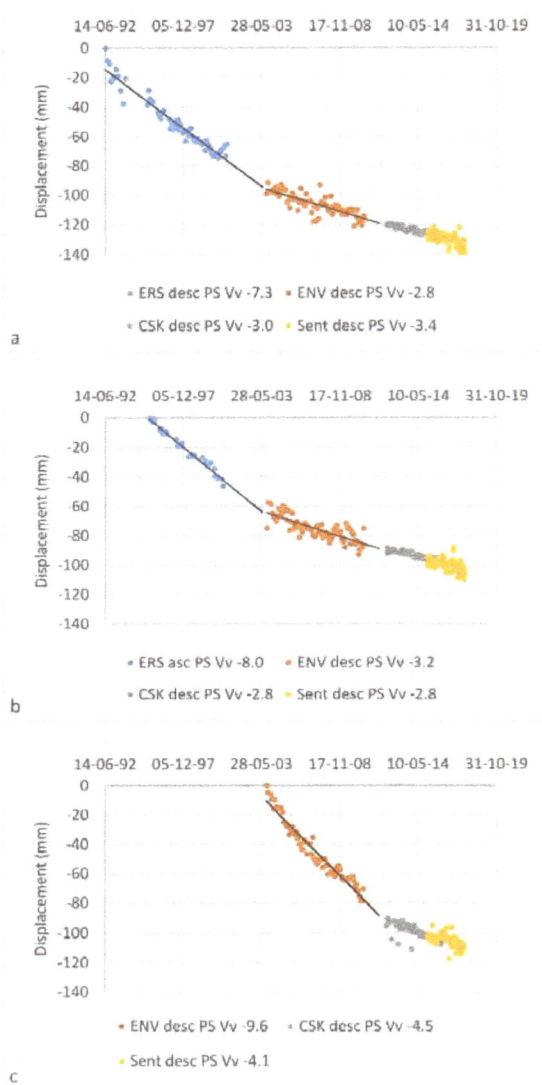

Figure 8. Time series of vertical displacements (Vv) calculated through Equation 3 applied to PSI results in the sectors 1 (**a**), 2 (**b**) and 3 (**c**) of the area showed in Figure 7b. To better follow the temporal evolution, the displacements derived by ENVISAT and COSMO –SkyMED (CSK) datasets are plotted starting from the linearly interpolated value (continuous lines) of the previous dataset. Note the similar results from PSI processing of CSK and Sentinel SAR data during the overlapping period of acquisition, which shows the high precision of the calculations.

The different behavior of sediments inside and outside the post-LGM fluvial incision can be clearly observed in the area included in the red square of Figure 9a. In this area a new overpass was built between 2012 (Figure 9b) and December 2014 (first Sentinel-1 acquisition). During the period between December 2014–July 2017, the east ramp of the overpass was affected by a displacement rate of about −12 mm/year, as measured through the PSI processing of Sentinel-1 SAR data (Figure 9c), while the west ramp has shown a lower rate (about 3.5 mm/year), which reflect the different geotechnical behavior between LGM and post-LGM deposits, locating the limit of the incision.

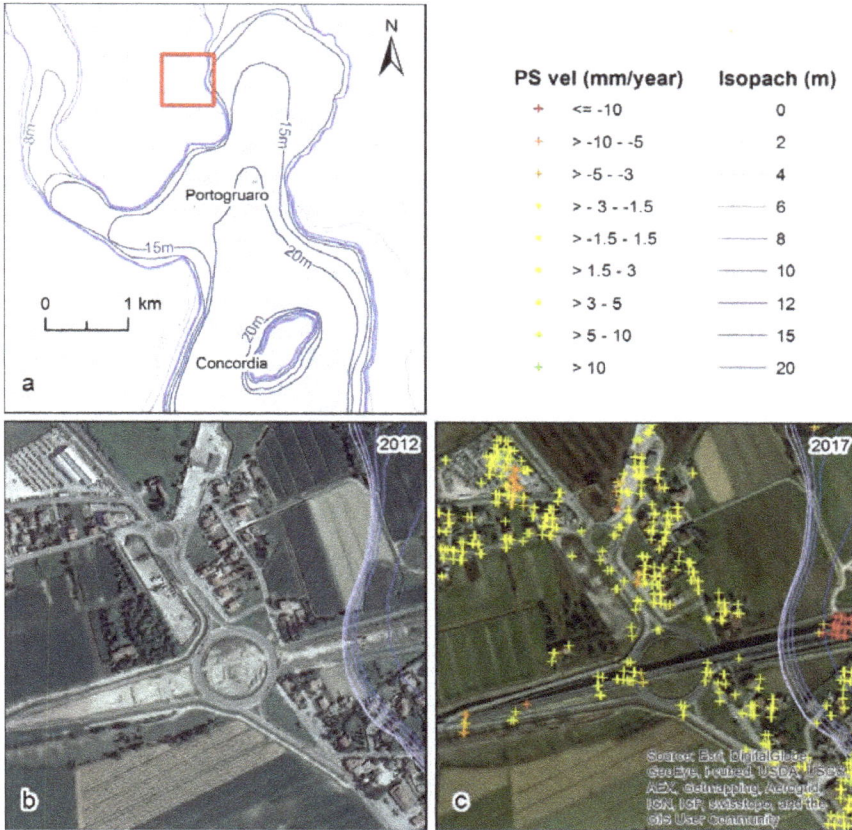

Figure 9. Land cover change before (**b**) and after (**c**) 2010 in the area indicated by the red square in (**a**) showing the different rate of subsidence outside and inside the post-LGM incision calculated through PSI processing of Sentinel-1 SAR data (**c**).

Further interesting results are outlined in Figure 10a, regarding a sector located slightly east of the historical center of Portogruaro. Also in this case, the limit of post-LGM fluvial incision can be easily identified as it is marked by the different displacement rates between past and recent deposits in all the interferometric datasets (Figure 10b–e). But the behavior of this sector is very different from the other subsiding areas because the PS velocities do not decrease over time. Taking into account all the PS inside the incision with a velocity < −1.5 mm/year, a mean displacement rate of −3.5, −2.6, −2.7, and 3.1 mm/year have been obtained by ERS, ENVISAT, COSMO, and Sentinel SAR data processing, respectively. These estimated velocities are not related to the load induced by the construction of new structures and infrastructures, indeed the land cover did not change during the observation period, as can be seen comparing Figure 10f,g.

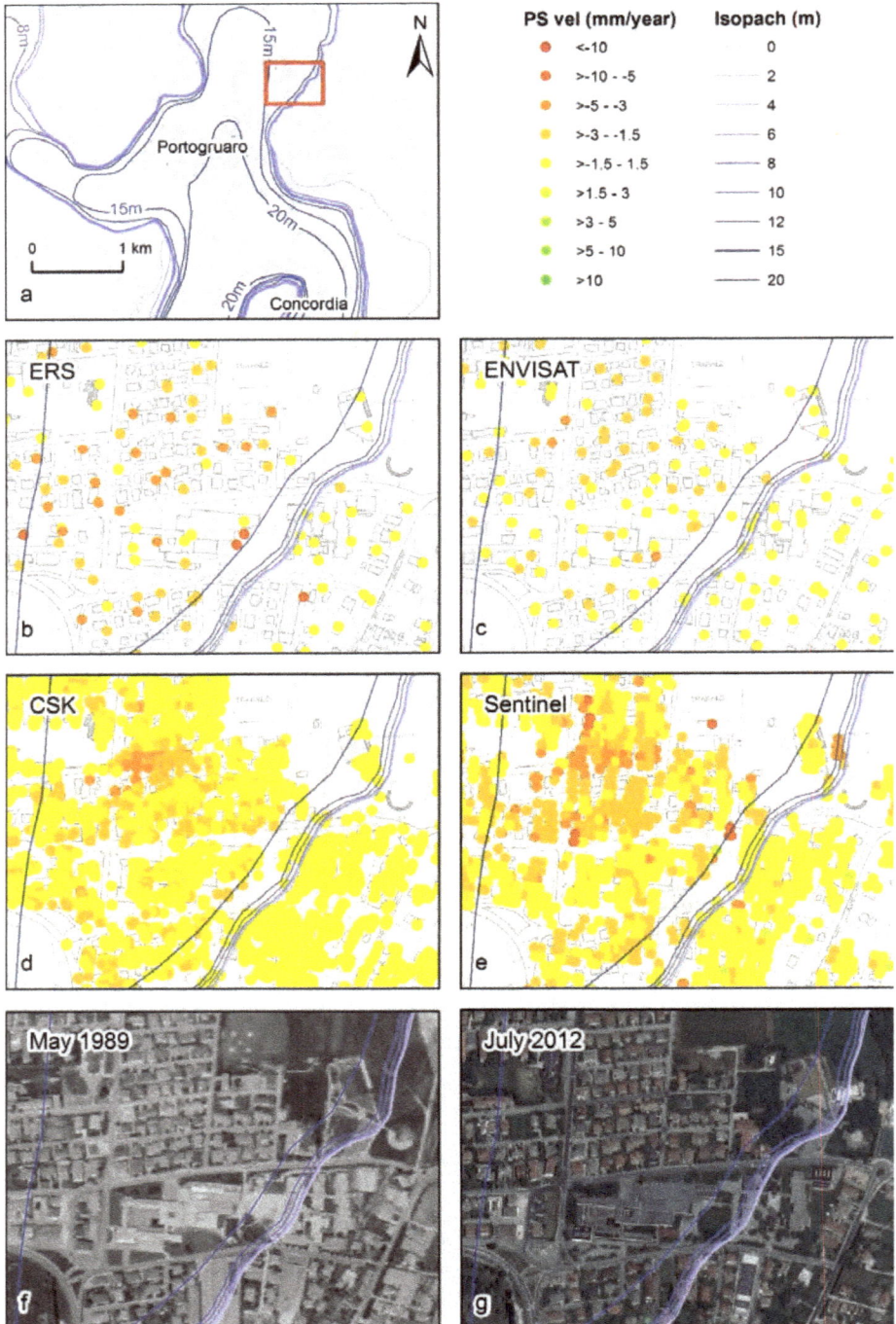

Figure 10. PS velocity maps of the area indicated by the red square in (**a**) ERS (**b**), ENVISAT (**c**), COSMO SKY-Med, (**d**) and Sentinel (**e**) interferometric data. The comparison between (**f**) and (**g**) shows that no land cover changes occurred during the observation period.

5. Discussion

The obtained results clearly show that interferometric data can help identify the border of post-LGM fluvial incision, due to the different rates of displacement between the LGM deposits of the ancient alluvial plain and younger incised valley fill. As it has been illustrated in the results section, recent deposits are affected by a rate of displacement up to −10 mm/year, while deposits of LGM plain can be considered stable. By combining ERS ascending and descending interferometric data, it can be seen that displacements are mainly vertical (Figure 5). This is a high favorable condition using PSI technique which may present some difficulties in evaluating the horizontal components of the movement, especially in the North-South direction.

In order to use interferometric techniques as a tool for geological and geomorphological mapping, it is essential to consider SAR data from different sensors covering a time span as long as possible. In our study area, ERS data are the best information to identify the presence of recent sediments (Figures 4a–b and 6), but also the other datasets helped to better delineate the shape of the post-LGM fluvial incision. In fact, ENVISAT, COSMO, and Sentinel interferometric data have shown sectors affected by subsidence in different periods, and sectors compared to ERS (Figure 4c–e). This is linked to the mechanism of ground displacement in the study area, which is mainly related to the construction of structures and infrastructures. Furthermore, such elements are essential in using PSI techniques, because they are one of the best reflectors of the incident RADAR signal and can be recognized as permanent scatterers in the processing.

Interferometric data from the different datasets allows monitoring the temporal evolution of subsidence and exploring the causes. The edification of new buildings induces a load in the soils triggering the clay deposit compaction (consolidation process [63,64]), with a rate of vertical displacement of up to −10 mm/year during the primary step, then the rate decreases to −4 mm/year and −3 mm/year in the secondary step (Figure 8). This behavior, which is similar to the one seen [65] in the city of Rome, has been clearly observed in urbanization occurring at the border of the incision in Figure 7. Here the consolidation process that is caused by the construction of new buildings can be entirely monitored through interferometry. Further important information from this study, is that the consolidation of the sediments filling the incision can last more than 25 years, in fact all the cases indicated in Figure 8 are still ongoing. This is a very important finding that suggests the possible use of interferometry to observe the consolidation process at the site scale, helping better define the mechanical behavior of soil, overpassing the limits of laboratory and in situ tests, which are performed on small samples, at specific points, and cannot take into account the variability of site conditions. Of course, laboratory and in situ tests remain mandatory for the evaluation of geo-mechanical properties of soils in the design of structures and infrastructures.

The above mentioned mechanism of subsidence limits the use of interferometry in the recognition of geological architecture of subsoil in the study area, because it depends on the development of new structures and infrastructures, it can then be applied only in urban areas and cannot provide information in scarcely urbanized sectors. Thus it is very difficult to evaluate the natural subsidence caused by the consolidation under lithostatic load, and, consequently, it seems to limit the use of interferometry in geological and geomorphological surveys. But this issue could be addressed considering L-band SAR data, which are more penetrating than C- and X-band, and are affected by a lower loss of coherence in forest and vegetated areas. These advantages have been recently shown by [17], evaluating ground displacements in the south sector of the Venetian-Friulian Plain by processing ALOS-PALSAR SAR data. Unfortunately, this kind of data was not available for our research. However, geological and geomorphological mapping is difficult in areas where the anthropic action has obliterated surficial evidences. These results show the potential of interferometric data in this new application field.

The mechanism of subsidence in our study area can also explain the good performance of ERS datasets in the identification of post-LGM incision, despite the lower PS density compared to other datasets. This area experienced a great urban expansion at the end of the 20th century. New urban settlements triggered the consolidation of post-LGM deposits and became the optimal target for the ERS acquisitions, which started at the beginning of the 1990s. In the following periods, urbanization was quite limited to new residential areas at the border of the main historical centers of Portogruaro and Concordia Sagittaria, and to the maintenance and improvement of the road network; such elements have been effectively detected and monitored by the ENVISAT, COSMO-SKY-Med and Sentinel-1 SAR sensors. Our case study shows that the use of interferometric data in geological and geomorphological surveys slightly depend on the characteristics of SAR data. The band and satellite revisiting time could influence the density of PSI points, but indeed do not limit the potential in defining the evolution of urbanization and the geological features of the area under investigation.

Certainly, the different rates of subsidence of pre- and post-LGM deposits are related to their geotechnical properties, in particular the degree of compressibility and permeability. In most cases the subsidence is linked to the loading of soft soils, but the are some exceptions as the area shown in Figure 10, where, although no land cover changes occurred in the last 30 years, ground displacements have been measured by all interferometric datasets. A similar situation has been observed in some localized sectors of the historical centers of Portogruaro and Concordia Sagittaria. According to the geological setting, the tectonic subsidence going on in the study area has a regional pattern and, between Tagliamento and Livenza rivers, it has an average rate of 0.4 mm/year. Thus, this long-term deformative process affects the whole study area and could not explain the differential vertical settlements characterizing the areas over the incised filled valley. This setting may mean the groundwater exploitation is a major cause of the differential subsidence.

From a hydrogeological point of view, in the study area, ten overlapping aquifers between 10 and about 500 m depth, have been recognized, and are composed of coarse deposits interbedded by clayey and silty layers [66,67]. The first aquifer is located between 10 and 25 m and is semi-confined, while the remaining are confined. A phreatic aquifer is present in the first ten meters, the water level is 2 m deep on average, and it has a seasonal fluctuation of about 1.5 m that is controlled by a pumping system to ensure groundwater does not reach the ground surface, which in some sectors has an elevation under the sea level. This shallow aquifer is recharged mainly by rainfall and the two rivers present in the area (Lemene and Reghene rivers). The most exploited aquifers by a number of water wells present in the area are the deepest ones, but quantitative data of exploitation are not available [67]. Then, it is not easy to evaluate the relationships between subsidence and groundwater exploitation. However, considering the ground displacements mapped in Figure 6, and their pattern, which perfectly matches the planform of the incised valley, it is evident that down lift movements are led by a surficial motivation. The depletion of deep aquifers can generate a large-scale subsidence, which could not induce a differential down lift in the very shallow deposits. Thus, the exploitation should also affect the surficial aquifers and, in particular, the unconfined one or that between 10–30 m, triggering the compaction of soft sub-soils with low displacement rates, due to the low permeability of the deposits [60,64].

6. Conclusions

In this paper we tested the use of multi-temporal differential interferometric synthetic aperture radar techniques as a tool for geological and geomorphological mapping, especially in urban areas where surficial evidences are often obliterated by the development of structures and infrastructures. To this end, the urbanized sector of the municipalities of Portogruaro and Concordia Sagittaria towns, located in the Venetian-Friulian Plain between Tagliamento and Piave rivers, has been considered, because its geological and geomorphological setting was already quite well known.

Interferometric data derived by permanent scatterers InSAR processing of SAR data acquired by ERS 1/2, ENVISAT, COSMO SKY-Med, and Sentinel-1 missions from 1992 to 2017, have been considered to evaluate the subsidence rate of the territory. Measuring the ground displacement rate allowed the identification, with high accuracy, the borders of a post-last glacial maximum incision filled by recent, unconsolidated lagoon and alluvial deposits, which are characterized by a subsidence rate much higher than the surrounding Pleistocene deposits of the external plain.

By monitoring the evolution of ground displacements inside the incision for a long period, the causes of the subsidence could be explored. The consolidation process, caused by the loading of post-LGM sediments after the construction of new buildings, and the development and maintenance of the road network, are the main causes of vertical movements. A primary consolidation step, with velocities of up to -10 mm/year, followed by a secondary step, with velocities about -3–4 mm/year, have been observed. In most of the cases, this is the mechanism of the subsidence, but in some sectors, where urban settlement changes have not been observed in the last 30 years, ground displacements seem to be related to the exploitation of the shallow phreatic aquifer. The ability to monitor the behavior of soils under loading for a long time, suggests that interferometry can be also used for the geo-mechanical characterization at the site scale, which represent a challenging use of the technique.

The obtained results show that DInSAR techniques can be effectively applied in geological and geomorphological surveys, mapping elongated features, with a length of kilometers and a width that is several hundreds of meters. Using these methods depends on the physical and mechanical properties of deposits and on the geological processes acting in the area under investigation. Processing SAR data, that is acquired by sensors with different wavelengths, covering a period of time as long as possible, is recommended.

Future research should regard the attempt to use interferometric techniques in geological and geomorphological mapping of scarce urbanized areas and the numerical simulation of physical phenomena occurring after the loading of soils and/or the exploitation of groundwater. To overcome the limits of interferometry, due to spatial and temporal de-correlations, the performance of different DInSAR techniques that make an accurate choice of processing parameters, could be tested; also the use of L-band SAR data could be considered. To validate the ex-post assumptions about the possible causes of subsidence, a hydro-mechanical numerical modelling that reconstructs the ground displacement pattern measured by interferometry should be performed on the basis of previous geotechnical and hydrogeological data, new geognostic and laboratory tests, and data on the water level changes of the shallow phreatic aquifer.

Author Contributions: Data curation, M.F., A.F., G.T. and M.M.; formal analysis, M.F., A.F., G.T. and M.M.; funding acquisition, M.F.; investigation, M.F., A.F., G.T. and M.M.; methodology, M.F., A.F. and G.T.; supervision, M.F. and A.F.; writing–original draft, M.F.; writing–review & editing, M.F., A.F. and G.T.

Funding: This research was funded by the Department of Geosciences at the University of Padua (Italy), grant number FLORIS_SID17_01, Principal Investigator Mario Floris.

Acknowledgments: ERS and ENVISAT interferometric data were provided by the Italian Ministry of the Environment and for Protection of the Land and Sea in the framework of the "Not Ordinary Plan of Remote Sensing" project (http://www.pcn.minambiente.it/mattm/en/). COSMO SKY-Med SAR data were provided by Italian Space Agency through an "Open Call for Science". Sentinel-1 SAR data were downloaded from Copernicus Open Access Hub (https://scihub.copernicus.eu/).

Conflicts of Interest: The authors declare no conflict of interest.

References

1. Sahu, P.; Sikdar, P.K. Threat of land subsidence in and around Kolkata City and East Kolkata Wetlands, West Bengal, India. *J. Earth Syst. Sci.* **2011**, *120*, 435–446. [CrossRef]
2. Amaş, I.; Mendes, D.A.; Popa, R.G.; Gheorghe, M.; Popovici, D. Long-term ground deformation patterns of Bucharest using multi-temporal InSAR and multivariate dynamic analyses: A possible transpressional system? *Sci. Rep.* **2017**, *7*, 43762. [CrossRef] [PubMed]
3. Castellazzi, P.; Arroyo-Domínguez, N.; Martel, R.; Calderhead, A.I.; Normand, J.C.L.; Gárfias, J.; Rivera, A. Land subsidence in major cities of Central Mexico: Interpreting InSAR-derived land subsidence mapping with hydrogeological data. *Int. J. Appl. Earth Obs. Geoinf.* **2016**, *47*, 102–111. [CrossRef]

4. Carbognin, L.; Gatto, P.; Mozzi, G.; Gambolati, G.; Ricceri, G. New trend in the subsidence of Venice. In Proceedings of the 2nd International Symposium on Land Subsidence, Anaheim, CA, USA, 13–17 December 1976; Rodda, J.C., Ed.; IAHS Publ.: Wallingford, UK, 1977; Volume 121, pp. 65–81.

5. Gatto, P.; Carbognin, L. The Lagoon of Venice: Natural environmental trend and man-induced modification. *Hydrol. Sci. Bull.* **1981**, *26*, 379–391. [CrossRef]

6. Teatini, P.; Tosi, L.; Strozzi, T.; Carbognin, L.; Wegmüller, U.; Rizzetto, F. Mapping regional land displacements in the Venice coastland by an integrated monitoring system. *Remote Sens. Environ.* **2005**, *98*, 403–413. [CrossRef]

7. Teatini, P.; Tosi, L.; Strozzi, T.; Carbognin, L.; Cecconi, G.; Rosselli, R.; Libardo, S. Resolving land subsidence within the Venice Lagoon by persistent scatterer SAR interferometry. *Phys. Chem. Earth* **2012**, *40–41*, 72–79. [CrossRef]

8. Tian, Y.; Liu-Zeng, J.; Luo, Y.; Li, Y.; Hu, Y.; Gong, B.; Liu, L.; Guo, P.; Zhang, J. Transient deformation during the Milashan Tunnel construction in northern Sangri-Cuona Rift, southern Tibet, China observed by Sentinel-1 satellites. *Sci. Bull.* **2018**, *63*, 1439–1447. [CrossRef]

9. Kong, S.M.; Kim, D.M.; Lee, D.Y.; Jung, H.S.; Lee, Y.J. Field and laboratory assessment of ground subsidence induced by underground cavity under the sewer pipe. *Geomech. Eng.* **2018**, *16*, 285–293. [CrossRef]

10. Choudhury, P.; Gahalaut, K.; Dumka, R.; Gahalaut, V.K.; Singh, A.K.; Kumar, S. GPS measurement of land subsidence in Gandhinagar, Gujarat (Western India), due to groundwater depletion. *Environ. Earth Sci.* **2018**, *77*, 770. [CrossRef]

11. Fernandez, J.; Prieto, J.F.; Escayo, J.; Camacho, A.G.; Luzón, F.; Tiampo, K.F.; Palano, M.; Abajo, T.; Pérez, E.; Velasco, J.; et al. Modeling the two- and three-dimensional displacement field in Lorca, Spain, subsidence and the global implications. *Sci. Rep.* **2018**, *8*, 14782. [CrossRef]

12. Tosi, L.; Teatini, P.; Carbognin, L.; Frankenfield, J. A new project to monitor land subsidence in the northern Venice coastland (Italy). *Environ. Geol.* **2007**, *52*, 889–898. [CrossRef]

13. Al-Halbouni, D.; Holohan, E.P.; Saberi, L.; Alrshdan, H.; Sawarieh, A.; Closson, D.; Walter, T.R.; Dahm, T. Sinkholes, subsidence and subrosion on the eastern shore of the Dead Sea as revealed by a close-range photogrammetric survey. *Geomorphology* **2017**, *285*, 305–324. [CrossRef]

14. Palamara, D.; Nicholson, M.; Flentje, P.N.; Baafi, E.Y.; Brassington, G.M. An evaluation of airborne laser scan data for coalmine subsidence mapping. *Int. J. Remote Sens.* **2007**, *28*, 3181–3203. [CrossRef]

15. Tomás, R.; Romero, R.; Mulas, J.; Marturià, J.J.; Mallorquí, J.J.; Lopez- Sanchez, J.M.; Herrera, G.; Gutiérrez, F.; González, P.J.; Fernández, J.; et al. Radar interferometry techniques for the study of ground subsidence phenomena: A re-view of practical issues through cases in Spain. *Environ. Earth Sci.* **2014**, *71*, 163–181. [CrossRef]

16. Yerro, A.; Corominas, J.; Monells, D.; Mallorquí, J.J. Analysis of the evolution of ground movements in a low densely urban area by means of DInSAR technique. *Eng. Geol.* **2014**, *170*, 52–65. [CrossRef]

17. Tosi, L.; Da Lio, C.; Strozzi, T.; Teatini, P. Combining L- and X-Band SAR Interferometry to Assess Ground Displacements in Heterogeneous Coastal Environments: The Po River Delta and Venice Lagoon, Italy. *Remote Sens.* **2016**, *8*, 308. [CrossRef]

18. Fiaschi, S.; Tessitore, S.; Bonì, R.; Di Martire, D.; Achilli, V.; Borgstrom, S.; Ibrahim, A.; Floris, M.; Meisina, C.; Ramondini, M.; et al. From ERS-1/2 to Sentinel-1: Two decades of subsidence monitored through A-DInSAR techniques in the Ravenna area (Italy). *GISci. Remote Sens.* **2017**, *54*, 305–328. [CrossRef]

19. Di Paola, G.; Alberico, I.; Aucelli, P.P.C.; Matano, F.; Rizzo, A.; Vilardo, G. Coastal subsidence detected by Synthetic Aperture Radar interferometry and its effects coupled with future sea-level rise: The case of the Sele Plain (Southern Italy). *J. Flood Risk Manag.* **2018**, *11*, 191–206. [CrossRef]

20. Milillo, P.; Giardina, G.; DeJong, M.J.; Perissin, D.; Milillo, G. Multi-Temporal InSAR Structural Damage Assessment: The London Crossrail Case Study. *Remote Sens.* **2018**, *10*, 287. [CrossRef]

21. Cerchiello, V.; Tessari, G.; Velterop, E.; Riccardi, P.; Defilippi, M.; Pasquali, P. Building damage risk by modelling interferometric time series. *IEEE Geosci. Remote Sens. Lett.* **2017**, *99*, 1–5. [CrossRef]

22. Tessitore, S.; Di Martire, D.; Calcaterra, D.; Infante, D.; Ramondini, M.; Russo, G. Multitemporal synthetic aperture radar for bridges monitoring. In *Remote Sensing Technologies and Applications in Urban Environments, Proceedings of SPIE, Warsaw, Poland, 11–13 September 2017*; Erbertseder, T., Zhang, Y., Chrysoulakis, N., Eds.; Volume 10431, Article number 104310C. [CrossRef]

23. Ouchi, K. Recent Trend and Advance of Synthetic Aperture Radar with Selected Topics. *Remote Sens.* **2013**, *5*, 716–807. [CrossRef]

24. Crosetto, M.; Monserrat, O.; Cuevas-González, M.; Devanthéry, N.; Crippa, B. Persistent Scatterer Interferometry: A review. *ISPRS J. Photogramm. Remote Sens.* **2016**, *115*, 78–89. [CrossRef]

25. Atzori, S.; Hunstad, I.; Chini, M.; Salvi, S.; Tolomei, C.; Bignami, C.; Stramondo, S.; Trasatti, E.; Antonioli, A.; Boschi, E. Finite fault inversion of DInSAR coseismic displacement of the 2009 L'Aquila earthquake (central Italy). *Geophys. Res. Lett.* **2009**, *36*, L15305. [CrossRef]

26. Merryman Boncori, J.P.; Papoutsis, I.; Pezzo, G.; Atzori, S.; Ganas, A.; Karastathis, V.; Salvi, S.; Kontoes, C.; Antonioli, A. The February 2014 Cephalonia earthquake (Greece): 3D deformation field and source modeling from multiple SAR techniques. *Seismol. Res. Lett.* **2015**, *86*, 1–14. [CrossRef]

27. Yu, C.; Li, Z.; Chen, J.; Hu, J.C. Small magnitude co-seismic deformation of the 2017 Mw 6.4 Nyingchi earthquake revealed by InSAR measurements with atmospheric correction. *Remote Sens.* **2018**, *10*, 684. [CrossRef]

28. Biggs, J.; Anthony, E.Y.; Ebinger, C.J. Multiple inflation and deflation events at Kenyan volcanoes, East African Rift. *Geology* **2009**, *37*, 979–982. [CrossRef]

29. Kobayashi, T.; Morishita, Y.; Munekane, H. First detection of precursory ground inflation of a small phreatic eruption by InSAR. *Earth Planet. Sci. Lett.* **2018**, *491*, 244–254. [CrossRef]

30. Tessari, G.; Beccaro, L.; Ippoliti, S.; Riccardi, P.; Floris, M.; Marzoli, A.; Ogushi, F.; Pasquali, P. Monitoring of Sakurajima volcano, Japan, with SAR data: From small displacement measurements to modeling and forecast. In Proceedings of the IGARSS, Valencia, Spain, 22–27 July 2018; pp. 3075–3078.

31. Hooper, A.; Bekaert, D.; Spaans, K.; Arikan, M. Recent advances in SAR interferometry time series analysis for measuring crustal deformation. *Tectonophysics* **2012**, *514–517*, 1–13. [CrossRef]

32. Fiaschi, S.; Closson, D.; Abou Karaki, N.; Pasquali, P.; Riccardi, P.; Floris, M. The complex karst dynamics of the Lisan Peninsula revealed by 25 years of DInSAR observations. Dead Sea, Jordan. *ISPRS J. Photogramm. Remote Sens.* **2017**, *130*, 358–369. [CrossRef]

33. Bacques, G.; de Michele, M.; Raucoules, D.; Aochi, H. The locking depth of the Cholame section of the San Andreas Fault from ERS2-Envisat InSAR. *Remote Sens.* **2018**, *10*, 1244. [CrossRef]

34. Takada, Y.; Sagiya, T.; Nishimura, T. Interseismic crustal deformation in and around the Atotsugawa fault system, central Japan, detected by InSAR and GNSS. *Earth Planets Space* **2018**, *70*, 32. [CrossRef]

35. Wasowski, J.; Bovenga, F. Investigating landslides and unstable slopes with satellite Multi Temporal Interferometry: Current issues and future perspectives. *Eng. Geol.* **2014**, *174*, 103–138. [CrossRef]

36. Tessari, G.; Floris, M.; Pasquali, P. Phase and amplitude analyses of SAR data for landslide detection and monitoring in non-urban areas located in the North-Eastern Italian pre-Alps. *Environ. Earth Sci.* **2017**, *76*, 85. [CrossRef]

37. Di Maio, C.; Fornaro, G.; Gioia, D.; Reale, D.; Schiattarella, M.; Vassallo, R. In situ and satellite long-term monitoring of the Latronico landslide, Italy: Displacement evolution, damage to buildings, and effectiveness of remedial works. *Eng. Geol.* **2018**, *245*, 218–235. [CrossRef]

38. Zhao, C.; Kang, Y.; Zhang, Q.; Lu, Z.; Li, B. Landslide identification and monitoring along the Jinsha River catchment (Wudongde reservoir area), China, using the InSAR method. *Remote Sens.* **2018**, *10*, 993. [CrossRef]

39. Rajendran, S.; Nasir, S. Capability of L-band SAR data in mapping of sedimentary formations of the Marmul region, Sultanate of Oman. *Earth Sci. Inform.* **2018**, *11*, 341–357. [CrossRef]

40. Antonioli, F.; Ferranti, L.; Fontana, A.; Amorosi, A.; Bondesan, A.; Braitenberg, C.; Fontolan, G.; Furlani, S.; Mastronuzzi, G.; Monaco, C.; et al. Holocene relative sea-level changes and vertical movements along the Italian and Istrian coastlines. *Quat. Int.* **2009**, *206*, 101–133. [CrossRef]

41. Carbognin, L.; Teatini, P.; Tosi, L.; Strozzi, T.; Vitturi, A.; Mazzuccato, A. Subsidenza. In *Atlante Geologico Della Provincia di Venezia*; Vitturi, A., Ed.; Cierre: Verona, Italy, 2011; pp. 519–529. ISBN 978-88-907207-0-3.

42. Fontana, A.; Bondesan, A.; Meneghel, M.; Toffoletto, F.; Vitturi, A.; Bassan, V. *Carta Geologica d'Italia alla Scala 1:50,000—Foglio 107 Portogruaro (Geological Map of Italy at 1:50,000 Scale—Sheet 107 Portogruaro)*; Infocartografica: Piacenza, Italy, 2012; 168p.

43. Bondesan, A.; Primon, S.; Bassan, V.; Vitturi, A. *Le unità Geologiche della Provincia di Venezia*; Cierre: Verona, Italy, 2008; 184p.

44. Fontana, A.; Mozzi, P.; Marchetti, M. Alluvial fans and megafans along the southern side of the Alps. *Sediment. Geol.* **2014**, *301*, 150–171. [CrossRef]

45. Zanferrari, A. Inquadramento geologico lineamenti strutturali. In *Carta Geologica d'Italia alla Scala 1:50,000—Foglio 107 Portogruaro (Geological Map of Italy at 1:50,000 Scale—Sheet 107 Portogruaro)*; Fontana, A., Ed.; Regione Veneto, Infocartografica: Piacenza, Italy, 2012; pp. 35–46.

46. Carminati, E.; Doglioni, C.; Scrocca, D. Apennines subduction-related subsidence of Venice (Italy). *Geophys. Res. Lett.* **2003**, *30*, 1–4. [CrossRef]

47. Fontana, A. *Evoluzione Geomorfologica della Bassa Pianura Friulana e sue Relazioni con le Dinamiche Insediative Antiche*; Enclosed Geomorphological Map of the Low Friulian Plain scale 1:50,000; Monografie Museo Friulano Storia Naturale: Udine, Italy, 2006; 288p.

48. Clark, P.; Dyke, A.; Shakun, J.; Carlson, A.; Clark, J.; Wohlfarth, B.; Mitrovica, J.; Hostetler, S.; McCabe, A. The Last Glacial Maximum. *Science* **2009**, *325*, 710–714. [CrossRef]

49. Monegato, G.; Ravazzi, C.; Donegana, M.; Pini, R.; Calderoni, G.; Wick, L. Evidence of a two-fold glacial advance during the last glacial maximum in the Tagliamento end moraine system (eastern Alps). *Quat. Res.* **2007**, *68*, 284–302. [CrossRef]

50. Fontana, A.; Mozzi, P.; Bondesan, A. Alluvial megafans in the Venetian-Friulian Plain (north-eastern Italy): Evidence of sedimentary and erosive phases during Late Pleistocene and Holocene. *Quat. Int.* **2008**, *189*, 71–90. [CrossRef]

51. Fontana, A.; Monegato, G.; Devoto, S.; Zavagno, E.; Burla, I.; Cucchi, F. Evolution of an Alpine fluvioglacial system at the LGM decay: The Cormor megafan (NE Italy). *Geomorphology* **2014**, *204*, 136–153. [CrossRef]

52. Mozzi, P.; Bini, C.; Zilocchi, L.; Becattini, R.; Mariotti Lippi, M. Stratigraphy, palaeopedology and palynology of Late Pleistocene and Holocene deposits in the landward sector of the lagoon of Venice (Italy), in relation to the Caranto level. *Alp. Mediterr. Quat.* **2003**, *16*, 193–210.

53. Fontana, A.; Vinci, G.; Tasca, G.; Mozzi, P.; Vacchi, M.; Bivi, G.; Salvador, S.; Rossato, S.; Antonioli, F.; Asioli, A.; et al. Lagoonal settlements and relative sea level during Bronze Age in Northern Adriatic: Geoarchaeological evidence and paleogeographic constraints. *Quat. Int.* **2017**, *439*, 17–36. [CrossRef]

54. Ferretti, A.; Prati, C.; Rocca, F. Nonlinear subsidence rate estimation using permanent scatterers in differential SAR interferometry. *IEEE Trans. Geosci. Remote Sens.* **2000**, *38*, 2202–2212. [CrossRef]

55. Ferretti, A.; Prati, C.; Rocca, F. Permanent Scatterers in SAR interferometry. *IEEE Trans. Geosci. Remote Sens.* **2001**, *39*, 8–20. [CrossRef]

56. Berardino, P.; Fornaro, G.; Lanari, R.; Sansosti, E. A new algorithm for surface deformation monitoring based on small baseline differential SAR interferograms. *IEEE Trans. Geosci. Remote Sens.* **2002**, *40*, 2375–2383. [CrossRef]

57. Rocca, F. Diameters of the Orbital Tubes in Long-Term Interferometric SAR Surveys. *IEEE Geosci. Remote Sens. Lett.* **2004**, *1*, 224–227. [CrossRef]

58. Crosetto, M.; Monserrat, O.; Adam, N.; Parizzi, A.; Bremmer, C.; Dortland, S.; Hanssen, R.F.; van Leijen, F.J. Final Report of the Validation of Existing Processing Chains in Terrafirma Stage 2, Terrafirma Project, ESRIN/Contract no. 19366/05/I-E. Available online: http://www.terrafirma.eu.com/product_validation.htm (accessed on 3 January 2019).

59. Samieie-Esfahany, S.; Hanssen, R.; van Thienen-Visser, K.; Muntendam-Bos, A. On the effect of horizontal deformation on InSAR subsidence estimates. In Proceedings of the FRINGE, Frascati, Italy, 30 November–4 December 2009; ESA: Frascati, Italy, 2010.

60. Raspini, F.; Cigna, F.; Moretti, S. Multi-temporal mapping of land subsidence at basin scale exploiting Persistent Scatterer Interferometry: Case study of Gioia Tauro plain (Italy). *J. Maps* **2012**, *8*, 514–524. [CrossRef]

61. Notti, D.; Herrera, G.; Bianchini, S.; Meisina, C.; García-Davalillo, J.C.; Zucca, F. A methodology for improving landslide PSI data analysis. *Int. J. Remote Sens.* **2014**, *35*, 2186–2214. [CrossRef]

62. Bonì, R.; Pilla, G.; Meisina, C. Methodology for Detection and Interpretation of Ground Motion Areas with the A-DInSAR Time Series Analysis. *Remote Sens.* **2016**, *8*, 686. [CrossRef]

63. Terzaghi, K.; Peck, R.B. *Soil Mechanics in Engineering Practice*; John Wiley & Sons: New York, NY, USA, 1967.

64. Lambe, T.W.; Whitman, R.V. *Soil Mechanics*; John Wiley & Sons: New York, NY, USA, 1979; 505p.

65. Stramondo, S.; Bozzano, F.; Marra, F.; Wegmuller, U.; Cinti, F.R.; Moro, M.; Saroli, M. Subsidence induced by urbanisation in the city of Rome detected by advanced InSAR technique and geotechnical investigations. *Remote Sens. Environ.* **2008**, *112*, 3160–3172. [CrossRef]

66. Dal Prà, A.; Gobbo, L.; Vitturi, A.; Zangheri, P. *Indagine Idrogeologica del Territorio Provinciale di Venezia (Hydrogeological Survey in the Territory of Venice Province)*; Venice Province: Venice, Italy, 2000.

67. Zangheri, P. Aspetti idrogeologici. In *Carta Geologica d'Italia alla Scala 1:50,000—Foglio 107 Portogruaro (Geological Map of Italy at 1:50,000 Scale—Sheet 107 Portogruaro)*; Fontana, A., Ed.; Regione Veneto, Infocartografica: Piacenza, Italy, 2012; pp. 139–146.

remote sensing

MDPI

Article

Measuring Urban Subsidence in the Rome Metropolitan Area (Italy) with Sentinel-1 SNAP-StaMPS Persistent Scatterer Interferometry

José Manuel Delgado Blasco [1,*], Michael Foumelis [2], Chris Stewart [3] and Andrew Hooper [4]

[1] Grupo de Investigación Microgeodesia Jaén (PAIDI RNM-282), Universidad de Jaén, 23071 Jaén, Spain
[2] BRGM–French Geological Survey, 45060 Orleans, France; M.Foumelis@brgm.fr
[3] Future Systems Department, Earth Observation Programmes, European Space Agency (ESA), 00044 Frascati, Italy; chris.stewart@esa.int
[4] Institute of Geophysics and Tectonics, University of Leeds, Leeds LS2 9JT, UK; A.Hooper@leeds.ac.uk
* Correspondence: j.dblasco@ujaen.es

Received: 30 November 2018; Accepted: 8 January 2019; Published: 11 January 2019

Abstract: Land subsidence in urban environments is an increasingly prominent aspect in the monitoring and maintenance of urban infrastructures. In this study we update the subsidence information over Rome and its surroundings (already the subject of past research with other sensors) for the first time using Copernicus Sentinel-1 data and open source tools. With this aim, we have developed a fully automatic processing chain for land deformation monitoring using the European Space Agency (ESA) SentiNel Application Platform (SNAP) and Stanford Method for Persistent Scatterers (StaMPS). We have applied this automatic processing chain to more than 160 Sentinel-1A images over ascending and descending orbits to depict primarily the Line-Of-Sight ground deformation rates. Results of both geometries were then combined to compute the actual vertical motion component, which resulted in more than 2 million point targets, over their common area. Deformation measurements are in agreement with past studies over the city of Rome, identifying main subsidence areas in: (i) Fiumicino; (ii) along the Tiber River; (iii) Ostia and coastal area; (iv) Ostiense quarter; and (v) Tivoli area. Finally, post-processing of Persistent Scatterer Inteferometry (PSI) results, in a Geographical Information System (GIS) environment, for the extraction of ground displacements on urban infrastructures (including road networks, buildings and bridges) is considered.

Keywords: urban subsidence; Copernicus Sentinel-1; Persistent Scatterer Interferometry; SNAP-StaMPS; Rome

1. Introduction

Since the launch of Copernicus Sentinel-1A on 8 April 2014, a new era of continuous monitoring using spaceborne Synthetic Aperture Radar (SAR) sensors has started. Sentinel-1 constitutes a significant improvement from previous European C-band SAR missions, European Remote Sensing (ERS) satellites and Environmental Satellite (ENVISAT), since it reduced the temporal revisit time from 35 to six days, at best using the two satellite segments A and B, with a large swath coverage of 250 km. The scientific communities as well as Earth Observation (EO) practitioners were thus given the means to extend the use of spaceborne SAR data to land applications.

In support to the EO community, the European Space Agency (ESA) continued developing appropriate tools for the utilization of the Copernicus Sentinel data. By evolving existing tools, such as the Next ESA SAR Toolbox (NEST) as well as integrating others, the SeNtinel Application Platform (SNAP) [1] becomes a multi-mission toolbox supporting both SAR and optical data processing.

Newly implemented on Sentinel-1, the Terrain Observation by Progressive Scans (TOPS) acquisition mode [2] required further development in terms of interferometric handling to ensure robust results. The SNAP TOPSAR) capabilities were made available to users at an early stage, just before the start of Sentinel-1 data dissemination, while SNAP TOPS Interferometric SAR (InSAR) development were first communicated at the ESA Fringe 2015 consultation meeting. Currently, TOPS InSAR processing is sufficiently documented [3–7] and SNAP remains a widely used end-to-end open source tool for processing of Sentinel-1 data.

Further development of SNAP was carried out to include exports to software packages supporting more advanced interferometric analysis, such as Stanford Method for Persistent Scatterers (StaMPS) [8]. StaMPS is a freely distributed package for research purposes with a large user community that incorporates Persistent Scatterer Interferometry (PSI) and Small Baseline methods to measure ground displacements from time series of SAR acquisitions. The idea was for the open source InSAR processor to be used together with StaMPS PSI, boosting the utilization of the Copernicus Sentinel-1 data for geohazard-related applications. The potential of SNAP-StaMPS integration has been already demonstrated at [9], for which the authors had also published a set of scripts to support utilization by the scientific community [10].

In this study we employ such open tools to analyze the urban deformation of the Rome metropolitan area for the first time using Sentinel-1 data, and combined ascending and descending orbits to depict the vertical urban deformation with special attention to different areas. Ground deformation analyses of Rome have already been undertaken in the past [11,12], but these have not been updated recently, and have not included Sentinel-1 SAR observations. Here we intend to update existing knowledge with contemporary information regarding ground deformation in the Rome metropolitan area using open data and tools. Previous relevant studies are indirectly used to verify our findings and to allow us to understand which subsidence patterns correspond to already identified phenomena and which new sources of deformation that would require further attention. Finally, a dedicated analysis was performed highlighting vertical displacements along urban infrastructures, including road networks, buildings and bridges.

1.1. Study Area

The study area includes the city of Rome and its surroundings, in the Lazio region of central Italy (Figure 1). The geology of the region is characterized by volcanic deposits (mainly pyroclastic tuff) from the Albano volcano district to the southeast and the Sabatino volcano district to the northwest, with alluvial sediments along the Tiber valley in between the two [13]. The topography gradually decreases from these two volcanic districts towards the Tiber, with valleys carved by fluvial erosion. The variability of heights in the Rome metropolitan plain does not exceed 100 m.

The southwest of the study area is dominated by the Tiber River delta and coastal plain. Both are comprised of alluvial sediments with a flat morphology [12]. The northeast of the study area includes the beginning of the Apennines mountain chain, comprising mainly sedimentary limestone and dolomite rocks [13].

Many cases of land subsidence have already been identified over the study area, and quantified through various InSAR techniques, a detailed review of which is presented in [14]. Subsidence of buildings on the alluvial sediments along the Tiber River in Rome has been measured using PSI, Small Baseline Subsets (SBAS), Interferometric Point Target Analysis (IPTA) and 4D SAR imaging techniques in many studies, e.g., by [11,15–17] and others, using ERS-1, 2 and ENVISAT ASAR data. The main cause of subsidence in this case is the weight of relatively recent construction on the unconsolidated alluvial material, especially in areas such as Grotta Perfetta, in the southwestern outskirts of the city [11–16].

Other studies, such as [14,18,19], focused on quantifying displacement which may affect the structural integrity of archaeological monuments in the historical center of the city, using SBAS, PSI and SqueeSAR with ERS-1 and -2, ENVISAT ASAR [18], Radarsat-1, 2 [19] and COnstellation of small Satellites for the Mediterranean basin Observation (COSMO-SkyMed) data [14].

Another type of ground displacement in the region of interest, which has been measured using InSAR techniques, has been identified in the Acque Albule Plain [20], in the northeastern part of the study area. Here a combination of groundwater extraction for mining and the presence of compressible soils has led to ground subsidence in the area. This has been quantified and studied with ERS and ENVISAT data using PSI and Quasi-PS InSAR (QPS) techniques by [21]. QPS is based on a different set of filtered interferograms (multi-master configuration) and is weighted by interferometric coherence.

More recently, subsidence affecting the area surrounding the Rome Fiumicino (FCO) airport, in particular over the third runway, has been studied with the PSI technique applied to ERS, ENVISAT and COSMO-SkyMed data by [12]. The authors showed how the varying rate of subsidence in the area correlates with the age of overlying man-made constructions and the nature of the underlying geology.

2. Materials and Methods

2.1. Open Source Toolboxes

This work has been carried out using the open source ESA SNAP and StaMPS software packages. The SNAP Graph Builder operator can be used to create processing chains which can be called using the batch mode Graph Processing Tool (GPT). We have exploited this utility to create the several templates necessary for creating single master TOPSAR coregistration and interferogram generation.

Finally, in order to fully automate the single master interferogram generation, we have developed and made available, based on well-designed SNAP graphs, a set of scripts called *"snap2stamps"*. These scripts enable automatic processing after setting some parameters in a configuration file. In fact, they are python wrappers which use the aforementioned templates based mainly on SNAP TOPSAR interferometric operators and whose outputs are compatible with StaMPS PSI chain. The *snap2stamps* scripts are available via the Zenodo repository [10]. Latest versions of the scripts (not verified by the developing team) can also be found on the GitHub repository (https://github.com/mdelgadoblasco/ snap2stamps). The authors had released a first version of the *snap2stamps* package in July 2018, which automates the TOPSAR single master Differential InSAR (DInSAR) processing, fully compatible with StaMPS PSI [9], allowing the creation of stacks of single master interferograms in batch mode, just by defining some simple settings such as project folder, subswath to process and defining the bounding box coordinates of the area of interest.

Additionally, for the removal of the Atmospheric Phase Screen (APS), we have employed the Toolbox for Reducing Atmospheric InSAR Noise (TRAIN) [22] and applied the linear approach (topography versus phase) integrated in the aforementioned open source package.

2.2. Data and Processing

For the data processing we have employed an ESA RSS CloudToolbox which is a Virtual Machine provided by the ESA Research and Service Support [23] with access to collocated Sentinel-1 data via Copernicus Data and Information Access Services (DIAS). This has the advantage of eliminating the data download time, as the data is locally accessible and ready to use. The resources employed were 8 vCPUs, 32 GB RAM and 1TB disk space, resulting in a total processing time of approximately 15 days, including the post-analysis of PSI results.

For the interferometric processing, the Advanced Land Observation Satellite (ALOS) World 3D (AW3D30) Digital Surface Model (DSM) [24], of 30 m spatial resolution, was utilized, while for examining the geolocation accuracy as well as the interpretation of PSI results we employed a very high resolution DSM (5 m/pixel), as extracted from CartoSat-1 satellite data [25].

2.2.1. Copernicus Sentinel-1 Data

We limited our analysis to Sentinel-1A data only (12-days repeat cycle) (Table 1), which is sufficient given the expected magnitude of ground displacements and the availability of a large number of acquisitions over the area of interest. Some details on the Sentinel-1 data employed for the processing are shown in Table 1. It should be noted that since Sentinel-1 products are not spatially synchronized, meaning that their starting and ending times may vary within each orbit, often more than one scene is required to fully cover our area of interest. This introduces additional storage and computational requirements, as consecutive scenes, for the same acquisition date, need to be downloaded and assembled into single products before proceeding with the interferometric processing. Our area of interest (AOI) and the extent of Sentinel-1 ascending (A117) and descending (D022) orbits is illustrated in Figure 1.

Among the different options, we have selected ascending and descending tracks, 117 and 022, respectively, for which the area of interest is mapped with comparable incidence angles. By ensuring combination of similar viewing geometry, i.e., sensitivity to vertical motion, we facilitate a more robust extraction of the vertical motion component, of interest for our investigation.

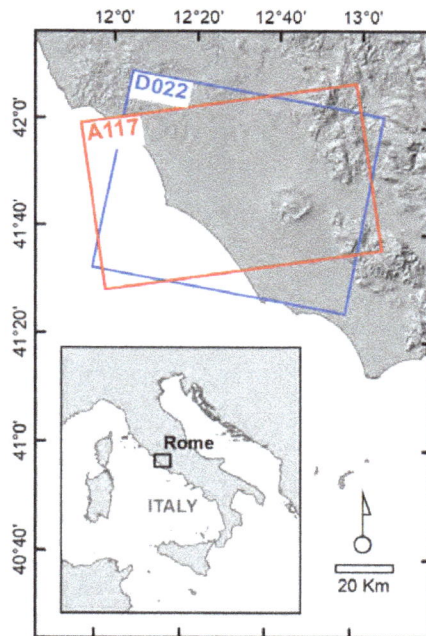

Figure 1. Area of interest and footprint of the selected Sentinel-1 master scenes for both ascending (A117) and descending (D022) tracks. ALOS World 3D DSM used as background.

Table 1. Sentinel-1 data employed for processing, with first and last image of each dataset, orbit pass, track and number of acquisitions.

Satellite	First Image	Last Image	Orbit Pass	Track	N Acquisitions
S1A	2015/03/24	2018/04/13	Descending	22	82
S1A	2015/03/30	2018/04/19	Ascending	117	87

2.2.2. SNAP-StaMPS PSI Processing

The PSI processing is split into two independent workflows: (i) single master DInSAR processing using ESA SNAP; and (ii) the PSI processing using StaMPS.

Firstly, the master scene is selected from the beginning of the data series, as Sentinel-1 has orbit control which guarantees any interferometric combination among the data. Additionally, as we want to obtain PS points over urban infrastructure, we expect that temporal baseline will not greatly affect the number of PS obtained. Master image splitting and update of orbit state vectors follow, also using the SNAP Graphical User Interface (GUI), to ensure proper selection of bursts covering our AOI. These steps are critical since they optimize time and resources for the rest of the processing. Table 2 details the parameters involved in master image splitting for burst selection over the AOI.

Table 2. Main characteristics of the selected Sentinel-1A master scene.

Track	Acquisition Date	Mean Inc. Angle (rad/degrees)	Sub-Swath	Polarization	Initial Burst	Last Burst
D022	2015/05/23	0.75/42.97	IW3	VV	5	8
A117	2015/08/09	0.67/38.39	IW2	VV	5	7

The next step involves generating all single master interferograms using the snap2stamps scripts, by following an automatic processing scheme implemented in four steps:

1. **Slave preparation.** In this step, the Sentinel-1 Single Look Complex (SLC) data are sorted by acquisition date while checking if SLC assembly (concatenation procedure) is necessary, depending on whether the defined AOI is covered by more than one scene per acquisition date.
2. **Slave splitting.** To enable processing in batch mode, the SNAP Graph Processing Tool (GPT) is used, which runs already-defined processing chains (graphs in xml format). For this step, the TOPSAR-Splitting and Apply Orbit operators are called, to update the annotated orbit information with more precise ones according to their availability (restituted or precise). These orbits are automatically downloaded by SNAP. The corresponding graph is illustrated in Figure 2, part A.
3. **Coregistration and interferogram computation.** This is the most computationally demanding step, as it performs the coregistration of the TOPSAR data (Back-geocoding with Enhanced Spectral Diversity [26] refinement) and produces the interferograms with the Flat-Earth and topographic phase contributions removed. Optionally, a finer subset can be applied over an AOI, as defined in the project configuration file. If no information is provided by the user, the full burst interferograms are generated. The outputs of this step are two debursted stacks of master-slave Single Look Complex (SLC) files and the master-slave interferogram. Supplementary data files required by StaMPS are also generated, including elevation band and orthorectified latitude and longitude coordinates as independent products. The graph employed for this step is shown in Figure 2, part B.
4. **Stamps export.** This is the final step of the single master DInSAR processing, which converts previous processing results into binary raster files compatible with StaMPS readers. Graph shown in Figure 2, part C.

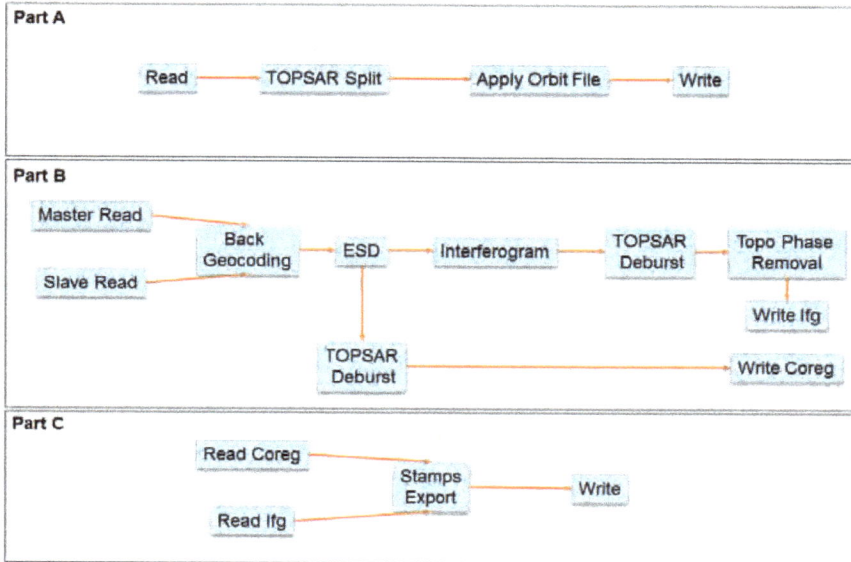

Figure 2. Schematic diagram presenting the different chains of the SentiNel Application Platform (SNAP) workflow to prepare the interferometric inputs for Stanford Method for Persistent Scatterers (StaMPS) Persistent Scatterer Interferometry (PSI) processing [9]. Part (**A–C**) illustrate the workflow employed for slave splitting, coregistration, interferometric computation and StaMPS export, respectively.

The following step involves the ingestion of SNAP exports into StaMPS using a specific script, called *mt_prep_snap*, available in the distribution. Subsequently, the StaMPS PSI processing chain is run from step 1 to 7 as described in the StaMPS User Manual [27].

In this case, we additionally applied the integrated TRAIN, using the linear tropospheric correction approach, to mitigate the topography-correlated atmospheric phase.

In order to properly merge the results from both ascending and descending tracks in subsequent post-processing steps, we selected the same reference point in both cases, corresponding to a permanent European Reference Frame (EUREF) Global Navigation Satellite System (GNSS) station (M0SE00ITA), located at the Aerospace Engineering Faculty of the University of Rome "La Sapienza". Based on the EUREF solution [28], the station seems stable, with no evident vertical motion (Vx = −0.7 ± 0.1 mm/yr, Vy = −1.3 ± 0.1 mm/yr and Vz =0.5 ± 0.1 mm/yr in ETRF2014) during the entire observation period.

We ran StaMPS PSI three times from the merging of the different patches (step 5) onwards, each time with different grid options [27]: (i) no merging; (ii) merging by 20 m grid; and (iii) merging by 40 m grid. For each run, the merging of PS candidates was performed with the same threshold selected for the phase noise filtering in StaMPS step 4.

2.2.3. Post-Processing

Having both ascending and descending PSI measurements, we combined them to calculate the vertical component for each individual Persistent Scatter (PS) point (see Figure 3) using Equation (1) and (2), as described in [29]:

$$
\begin{bmatrix} d_{LOS}{}^{asc} \\ d_{LOS}{}^{desc} \end{bmatrix} = A \begin{bmatrix} d_{up} \\ d_{h_{ald}} \end{bmatrix} \tag{1}
$$

with

$$A = \begin{bmatrix} \cos\theta^{asc} & \frac{\sin\theta^{asc}}{\cos\Delta\alpha} \\ \cos\theta^{desc} & \sin\theta^{desc} \end{bmatrix} \tag{2}$$

where d_{LOS} is displacement along Line-Of-Sight (LOS). d_{up} is the vertical displacement. d_{hald} is the projection of horizontal displacement in descending azimuth look direction (ALD). θ is the incident angle. $\Delta\alpha$ is the satellite heading, difference between ascending and descending orbit.

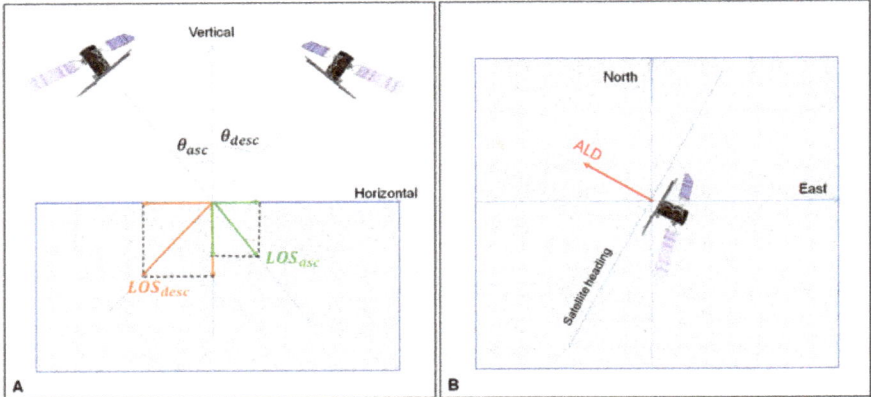

Figure 3. Ascending and descending decomposition in vertical and horizontal components (**A**) and the Azimuth Look Direction (ALD) for the descending orbit pass (**B**).

The combination of ascending and descending PS points was performed in the vector domain avoiding any rasterization option, as described in [30]. The method combines PS targets based on their geographic proximity, while attributes transfer and decomposition of motion is done within the features geodatabases. This leads to a higher number of final PS points and reduction of error budget introduced by spatial interpolation and rasterization procedures. PS points of one geometry having no neighbouring targets by the opposite geometry, within a defined search radius, are being excluded. The selection of the maximum search radius for the combination, in our case 40 m, was based on the statistical analysis of the distances between PS targets from the independent LOS solutions.

After obtaining the vertical component of the deformation rates, by using GIS capabilities, we overlaid the deformation information over buildings, roads, highways and railways vector layers to calculate the maximum observed deformation for each of these elements. In such a way, we were able to provide subsidence information over critical infrastructure (roads, bridges etc.) as well as on individual building blocks. For the special case of the road networks, and in order to depict the spatial variability of motion along them, a segmentation procedure was applied considering a distance of 20 m for each segment. The displacement value is then calculated based on PS points located at a specified distance across each road segment. To reduce overlaps between successive segments, we considered a square buffering option, i.e., buffer zones do not exceed the segments' start and end points.

Since StaMPS does not update PS heights, inaccuracies in the geolocation of PS targets over urban environments, especially for high buildings, might occur. To compensate for the above mentioned issue, during the calculation of the deformation statistics for each vector element (road, buildings etc.) a 20 m buffer was considered. The choice of the buffer distance was based on the visual interpretation of the results.

3. Results

We have obtained the average PSI LOS deformation rates for both ascending and descending tracks (Figure 4). An indicator of the compatibility of solutions between acquisition geometries is the standard deviation of the mean LOS deformation velocities, which correspond to 1.04 mm/yr and 1.17 mm/yr for A117 and D022 tracks, respectively.

Given the difference in area covered by each track, the relatively larger number of PS points in the descending solution could be explained (Table 3). Apart from the effect of area coverage, it seems that the overall numbers of PS points is comparable. We attribute this to the common observation period and incidence angles considered for both ascending and descending datasets.

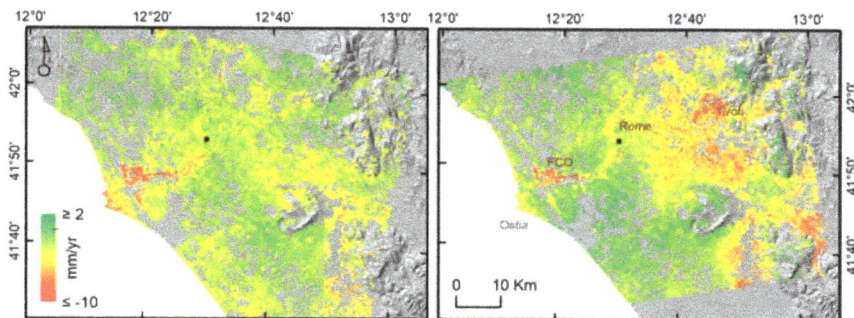

Figure 4. Sentinel-1 average Line-Of-Sight (LOS) deformation rate maps over the period March 2015–April 2018 for descending (**left**) and ascending (**right**) acquisition geometries. Positive values indicate motion towards the sensor or uplift, whereas negative values motion away from the sensor or subsidence. Selected reference point (M0SE00ITA EUREF station) is shown as square. ALOS World 3D Digital Surface Model (DSM) as background.

After decomposing the ascending and descending LOS measurements, we obtained the vertical deformation rate map presented in Figure 5. It can be easily seen that various deformation patterns exist, attributed mainly to the different subsidence mechanisms acting in the metropolitan area of Rome. There are several areas undergoing significant subsidence, such as (i) FCO airport; (ii) along the Tiber River; (iii) the coastal zone of Ostia; (iv) Ostiense quarter within Rome and; (v) Tivoli area, while the rest of the region exhibits relatively low ground deformation rates. In the following sections we provide a more detailed analysis of the PSI results.

As we ran StaMPS using different merging options, we obtained different numbers of PS points for each solution. In Table 3, we summarize the total points obtained for each merging configuration and those remaining after the vertical decomposition. It should be noted that, for the decomposition, only the overlapping area between ascending and descending results is exploited.

Table 3. Total number of PS points obtained by StaMPS processing (LOS) and after the decomposition to actual motion component (Vertical).

Orbit	No Merging LOS/Vertical	20 m LOS/Vertical	40 m LOS/Vertical
Ascending (A117)	1065328/947386	486188/418481	264024/211999
Descending (D022)	1342924/1061976	580578/439738	311615/217237

For the full resolution datasets, we obtained over 1 million point targets for both orbits (Table 3), a fact which poses some difficulties in handling the dataset for post analysis purposes. The decision to reduce the initial number of PS points or not depends actually on the application at hand. For example, while working on infrastructure monitoring, it may be relevant to maintain full resolution results, as the number of points decreases rapidly after merging. For the needs of our work, the 20 m merged solutions offered a reasonable trade-off between density of PS targets and computational requirements for post-processing.

The obtained PS density after the decomposition, calculated at 200 m grid, is shown in Figure 5. The increase of density over the built-up area of Rome, reaching 1700 points/km^2 in the center of the city, is evident, while several rural urban centers also exhibit relatively high PS densities varying from 300 to 600 points/km^2. The locations of ascending and descending PS targets over an urban fabric of moderate density are presented (Figure 6), showing the advantages of the applied decomposition approach in retaining larger numbers of PS targets that allows proper characterization of on-going deformation phenomena. The final average velocity map of vertical deformation over the study area is shown in Figure 5, where the different areas with higher subsidence can be clearly identified.

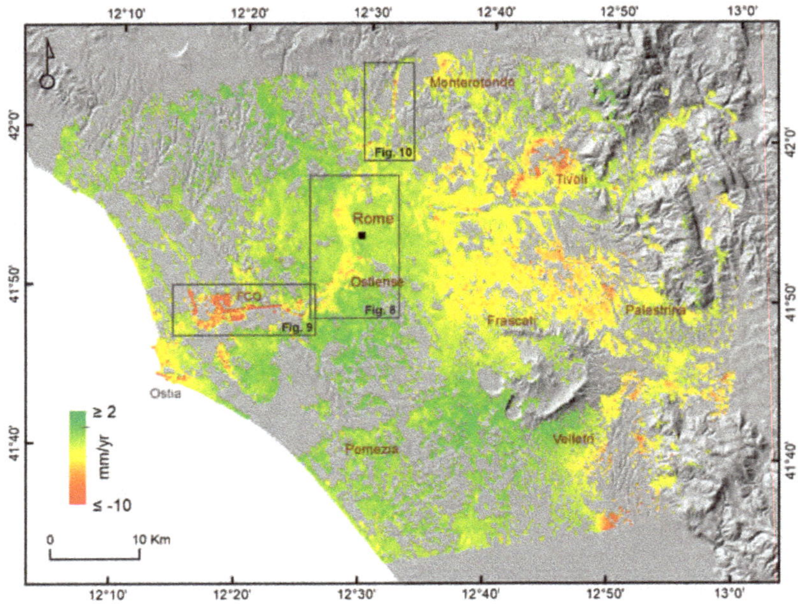

Figure 5. Figure 5. Sentinel-1 average vertical motion rates for the period March 2015–April 2018. For the decomposition, PS points at full resolution were spatially down sampled by a window of 40 m radius (see Section 2.2.2). Selected reference point (M0SE00ITA EUREF station) is marked by a black square. ALOS World 3D DSM as background. The locations of other figures are also shown.

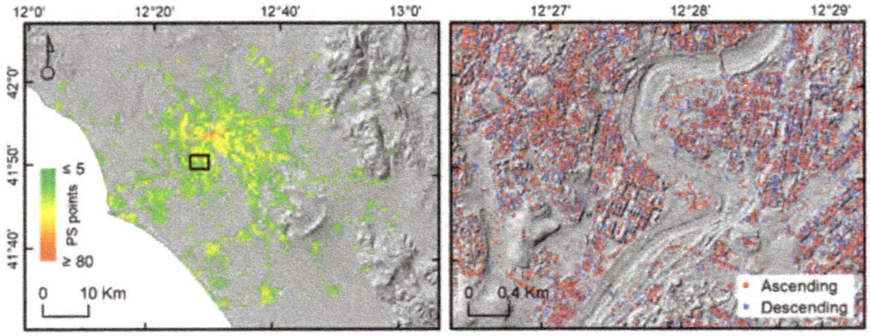

Figure 6. Sentinel-1 PS density over a 200 m resolution grid (**left**) and PS locations for both ascending and descending geometries at full resolution. Black square indicates the location of the zoom image shown on the right. ALOS World 3D DSM (**left**) and CartoSat-1 DSM (**right**) as backgrounds.

3.1. Critical Urban Infrastructures: Global Road Network

As mentioned in Section 2, we separated the vertical deformation of the different elements such as roads, highways, railways and buildings using the extracted OpenStreetMap shapefile layers available in [31], and here in Figure 7 we show the deformation for the different roads and highways over our AOI. Significant deformation is revealed near the Fiumicino area, along the Tiber River and in the eastern part outside the Grande Raccordo Anulare (GRA). However, other roads show stable behavior or irrelevant deformation.

Systematic monitoring of human infrastructure, particularly critical for communication and transportation (highways, railways, roads, bridges, viaducts) or providing resources (electricity plants, dikes, dams) can be used for maintenance planning activities as well as for infrastructure risk assessment.

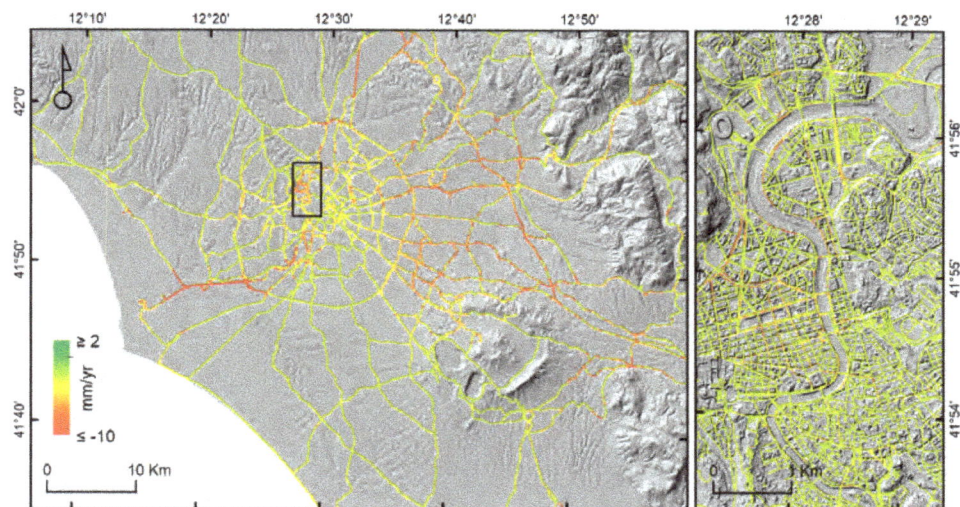

Figure 7. Average vertical motion rates along motorways as well as primary and secondary road networks (**left**) and a detail over Rome city, including tertiary, residential and pedestrian roads (**right**). ALOS World DSM (AW3D30) (**left**) and CartoSat-1 DSM (**right**) as backgrounds.

3.2. Subsidence along the Tiber River

This type of subsidence phenomena is quite common on areas constructed over alluvial deposits along river floodplains.

As mentioned in Section 1.1, the subsidence of buildings on the alluvial sediments along the Tiber River in Rome has been studied using similar techniques. Some notable studies include [11–17]. These all used ERS-1, 2 and ENVISAT ASAR data. The main cause of subsidence in this case is the weight of relatively recent construction on the unconsolidated alluvial material [11–16]. The older constructions in the city center of Rome on the other hand display less movement, as their position has consolidated over time. Findings from the PanGeo project [17], and the studies mentioned above, reported subsidence along the Tiber River and its tributaries determined using PSI and other techniques applied to ERS SAR and ENVISAT ASAR data acquired from 2002 to 2005.

Similar patterns over these alluvial deposits are present in our results (Figure 8). It is difficult to compare the deformation velocities between the studies, due to the different datasets employed and periods analyzed. We measure strong subsidence in several areas such as Ostiense and Santa Victoria quarter, with maximum vertical deformation rate of −7.2 mm/yr. Also remarkable is the subsidence along the Tiber River and its tributaries (with more than 52k PS points), with a maximum deformation rate of −8.7 mm/yr and an average of −1.4 mm/yr.

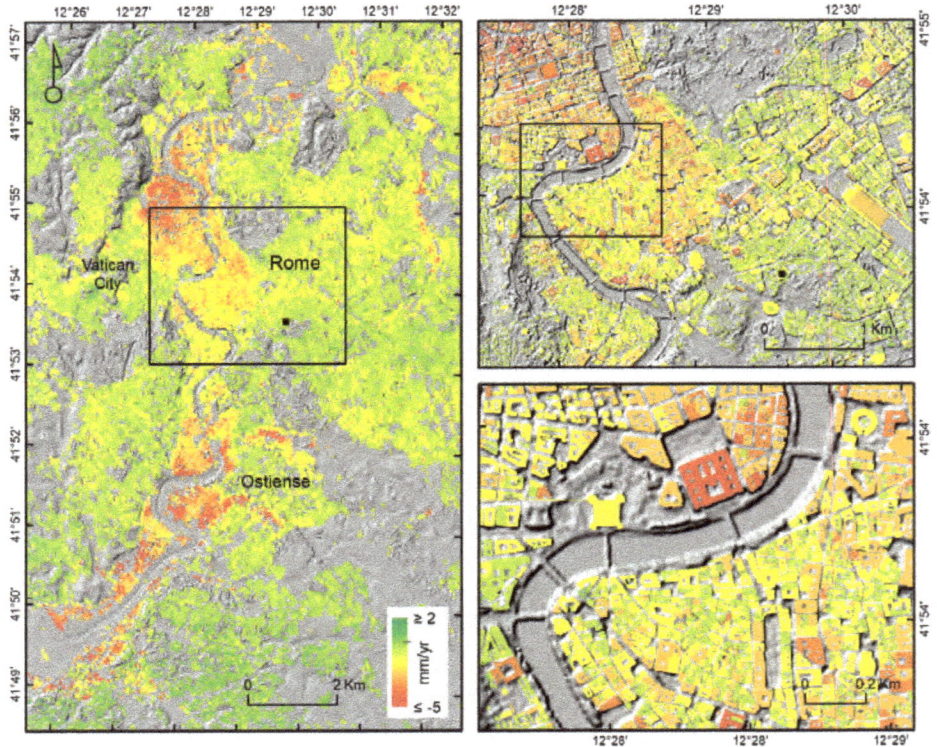

Figure 8. Average vertical deformation rates along the Tiber River and its tributaries (**left**) and detailed views presenting the estimated maximum deformation rates per building block within the center of Rome (**right**). The selected reference point (M0SE00ITA EUREF station) is shown in the black square. CartoSat-1 DSM as background.

3.3. Fiumicino Airport and Ostia Coastal Region

The deformation over the Fiumicino area is also known, and consistent with previous studies (e.g., [12]), in which similar subsidence patterns are found. There are some points to highlight on the deformation over this area: (i) the airport runway oriented north-south; (ii) the highway; and (iii) the harbor area.

Regarding the runway, part of the track is located over an ancient lake. This has a different degree of consolidation compared to the other part, located over alluvial deposits. A very different behavior is thus shown between the two parts [12]. Also, the highway from Rome towards Fiumicino airport suffers from a high rate of vertical subsidence. This poses a higher risk than comprehensive subsidence over the entire area as large variations in deformation rate can lead to cracks in infrastructures.

Figure 9 illustrates the spatial distribution of the vertical deformation on the Fiumicino area, where the Rome-Fiumicino highway deformation is highlighted in (A), the full time series deformation of a smaller area for both ascending and descending orbits is shown in (B) and in (C) the total accumulated vertical deformation of the portion of highway inside the black ellipse found in (A). Figure 9C has been obtained by considering the linear velocity of the vertical motion for each PS point inside the dashed lines in Figure 9A from West to East. The horizontal axis is the longitude and vertical axis refers to the total accumulated vertical motion. There are visible transitions of blue to red and vice versa between adjacent points, where blue corresponds to no motion or the PS point in that position, as there is not always a PS point per each 20 m of highway.

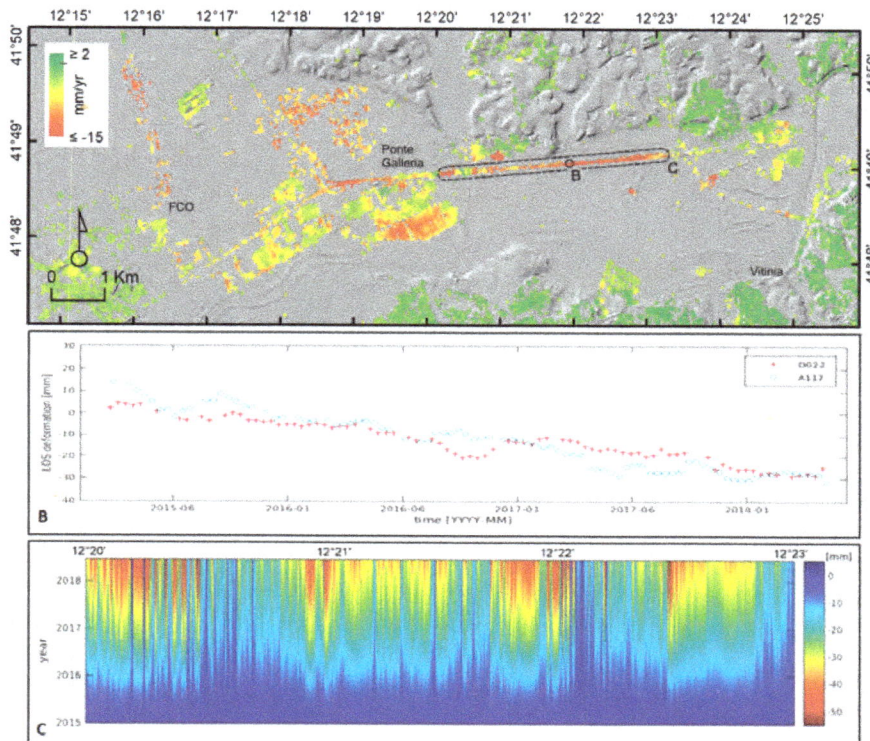

Figure 9. Spatial distribution of average vertical deformation rates at Fiumicino (FCO) airport area overlain on CartoSat-1 DSM (**A**). PSI LOS displacement time series for selected point target (**B**) and cumulative motion plot for the Roma-Fiumicino highway (**C**) are also shown.

Moving from West to East, the first area with strong vertical deformation is near the Mediterranean coast of Ostia-Fiumicino, where more than 20k PS are obtained, measuring a maximum deformation rate of −9 mm/yr. Further East, the nearest area with strong deformation is the Fiumicino airport, where only on the airport North–South oriented runway are located 270 PS points, with a maximum vertical deformation of −14.8 mm/yr, and an average deformation along the track of −6.6 mm/yr. Next to the Fiumicino airport, in the Ponte Galleria industrial area, with more than 3700 PS points, the maximum subsidence value is less than −19 mm/yr while the average vertical deformation is less than −5 mm/yr. Continuing towards Rome city, the highway Rome-Fiumicino also suffers strong vertical deformation, having a maximum deformation of −19 mm/yr and an average vertical deformation of almost −7 mm/yr.

3.4. Other Cases of Strong Displacement Patterns

Other important patterns are found over highways and roads, for which it is worth highlighting that the areas where strong transitions between subsidence rates occur are generally more dangerous as these are responsible for cracks in infrastructures. Hence, these areas with strong variations in subsidence behavior, and with high spatial frequency are the ones we would suggest to closely monitor, as well as the areas with significant subsidence, as continuous subsidence may also pose a risk.

In Figure 10, we show an example of the road subsiding while the bridges seem to remain more stable. Specifically, near Settebagni, on the "A1 Diramazione Roma Nord" there are more than 700 PS, measuring a maximum vertical deformation of −7.8 mm/yr with an average of around −4 mm/yr.

Finally, on the Eastern part of Rome (see Figure 6), we find Tivoli and Tivoli Terme, with more than 17k PS, measuring a maximum vertical deformation of −12 mm/yr, with an average of −3 mm/yr, similar to values obtained in [20].

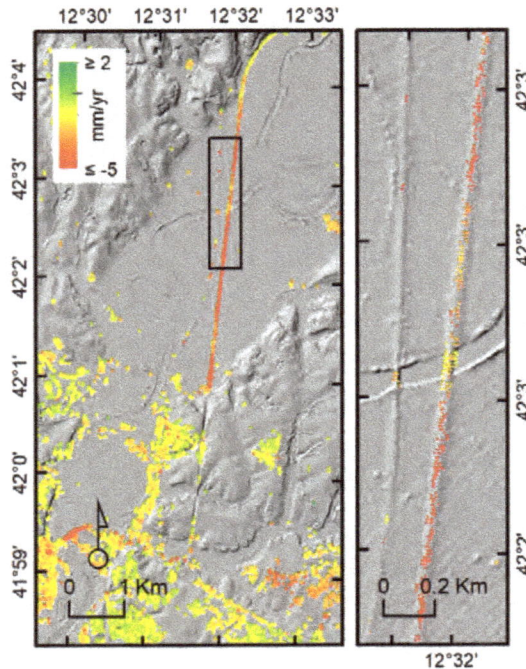

Figure 10. Deformation patterns at the "A1 Diramazione Roma Nord" north of Rome, showing subsidence of alluvial deposits and relative stability of constructed bridges. Geolocation accuracy of the obtained PS results can be visually assessed based on the overlap on the CartoSat-1 DSM in the background.

4. Discussion

We present ground deformation in the Rome metropolitan area for the first time, using Copernicus Sentinel-1 mission data and open source toolboxes, paving the way for the broader utilization of the proposed chain by EO practitioners in geohazards applications.

Despite the availability of additional satellite data, limiting the analysis to Sentinel-1A acquisitions (repeat of 12 days) was considered sufficient, given the relatively low deformation rates in the area and the expected linear behavior of motion in time.

The area exhibits diverse deformation patterns, the spatial expression of which indirectly suggests, at least for some cases, the underlying deformation mechanism. The demonstrated quality of interferometric products obtained by Sentinel-1, both in terms of spatial density and uncertainties of displacement estimates is of key importance for the interpretation of motion and the phenomena involved. In our case, the density of the results reaches 1700 point/km^2 (~70 PS targets on a 200 m × 200 m grid) for dense urban centers, 300–600 point/km^2 for the suburban environments, while for the entire area of interest we calculated (including areas with no PS) on average 250 point/km². It is worth mentioning that we exploit only the co-pol. (VV) channel of Sentinel-1, yet by considering both polarizations (VV+VH) an increase in the number of PS targets is expected. As a result of the high PS density in built-up areas, and thus, availability of multiple PS targets within each building block, it is possible to differentiate between deformation of the buildings themselves from those related to soil foundations. Yet, given the uncertainties in the location of the PS points and the resolution of Sentinel-1, such separation of motion should be handled with care.

Among the most pronounced deformation patterns is the subsidence at Fiumicino airport reaching −20 mm/yr and along the Tiber River and its tributaries with motion in the order of −5mm/yr. Several other cases with a magnitude of motion worth mentioning, in the range of −12 to −3 mm/yr, are found along the coastal zone of Ostia as well as in the Tivoli region. The presence of relatively high deformation gradients for the above mentioned sites is well documented in several studies [11–18,20]. A more detailed analysis is provided in the PanGeo project report about the Geohazard description of Rome [17]. In [20], an analysis integrating geological and hydrogeological modelling provided insights on the relation between ground displacement, variations in the groundwater table and geotechnical properties of the subsoil for the specific case of Fiumicino area. For most of the cases, ground deformation can be attributed to the local geological conditions, such as the compaction of soft sediments, whereas loading by urban constructions should be one of the major reinforcing factors. However, further investigation is required to characterize the on-going subsidence induced phenomena and their temporal evolution compared to past ground displacement measurements.

Computed average vertical displacement rates for the various lithological types, as described in the 1/25,000 scale geological map of Lazio [24] are shown in (Figure 11). As expected, unconsolidated deposits (e.g., sands, clays and other alluvial material) show higher subsidence rates compared to basement formation such as marls, limestones and dolomites.

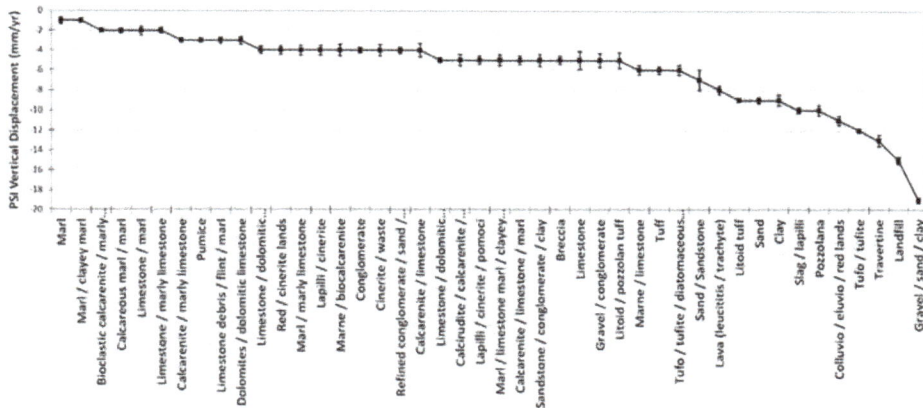

Figure 11. Maximum subsidence rate per soil lithological type in the broader metropolitan area of Rome [31]. Sorted from lower to higher maximum subsidence value.

Deformation over urban infrastructures was also detected, with the case of the GRA, the ring highway surrounding the city of Rome, showing subsidence rates of up to −7.8 mm/yr. In fact, the proposed segmentation approach enabled the localization of the deformation along the entire road network of the Rome metropolitan area, a valuable option to support planners.

The above-mentioned findings were verified by inter-comparison with previous studies. Further validation activities were not considered necessary, since InSAR techniques have already undergone a long period of validation [32], confirming their capacity to measure surface motion. Actually, the majority of the observed subsidence has been already reported in the past [11–18].

However, the updated status as provided by Sentinel-1 is of significant importance, since the continuation of ground deformations without addressing any mitigation actions may lead to undesirable consequences in the future.

Finally, our ability to detect and measure ground movements with substantial accuracy within a short time should be underlined. Comparable studies in the past would require long observation periods to collect sufficient satellite data to obtain robust solutions. At the same time, our capacity to actually monitor phenomena has been increased by the systematic availability of the Copernicus Sentinel-1 data.

5. Conclusions

We have demonstrated the utilization of open and free data from the Copernicus Sentinel-1 mission using open source toolboxes for advanced interferometric processing. We provide details on a dedicated package for the automation of SNAP-StaMPS PSI processing, which we make available to EO practitioners in response to the growing need for EO-based solutions for monitoring geohazards. The integration of such a chain on a cloud processing environment will enable EO practitioners to respond to the ever-increasing volume of satellite data and high processing capacity requirements.

We verified the results by inter-comparison with previously published studies. To facilitate openness, we have made the PSI measurements over Rome available online, encouraging further analysis and interpretation as well as promoting collaboration between research communities.

Supplementary Materials: The following are available online at http://www.mdpi.com/2072-4292/11/2/129/s1, Sentinel-1 PS LOS displacement rates over the period April 2015–May 2018 in WGS84 projection are provided in Environmental Systems Research Institute (ESRI) shapefile format for both ascending and descending datasets. Each record contains latitude and longitude coordinates in decimal degrees, record identifier, average LOS deformation rates and standard deviation of average LOS deformation rates, both in mm/yr.

Author Contributions: Conceptualization and methodology, by all authors; software, J.M.D.B., A.H. and M.F.; formal analysis and investigation, J.M.D.B., M.F. and C.S.; writing—review and editing, by all authors; visualization, M.F. and J.M.D.B.

Funding: This research was partly funded by BRGM (www.brgm.fr), in the form of an internal research project.

Acknowledgments: The authors would like to acknowledge the ESA Research and Service Support for the provisioning of the employed computer resources with collocated Copernicus Sentinel-1 data. The suggestions and recommendations by anonymous reviewers that contributed to improving the manuscript are also acknowledged. ALOS World 3D DSM (AW3D30) was downloaded from The Japan Aerospace Exploration Agency (JAXA) and CartoSat-1 DSM was obtained from ESA through the Category-1 research project with ID 42428. Extracted OpenStreetMap layers over Lazio region downloaded from [31].

Conflicts of Interest: The authors declare no conflict of interest.

References

1. Veci, L.; Lu, J.; Prats-Iraola, P.; Scheiber, R.; Collard, F.; Fomferra, N.; Engdahl, M. The Sentinel-1 Toolbox. In Proceedings of the IEEE International Geoscience and Remote Sensing Symposium (IGARSS), Quebec City, QC, Canada, 13–18 July 2014; pp. 1–3.
2. De Zan, F.; Guarnieri, A.M. TOPSAR: Terrain observation by progressive scans. *IEEE Trans. Geosci. Remote Sens.* **2006**, *44*, 2352–2360. [CrossRef]
3. Prats-Iraola, P.; Nannini, M.; Yague-Martinez, N.; Pinheiro, M.; Kim, J.-S.; Vecchioli, F.; Minati, F.; Costantini, M.; Borgstrom, S.; De Martino, P.; et al. Interferometric investigations with the Sentinel-1 constellation. In Proceedings of the 2017 IEEE International Geoscience and Remote Sensing Symposium (IGARSS), Fort Worth, TX, USA, 23–28 July 2017; pp. 5537–5540.
4. Prats-Iraola, P.; Nannini, M.; Yague-Martinez, N.; Scheiber, R.; Minati, F.; Vecchioli, F.; Costantini, M.; Borgstrom, S.; De Martino, P.; Siniscalchi, V.; et al. Sentinel-1 tops interferometric time series results and validation. In Proceedings of the 2016 IEEE International Geoscience and Remote Sensing Symposium (IGARSS), Beijing, China, 10–15 July 2016; pp. 3894–3897.
5. GUNCE, H.B.; SAN, B.T. Measuring Earthquake-Induced Deformation in the South of Halabjah (Sarpol-e-Zahab) Using Sentinel-1 Data on November 12, 2017. *Proceedings* **2018**, *2*, 346. [CrossRef]

6. Ganas, A.; Elias, P.; Bozionelos, G.; Papathanassiou, G.; Avallone, A.; Papastergios, A.; Valkaniotis, S.; Parcharidis, I.; Briole, P. Coseismic deformation, field observations and seismic fault of the 17 November 2015 M = 6.5, Lefkada Island, Greece earthquake. *Tectonophysics* **2016**, *687*, 210–222. [CrossRef]

7. Jelének, J.; Kopačková, V.; Fárová, K. Post-Earthquake Landslide Distribution Assessment Using Sentinel-1 and-2 Data: The Example of the 2016 Mw 7.8 Earthquake in New Zealand. *Proceedings* **2018**, *2*, 361. [CrossRef]

8. Hooper, A.; Bekaert, D.; Spaans, K.; Arıkan, M. Recent advances in SAR interferometry time series analysis for measuring crustal deformation. *Tectonophysics* **2012**, *514–517*, 1–13. [CrossRef]

9. Foumelis, M.; Delgado Blasco, J.M.; Desnos, Y.-L.; Engdahl, M.; Fernandez, D.; Veci, L.; Lu, J.; Wong, C. ESA SNAP-StaMPS Integrated Processing for Sentinel-1 Persistent Scatterer Interferometry. In Proceedings of the 2018 IEEE International Geoscience and Remote Sensing Symposium, Valencia, Spain, 22–27 July 2018; pp. 1364–1367.

10. Delgado Blasco, J.M.; Foumelis, M. Automated SNAP Sentinel-1 DInSAR Processing for StaMPS PSI with Open Source Tools. Available online: https://zenodo.org/record/1322353 (accessed on 29 July 2018).

11. Stramondo, S.; Bozzano, F.; Marra, F.; Wegmuller, U.; Cinti, F.R.; Moro, M.; Saroli, M. Subsidence induced by urbanisation in the city of Rome detected by advanced InSAR technique and geotechnical investigations. *Remote Sens. Environ.* **2008**, *112*, 3160–3172. [CrossRef]

12. Bozzano, F.; Esposito, C.; Mazzanti, P.; Patti, M.; Scancella, S. Imaging Multi-Age Construction Settlement Behaviour by Advanced SAR Interferometry. *Remote Sens.* **2018**, *10*, 1137. [CrossRef]

13. Funiciello, R.; Heiken, G.; De Rita, D. *I Sette Colli: Guida Geologica a Una Roma Mai Vista*; Raffaello Cortina Editore: Milano, Italy, 2006.

14. Cigna, F.; Lasaponara, R.; Masini, N.; Milillo, P.; Tapete, D. Persistent scatterer interferometry processing of COSMO-skymed stripmap HIMAGE time series to depict deformation of the historic centre of Rome, Italy. *Remote Sens.* **2014**, *6*, 12593–12618. [CrossRef]

15. Manunta, M.; Marsella, M.; Zeni, G.; Sciotti, M.; Atzori, S.; Lanari, R. Two-scale surface deformation analysis using the SBAS-DInSAR technique: A case study of the city of Rome, Italy. *Int. J. Remote Sens.* **2008**, *29*, 1665–1684. [CrossRef]

16. Fornaro, G.; Serafino, F.; Reale, D. 4-D SAR imaging: The case study of Rome. *IEEE Geosci. Remote Sens. Lett.* **2010**, *7*, 236–240. [CrossRef]

17. Comerci, V.; Cipolloni, C.; di Manna, P.; Guerrieri, L.; Vittori, E.; Bertoletti, E.; Ciuffreda, M.; Succhiarelli, C. PanGeo: Enabling Access to Geological Information in Support of GMES-D7.1.26 Geohazard Description for Rome. 2012. Available online: http://www.pangeoproject.eu/pdfs/english/rome/Geohazard-Description-rome.pdf (accessed on 19 September 2018).

18. Zeni, G.; Bonano, M.; Casu, F.; Manunta, M.; Manzo, M.; Marsella, M.; Pepe, A.; Lanari, R. Long-term deformation analysis of historical buildings through the advanced SBAS-DInSAR technique: The case study of the city of Rome, Italy. *J. Geophys. Eng.* **2011**, *8*, S1. [CrossRef]

19. Tapete, D.; Fanti, R.; Cecchi, R.; Petrangeli, P.; Casagli, N. Satellite radar interferometry for monitoring and early-stage warning of structural instability in archaeological sites. *J. Geophys. Eng.* **2012**, *9*, S10–S25. [CrossRef]

20. Bozzano, F.; Esposito, C.; Franchi, S.; Mazzanti, P.; Perissin, D.; Rocca, A.; Romano, E. Analysis of a Subsidence Process by Integrating Geological and Hydrogeological Modelling with Satellite InSAR Data. In *Engineering Geology for Society and Territory-Volume 5*; Springer: New York, NY, USA, 2015; pp. 155–159.

21. Perissin, D.; Wang, T. Repeat-pass SAR interferometry with partially coherent targets. *IEEE Trans. Geosci. Remote Sens.* **2012**, *50*, 271–280. [CrossRef]

22. Bekaert, D.P.S.; Walters, R.J.; Wright, T.J.; Hooper, A.J.; Parker, D.J. Statistical comparison of InSAR tropospheric correction techniques. *Remote Sens. Environ.* **2015**, *170*, 40–47. [CrossRef]

23. Delgado Blasco, J.M.; Sabatino, G.; Cuccu, R.; Rivolta, G.; Pelich, R.; Matgen, P.; Chini, M.; Marconcini, M. Support for Multi-temporal and Multi-mission data processing: The ESA Research and Service Support. In Proceedings of the 2017 9th International Workshop on the Analysis of Multitemporal Remote Sensing Images (MultiTemp), Brugge, Belgium, 27–29 June 2017.

24. Takaku, J.; Tadono, T.; Tsutsui, K.; Ichikawa, M. VALIDATION of "aW3D" GLOBAL DSM GENERATED from ALOS PRISM. In Proceedings of the ISPRS Annals of the Photogrammetry, Remote Sensing and Spatial Information Sciences, Prague, Czech Republic, 12–19 July 2016.

25. D'Angelo, P.; Lehner, M.; Krauss, T.; Hoja, D.; Reinartz, P. Towards automated DEM generation from high resolution stereo satellite images. *Int. Arch. Photogramm. Remote Sens. Spat. Inf. Sci.* **2008**, *37*, 1137–1342.

26. Scheiber, R.; Moreira, A. Coregistration of interferometric SAR images using spectral diversity. *IEEE Trans. Geosci. Remote Sens.* **2000**, *38*, 2179–2191. [CrossRef]

27. Hooper, A.J.; Bekaert, D.; Hussain, E.; Spaans, K. *StaMPS/MTI Manual*; School of Earth and Environment: Leeds, UK, 2018.

28. EUREF Permanent GNSS Network. M0SE00ITA (Roma, Italy). Available online: http://www.epncb.oma.be/_productsservices/coordinates/crd4station.php?station=M0SE00ITA (accessed on 1 October 2018).

29. Samieie-Esfahany, S.; Hanssen, R.F.; Van Thienen-Visser, K.; Muntendam-Bos, A.; Systems, S. On the effect of horizontal deformation on insar subsidence estimates. In Proceedings of the 2009 Workshop on Fringe, Frascati, Italy, 30 November–4 December 2009.

30. Foumelis, M. Vector-based approach for combining ascending and descending persistent scatterers interferometric point measurements. *Geocarto Int.* **2018**, *33*, 38–52. [CrossRef]

31. Estratti OpenStreetmap: Regione Lazio. Available online: https://osm-estratti.wmflabs.org/estratti/ (accessed on 16 September 2018).

32. Crosetto, M.; Monserrat, O.; Cuevas-González, M.; Devanthéry, N.; Crippa, B. Persistent Scatterer Interferometry: A review. *ISPRS J. Photogramm. Remote Sens.* **2016**, *115*, 78–89. [CrossRef]

![remote sensing logo] *remote sensing*

MDPI

Article

Using TSX/TDX Pursuit Monostatic SAR Stacks for PS-InSAR Analysis in Urban Areas

Ziyun Wang [1], Timo Balz [1,*], Lu Zhang [1], Daniele Perissin [2] and Mingsheng Liao [1]

[1] State Key Laboratory of Information Engineering in Surveying, Mapping and Remote Sensing (LIESMARS), Wuhan University, Wuhan 430072, China; ziyunwang@whu.edu.cn (Z.W.); luzhang@whu.edu.cn (L.Z.); liao@whu.edu.cn (M.L.)

[2] RASER Limited, Hong Kong, China; daniele.perissin@sarproz.com

* Correspondence: balz@whu.edu.cn; Tel.: +86-27-6877-9960

Received: 7 November 2018; Accepted: 18 December 2018; Published: 24 December 2018

Abstract: Persistent Scatterer Interferometry (PS-InSAR) has become an indispensable tool for monitoring surface motion in urban environments. The interferometric configuration of PS-InSAR tends to mix topographic and deformation components in differential interferometric observations. When the upcoming constellation missions such as, e.g., TanDEM-L or TWIN-L provide new standard operating modes, bi-static stacks for deformation monitoring will be more commonly available in the near future. In this paper, we present an analysis of the applicability of such data sets for urban monitoring, using a stack of pursuit monostatic data obtained during the scientific testing phase of the TanDEM-X (TDX) mission. These stacks are characterized by extremely short temporal baselines between the TerraSAR-X (TSX) and TanDEM-X acquisitions at the same interval. We evaluate the advantages of this acquisition mode for urban deformation monitoring with several of the available acquisition pairs. Our proposed method exploits the special properties of this data using a modified processing chain based on the standard PS-InSAR deformation monitoring procedure. We test our approach with a TSX/TDX mono-static pursuit stack over Guangzhou, using both the proposed method and the standard deformation monitoring procedure, and compare the two results. The performance of topographic and deformation estimation is improved by using the proposed processing method, especially regarding high-rise buildings, given the quantitative statistic on temporal coherence, detectable numbers, as well as the PS point density of persistent scatters points, among which the persistent scatter numbers increased by 107.2% and the detectable height span increased by 78% over the standard processing results.

Keywords: pursuit monostatic; PS-InSAR; urban monitoring; skyscrapers

---

1. Introduction

Persistent Scatterer Interferometry (PS-InSAR) is a multi-temporal interferometric method that has been widely used for urban monitoring. Various cities use PS-InSAR as a standard monitoring tool for monitoring slow deformation, including ground [1–4], single buildings [5–7], and infrastructures [8,9]. The upcoming bi-static sensor missions promise even higher performance in global monitoring. In 2010, the TanDEM-X (TerraSAR-X Add-on for Digital Elevation Measurements) mission was launched. Together with TerraSAR-X (TSX), the constellation operates in a bistatic mode, with the primary goal of producing a highly accurate global digital elevation model (DEMs) of the Earth [10]. A secondary goal of the program is to demonstrate potential new applications of TanDEM-X (TDX) [11,12] and innovative new operating modes, including the bistatic, monostatic pursuit, and alternating bistatic modes, during the commissioning and scientific research phases of the mission [13,14]. Among those operation modes, the pursuit monostatic (PM) data is of interest for urban PS-InSAR applications that address slow motion, especially subsidence in urban areas.

In the PM mode, two satellites fly in the same orbit, one behind another, with an increased along-track baseline of 20 km, which is equal to a temporal baseline of 10 s. This enlarged distance allows both satellites to transmit and receive signals independently; thus, for each target scene, an independent image pair can be obtained within a time interval of 10 s. With this extremely short along-track temporal baseline between acquisitions within the PM mode, various applications on the datasets were conducted [15,16], and qualities were analyzed [17,18]. Few applications are found on urban areas [19], while most publications about PM data applications are focused on Moving Target Indication (MTI) [20,21], with examples drawn from sea surface derivation [22], and oil and ship detection [23]. Another example demonstrate the use of PM data for agricultural monitoring [24]. Many of the methods adopted in those publications are designed for change detection (CD), and are focused on minor changes of the targets occurring over short time intervals. Research on mono-static pursuit data for surface motion estimation, however, is limited; we assume that this is due to the limited availability of such stacks.

Upcoming Synthetic Aperture Radar (SAR) missions show similar constellations to TerraSAR-X and TanDEM-X; for example, TanDEM-L [25–27] or TWIN-L [28], will increase the availability of stacks of bi-static or PM image pairs considerably. In preparation for this, we use the technique of persistent scatter (PS) SAR interferometry for surface motion monitoring in urban areas. Although PM data allows for standard procedures without considering bi-static processing steps, processing PM data stacks with the standard PS-InSAR method does not make full use of the extremely short 10 s temporal baselines. Interferometric configuration in the standard PS-InSAR method tends to mix the topographic and deformation components in differential phase observations during estimations, especially in areas crowded with high-rise buildings; however, the interferometric phase generated by using only PM pairs is almost free from long-term deformation and atmospheric delay; hence, it is suitable for retrieving topographic residuals of urban areas, and the deformation estimation can thus be improved, especially regarding high-rise buildings, where spatially uncorrelated topographic residuals are commonly observed. Addressing this issue, we developed a new processing chain for PM data stacks, based on interferometry of PM pairs. The method will, in principle, be also applicable to bi-static stacks from the TanDEM-X mission, as well as future space-borne systems.

The paper is arranged as follows: in Section 2, the methodology of the processing is explained; in Section 3, the results of urban monitoring using pursuit monostatic spotlight acquisitions are demonstrated; in Section 4, the results are analyzed, and finally in Section 5, conclusions are drawn.

2. Materials and Methods

Principles of the standard PS-InSAR method are first stated in this section; after that, the proposed method will be presented based on the modification of the standard processing chain.

2.1. Standard PS-InSAR

The core of the standard PS-InSAR method is to conduct a time-series analysis of PS that can preserve stable backscattering over time. PS points can be selected based on several adoptable indexes, such as amplitude stability or temporal coherence. Selected PS points are connected to form a network; the deformation and topography differences for each arc are derived. The differential interferometric phase between the neighboring PS points is noted as:

$$\Delta\phi_{diff} = \Delta\phi_{defo} + \Delta\phi_{topo} + \Delta\phi_{atmos} + \Delta\phi_{noise} \tag{1}$$

$\Delta\phi_{diff}$ express the differential interferometric phase, while ϕ_{defo}, $\Delta\phi_{topo}$, and $\Delta\phi_{atmos}$ stand for the deformation, topographic, and atmospheric components of the differential interferometric phase. $\Delta\phi_{noise}$ is the residual noise. The deformation component describes the deformation velocity difference between the neighboring PS points in the same arc. Both linear and non-linear deformation can be contained in this term; for example, urban subsidence [1] for linear deformation, and periodic thermal deformation [7] for non-linear deformation. The topographic component describes the residual topographic phase between neighboring PS points in the same arc. The atmospheric component describes phase change caused by phase delay during transmission, e.g. due to water vapor differences in the atmosphere. These three interferometric phase components, are main elements in the differential phase, together with non-linear deformation and noise, the equation can be extended as:

$$\Delta\phi_{diff} = \frac{4\pi}{\lambda}T\Delta v + \frac{4\pi}{\lambda}\frac{B_{\perp}}{\rho\sin\theta}\Delta h + \Delta\phi_{atmos} + \frac{4\pi}{\lambda}\Delta D_{non-linear} + \Delta\phi_{noise} \qquad (2)$$

In this equation, λ is the wavelength, T is the temporal baseline, ρ is the slant range, and θ is the incident angle. $\Delta\phi_{diff}$ is the differential observation; the linear deformation velocity Δv and the residual topographic phase Δh are the unknowns. If the actual differential phase is bigger than 2π, phase ambiguities also need to be considered in the resolution process, which are also unknowns. The atmospheric component $\Delta\phi_{atmos}$ can be mitigated by spatial filtering, or optionally supported by introducing an external water vapor mapping product. It can also be estimated within standard PS-InSAR, which will be mentioned in the later paragraphs. The atmospheric component $\Delta\phi_{atmos}$, non-linear deformation component $\Delta D_{non-linear}$, and noise term $\Delta\phi_{noise}$ are considered as residual phase ε, as we can see in:

$$\varepsilon = \Delta\phi_{atmos} + \frac{4\pi}{\lambda}\Delta D_{non-linear} + \Delta\phi_{noise} \qquad (3)$$

Nevertheless, distances between PS candidates (PSC) within the PSC network can be controlled within certain limits, and the area of triangles in this network can be restricted to be small enough either, hence the residual phase ε of differential observations can normally be considered to be no bigger than π; therefore, phase ambiguities can be assumed to be zero. If the number of images in one interferograms stack is M, differential observations generated between the neighboring PS points x and y can be formulated as:

$$\Delta\phi^k_{x-y,ifg} = \left[\alpha^k, \beta^k\right] \cdot \begin{bmatrix} \Delta v_{x-y} \\ \Delta h_{x-y} \end{bmatrix} + \varepsilon \; (k = 1, 2, \ldots, M) \qquad (4)$$

where $\Delta\phi^k_{x-y,ifg}$ is the differential interferometric phase generated between x and y, k is the index of interferograms, α^k and β^k are the coefficients of mean velocity Δv_{x-y} and the residual topographic phase Δh_{x-y}; Δv_{x-y} and Δh_{x-y} are unknowns, and they have to be determined under maximized temporal coherence γ.

To solve for the mean velocity Δv_{x-y} and the residual topographic phase Δh_{x-y} in Equation (4), periodogram spectral analysis [29] is applied in the standard PS-InSAR technique. The method can transform the residual phase into a frequency domain in space $\{\Delta h, \Delta v\}$, and search for the location of single PS points that could maximize the temporal coherence γ. This method is based on the original wrapped interferograms, and the performance of this method depends on the temporal and spatial distribution of baselines; with an increased number and a more even distribution of the temporal baseline, the performance could be improved.

The atmospheric component is also estimated, together with mean velocity Δv_{x-y} and the residual topographic phase Δh_{x-y}. After estimation of the topographic error and deformation for each point along the network based on a previously selected sparse PSC network, their topography and deformation can be derived, starting from a selected reference point. The remaining phase residuals caused by atmospheric effects and are subsequently used to estimate the atmospheric phase screen (APS) of each interferogram. The atmospheric components in differential phases can be extracted through low-pass filtering in the spatial domain, and high-pass filtering in the temporal domain [30]. On this basis, the APS of each interferogram can be obtained by interpolation, using various methods, for instance, kriging. After removing the APS, the deformation and topographic error for every PSC in the stack are estimated relative to the previously selected reference point.

After determination of Δh and Δv, the topography and deformation can be derived for each PS point. Topographic components for each PS point can be resolved within an indirect adjustment model, and can be abbreviated as:

$$\Delta h_{M \times 1} = C_{M \times N} \times h_{N \times 1} \tag{5}$$

where M is the number of PS connection arcs, N is the number of selected PS points, Δh is the residual topographic observation, h is the unknown PS topography, and C is the coefficient matrix describing the relationship between residual topographic phase Δh between neighboring PS points and the topography h for each PS point, which is also what we referred as PS height. In response to least-square principle, the solution for this function is:

$$h_{N \times 1} = \left(C_{N \times M}^T P_{M \times M} C_{M \times N} \right)^{-1} C_{N \times M}^T P_{M \times M} \, \Delta h_{M \times 1} \tag{6}$$

In Equation (6), P is weight matrix. On the other hand, deformation pattern may not necessarily be linear. As we consider the whole scene, the topographic phase between neighboring PS points may also lead to phase wrapping in the spatial domain. In this context, the topographic component is considered as an error source and deformation is extracted based on the resumed interferometric phase after removal of the topographic components. Phase unwrapping is applied to the interferograms, and various algorithms can be chosen, for instance, Branchcut [31], SNAPU [32], etc. The unwrapped interferometric phase is noted as:

$$\phi_{x,uw}^k = \phi_{x,defo}^k + \hat{\phi}_{x,topo}^k + \hat{\phi}_{x,atmos}^k + \hat{\phi}_{x,noise}^k \tag{7}$$

where $\phi_{x,uw}^k$ expresses the unwrapped phase, $\phi_{x,defo}^k$ expresses the deformation component, and $\hat{\phi}_{x,topo}^k$, $\hat{\phi}_{x,atmos}^k$ and $\hat{\phi}_{x,noise}^k$ express the topographic, atmospheric components as well as the noise term of the unwrapped phase, respectively. After removal of the topographic component and the estimated atmospheric component from the interferometric phase, the residual phase can be considered as deformation signal. Converting the deformation phase $\hat{\phi}_{x,topo}^k$ into the slant-range deformation distance d, $d = \phi_{defo} \cdot \lambda / 4\pi$ together with the temporal baseline t, according to least-square principle, the deformation velocity within each time interval can be resolved as:

$$v = \left(t' P t \right)^{-1} P d \tag{8}$$

In this equation, v is the expected deformation velocity, P is the weight matrix, d is the slant range deformation distance, and t is the temporal baseline.

In practical image processing, the processing chain is implemented using SARProZ (Daniele Perissin, RASER Limited, Hong Kong) [33], in which the standard PS-InSAR can be divided into four parts that are listed below:

- Preliminary processing: image import; master selection; co-registration; resampling; reflectivity map generation; amplitude stability index generation;
- Preliminary geocoding: ground control point (GCP) selection; external DEM input; projection of external DEM and synthetic amplitude in SAR coordinates; Mask for PSCs;
- APS estimation: interferometric formation using star graph; select a subset of PSCs according to a certain threshold; triangulation of sparse PSC network and generation of differential phase; deformation and topographic estimation based on given search space; selection of reference point; inversion of initial APS;
- Sparse estimation: interferometric formation using star graph; selection of all PSCs based on a certain threshold; Input of initial APS, differential phase and reference points; deformation and topographic estimations based on the given search space for the PSCs.

However, applying the standard PS-InSAR approach to a PM data stack does not provide the best possible solution. As mentioned in the first section, there are two kinds of temporal baselines in PM data stacks: normal temporal baselines between repeat-pass intervals of the same satellite, which is 11 days for our case, and the extremely short 10 s temporal baselines between the acquisition times of TSX and TDX satellites at the same repeat-pass interval. These short temporal baselines are also a feature of this PM data. The interferometric phase generated with 10 s pairs are less likely to contain deformation components and also are less affected by atmospheric delay, consequently making these interferometric phases contain mainly the topographic phase component. As the deformation and residual topographic phase are estimated together in standard PS-InSAR, an external estimated height with the 10 s pairs could provide more accurate residual topography values for deformation estimation, thus allowing for a better performance of urban target monitoring.

2.2. A PS-InSAR Solution for Mono-Static Pursuit Datasets

To explore the full potential of these interferometric phases generated with 10 s pairs, and to refine the performance of deformation estimation, we propose a new processing chain that separates topographic error estimation from velocity estimation. The purpose is to use the 10 s pairs for topographic error estimation so that influences of un-modeled deformation and APS can be largely avoided. We assume that the retrieved topographic error would be better estimated with this modification; when measuring deformation and calculating APS, we use these precise topographic error estimations; in this way, the precision of the overall processing will be increased. This new processing chain is especially advantageous for areas with high-rise buildings, where large spatially uncorrelated topography errors occur.

Our proposed processing chain is similar with the standard PS-InSAR processing chain mentioned in Section 2.1. The preliminary processing steps and geocoding steps are the same. We read the SAR image data stack, select a common master image, and co-register all slave images to this master and resample. The reflectivity map and amplitude stability index are generated for each pixel based on this co-registered data stack. External Digital Elevation Models (DEM) are used, in order to simulate the spatial correlated topographic component. After this conventional preprocessing, APS and sparse estimations are implemented by using a modification of the original PS-InSAR processing chain.

In the proposed processing chain, both APS and sparse estimation are processed in two rounds; the first round aims for the residual topographic estimation, and the second round for the deformation. In round one, the interferometric formation of the PM data are connected only between the 10 s pairs, as shown in the real example used in our experiment listed below.

According to Figure 1, the interferometric phases are generated between SAR acquisitions derived from TSX and TDX satellites that fly within a temporal baseline of 10 s. Topographic errors, instead of deformation velocities, are estimated in this formation, since the generated interferometric phases are free from atmospheric delay. Furthermore, as we are working in urban areas, the temporal decorrelation for man-made targets is negligible, making this an ideal configuration for height estimation. Nevertheless, given that interferometric phases are less affected by atmospheric delay or by long-term deformation, regular spatial correlated error such as orbital error may still have some impact on the differential phase generated from these interferometric phases.

Once estimation for the residual topographic error is completed, we reprocess the stack in round two, connecting the interferograms in time. In our experiments, we used a standard single master approach, and as shown in Figure 2, all of the slave images are connected to a common master image in a star graph. However, other approaches, e.g., SBAS-like, are also possible.

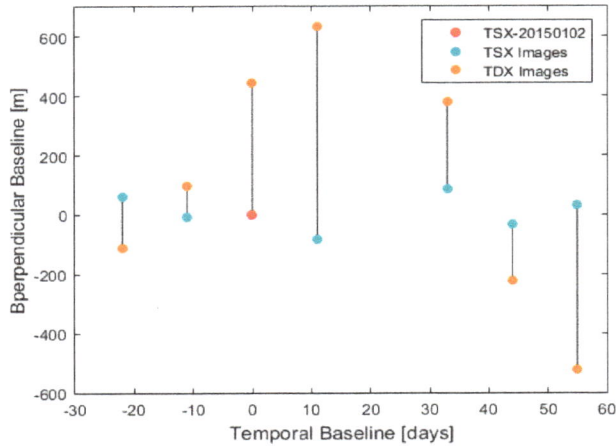

Figure 1. Interferometric formation of the proposed method for height estimation from TerraSAR-X (TSX) and TanDEM-X (TDX) image pairs.

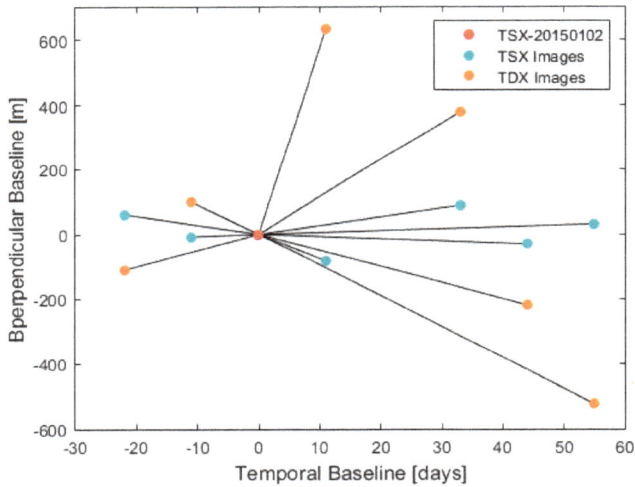

Figure 2. Interferometric formation of the proposed method for deformation estimation.

Deformation estimation is conducted in this round, considering the residual topographic phase derived previously in the first round; after removing this component, the deformation phase is unwrapped and the mean velocity of each PS point is resolved according to Equation (9). In summary, the processing chain of the proposed method is listed below:

- Preliminary processing: image import; master selection; co-registration; resampling; reflectivity map generation; amplitude stability index generation;
- Preliminary geocoding: GCP selection; external DEM input; projection of external DEM and synthetic amplitude in SAR coordinates; Mask for PSCs;
- First APS estimation: interferometric formation using only 10 s PM pairs; selection of a subset of PSCs according to a certain threshold; triangulation of sparse PSC network and generation of the differential phase; topographic estimation only based on a given search space; selection of the reference point; inversion of the initial APS;

- First sparse estimation: interferometric formation using only 10 s PM pairs; selection of all PSCs based on a certain threshold; input of initial APS, differential phase, and reference point; topographic estimation only based on the given search space for PSCs;
- Second APS estimation: interferometric formation using a star graph; selection of a subset of PSCs according to a certain threshold; triangulation of sparse PSC networks and generation of a differential phase; deformation estimation based on residual phase components derived in the first round within a given search space; selection of the reference point; inversion of the initial APS;
- Second Sparse estimation: interferometric formation using star graph; selection of all PSCs based on a certain threshold; Input of initial APS, differential phase and reference point; deformation estimation based on the residual phase component derived in first round within a given search space for PSCs;

The specific flow charts of both the standard and our proposed processing chains are briefly shown in Figure 3.

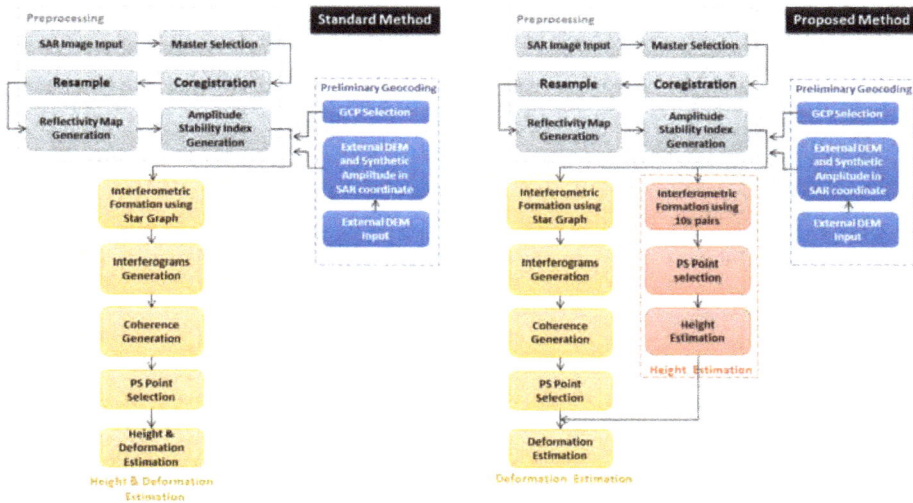

Figure 3. Flow charts of the standard processing chain (**left**) and the proposed processing method (**right**) for PS-InSAR.

3. Experiment Results

Both processing chains mentioned in Figure 3 are tested with a pursuit-monostatic dataset acquired over our study area in Guangzhou. The results are presented separately in this section. Specific information on our study area and dataset information are also demonstrated below.

3.1. Study Area and Dataset Information

Guangzhou is one of the most developed cities in China, where rapid economic development brings many new buildings in recent years. Our study area is displayed in the yellow square shown in Figure 4, which covers the central area of the Zhujiang New Town Central Business District (CBD).

This is an extremely dense urban area with dozens of skyscrapers. The landmark buildings in this area include the CTF Finance Centre (539.2 m), known as Canton East Tower, and Guangzhou International Finance Centre (440.75 m), known as West Tower. These two are the tallest two buildings in this area. Other tall buildings such as Universal Plaza (318 m), Guangcheng International mansion

(360 m), Pearl River Tower (309 m), Leatop Plaza (303 m), and Fuli Ying Kai Plaza (296.4 m), are among the top 10 highest buildings in Guangzhou. They are marked with pentagons in Figure 4. Also, the Guangzhou Opera House, the Guangdong Museum, and the Haixinsha Asian Games Theme Park are located in the south of Zhujiang New Town. These are located along the two sides of the so-called Automated People Mover System (APM), together with the skyscrapers, as we can see in the blue north–south traffic line in that passes though the central area of Zhujiang New Town CBD. Therefore, this area is marked as the central part of our test site, as shown in the orange region in Figure 4.

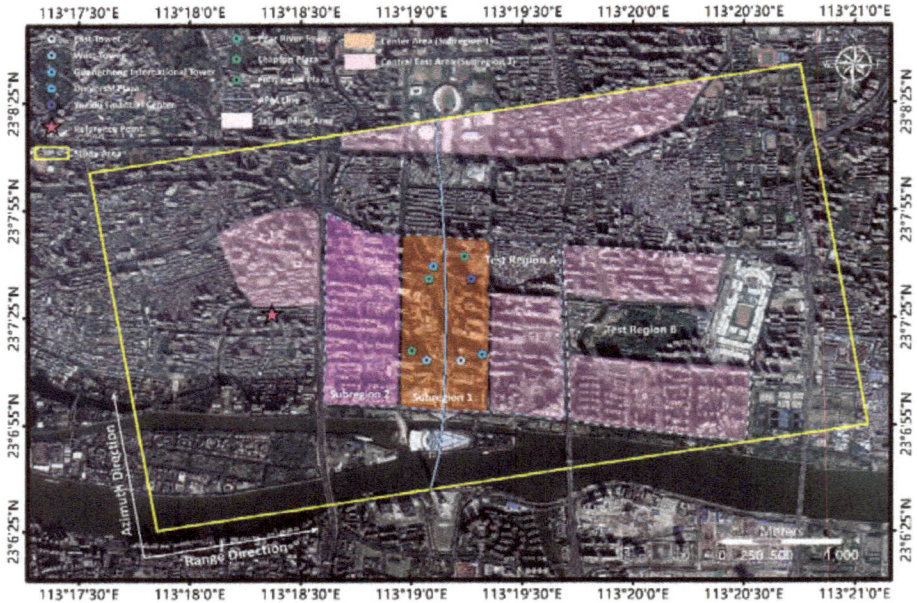

Figure 4. Study area of the staring spotlight monostatic pursuit stack over Guangzhou.

Apart from the central area, there are also tall buildings around the central area. The general distribution of the buildings with different heights is very different in our study area, though according to the approximate average heights of buildings, the study area could be divided into several sub-regions, if one street block is counted as a single unit. Here, we sorted the sub-regional urban surface into three categories. The first one is the central CBD marked in orange; with many skyscrapers, the average height of this area is very high. The second one is the newly built residential areas, where many tall buildings with height limits are distributed. These residual areas include urban communities, shopping districts, and public service buildings, such as schools, hospitals, and tax bureaus. Due to the height limit of 100 m according to the definition of "high-rise buildings" given by the General code for civil building design (GB50352-2005), these residual buildings, marked in the pink regions, are typically not higher than 100 m. The rest of the test site are old residual areas in Guangzhou with building heights not exceeding 50 m; residual apartments with eight floors about 24 m are the most common.

Based on the three categories, we select two small regions in our test site, i.e., test region A and B, marked with dashed boxes in Figure 4 for simple analysis in the experiment. These two regions cover the first two typical types of urban areas, as mentioned above, and demonstrate the different features in height distribution. Two subsets, the central area and central east area in test region A, are further selected for their high density of tall buildings, named subregions 1 and 2, shown as purple and orange areas, respectively, in Figure 4.

The Pursuit Monostatic data stack used in the experiment contains 14 ascending staring spotlight (ST) images, i.e., seven TSX/TDX mono-static pursuit pairs during 11 December 2014 to 26 February 2015. Other specific parameters about the dataset are listed in Table 1.

Table 1. Dataset information of pursuit monostatic data in Guangzhou.

Operation Mode	Pursuit Mono-Static	
Image Mode	Staring Spotlight (ST)	
Look Angle	43.4°	
Polarization	VV	
Direction	Ascending	
Resolution	0.24 m (azimuth) 0.80 m (ground range)	
Azimuth Extent	3 km	
Range Extent	5.5 km	
	Acquisition Time	Temperature (°C)
	20141211	14.5
	20141222	9.1
External Information	20150102	11.6
	20150113	11.2
	20150204	12.4
	20150215	16.7
	20150226	22.0

3.2. Experimental Results

Experimental results obtained in processing these pursuit monostatic data using both standard and proposed processing chains are displayed in the following paragraphs.

3.2.1. Processing Results from the Standard Method

The Pursuit Monostatic dataset in the Guangzhou test site is first processed with the standard processing method. Preprocessing and preliminary geocoding are accomplished according to the steps listed in Section 2.2. Interferometric formations of SAR images are configured in a star graph, using a single master as shown in Figure 2.

After generating interferograms and coherence, initial PS candidates are selected based on the amplitude stability, index with a threshold of 0.2. Although 0.3 is usually sufficient for urban monitoring, we set this threshold to be slightly lower, because the image resolution of the staring spotlight data is high enough to ensure abundant reliable PS points in the results. In this result, we select 740,105 PSC according to our threshold; the temporal coherence of these PSC points after processing is shown in Figure 5.

In this result, the PSCs generally demonstrate good temporal coherence, as we can see in Figure 5A, where most of PSC coherence values are distributed in the numerical interval [0.6, 1], especially PSCs from the ground and from lower parts of building façades. Their temporal coherence is mostly distributed in [0.8, 1], as the red points shown in Figure 5. Nevertheless, we can still find some relative unstable PSCs, shown in yellow in Figure 5A. These yellow points are mainly distributed along the façades of high-rise buildings, especially in the central area and its surrounding regions marked in Figure 4, as we can see in the enlarged coherence map in test regions A and B, as shown in Figure 5B,C. The triangle icons in Figure 5A,B represent skyscrapers with heights above or around 300 m. By comparing the coherence maps of the test regions A and B, we can find some unstable PSCs that have clear shapes of buildings edges, but there is a shift between the actual locations and the derived locations from standard processing results. These unstable PSCs are referred to "shifted PSCs" in our experience, and their shifts in location can be found around skyscrapers in Figure 5B. In Figure 5C, where most of the buildings are residential buildings that have a height limit of 100 m, much less shifted PSCs are found.

After selecting the initial PSCs according to the amplitude stability index, the estimation of deformation and topographic residuals are implemented, together within a total resolving of the differential phases, based on the least-square principles, according to Equation (4). Deformation and topographic residuals of PSCs are accomplished by setting pre-defined search spaces between −100 to 100 mm/year for linear deformation, and −300 m to 300 m for topographic residuals, since the vast majority of PSCs are expected to fall into this interval. 343,965 PS points are extracted, with temporal coherences of larger than 0.97 and their estimated topographic residuals, or residual height maps, are shown in Figure 6.

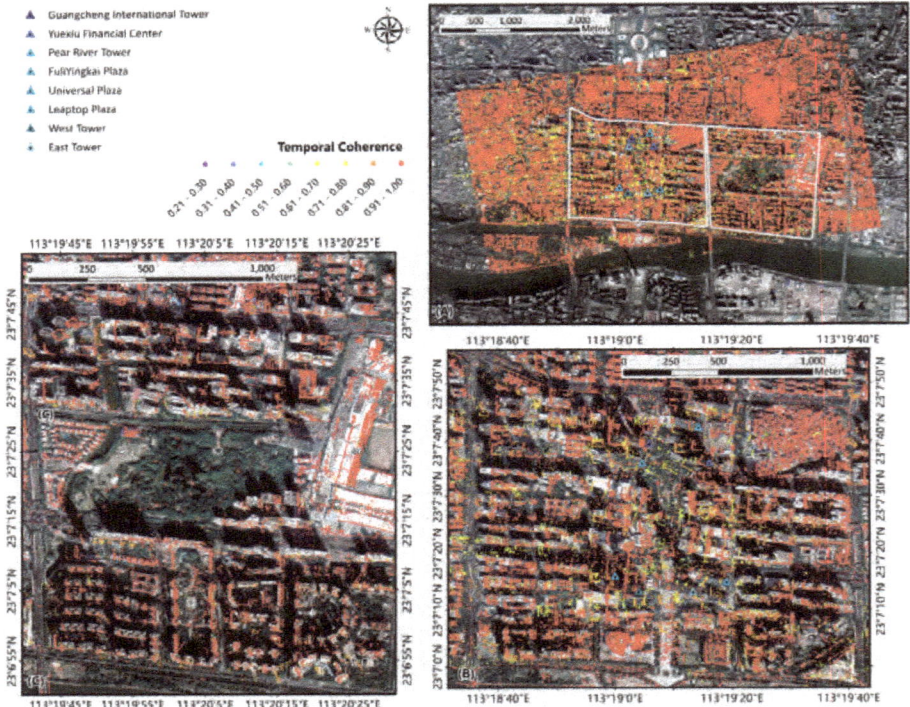

Figure 5. Temporal coherence map of the selected persistent scatterer candidates in the standard processing result: (**A**) The whole study area; (**B**) Subset of the map in test region A; (**C**) Subset of the map in test region B.

The resulted PS point residual heights range from −288.8 m to 168.5 m, with a mean value of 27.3 m. As we can see from Figure 7, a majority of the PS points detected in this result are points with low heights, since ground points, i.e., green points, are dominant in the map. High PS points can also be found in high-rise building areas, as we can see in the test regions A and B; however, the detectable height is limited, and only found in the lower part or the bottom of the buildings; this is clearer in the 3D demonstration of the PS-InSAR result in test region A, as shown in Figure 7.

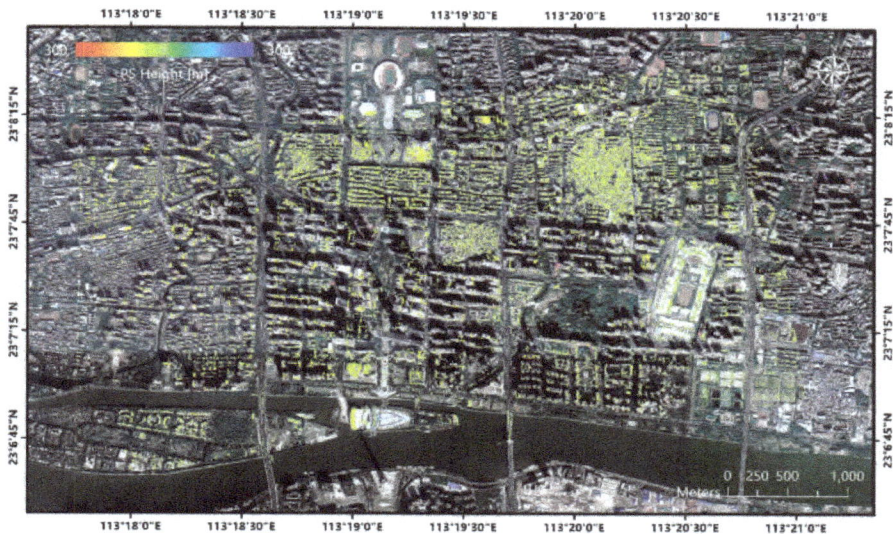

Figure 6. Residual height map for pursuit monostatic (PM) data in Guangzhou using the standard PS-InSAR method.

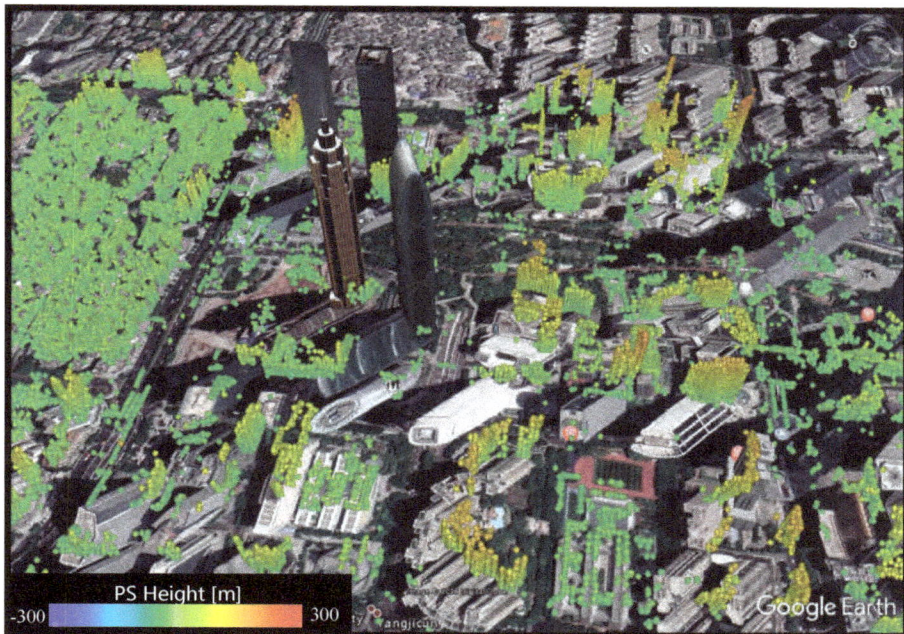

Figure 7. 3D demonstration of the highly temporal coherent persistent scatterers in test region A using standard PS-InSAR processing.

The deformation map is shown in Figure 8. A total of 343,963 PS points were selected with a temporal coherence bigger than 0.85 and amplitude stability index smaller than 0.2. As we can see from

Figure 8, most ground points are stable, while in the southeastern part of our test site, a subsidence pool can be found.

Figure 8. Linear deformation map generated from standard processing results using PS points with a temporal coherence bigger than 0.85.

Clear structured PS point clusters can be found and they show significant negative motion, i.e., motion away from the sensor in the line-of-sight direction. The shape of these point clusters indicate that they are part of high-rise buildings, but they show a shift between the location of these point clusters and normal PS points of the same structure; as mentioned in previous paragraphs, these clusters are referred as "shifted PS" and their presence is assumed to be caused by geocoding error, with an inaccurate estimation of topographic residuals. A clearer view of this deformation result is shown below, in which a 3D deformation map of a small area in test region A is displayed.

We can see more clearly that in Figure 9, thermal expansions are also visible on high-rise buildings where velocity is accumulated along with height. An estimation with seasonal deformation could be applied to remove the thermal deformation, as mentioned in Crosetto et al. [7]; however, the mono-static pursuit data has only seven pairs, covering a time span of just three months, which is not enough to model the temperature-related deformation correctly.

3.2.2. Processing Results from Our Proposed Method

After processing the pursuit monostatic data using the standard processing chain, we continue to process these data using the proposed processing chain. Preprocessing and preliminary geocoding are also accomplished according to the steps listed in Section 2.2. Topographic residuals and deformation estimation steps are accomplished in separated processing rounds. Topographic residuals are accomplished in the first round, with interferometric formation set in accordance with Figure 1.

In this test, PSCs are selected with the same amplitude stability index used in the standard processing steps; the topographic component of PSCs is retrieved in a least square process, assuming that only topographic residuals are left in the interferometric phase. Meanwhile, the temporal coherence of the PSCs is also generated, as shown in Figure 10.

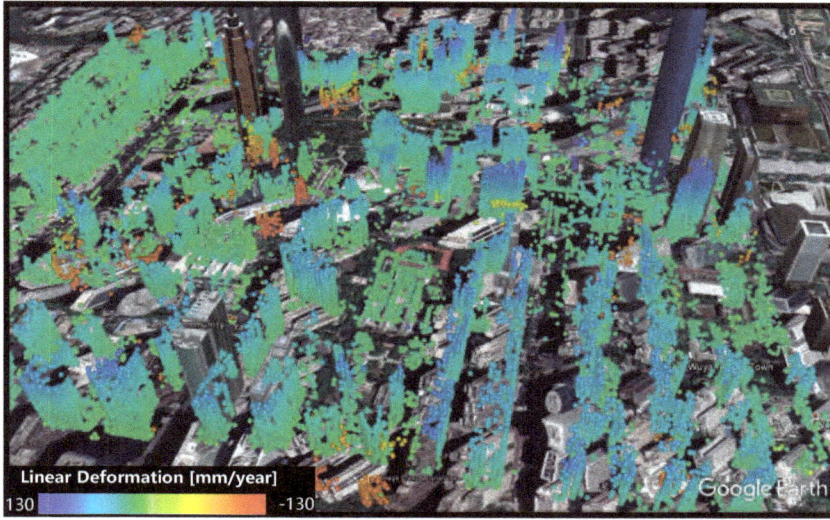

Figure 9. 3D-view of the estimated linear deformation trends visualized with Google Earth for the standard processing method.

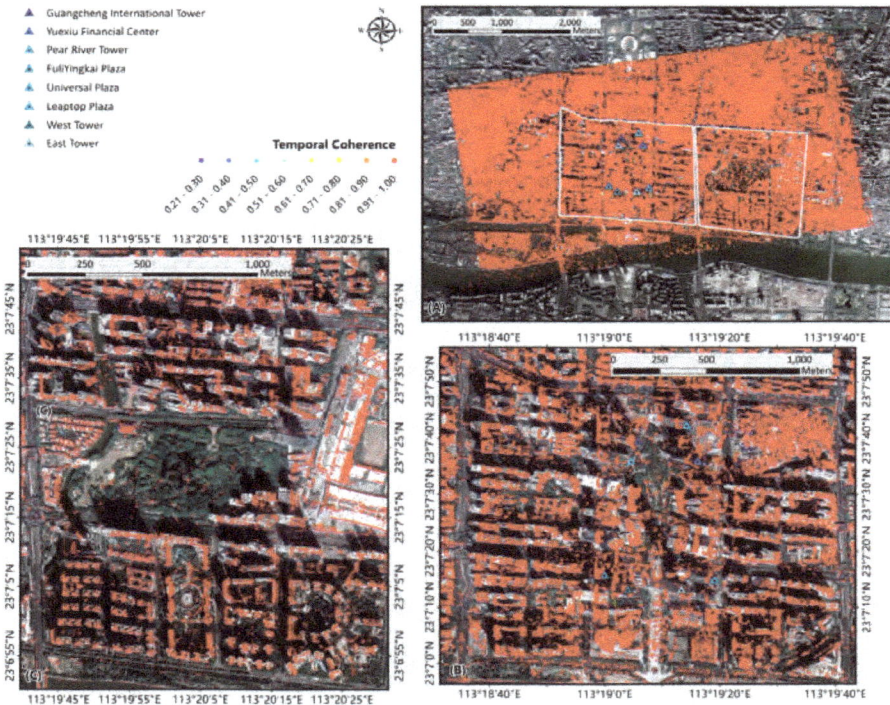

Figure 10. Temporal coherence map of the selected PSC in the standard processing result: (**A**) The whole study area (**B**) Subset of the map in test region A (**C**) Subset of the map in test region B.

Figure 10 shows that an overwhelming majority of PS points demonstrate very high temporal coherence distributed in the numerical interval [0.9, 1], even in areas with many high-rise buildings in test regions A and B, as displayed in the enlarged coherence map in Figure 10B,C. Also, we do not find shifted PSCs in this result. The point number and general stability of the PSCs are largely improved in our proposed processing results.

Residual height map generated in the proposed processing chain is displayed in Figure 11 and visualized in 3D in Figure 12, with a pre-defined search span for height estimation set between −300 m to 300 m.

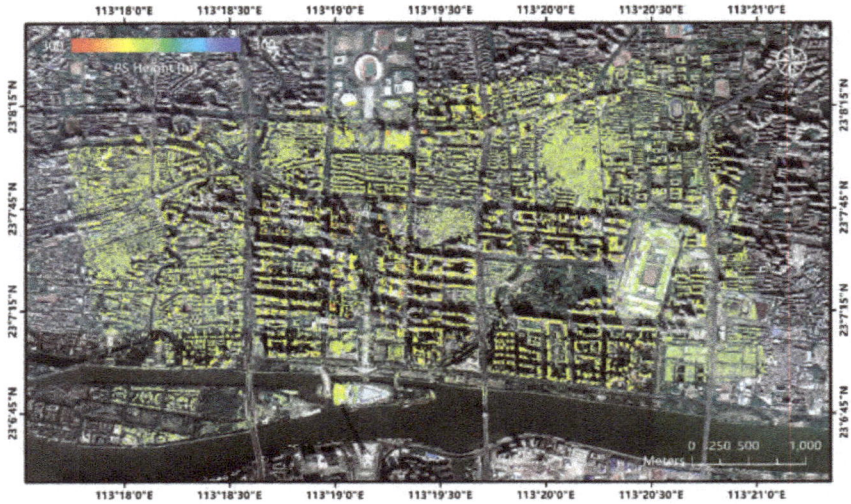

Figure 11. Residual height map for pursuit mono-static data in Guangzhou, using a proposed processing chain.

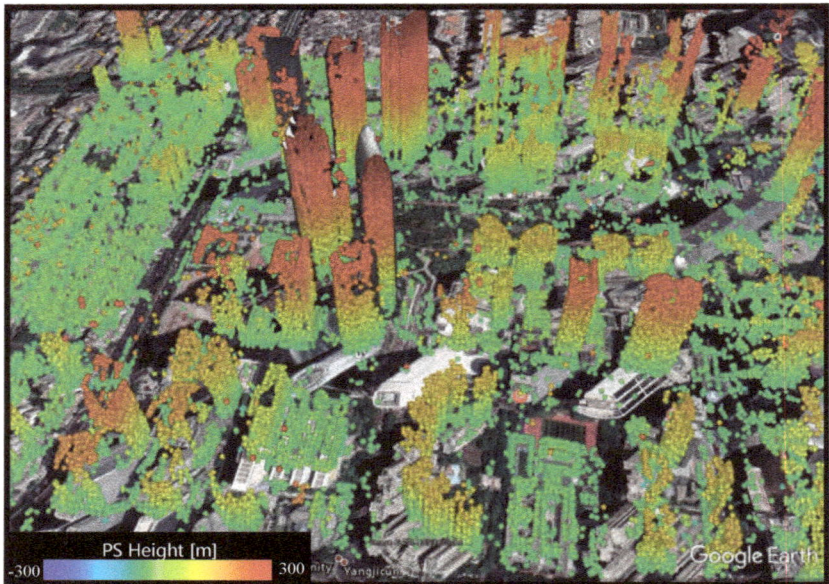

Figure 12. 3D-visualization of PS-InSAR results, using the proposed processing method.

A total of 712,760 PS points were extracted with the same coherence threshold of 0.97 being used for displaying the results from standard processing. As we can see in Figure 11, the PS heights in the high-rise building areas are fully detected. The detectable height limit is largely improved in this result, ranging from −294.9 m to 300 m, with an average height of 38.1 m. For skyscrapers, the height estimation is also largely improved, as the detectable PS points on the buildings are increased from the ground or the bottom to the full façades, as we can see in the 3D image of test region A below.

These PS heights are restored and used in the following deformation estimation. The interferometric formation is set according to Figure 2, where the identical star graph is adopted. Deformation estimation is thus accomplished based on pre-derived topographic residuals. Consequently, the linear deformation results are shown in Figure 13, where 465,564 PS points with a temporal coherence larger than 0.85 are selected, with an amplitude stability index smaller than 0.2.

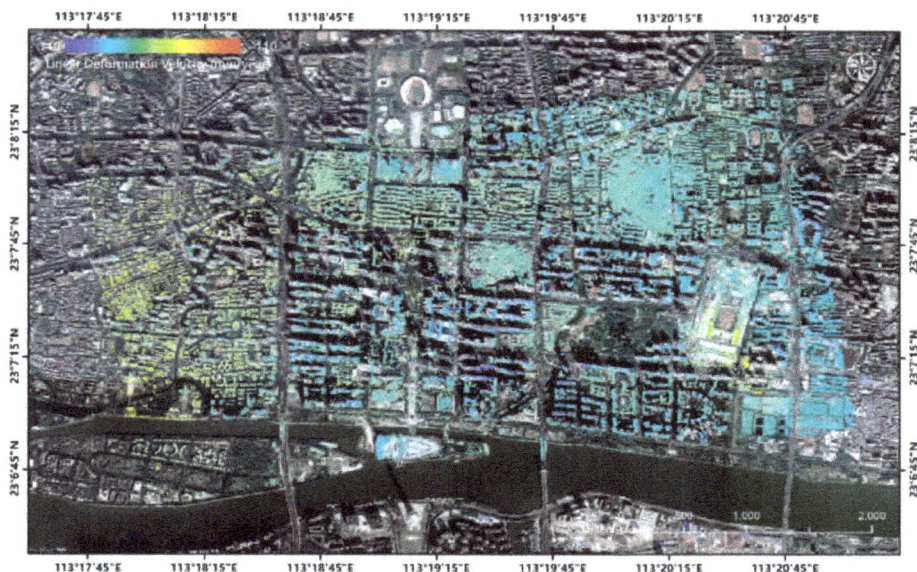

Figure 13. Linear deformation map generated from the proposed processing results, using PS points with a temporal coherence of larger than 0.85.

Deformation patterns in Figure 13 shows similar results compared to Figure 8, with a subsidence pool detected in southeastern area of our test site. This subsidence signal can be more clearly observed in Figure 13, as we can see that the shifted PS clusters in Figure 9 disappear in our proposed processing results. As we assume, this is due to the large reduction of topographic errors during the height estimation using only PM pairs. This improvement of topographic residual estimations further improves the deformation estimation accuracy. Although we still cannot estimate the temperature-related motion with only seven TSX and TDX pairs, with increased detectable height limits in our proposed processing method, the thermal deformation is also improved. For the same single building, the detected deformation velocity grows larger than that in the standard processing results, especially for skyscrapers in the central area, as we see in 3D demonstration of PS deformation in Figure 14.

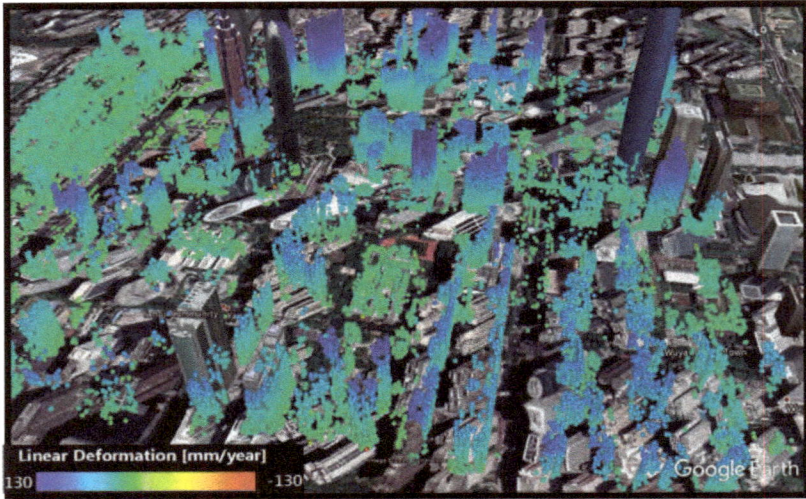

Figure 14. 3D view of the estimated linear deformation trends visualized with Google Earth for the proposed method.

4. Discussion

To evaluate the performance of these two processing chains, the results of the two processing chains are compared and analyzed in this section, on the basis of quantitative statistics. PS point properties are discussed in three aspects, including temporal coherence, and topographic and deformation estimation.

4.1. Temporal Coherence

The temporal coherence of PSCs selected using the same threshold in both standard and proposed processing chains are analyzed according to the histogram shown in Figure 15. A histogram of PS points for temporal coherence in both standard and proposed processing results.

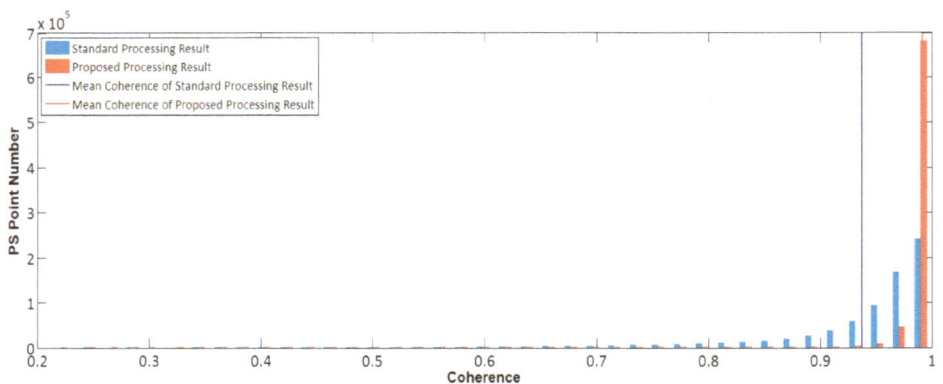

Figure 15. Histogram of PS points for temporal coherence in both standard and proposed processing results.

The histogram was generated, taking 0.02 as bin distance. The blue and red bars express the PS numbers in each bin; Blue and red vertical lines express the mean height value of the total PS heights, for standard and proposed processing results respectively. As we can infer from Figure 15, the persistent scatterers from the proposed method demonstrate a steeper distribution in the histogram, with quicker increase in PS numbers in high coherence intervals, compared to PSs derived from the standard processing result. The mean temporal coherence of the proposed processing result, 0.99, is also bigger than the mean temporal coherence, 0.94, of the standard processing result, which indicates a generally higher stability of PSs in the proposed processing results, as well as less disturbance in the phase observations. Topography estimation using these PSs is more likely to obtain accurate results.

4.2. Topography Estimation

Topography estimation results are also quantitatively analyzed, mainly focusing on PS point numbers and density.

4.2.1. PS Point Numbers

Histogram of PS point heights for both standard and proposed processing results are presented with a distance bin of 5 m, as shown in Figure 16.

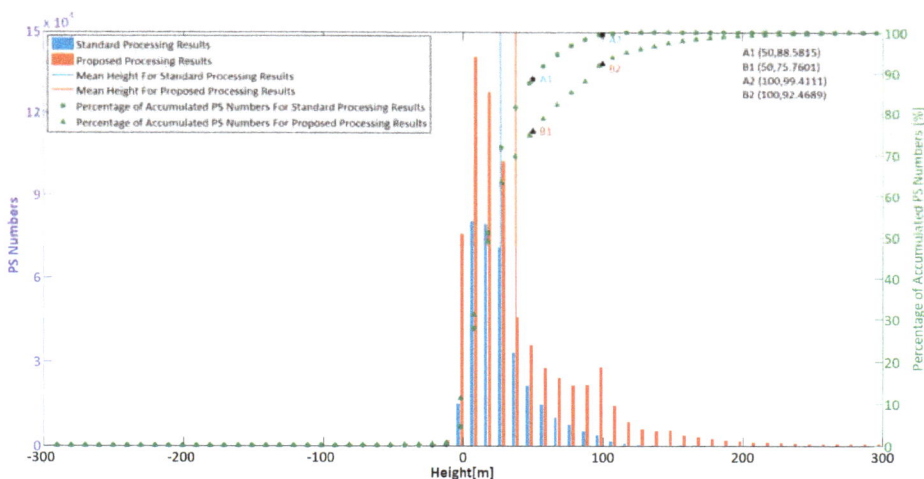

Figure 16. Histogram of PS point heights for both the standard and proposed processing results.

The blue and red bars represent PS point numbers in each bin of PS height for the standard and proposed processing results, respectively. The blue and red vertical lines represent the mean height value of the total PS points in stand and proposed processing results, respectively. The green dotted lines express the accumulated PS numbers for the standard processing results as a percentage; meanwhile, the green triangle lines express the accumulated PS numbers for the proposed processing results as a percentage. As we can infer from Figure 16, for PS points selected with the same temporal coherence threshold, more PS points are detected in each height bin in the proposed processing results, than in the standard processing results, with an overall rate of increase of 107.2%, and 663% in the 100 m bin. Also, the detected PS height span is increased from the standard processing results to the proposed processing results, from 168.5 m to 300 m, increased by 78%. The majority of PS heights are distributed in interval from 0 to 50 m.

The percentage curves of accumulated PS numbers, marked in green circle and triangle icons in Figure 16, also reveals that more than 88% of the PS points are distributed in height intervals between 0 to 50 m in standard processing results, as marked with A1, which are assumed to be ground or low building points. About 99.4% of the total PS points have height values below 100 m, marked in A2;

and only around 0.6% percent of the PS points are detected over targets with heights bigger than 100 m. On the other hand, for proposed processing results, 76% percent of the detected heights are below 50 m, as marked with B1, and 92.4% of the PS points are below 100 m, marked with B2; the other 7.6% of the detected PS heights are between 100 m and 300 m. There is also a local peak of PS numbers with a height value around 100 m; this is in accordance with the emergence of high-rise buildings, which are built under the height limit of 100 m. These increased number of PS points with heights around 100 m are assumed to be building roofs. This comparison shows the limited detection ability of standard processing method on high-rise buildings, comparing to proposed processing results, especially for high-rise buildings.

4.2.2. Persistent Scatterer Density

Apart from the general increase of the PS points, the PS point density is further estimated based on the statistics of the variance and mean value of a PS point. This variance is defined as the square deviation of the estimated height values within a patch centered on each PS point with extension of 10 m; the mean value of a PS point is defined as the mean height value of all PS points within the patch. The statistical results of variance and mean value, as well as the differences between these two, are calculated and normalized. The sub region 1 and 2 are selected as our test regions as we can see in Figure 4. The Statistical results of these values are listed below.

Figures 17 and 18 demonstrate variance and mean values, as well as the difference histogram of these two indexes. The variance of the PS points in the proposed processing results increases in our test region, compared to the standard processing results, as most of the variance difference is positive in the histograms; the detectable locations are also expanding in the proposed processing method. As we know, façade PS points are vertically distributed on buildings, so that if the variance gets bigger, it could be either caused by an increased point density within a certain height limit, or an increased detected height limit. In this context, the mean value of the test region is also presented. The mean value of the test region is slightly increased; as we infer from the mean value difference histogram from both Figures 17 and 18, the difference value is distributed around 0, but it has a larger positive span. As we can see in the distribution map, PS points with an increased mean value are mostly derived from building façades. Hence, we can infer from these statistics of the variance and mean value that the point density on the building façades is increasing. Besides, in the statistical result of the central area of Zhujiang CBD, increased variances are detected. For the skyscrapers in the central area, the variance and mean value map increases, which is a hint towards the increase of the point density and the detected heights, which is also in accordance with the 3D visualization of these areas in Figures 7 and 12.

4.3. Deformation Estimation

Comparisons of surface motion estimated from the standard method and our proposed methods are also addressed, as we can see in Figures 8 and 13. Similarities can be found in these two results: first, for PS near the ground, both results show good performance, as we can see in the northeastern part of our results. The traffic lines can be clearly seen in the northeastern corner of our test site. Second, the same subsidence patterns can be detected in the southeastern part of our study area. A linear ramp could be detected in the deformation maps, which is assumed to be caused by orbital error, such orbital effects can occur due to the limited number of the images, the special nature of the data acquired, and the baseline distribution during the science phase of the project.

For deformation results, the main improvement from our proposed processing method lies in the removal of the shifts and refinements of thermal deformation in the high-rise building area. Since the thermal deformation in the high-rise buildings accumulated from the bottom to the top of buildings, a linear relationship exists between the thermal deformation and the PS height. This linear relation is also due to the short overall time-span of the data acquisition. For data stacks covering a longer time period, we would rather expect a sinusoidal behavior. However, in our example, a linear regression analysis is accomplished, in order to evaluate this refinement. The results are demonstrated below.

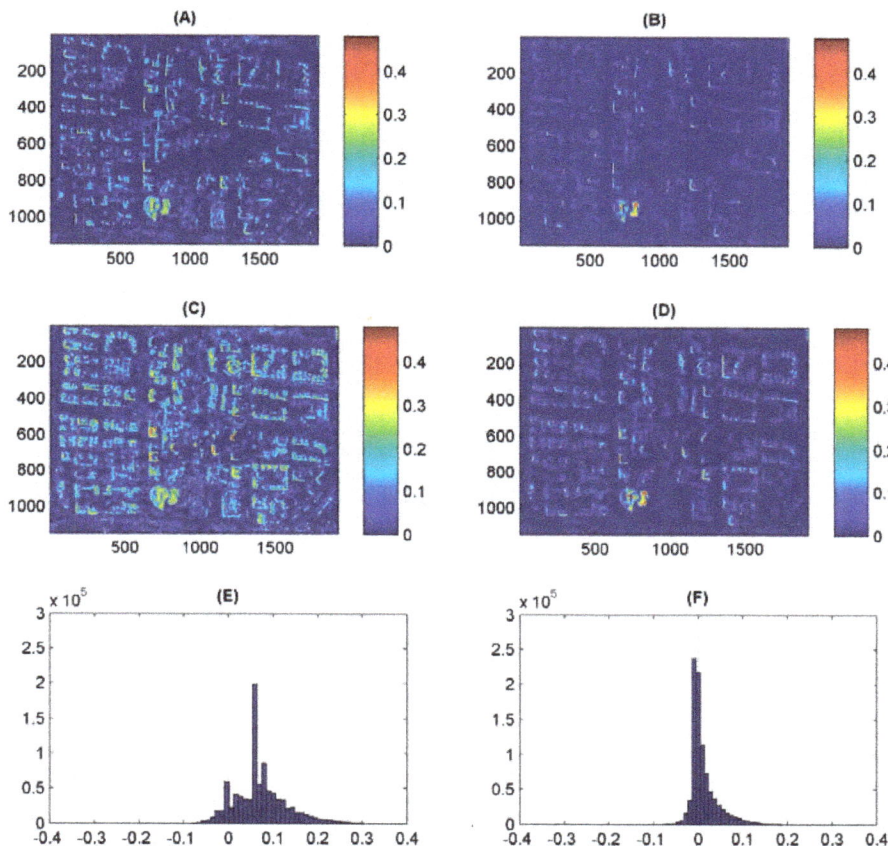

Figure 17. Statistics result of the two methods in sub-region 1: (**A**) variance of the standard processing method (**B**) mean height of the standard processing method (**C**) variance of the proposed processing method (**D**) mean height of the proposed processing method (**E**) Histogram of the variance difference to the standard processing method (**F**) Histogram of the mean height difference to the standard processing method.

As we can see in Figure 19, there is a clear linear relationship in the regression result in sub-region 1 and 2, with deformation and height in the central area. It can be inferred that the deformation changes faster along with height in the proposed processing, as we can see in the specific slope in Figure 19A,C, compared to Figure 19B,D. The reason for this increase in the linear slope is the removal of shifts, as well as an increased number of PS points detected on high façades, which increases the detectable span of the temperature-related deformation, thus enlarging the proportion of larger deformations in the results. The absolute value of this deformation could be corrected if we had more image pairs covering a longer period. Furthermore, we can see that we can estimate points at much higher positions correctly.

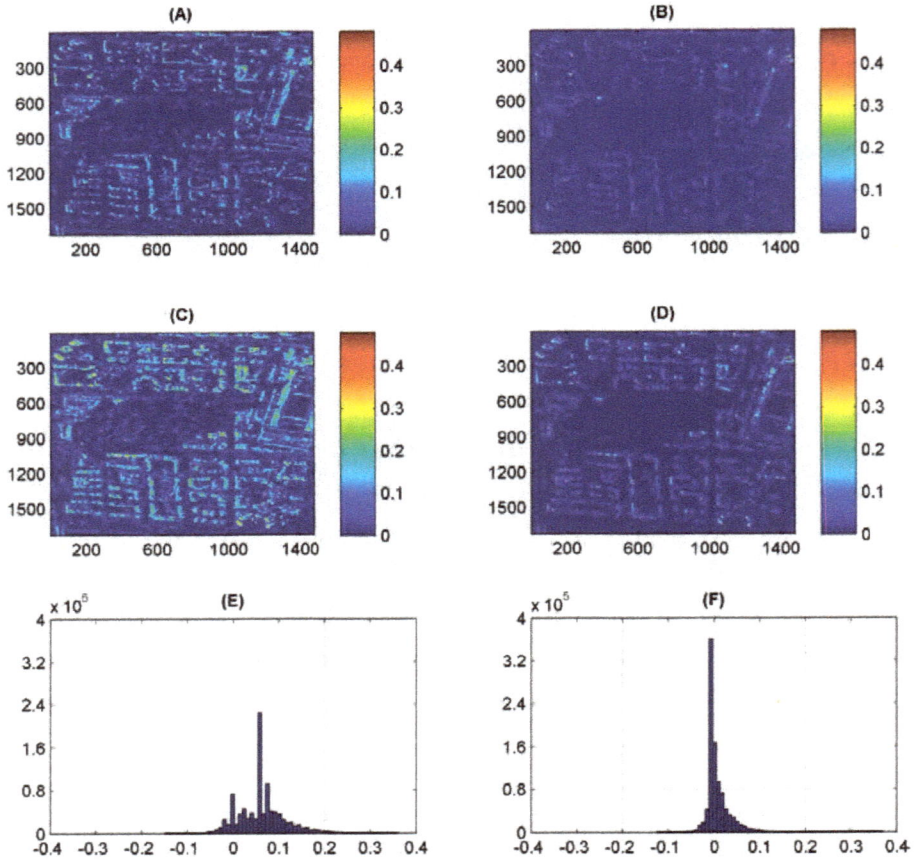

Figure 18. Statistical result of the two methods in sub-region 2: (**A**) variance of the standard processing method (**B**) mean height of the standard processing method (**C**) variance of the proposed processing method (**D**) mean height of the proposed processing method (**E**) Histogram of the variance difference to the standard processing method (**F**) Histogram of the mean height difference to the standard processing method.

However, when processing the deformation, influences from the atmosphere are unavoidable. For ground subsidence monitoring, the main deformation trends can be found with the standard PS-InSAR processing, as well as our proposed method. We can see differences and reduced noise in our approach. For monitoring high-rise buildings, and especially for skyscrapers, shifts during estimation are refined, and the linear relationship between the deformation and PS height are better restored. However, the number of images, and especially of image pairs, is very limited in both stacks, which leads to imperfect results, as the number of images is too low to successfully estimate temperature-related deformation. Nevertheless, we demonstrate improvements when using our proposed approach, and its practical significance for mono-static pursuit and bi-static data stack processing.

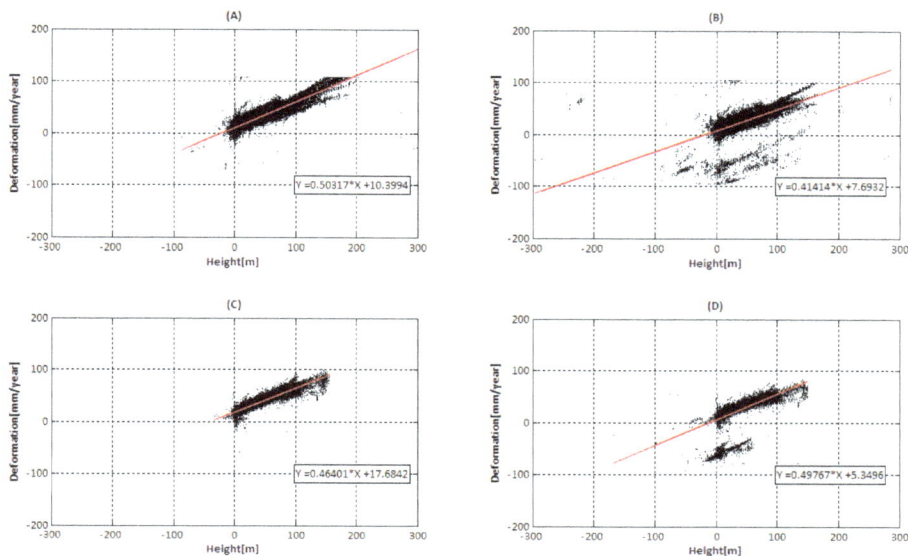

Figure 19. Linear regression results between thermal deformation and PS height: (**A**) the proposed processing result in sub-region 1, (**B**) the standard processing result in sub-region 1, (**C**) the proposed processing result in sub-region 2, (**D**) the standard processing result in sub-region 2.

5. Conclusions

Urban monitoring is an important issue in China. With upcoming SAR missions, e.g., TanDEM-L and Twin-L, the application of innovative SAR products in urban monitoring has to be explored. Pursuit monostatic data is tested in this paper as an example, but the results can also be used for bi-static stacks. A direct application of this data to standard PS-InSAR shows only a restricted ability for urban monitoring and does not make full use of the extremely short temporal baselines, which are the characteristic of the pursuit monostatic pairs. In this context, a new processing method is proposed on basis of the standard PS-InSAR method in this paper. Both proposed and standard PS-InSAR processing methods are tested with pursuit monostatic staring spotlight data in Guangzhou. The detectable PS numbers and height span using the proposed processing method are largely increased by 107.2% and 78%, respectively. For surface motion estimations near the ground surface, both methods achieve similar performances. For deformation estimation in high-rise buildings, shifts that occurred in the standard processing results are fixed with the proposed processing. Based on the experimental results mentioned above, we conclude that the proposed processing method demonstrates better ability in urban monitoring, and it is a more suitable choice for monostatic pursuit and bi-static data.

Author Contributions: Funding acquisition, M.L.; Investigation, Z.W.; Methodology, T.B.; Software, D.P.; Supervision, T.B. and L.Z.; Writing—original draft, Z.W.; Writing—review & editing, T.B. and L.Z.

Funding: This work was supported by the Natural Science Foundation of China under the Grant 61331016 and 41571435.

Acknowledgments: The TerraSAR-X and TanDEM-X data was provided by DLR via NTI_INSA6712.

Conflicts of Interest: The authors declare no conflict of interest.

References

1. Ferretti, A.; Prati, C.; Rocca, F. Nonlinear subsidence rate estimation using permanent scatterers in differential SAR interferometry. *IEEE Trans. Geosci. Remote Sens.* **2000**, *38*, 2202–2212. [CrossRef]

2. Lanari, R.; Zeni, G.; Manunta, M.; Guarino, S.; Berardino, P.; Sansosti, E. An integrated SARGIS approach for investigating urban deformation phenomena: A case study of the city of Naples, Italy. *Int. J. Remote Sens.* **2004**, *25*, 2855–2862. [CrossRef]

3. Colesanti, C.; Mouelic, S.L.; Bennani, M.; Raucoules, D.; Carnec, C.; Ferretti, A. Detection of mining related ground instabilities using the Permanent Scatterers technique—A case study in the east of France. *Int. J. Remote Sens.* **2005**, *26*, 201–207. [CrossRef]

4. Vallone, P.; Giammarinaro, M.S.; Crosetto, M.; Agudo, M.; Biescas, E. Ground motion phenomena in Caltanissetta (Italy) investigated by InSAR and geological data integration. *Eng. Geol.* **2008**, *98*, 144–155. [CrossRef]

5. Gernhardt, S.; Adam, N.; Eineder, M.; Bamler, R. Potential of very high resolution SAR for persistent scatterer interferometry in urban areas. *Ann. GIS* **2010**, *16*, 103–111. [CrossRef]

6. Gernhardt, S.; Bamler, R. Deformation monitoring of single buildings using meter-resolution SAR data in PSI. *ISPRS J. Photogramm. Remote Sens.* **2012**, *73*, 68–79. [CrossRef]

7. Crosetto, M.; Monserrat, O.; Cuevas-González, M.; Devanthéry, N.; Luzi, G.; Crippa, B. Measuring thermal expansion using X-band persistent scatterer interferometry. *ISPRS J. Photogramm. Remote Sens.* **2015**, *100*, 84–91. [CrossRef]

8. Qin, X.; Zhang, L.; Yang, M.; Luo, H.; Liao, M.; Ding, X. Mapping surface deformation and thermal dilation of arch bridges by structure-driven multi-temporal DInSAR analysis. *Remote Sens. Environ.* **2018**, *216*, 71–90. [CrossRef]

9. Qin, X.; Zhang, L.; Ding, X.; Liao, M.; Yang, M. Mapping and Characterizing Thermal Dilation of Civil Infrastructures with Multi-Temporal X-Band Synthetic Aperture Radar Interferometry. *Remote Sens.* **2018**, *10*, 941. [CrossRef]

10. Schulze, D.; Zink, M.; Krieger, G.; Böer, J.; Moreira, A. TANDEM-X mission concept and status. In Proceedings of the FRINGE, Frascati, Italy, 30 November–4 December 2009.

11. Fritz, T.; Rossi, C.; Yague-Martinez, N.; Rodriguez-Gonzalez, F.; Lachaise, M.; Breit, H. Interferometric processing of TanDEM-X data. In Proceedings of the IEEE International Geoscience and Remote Sensing Symposium (IGARSS), Vancouver, BC, Canada, 24–29 July 2011; pp. 2428–2431.

12. Fritz, T.; Breit, H.; Rossi, C.; Balss, U.; Lachaise, M.; Duque, S. Interferometric processing and products of the TanDEM-X mission. In Proceedings of the IEEE International Geoscience and Remote Sensing Symposium (IGARSS), Munich, Germany, 22–27 July 2012; pp. 1904–1907.

13. Hajnsek, I.; Krieger, G.; Papathanassiou, K.; Baumgartner, S.; Rodriguez-Cassola, M.; Prats, P. TanDEM-X: First scientific experiments during the commissioning phase. In Proceedings of the 2010 8th European Conference on Synthetic Aperture Radar (EUSAR), Aachen, Germany, 7–10 June 2010; pp. 1–3.

14. Bachmann, M.; Hofmann, H. Challenges of the TanDEM-X commissioning phase. In Proceedings of the 2010 8th European Conference on Synthetic Aperture Radar (EUSAR), Aachen, Germany, 7–10 June 2010; pp. 1–3.

15. Prats, P.; Scheiber, R.; Mittermayer, J.; Wollstadt, S.; Baumgartner, S.V.; López-Dekker, P.; Schulze, D.; Steinbrecher, U.; Rodríguez-Cassolà, M.; Reigber, A.; et al. TanDEM-X experiments in pursuit monostatic configuration. In Proceedings of the 2012 9th European Conference on Synthetic Aperture Radar (EUSAR), Nürnberg, Germany, 23–26 April 2012; pp. 159–162.

16. Lumsdon, P.; Schlund, M.; von Poncet, F.; Janoth, J.; Weihing, D.; Petrat, L. An encounter with pursuit monostatic applications of TanDEM-X mission. In Proceedings of the IEEE International Geoscience and Remote Sensing Symposium (IGARSS), Milan, Italy, 26–31 July 2015; pp. 3187–3190.

17. Bello, J.L.B.; Martone, M.; Gonzalez, C.; Kraus, T.; Bräutigam, B. First interferometric performance analysis of full polarimetric TanDEM-X acquisitions in the pursuit monostatic phase. In Proceedings of the IEEE International Geoscience and Remote Sensing Symposium (IGARSS), Milan, Italy, 26–31 July 2015; pp. 1242–1245.

18. Bueso-Bello, J.L.; Martone, M.; Prats-Iraola, P.; González-Chamorro, C.; Kraus, T.; Reimann, J.; Jäger, M.; Bräutigam, B.; Rizzoli, P.; Zink, M. Performance Analysis of TanDEM-X Quad-Polarization Products in Pursuit Monostatic Mode. *IEEE J. Sel. Top. Appl. Earth Obs. Remote Sens.* **2017**, *10*, 1853–1869. [CrossRef]

19. Shi, Y.L.; Wang, Y.Y.; Kang, J.; Lachaise, M.; Zhu, X.X.; Bamler, R. 3D Reconstruction from Very Small TanDEM-X Stacks. In Proceedings of the EUSAR 2018: 12th European Conference on Synthetic Aperture Radar, Aachen, Germany, 4–7 June 2018.

20. Sjögren, T.; Vu, V. Detection of slow and fast moving targets using hybrid CD-DMTF SAR GMTI mode. In Proceedings of the 2015 IEEE 5th Asia-Pacific Conference on Synthetic Aperture Radar (APSAR), Singapore, 1–4 September 2015; pp. 818–821.

21. Sjögren, T.; Vu, V.; Mats, P. Experimental result for SAR GMTI using monostatic pursuit mode of TerraSAR-X and TanDEM-X on Staring Spotlight images. In Proceedings of the EUSAR 2016: 11th European Conference on Synthetic Aperture Radar, Hamburg, Germany, 6–9 June 2016; pp. 1–4.

22. Ren, Y.; Li, X.M. Derivation of sea surface current fields using TanDEM-X pursuit monostatic mode data. In Proceedings of the International Geoscience and Remote Sensing Symposium (IGARSS), Beijing, China, 10–15 July 2016; pp. 4019–4022.

23. Velotto, D.; Nunziata, F.; Bentes, C.; Migliaccio, M.; Lehner, S. Investigation of the experimental TanDEM-X pursuit monostatic mode for oil and ship detection. In Proceedings of the International Geoscience and Remote Sensing Symposium (IGARSS), Beijing, China, 10–15 July 2016; pp. 4023–4026.

24. Kemp, J.; Burns, J. Agricultural monitoring using pursuit monostatic TanDEM-X coherence in the Western Cape, South Africa. In Proceedings of the EUSAR 2016: 11th European Conference on Synthetic Aperture Radar, Hamburg, Germany, 6–9 June 2016; pp. 1–4.

25. Krieger, G.; Hajnsek, I.; Papathanassiou, K.; Eineder, M.; Younis, M.; De Zan, F.; Prats, P.; Huber, S.; Werner, M.; Fiedler, H.; et al. The tandem-L mission proposal: Monitoring earth's dynamics with high resolution SAR interferometry. In Proceedings of the Radar Conference, Pasadena, CA, USA, 4–8 May 2009; pp. 1–6.

26. Moreira, A. A golden age for spaceborne SAR systems. In Proceedings of the 2014 20th International Conference on Microwaves, Radar, and Wireless Communication (MIKON), Gdansk, Poland, 16–18 June 2014; pp. 1–4.

27. Huber, S.; Villano, M.; Younis, M.; Krieger, G.; Moreira, A.; Grafmueller, B.; Wolters, R. Tandem-L: Design concepts for a next-generation spaceborne SAR system. In Proceedings of the EUSAR 2016: 11th European Conference on Synthetic Aperture Radar, Hamburg, Germany, 6–9 June 2016; pp. 1–5.

28. Wang, Y. Twin-L SAR: Terrain-Wide Swath Interferometric L-Band SAR. In Proceedings of the International Society for Photogrammetry and Remote Sensing (ISPRS) Geospatial Week 2017, Wuhan, China, 18–22 September 2017.

29. Marple, S.L. Digital spectral analysis: With applications. *J. Acoust. Soc. Am.* **1998**, *86*, 2043. [CrossRef]

30. Hanssen, R.F. *Radar Interferometry: Data Interpretation and Error Analysis*; Springer Science & Business Media: Berlin, Germany, 2001; Volume 2.

31. Goldstein, R.M.; Zebker, H.A.; Werner, C.L. Satellite radar interferometry—Two-dimensional phase unwrapping. *Radio Sci.* **1988**, *23*, 713–720. [CrossRef]

32. Chen, C.W.; Zebker, H.A. Two-dimensional phase unwrapping with use of statistical models for cost functions in nonlinear optimization. *J. Opt. Soc. Am. A Opt. Image Sci. Vis.* **2001**, *18*, 338–351. [CrossRef] [PubMed]

33. Perissin, D.; Wang, Z.; Lin, H. Shanghai subway tunnels and highways monitoring through Cosmo-SkyMed Persistent Scatterers. *ISPRS J. Photogramm. Remote Sens.* **2012**, *73*, 58–67. [CrossRef]

![remote sensing logo] *remote sensing*

MDPI

Article

A Persistent Scatterer Interferometry Procedure Based on Stable Areas to Filter the Atmospheric Component

Michele Crosetto [1,*], Núria Devanthéry [1], Oriol Monserrat [1], Anna Barra [1], María Cuevas-González [1], Marek Mróz [2], Joan Botey-Bassols [3], Enric Vázquez-Suñé [3] and Bruno Crippa [4]

[1] Centre Tecnològic de Telecomunicacions de Catalunya (CTTC), Division of Geomatics, Av. Gauss 7, E-08860 Castelldefels (Barcelona), Spain; nuria.devanthery@gmail.com (N.D.); omonserrat@cttc.cat (O.M.); abarra@cttc.cat (A.B.); mcuevas@cttc.cat (M.C.-G.)
[2] Institute of Geodesy, University of Warmia and Mazury in Olsztyn, ul. Oczapowskiego 1, 10-719 Olsztyn, Poland; marek.mroz@uwm.edu.pl
[3] Institute of Environmental Assessment and Water Research (IDAEA), CSIC, c/Jordi Girona 18, 08034 Barcelona, Spain; w.jbotey@gmail.com (J.B.-B.); enric.vazquez@idaea.csic.es (E.V.-S.)
[4] Department of Earth Sciences, University of Milan, Via Cicognara 7, I-20129 Milan, Italy; buno.crippa@unimi.it
* Correspondence: mcrosetto@cttc.cat; Tel.: +34-93-645-2900

Received: 6 October 2018; Accepted: 8 November 2018; Published: 10 November 2018

Abstract: This paper describes a Persistent Scatterer Interferometry (PSI) procedure to monitor the land deformation in an urban area induced by aquifer dewatering and the consequent drawdown of the water table. The procedure, based on Sentinel-1 data, is illustrated considering the construction works of Glories Square, Barcelona (Spain). The study covers a period from March 2015 to November 2017, which includes a dewatering event in spring 2017. This paper describes the proposed procedure, whose most original part includes the estimation of the atmospheric phase component using stable areas located in the vicinity of the monitoring area. The performances of the procedure are analysed, characterising the original atmospheric phase component and the residual one that remains after modelling the atmospheric contribution. This procedure can work with any type of deformation phenomena, provided that its spatial extension is sufficiently small. The quality of the obtained time series is illustrated discussing different deformation results, including a validation result using piezometric data and a thermal expansion case.

Keywords: SAR; Sentinel-1; differential SAR interferometry; atmospheric component; modelling; deformation time series; validation

1. Introduction

Deformation monitoring of urban areas is an important tool for city management and asset maintenance. An important application is the monitoring of the deformation caused by construction works that involve aquifer dewatering, which can affect buildings and infrastructures. In this paper we describe one of this type of application: the monitoring of the deformation associated with the construction works related to the transformation process of the Glories Square, located in the centre of Barcelona (Spain). These works involve the construction of underground tunnels, which requires aquifer dewatering. They are monitored with a set of in situ measurements, e.g., inclinometers, topographic surveys, levelling, which are mainly located in the area of the Glories Square. Such measurements are complemented with Persistent Scatterer Interferometry (PSI) observations, which aim at achieving a global view of the deformation phenomena occurring in the square and, especially, its surroundings, where in situ measurements are not available.

The PSI monitoring is based on C-band data acquired by the Sentinel-1A sensor. The Sentinel-1 mission offers significant improvements, with respect to previous European Space Agency missions, in terms of revisiting time, spatial coverage and quality of the Synthetic Aperture Radar (SAR) imagery. The Interferometric Wide Swath (IWS) acquisition mode of Sentinel-1 images provides a 250-km swath with a repeat cycle of 6 days, considering both 1A and 1B. The Sentinel-1 mission has been especially designed for massive wide-area monitoring. An important advantage of the Sentinel-1 mission is that the data are freely available for both scientific and commercial applications. Several studies based on Sentinel-1 data have been devoted to urban deformation monitoring. Relevant examples include the monitoring of Mexico City [1], Madrid [2], Wuhan [3], Shanghai [4], Ravenna [5], the San Francisco Bay Area [6], Beijing [7], the Lanzhou New District [8], Florence [9], and Istanbul [10]. The monitoring of a slow-moving urban landslide area is described in [11], while the monitoring of sink-holes in urban areas is described in [12]. In [13] a high-speed railway bridge is studied, while in [14] the monitoring of a network of roads and railways is focused on.

This paper is organised as follows. Section 2 examines some important characteristics of PSI monitoring in urban areas. Section 3 describes the PSI approach proposed by the authors to monitor the area of interest. Section 4 discusses the results achieved over the area of interest. Section 5 includes the conclusions of this work.

2. Urban PSI Monitoring

Several PSI deformation monitoring approaches have been proposed in the last two decades, see [15] for a review. The PSI approach usually used by the authors is described in [16]. However, in some cases, it is possible to tailor the PSI data processing and analysis to the specific characteristics of the application at hand. This was the case in the analysis of the Glories Square area, which is described in this work. In the following, we briefly describe the key characteristics of the proposed processing.

The monitoring of the Glories Square and its neighbouring areas primarily concerns buildings and infrastructures. In order to obtain measurement points (hereafter called Persistent Scatterers, PSs) over such elements, it is important to properly model the Residual Topographic Error (RTE) component of the PSI observations for two reasons: for PSI modelling purposes and to properly geocode the PSI results. A second aspect regards the thermal expansion displacements. Some approaches have been proposed to explicitly model and estimate such displacements, e.g., see [17]. It is worth mentioning that these displacements can be neglected if one could filter out the PSs characterised by high RTE values (e.g., those above 10 m). However, in this work, this approach was discarded because the PSs located on the top of the buildings are of major interest, and we decided to keep the thermal displacements together with the ground deformation. This has the advantage of performing a standard PSI analysis, i.e., avoiding the computational costs of explicitly modelling the thermal expansion component. The analysis of the thermal expansion displacements and their separation from the deformation was carried out during data interpretation.

The third and most important aspect concerns the atmospheric phase component. Most of the PSI approaches use sets of spatial and temporal filters to estimate the atmospheric phase component [18–20]. The critical issue is to separate two low-pass spatial contributions: the deformation and the atmospheric components. The assumption usually employed is that the former one is temporally correlated, while the latter component is not correlated in time. In the targeted application this approach has two limitations. The first one is the difficulty to properly separate subtle deformation from the atmospheric contribution, with the risk to underestimate the former one. The second limitation is that no prior information is available about the dewatering plans and hence on the expected ground deformation. For instance, water pumping can be intermittent, implying a series of ups and downs of the ground, which cannot be smooth in time. For these reasons, we decided to adopt an alternative approach, which does not make use of spatio-temporal filters and, more importantly, does not make any assumption associated with such filters. We use a different philosophy based on stable areas, which allows us to estimate the atmospheric component without making any assumption concerning the deformation at hand. This has two advantages: we can correctly estimate sudden deformation, without the disadvantage of filtering out the high-frequency temporal components of the deformation; in addition, we can avoid the subjective decisions often associated with filtering. We use a similar procedure to process interferometric ground-based SAR data [21].

This approach requires the availability of known stable areas in the vicinity of the area of interest. This needs external information, which however can be validated during the data analysis. The stable areas are used to estimate the atmospheric component, which is then predicted and removed from the PSI observations over the area of interest. Several approaches can be used for this purpose, e.g., kriging, least squares collocation (e.g., see [22]), polynomials, etc.

3. The Proposed PSI Approach

In this section we detail, step by step, the proposed PSI approach.

1. Acquisition of a set of N interferometric Sentinel-1 SAR images that cover the area of interest. In this work, a minimum of 25 IWS images were used.
2. Precise co-registration of the entire burst stack that covers the area of interest. This is based on the information provided by the precise orbits associated with the images.
3. Generation of two redundant networks of interferograms: full-resolution (pixel footprint: 4 by 14 m) and 10 in range by 2 in azimuth (10 × 2) multi-look (pixel footprint: 40 by 28 m).
4. Candidate PS selection using the Amplitude Dispersion index [18].
5. The 2 + 1D phase unwrapping of the redundant 10 × 2 multi-look interferograms, see for details [16].
6. Identification of stable areas in the surroundings of the area of interest.
7. Estimation of the atmospheric phase component over the stable areas. In the current implementation of the monitoring, this step is performed assuming a linear phase model. The residuals of such models are used to validate the hypothesis regarding the stable areas.
8. Prediction and removal of the estimated atmospheric component from the original single-look interferograms.
9. Using the atmospheric-free single-look interferograms, estimation of linear deformation velocity and RTE using the periodogram, see [23].
10. Removal of the RTE from the atmospheric-free single-look interferograms.
11. The 2 + 1D phase unwrapping of single-look (RTE- and atmospheric-free) interferograms.
12. Generation of the deformation time series and estimation of the deformation velocity.
13. Geocoding of the results: the deformation velocity and the deformation time series.

4. Data Description

The study area concerns the surroundings of the Glories Square in Barcelona, see Figure 1. This area is located in the Besòs river delta, that is in a sedimentary environment with (sub-)horizontal layers. Below a 6-m thick anthropogenic deposits layer, two stratigraphic levels can be identified: quaternary units and tertiary (Pliocene) units. The quaternary materials are mainly clays with sand and gravel interbeds. The tertiary units are composed of medium to coarse sands from NE to SE and from SW to NW around the Glories Square, and mainly of grey marls from NW to NE and from SE to SW. The processed area is bordered by the red line shown in Figure 1, while the main area of interest has a perimeter indicated by a yellow line. In order to define this area, based on aquifer hydraulic properties and piezometers located in the area of interest, it was assumed that water pumping can have a maximum influence area (i.e., the area where changes in the water table can have effect on the surface) with a radius of 1 km, centred in the middle of the Glories Square. The remaining areas, i.e., outside the influence area, were considered stable. In this specific case there were previous PSI processing results that indicate the stability of this area. However, as it is discussed later, the hypothesis of the stable area can be validated during data processing and analysis. The analysed dataset includes

78 descending IWS SLC Sentinel-1A images, which cover the period from March 2015 to November 2017, see Table 1. The dataset includes 1813 interferograms, which were generated using all possible image combinations, with a limit of one year for the temporal baselines.

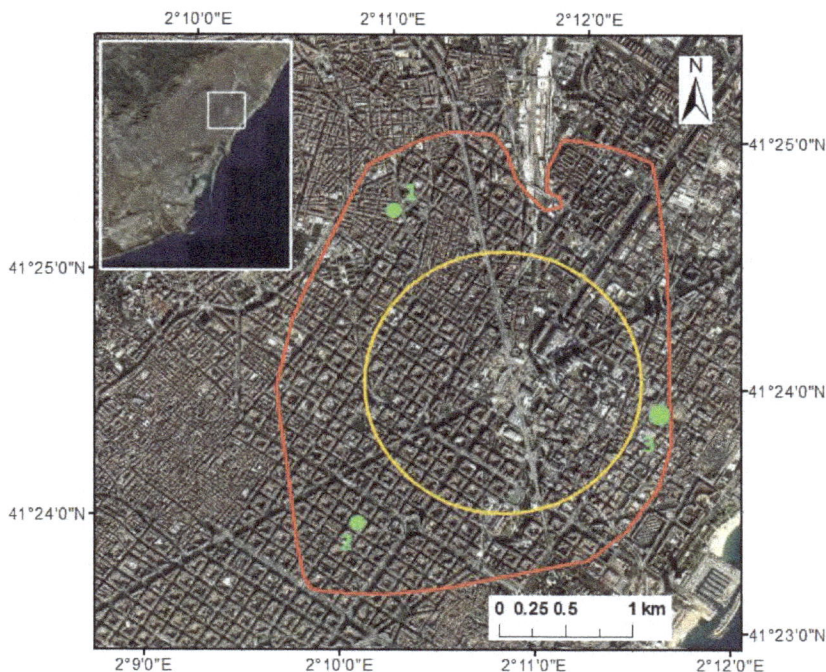

Figure 1. Study area (included in the red perimeter) and main area of interest (included in the yellow circle), which is the maximum area potentially affected by the water pumping activities. The area outside the yellow circle is considered stable. The figure inset shows the city of Barcelona.

Table 1. Dates of the 78 descending IWS images used in this work. Relative orbit: 110.

#	Date	#	Date	#	Date	#	Date
1	20150306	21	20151219	41	20160827	61	20170424
2	20150318	22	20151231	42	20160908	62	20170506
3	20150330	23	20160112	43	20160920	63	20170518
4	20150411	24	20160124	44	20161002	64	20170530
5	20150505	25	20160205	45	20161014	65	20170611
6	20150517	26	20160217	46	20161026	66	20170623
7	20150529	27	20160229	47	20161107	67	20170705
8	20150610	28	20160312	48	20161119	68	20170717
9	20150704	29	20160324	49	20161201	69	20170729
10	20150716	30	20160405	50	20161213	70	20170810
11	20150728	31	20160417	51	20161225	71	20170822
12	20150809	32	20160429	52	20170106	72	20170903
13	20150821	33	20160511	53	20170118	73	20170915
14	20150914	34	20160523	54	20170130	74	20170927
15	20150926	35	20160604	55	20170211	75	20171009
16	20151008	36	20160628	56	20170223	76	20171021
17	20151101	37	20160710	57	20170307	77	20171102

Table 1. *Cont.*

#	Date	#	Date	#	Date	#	Date
18	20151113	38	20160722	58	20170319	78	20171114
19	20151125	39	20160803	59	20170331		
20	20151207	40	20160815	60	20170412		

5. PSI Results over the Test Area

First of all, we analysed the characteristics of the output of step 5, described in the previous section. This output includes a set of 78 of 10 x 2 multi-look unwrapped phase images, where the first image values are set to zero because they represent the temporal reference. The objective of the analysis focused on the stable areas was the inspection of the characteristics of the atmospheric phase component. We assumed that, in the stable areas, this component is the only one spatially correlated (i.e., the RTE phase component, the thermal phase component and the phase noise are assumed to be spatially uncorrelated).

The data were analysed using the empirical autocorrelation function, EAF, $C(d_K)$, whose values are second order statistics [24]:

$$C(d_K) = \frac{1}{n_T} \cdot \sum_{i=1}^{n_T} \left\{ [M(P_i) - m_M] \cdot \frac{1}{n_J} \cdot \sum_{j=1}^{n_J} [M(P_j) - m_M] \right\} \tag{1}$$

where $d_K = K \cdot \Delta$ is the distance from the origin; K is an integer number; Δ is the function step; $n_T = l \cdot m$ is the total image pixel number, given by the number of line times the number of columns; $M(P_i)$ is the image value in the pixel P_i; m_M is the mean image value; and n_J is the total number of pixels P_j that, for a given P_i, satisfy the condition: $(K-1) \cdot \Delta < \|P_i - P_j\| < K \cdot \Delta$. Two examples of EAF are shown in Figure 2. From this function, the following information can be derived:

(1) σ_{tot}, the total standard deviation of the phase image;
(2) σ_{corr}, the standard deviation of the spatially correlated part of the phase image;
(3) σ_{noise}, the standard deviation of the spatially uncorrelated part of the phase image;
(4) L_{corr}, the correlation length, i.e., the distance from the origin where the EAF has a correlation which is half of that in the origin.

Figure 2. Example of correlation drop. Normalised EAFs (i.e., EAF divided by σ^2_{corr}) of the phase image #14: before the atmospheric correction (**red**) and residual phase image after the correction (**green**).

The main results of the analysis are summarised below:

- A number of 24 images out of 77 (i.e., 31.1% of the total) have σ_{corr} <0.4 rad. This indicates a rather weak atmospheric component. In terms of displacement this corresponds to a standard deviation below 1.76 mm. A number of 14 images out of 24 were acquired during winter or late autumn: this confirms that this is the period of the year when less atmospheric turbulence occurs. With the exception of one image, with L_{corr} = 680 m, this group is characterised by zero or negligible L_{corr}.

- A number of 26 images out of 77 (i.e., the 33.8% of the total) have σ_{corr} between 0.4 and 0.5 rad. With the exception of one image, with L_{corr} = 833 m, this group is characterised by moderated L_{corr} values, in the range between 20 and 50 m.

- Finally, 27 out of 77 images (i.e., the 35% of the total) have σ_{corr} above 0.5 rad, see their main characteristics in Table 2. A number of 18 images out of 27 were acquired in summer or late spring: this is the period when there is maximum atmospheric turbulence. With the exception of 7 images, which have negligible L_{corr} values, all the remaining images have correlation lengths above 50 m, with a maximum value of 1785 m.

Table 2. Main characteristics of the 27 phase images that are affected by the strongest atmospheric phase component. Values before atmospheric correction (orig) and after the correction (res).

Image #	s_tot_orig	s_corr_orig	s_tot_res	s_corr_res	L_corr_orig	L_corr_res
34	0.63	0.50	0.60	0.47	204	192
68	0.71	0.51	0.69	0.49	24	0
45	0.67	0.50	0.62	0.44	27	18
62	0.64	0.51	0.64	0.51	255	183
40	0.67	0.54	0.48	0.32	655	0
39	0.69	0.54	0.63	0.47	26	18
4	0.78	0.54	0.67	0.44	0	0
72	0.67	0.54	0.49	0.33	417	0
48	0.69	0.55	0.62	0.48	417	18
16	0.72	0.56	0.54	0.35	281	0
69	0.69	0.56	0.52	0.36	468	0
37	0.74	0.59	0.55	0.36	24	0
3	0.71	0.60	0.64	0.52	204	18
24	0.71	0.61	0.40	0.22	1785	0
42	0.80	0.63	0.62	0.41	28	0
76	0.76	0.63	0.59	0.43	553	37
35	0.77	0.67	0.67	0.55	765	329
23	0.76	0.69	0.47	0.35	1029	18
36	0.83	0.72	0.55	0.38	1029	0
57	0.99	0.75	0.81	0.53	26	0
41	0.94	0.80	0.59	0.40	1122	0
67	0.97	0.82	0.64	0.45	842	0
14	1.01	0.92	0.53	0.37	1496	0
64	1.16	0.95	1.07	0.86	51	18
70	1.17	1.08	0.60	0.46	1658	18
12	1.33	1.21	1.10	0.97	468	384
44	1.40	1.30	0.98	0.85	638	185

As mentioned in the previous section, the atmospheric component was estimated over the stable areas using a linear phase model, which represents an easy-to-implement and robust modelling approach over small areas. The original 78 phase images, the atmospheric models (the linear planes) and the residual phase images after removing the models are shown in Figure 3. The hypothesis of the stable area can be validated by analysing the latter ones. In fact, a deformation signal would typically appear as a persistent pattern in consecutive images. This is not the case in the analysed dataset. The main results of the EAF analysis of the residual phase images displayed in Table 2 are:

- Compared to the original data, the average reduction of σ_{corr} is 30.5%. The most relevant result is a drop of 90% of the average L_{corr} values. This is an important indicator to judge the goodness of the proposed method. An example of correlation drop is shown in Figure 2: in this case the L_{corr} values drop from 1496 m to zero.

- It is worth noting from Table 2 that there are five images where the L_{corr} values remain quite high (i.e., above 180 m) after removing the atmospheric component (images 12, 34, 35, 44, and 62). In two cases (images 12 and 44), the corresponding σ_{corr} values are also high (0.85 and 0.97 rad). These two images represent a case where there is an important atmospheric component, which cannot be modelled by a linear atmospheric model. In this case there are two options: (i) discarding the images, especially if the dataset is big enough; (2) if the images cannot be eliminated, the deformation time series have to be interpreted with attention: the time series values in correspondence of the two images may contain spikes.

- The remaining images have σ_{corr} values ranging between 0.22 and 0.55 rad. These values indicate the dispersion of the residual atmospheric signal, which affect the corresponding deformation time series. In terms of displacements, the standard deviations of such a signal, range from 0.97 mm (best case) to 2.42 mm (worst case): these values seem to be acceptable for the purpose of the application at hand, as discussed later.

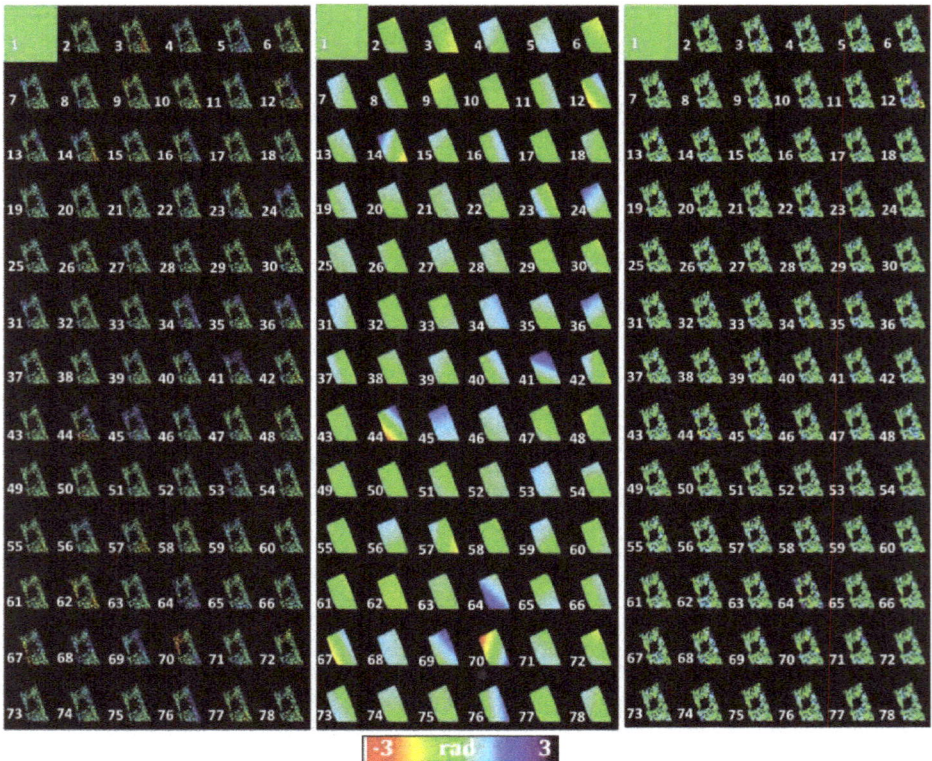

Figure 3. Atmospheric component estimation using stable areas. Original phases that cover an area of approximately 16 km^2 (**left**). The black circles show the 1-km radius area of interest. Estimated linear atmospheric components (**middle**). Residual phase after removing the linear atmospheric component (**right**). The black colour means no-used data. The first image is set to zero (green colour) because it is used as reference image.

The last part of the analysis concerned the 1-km radius area of interest (influence area, Figure 1). The objective was to assess the effects of the residual atmospheric component in the deformation time series (computed in the step 12 of the proposed PSI procedure). The analysis concerned the stable PSs of the main area of interest: they were selected by only considering those points that have an absolute deformation velocity below 1 mm/yr. The time series were analysed using the EAF: all the 78 images were included in the analysis (no images were discarded). These are the main results:

- The average σ_{tot} of all the time series is 1.90 mm. Only 127 out of 3862 time series have σ_{tot} above 3 mm: this represents 3.3% of the PSs. It is worth observing that the σ_{tot} includes, among others, the residual (non-modelled) atmospheric effects and the noise of the observations, which depends on the PS quality.
- As expected, the L_{corr} is close to zero for the great majority of time series. This is key to detect subtle (temporally correlated) deformation using the time series.

6. Deformation Results and Discussion

We discuss in the following some examples of time series, to illustrate the goodness of the proposed procedure. Figure 4 shows the line-of-sight (LOS) deformation time series of a point located close to the Glories Square. In this figure, during the period from March to June 2017, there is a terrain subsidence (up to about -10 mm), followed by an uplift to roughly recover to the original height. It is worth underlining that this result was achieved using exclusively SAR data, without any additional information. In the interpretation of the data, it was identified that this behaviour was due to aquifer dewatering, and the subsequent recovery of the water level due to the stop of the pumping activity. This is evident from the piezometer data plotted in Figure 4, which were acquired in a location close to the point, and which match very well with the estimated deformation time series. This represents an example of validation of the PSI results.

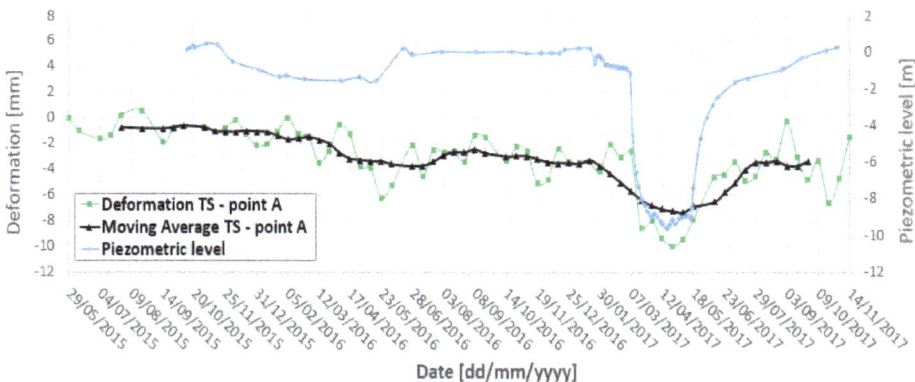

Figure 4. Example of time series validation: the deformation time series (green) is strongly correlated with the piezometric data of the same area. The black deformation time series represents a solution based on spatio-temporal filters, see [16]: a 96-day moving average was used. The location of the point and the piezometer is shown in Figure 5. The deformation values refer to the radar line-of-sight (LOS).

Figure 5. Examples of LOS accumulated deformation maps corresponding to the maximum of the displacement (12 April 2017, above) and to the recovery of the displacements (3 September 2017, below). In the above image, the three rectangles (grey-zone 1, orange-zone 2 and pink-zone 3) show the location of the three zones shown in Figure 6. The green circle A shows the location of the point considered in Figure 4. The white circle shows the location of the piezometer.

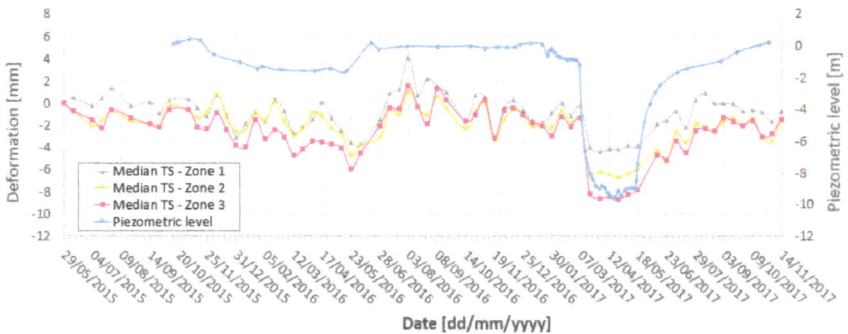

Figure 6. Examples of LOS deformation time series of three zones located in the deformation area shown in Figure 5. The time series display the median values of the points contained in each zone. The blue time series concerns the piezometric data.

To illustrate the performance of the proposed procedure, in Figure 4 the above time series is compared with a solution based on spatio-temporal filters, see the procedure described in [16]. In this case we used a 96-day moving average. In this example, it is evident that the first deformation time series matches the sudden drop of the piezometric data, while the second solution is basically missing such a drop, providing a biased temporal low-pass solution. One may observe that in the time series there are several peaks that are not linked to a particular period of dewatering. They have an amplitude of ± 2–3 mm with respect to the global trend of the time series. They are not due to thermal expansion: they are examples of the residual atmospheric effects, discussed earlier in this paper.

Figure 5 shows the LOS accumulated deformation maps corresponding to the maximum of the displacement (12 April 2017) and to the recovery of the displacements (3 September 2017). From the first map it is possible to assess the actual dimension of the area affected by ground deformation induced by dewatering. Figure 6 shows some additional examples of LOS deformation time series of three zones located in the deformation area shown in Figure 5. The time series display the median values of the points contained in each zone. One may appreciate that there is a good agreement between the measured time series and the piezometric data. Finally, Figure 7 shows the displacement time series of PSs located outside the maximum potential influence area. The location of such points is shown in Figure 1. The time series correspond to basically stable areas, with most of the time series values between ±2 mm. The time series include some spikes, with absolute values up to 6 mm, which are mainly due to residual atmospheric effects.

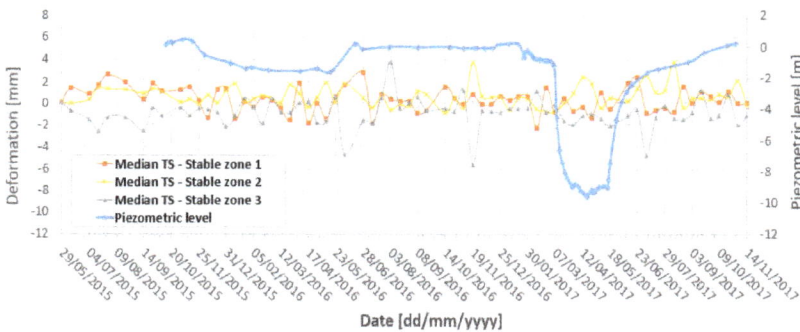

Figure 7. Example of LOS deformation time series related to three zones located in the stable area shown in Figure 1. The blue line shows the piezometric data. The plotted values represent the median deformation measured within the 3 green circles from Figure 1.

Another example of deformation time series is shown in Figure 8. In the same figure, the temperatures of the scenes are plotted in correspondence to the days of acquisition of the SAR images. One may appreciate a strong correlation between deformation and temperature. This is clearly a displacement behaviour induced by thermal expansion. This type of result is possible only if an appropriate estimation of the atmospheric phase contribution is carried out: this confirms the goodness of the procedure proposed in this work.

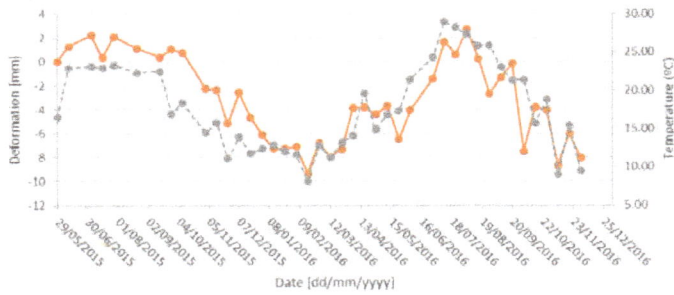

Figure 8. LOS displacement time series (**orange line**) of one single PS and plot of the corresponding temperatures (**grey line**). This is a clear example of thermal expansion displacements.

This paper has focused on a PSI procedure to monitor the land deformation associated with construction works that involve water pumping and hence lowering of the water table. The case study focuses on the construction works related to the transformation process of the Glories Square, located in the centre of Barcelona (Spain). The PSI monitoring was based on C-band data acquired by Sentinel-1 sensors. The used PSI procedure was tailored to the specific characteristics of the application at hand. The performances of the proposed procedure were analysed. This involved the characterization of the original atmospheric phase component of 78 Sentinel-1 SAR images using the EAF and, in particular, the standard deviation of the spatially correlated part of the phase images, σ_{corr}, and their correlation length, L_{corr}. In the analysed case study, the atmospheric component was estimated and removed from the original PSI observations using a linear phase model. The same EAF analysis was carried out on the residual atmospheric phase component. The most important aspect is a drop of 90% of the average L_{corr} values: this indicates that the proposed atmospheric estimation procedure works properly. However, in the analysed case, at least two images (2.5% of the images) show relatively high σ_{corr} and L_{corr} values: they represent two cases where a strong atmospheric component cannot be properly modelled by the linear atmospheric model used. However, this 2.5% does not compromise the quality of the estimated deformation time series. An EAF analysis of such time series was carried out: the average σ_{tot} of the time series of stable points is 1.90 mm. Only 3.3% of the time series have σ_{tot} above 3 mm. The quality of the time series was further illustrated considering different deformation result examples. The most important one concerns a validation result, where the PSI estimated deformation matches well with external piezometric data. This result is confirmed by other time series coming from different locations in the main deformation area. A final example regards a thermal expansion case. The strong correlation with the temperature is only possible with an appropriate estimation of the atmospheric phase contribution: this confirms the goodness of the approach proposed in this work.

7. Conclusions

The most original aspect of this procedure is the estimation of the atmospheric phase component using stable areas located in the vicinity of the monitoring area. This approach overcomes some important limitations of PSI techniques that use sets of spatial and temporal filters. In particular, it avoids the assumption that the low-pass spatial deformation is temporally correlated. This implies that, with the proposed procedure, sudden deformation can properly be estimated. In addition to the PSI approach described in this paper, the authors routinely apply this procedure to process interferometric ground-based SAR data.

The proposed PSI procedure was developed and tested for a specific application, i.e., the monitoring of urban land deformation due to water extraction. However, the same procedure can work with any type of deformation phenomena, provided that its spatial extension is sufficiently small. This is an important characteristic of the proposed procedure: it works only over relatively small areas,

Remote Sens. **2018**, *10*, 1780

where the atmospheric component, estimated over the stable areas, can be interpolated in the area of interest. The larger the area of interest, the bigger will be the error in the estimation of the atmospheric component over this area (i.e., the larger will be the residual atmospheric component). It is worth noting that the interpolation of the atmospheric component is only possible if stable areas surround the area of interest. In some applications this cannot be possible: in this case the atmospheric component needs to be extrapolated. This implies larger errors in the estimation of the atmospheric component.

Author Contributions: N.D., O.M., and B.C. developed the software tools used in this work. A.B. and M.C.-G. were involved in the PSI data processing. N.D., B.C., M.M., and M.C. were involved in the PSI data analysis. J.B.-B. and E.V.-S. contributed to data interpretation. M.C. coordinated the entire work and wrote the main paper draft. All authors contributed to the paper editing.

Funding: This work was partially funded by BIMSA (Barcelona d'Infraestructures Municipals, SA) of the Municipality of Barcelona, through the contract number 124.1215.349. In addition, it was partially funded by the Spanish Ministry of Economy and Competitiveness through the DEMOS project "Deformation monitoring using Sentinel-1 data" (Ref: CGL2017-83704-P).

Conflicts of Interest: The authors declare no conflict of interest.

References

1. Sowter, A.; Amat, M.B.C.; Cigna, F.; Marsh, S.; Athab, A.; Alshammari, L. Mexico City land subsidence in 2014–2015 with Sentinel-1 IW TOPS: Results using the Intermittent SBAS (ISBAS) technique. *Int. J. Appl. Earth Obs. Geoinf.* **2016**, *52*, 230–242. [CrossRef]

2. Bakon, M.; Marchamalo, M.; Qin, Y.; García-Sánchez, A.J.; Alvarez, S.; Perissin, D.; Papco, J.; Martínez, R. Madrid as Seen from Sentinel-1: Preliminary Results. *Procedia Comput. Sci.* **2016**, *100*, 1155–1162. [CrossRef]

3. Zhou, L.; Guo, J.; Hu, J.; Li, J.; Xu, Y.; Pan, Y.; Shi, M. Wuhan surface subsidence analysis in 2015–2016 based on sentinel-1a data by SBAS-InSAR. *Remote Sens.* **2017**, *9*, 982. [CrossRef]

4. Yu, L.; Yang, T.; Zhao, Q.; Liu, M.; Pepe, A. The 2015–2016 Ground Displacements of the Shanghai coastal area Inferred from a combined COSMO-SkyMed/Sentinel-1 DInSAR Analysis. *Remote Sens.* **2017**, *9*, 1194. [CrossRef]

5. Fiaschi, S.; Tessitore, S.; Bonì, R.; Di Martire, D.; Achilli, V.; Borgstrom, S.; Ibrahim, A.; Floris, M.; Meisina, C.; Ramondini, M.; Calcaterra, D. From ERS-1/2 to Sentinel-1: two decades of subsidence monitored through A-DInSAR techniques in the Ravenna area (Italy). *GIsci Remote Sens.* **2017**, *54*, 305–328. [CrossRef]

6. Shirzaei, M.; Bürgmann, R.; Fielding, E.J. Applicability of Sentinel-1 Terrain Observation by Progressive Scans multitemporal interferometry for monitoring slow ground motions in the San Francisco Bay Area. *Geophys. Res. Lett.* **2017**, *44*, 2733–2742. [CrossRef]

7. Du, Z.; Ge, L.; Ng, A.H.M.; Xiaojing, L.; Li, L. Mapping land subsidence over the eastern Beijing city using satellite radar interferometry. *Int. J. Digit. Earth* **2018**, *11*, 504–519. [CrossRef]

8. Chen, G.; Zhang, Y.; Zeng, R.; Yang, Z.; Chen, X.; Zhao, F.; Meng, X. Detection of Land Subsidence Associated with Land Creation and Rapid Urbanization in the Chinese Loess Plateau Using Time Series InSAR: A Case Study of Lanzhou New District. *Remote Sens.* **2018**, *10*, 270. [CrossRef]

9. Del Soldato, M.; Farolfi, G.; Rosi, A.; Raspini, F.; Casagli, N. Subsidence Evolution of the Firenze–Prato–Pistoia Plain (Central Italy) Combining PSI and GNSS Data. *Remote Sens.* **2018**, *10*, 1146. [CrossRef]

10. Aslan, G.; Cakır, Z.; Ergintav, S.; Lasserre, C.; Renard, F. Analysis of Secular Ground Motions in Istanbul from a Long-Term InSAR Time-Series (1992–2017). *Remote Sens.* **2018**, *10*, 408. [CrossRef]

11. Béjar-Pizarro, M.; Notti, D.; Mateos, R.M.; Ezquerro, P.; Centolanza, G.; Herrera, G.; Bru, G.; Sanabria, M.; Solari, L.; Duro, J.; Fernández, J. Mapping Vulnerable Urban Areas Affected by Slow-Moving Landslides Using Sentinel-1 InSAR Data. *Remote Sens.* **2017**, *9*, 876. [CrossRef]

12. Kim, J.W.; Lu, Z.; Degrandpre, K. Ongoing deformation of sinkholes in Wink, Texas, observed by time-series Sentinel-1a SAR interferometry (preliminary results). *Remote Sens.* **2016**, *8*, 313. [CrossRef]

13. Huang, Q.; Crosetto, M.; Monserrat, O.; Crippa, B. Displacement monitoring and modelling of a high-speed railway bridge using C-band Sentinel-1 data. *ISPRS J. Photogramm. Remote Sens.* **2017**, *128*, 204–211. [CrossRef]

14. North, M.; Farewell, T.; Hallett, S.; Bertelle, A. Monitoring the Response of Roads and Railways to Seasonal Soil Movement with Persistent Scatterers Interferometry over Six UK Sites. *Remote Sens.* **2017**, *9*, 922. [CrossRef]

15. Crosetto, M.; Monserrat, O.; Cuevas-González, M.; Devanthéry, N.; Crippa, B. Persistent Scatterer Interferometry: a review. *ISPRS J. Photogramm. Remote Sens.* **2016**, *115*, 78–89. [CrossRef]

16. Devanthéry, N.; Crosetto, M.; Monserrat, O.; Cuevas-González, M.; Crippa, B. An approach to Persistent Scatterer Interferometry. *Remote Sens.* **2014**, *6*, 6662–6679. [CrossRef]

17. Monserrat, O.; Crosetto, M.; Cuevas, M.; Crippa, B. The Thermal Expansion Component of Persistent Scatterer Interferometry Observations. *IEEE Geosci. Remote Sens. Lett.* **2011**, *8*, 864–868. [CrossRef]

18. Ferretti, A.; Prati, C.; Rocca, F. Nonlinear subsidence rate estimation using permanent scatterers in differential SAR interferometry. *IEEE Trans. Geosci. Remote Sens.* **2000**, *38*, 2202–2212. [CrossRef]

19. Ferretti, A.; Prati, C.; Rocca, F. Permanent scatterers in SAR interferometry. *IEEE Trans. Geosci. Remote Sens.* **2001**, *39*, 8–20. [CrossRef]

20. Berardino, P.; Fornaro, G.; Lanari, R.; Sansosti, E. A new algorithm for surface deformation monitoring based on small baseline differential SAR interferograms. *IEEE Trans. Geosci. Remote Sens.* **2002**, *40*, 2375–2383. [CrossRef]

21. Crosetto, M.; Monserrat, O.; Luzi, G.; Cuevas-González, M.; Devanthéry, N. Discontinuous GBSAR deformation monitoring. *ISPRS J. Photogramm. Remote Sens.* **2014**, *93*, 136–141. [CrossRef]

22. Crosetto, M.; Tscherning, C.C.; Crippa, B.; Castillo, M. Subsidence Monitoring using SAR interferometry: reduction of the atmospheric effects using stochastic filtering. *Geophys. Res. Lett.* **2002**, *29*, 26–29. [CrossRef]

23. Biescas, E.; Crosetto, M.; Agudo, M.; Monserrat, O.; Crippa, B. Two radar interferometric approaches to monitor slow and fast land deformations. *J. Surv. Eng.* **2007**, *133*, 66–71. [CrossRef]

24. Crosetto, M.; Moreno Ruiz, J.A.; Crippa, B. Uncertainty propagation in models driven by remotely sensed data. *Remote Sens. Environ.* **2001**, *76*, 373–385. [CrossRef]

remote sensing

MDPI

Article

Displacement Monitoring and Health Evaluation of Two Bridges Using Sentinel-1 SAR Images

Qihuan Huang [1,*], Oriol Monserrat [2], Michele Crosetto [2], Bruno Crippa [3], Yian Wang [1], Jianfeng Jiang [1] and Youliang Ding [4]

[1] School of Earth Sciences and Engineering, Hohai University, JiangNing District, Nanjing 211100, China; wang.yi.an@hhu.edu.cn (Y.W.); jfjiang@hhu.edu.cn (J.J.)
[2] Centre Tecnològic de Telecomunicacions de Catalunya (CTTC), Geomatics Division, Av. Gauss 7, E-08860 Castelldefels, Spain; omonserrat@cttc.cat (O.M.); mcrosetto@cttc.cat (M.C.)
[3] Department of Earth Sciences, Section of Geophysics, University of Milan, Via Cicognara 7, I-20129 Milan, Italy; bruno.crippa@unimi.it
[4] Key Laboratory of C&PC Structures of the Ministry of Education, Southeast University, Nanjing 210096, China; civilchina@hotmail.com
* Correspondence: InSAR@hhu.edu.cn; Tel.: +86-025-83786961

Received: 2 August 2018; Accepted: 23 October 2018; Published: 30 October 2018

Abstract: Displacement monitoring of large bridges is an important source of information concerning their health state. In this paper, a procedure based on satellite Persistent Scatterer Interferometry (PSI) data is presented to assess bridge health. The proposed approach periodically assesses the displacements of a bridge in order to detect abnormal displacements at any position of the bridge. To demonstrate its performances, the displacement characteristics of two bridges, the Nanjing-Dashengguan High-speed Railway Bridge (NDHRB, 1272 m long) and the Nanjing-Yangtze River Bridge (NYRB, 1576-m long), are studied. For this purpose, two independent Sentinel-1 SAR datasets were used, covering a two-year period with 75 and 66 images, respectively, providing very similar results. During the observed period, the two bridges underwent no actual displacements: thermal dilation displacements were dominant. For NDHRB, the total thermal dilation parameter from the PSI analysis was computed using the two different datasets; the difference of the two computations was 0.09 mm/°C, which, assuming a temperature variation of 30 °C, corresponds to a discrepancy of 2.7 mm over the total bridge length. From the total thermal dilation parameters, the coefficients of thermal expansion (CTE) were calculated, which were $11.26 \times 10^{-6}/°C$ and $11.19 \times 10^{-6}/°C$, respectively. These values match the bridge metal properties. For NYRB, the estimated CTE was $10.46 \times 10^{-6}/°C$, which also matches the bridge metal properties $(11.26 \times 10^{-6}/°C)$. Based on a statistical analysis of the PSI topographic errors of NDHRB, pixels on the bridge deck were selected, and displacement models covering the entire NDHRB were established using the two track datasets; the model was validated on the six piers with an absolute mean error of 0.25 mm/°C. Finally, the health state of NDHRB was evaluated with four more images using the estimated models, and no abnormal displacements were found.

Keywords: SAR interferometry; displacement monitoring; Sentinel-1; permanent scatterers; thermal dilation; health monitoring

---

1. Introduction

The long-term millimeter-level displacement monitoring of man-made structures, such as dams, embankments, bridges, and railways, is a promising field of application for satellite Persistent Scatterer Interferometry (PSI). This technique offers the advantages of wide area coverage, high sensitivity to small deformations, and day and night and all-weather operation, which makes it suitable for man-made structural health monitoring. A review of the PSI technique is provided in Reference [1].

PSI monitoring of man-made structures has usually been based on high-resolution SAR data. Relevant examples include dam monitoring using ALOS PALSAR data [2,3], and several works based on TerraSAR-X data [4–7]. A combination of ALOS PALSAR and TerraSAR-X data is described in Reference [8]. RADARSAT-2 images were used for railway monitoring [9]. An example based on COSMO-SkyMed imagery is described in Reference [6]. As far as bridge monitoring is concerned, the X-band images are the extensively used PSI data [7,10–12]. This is mainly due to the high spatial resolution of X-band data, and their high sensitivity to displacements with respect to the C- and L-band. However, with the advent of Sentinel-1 SAR sensors, this has changed slightly. The main reasons for this are that the resolution is still high (with a footprint of 4 by 14 m) and the quality of the signal is good enough to measure millimeter displacements [13]. Moreover, the spatial coverage of a single Sentinel-1 image (250 by 180 km) and their free availability suggest a great advantage with respect to X-band data in terms of costs.

In this study, we have focused on C-band Sentinel-1 data, taking advantage of open access SAR data. This study is a continuation of the work described in Reference [13]. The main improvements in this study can be found at different levels. From the methodological point of view, the new approach includes: the assessment of the initial conditions of the bridge by using the extended PSI model [14]; the removal of the topographic phase error, which is an important error source during the health evaluation phase; and the evaluation of the displacements along the bridge, instead of focusing only on the piers. From the point of view of the analysis, we have added the analysis of two independent datasets for each bridge in order to cross-validate the results, and we have described a procedure to evaluate the sensitivity of PSI for different tracks, in order to find the best one. Finally, it is worth noting that the work shows the applicability of the approach to different bridges, by adding the results over a second bridge.

In Section 2, the main steps of bridge health evaluation are described. Section 3 provides general information concerning the bridges analyzed and the Sentinel-1 datasets. Section 4 presents the analysis of the sensitivity of the PSI measurements to the longitudinal displacements of the bridges. Section 5 describes the main issues related to the data processing. Section 6 describes the SAR interferometric results obtained, and Section 7 shows some examples of the bridge health evaluation. Section 8 includes the discussion and main conclusions.

2. A Bridge Health Evaluation Procedure

A large number of long bridges have been built in the last few decades. Maintaining the safety of these bridges is crucial. To monitor the evolution of the condition of a bridge, to locate and repair damages and also to perform a reliability assessment, a long-term structural health monitoring (SHM) system is generally installed on the bridges [15,16]. An SHM system is a tool for engineers and managers to plan and evaluate the maintenance operations on a structure. Long-term monitoring data collected from the SHM system can be used, e.g., to evaluate the vibrations of the main girder, the static performance of steel truss arc, the movement of piers, and the fatigue of the steel deck [17]. The SHM system for bridge health evaluation has the advantage of high temporal resolution, while its spatial resolution depends on the number of point sensors mounted on the bridge. Spatial resolution can, in some cases, be improved using the PSI technique, which is characterized by spatially dense measurement points. In the following sections, we describe a health evaluation procedure, focused on thermal dilation displacements of the bridge.

The key idea of the procedure is to: (i) use a set of SAR images to model the thermal dilation behavior of a given bridge; and then (ii) use additional SAR images to monitor the temporal evolution of the bridge. Figure 1 illustrates the flow chart of the procedure, which is composed of three parts highlighted in different colors. The procedure can be used to monitor the thermal dilation displacements of the entire bridge, exploiting the dense set of measurements provided by the PSI technique. The main steps of the procedure are described below.

1.   Collect N SAR images over the bridge of interest, acquired at times t_1 to t_N. Acquire, for each image, the temperature of the given scene at the time of acquisition of each image: T_1 to T_N.
2.   Generation of a redundant network of M interferograms from the N collected images (M >> N) [18], and calculation of the displacement time series using the traditional PSI method.

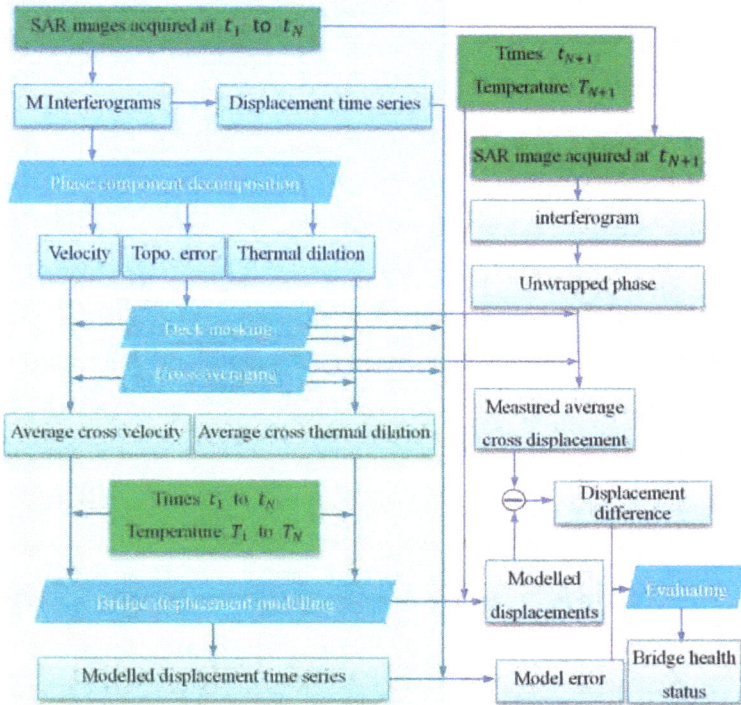

Figure 1. Flow chart of the bridge health evaluation procedure.

1.   The extended PSI model described in Reference [14] is used to estimate the main PSI phase components. This involves the following steps:

     (a)   *Pixel selection.* In the SAR images, only those points characterized by low noise levels are selected using the amplitude dispersion index [19].
     (b)   *Pixels connection.* The selected pixels are connected by edges (Figure 2). For each interferogram k and edge e, the phase difference $\Delta\Phi^k(e)$ is derived. Let us call this difference $\Delta\Phi^k_{obs}$.
     (c)   *Phase modeling and parameter estimation.* For each phase difference, we can write:

$$\Delta\varepsilon^k = \Delta\Phi^k_{obs} - \Delta\Phi^k_m(\Delta v, \Delta te, \Delta Th) \tag{1}$$

where $\Delta\varepsilon^k$ is the differential phase residual associated with a given edge e, while $\Delta\Phi^k_m(\Delta v, \Delta te, \Delta Th) = \frac{4\pi}{\lambda}\Delta T^k\Delta v + \frac{4\pi}{\lambda}\frac{B^k_\perp}{R^k\sin\theta}\Delta te + \frac{4\pi}{\lambda}\Delta Temp^k\Delta Th$ is the modeled differential phase. Δv, Δte and ΔTh are the differential unknowns associated with the edge e; Δv is the differential deformation velocity; Δte is the differential topographic error; and ΔTh is the so-called differential thermal dilation parameter; ΔT^k and $B_\perp^k$ are the temporal and perpendicular baseline of the interferogram k; $\Delta Temp^k$ is the temperature difference between the acquisitions of the two images of the interferogram k; R^k and θ^k are the slant range and incidence angle of the interferogram k; and λ is the radar wavelength. To estimate the three unknowns for each edge e, the following function is maximized numerically:

$$\gamma = \frac{1}{M} \sum_{k=1}^{M} \exp\left(j\cdot(\Delta\Phi_{obs}^k - \Delta\Phi_m^k(\Delta v, \Delta te, \Delta Th))\right) \tag{2}$$

where γ is a goodness of fit parameter, which indicates the quality of the estimation of the three unknowns; and M is the number of interferograms.

(d)   *Phase component reconstruction.* This step involves the integration of the differential unknowns Δte, Δv, and ΔTh. A minimum number of edges associated with a single pixel is set during the integration.

2.   Bridge displacement modeling and error estimation based on the estimated phase components. This involves the following steps:

(a)   *Bridge deck masking.* This step is based on the statistic result of the topographic errors achieved in step 3(d). Assuming the bridge deck is flat, a mask is built to select pixels on the bridge deck

(b)   *Cross averaging.* Instead of using the displacement measurements along the bridge longitudinal profile as in Refence [13], we average the above selected pixels in the cross-bridge direction. Therefore, robust and accurate displacement measurements along the longitudinal direction of the bridge are measured.

(c)   *Bridge displacement modeling.* Considering the cross averaged phase components on the bridge deck, the following displacement model can be established:

$$d_{Long} = \Delta Temp \cdot \Delta Th + \Delta v \cdot \Delta t \tag{3}$$

where d_{Long} is the modeled longitudinal displacement, $\Delta Temp$ and Δt are the temperature and temporal difference, respectively, and ΔTh and Δv are the thermal dilation and linear velocity parameters along the bridge estimated by PSI.

(d)   *Bridge displacement model error estimation.* With the acquisition time t_1 to t_N and the temperature T_1 to T_N, the model error, measured by the standard deviation of the differences between the cross-average value of the modeled displacements and the displacement time series achieved in step 2, is estimated.

3.   Bridge health evaluation. The idea for this evaluation is based on the hypothesis testing of the displacement differences, which are calculated between the upcoming measurements and the modeled ones, similar to Deviation Index DI1 described in Reference [20]. This includes the following two steps:

(a)   *Differential displacement estimation.* Let us assume that a new SAR image is acquired at t_{N+1}, with a temperature of T_{N+1}. Then, more interferograms are generated with the image t_{N+1}, and the cross-average displacements of the entire bridge deck are evaluated through interferogram phase unwrapping, bridge deck masking, and cross averaging. Such displacements are then compared with the modeled ones, and their difference is calculated:

$$\text{Diff}_{\text{Long}}^{k,k+1} = \text{Interf}_{\text{Long}}^{k,k+1} - \text{Model}_{\text{Long}}^{k,k+1} \tag{4}$$

where $\text{Model}_{\text{Long}}^{k,k+1'} = \Delta\text{Temp}^{k,k+1}\cdot\Delta\text{Th} + \Delta v\cdot\Delta\text{Tt}^{k,k+1}$ is the modeled longitudinal displacement, while $\text{Interf}_{\text{Long}}^{k,k+1}$ is the measured one.

(b)   *Bridge health evaluation.* The differences between the measured and modeled displacements are assessed using the procedure described in Reference [13], and the confidence interval is given as twice the estimated model error. A positive evaluation is when the measured displacements are within the confidence interval (i.e., the bridge shows a good behavior, or the displacements are within the design parameters of the bridge). Otherwise, a detailed analysis of the bridge, and especially of the bearings, is required.

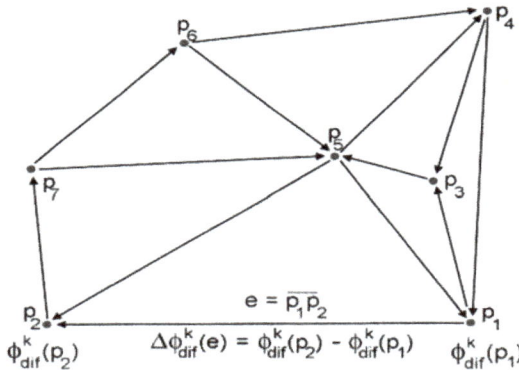

Figure 2. Scheme of the selected pixels connection.

3. Description of the Test Sites and Datasets

3.1. The NDHRB

Nanjing-Dashengguan High-speed Railway Bridge (NDHRB) is located in the Nanjing section of the middle and lower reaches of the Yangtze River, in China. This bridge (see the blue rectangle in Figure 3a) is the world's longest span high-speed railway bridge and the largest bridge with the heaviest design load ever built [17]. The structure of the NDHRB includes an orthogonal steel deck system; the heights of each part are highlighted in Figure 3c. The bridge includes six tracks: two tracks of the Beijing-Shanghai high-speed line; two tracks of the Shanghai-Chengdu railway lines; and two tracks of the Nanjing Metro. For more details, see Reference [17]. The bridge is supported by six sliding bearings (4#, 5#, 6#, 8#, 9#, 10#) on the two sides of the bridge and a fixed bearing (7#) located in the center of the bridge. The deck cross-section of NDHRB is shown in Figure 1d. The main structure of the bridge was built using three types of steel: Q345qD, Q370qE, and Q420qE. Their coefficients of thermal expansion (CTE) are $16.0 \times 10^{-6}/°C$, $13.0 \times 10^{-6}/°C$, and $13.0 \times 10^{-6}/°C$, respectively.

3.2. The NYRB

Nanjing-Yangtze River Bridge (NYRB) (the yellow rectangle in Figure 3a), connects the Beijing-Shanghai railway and the Nanjing-Yangzhou national highway. It is the first highway-cum-railway bridge (the upper layer is a highway, and the lower layer is a railway) built in China. It was opened on 29 December 1968, after ten years of construction. The main bridge is 1576-m long (128 m, plus 160 m by 9): it includes a simply-supported steel truss girder, with a span of 128 m, and 9 remaining continuous steel truss girders of 160 m, where every three spans are united as a bridge segment (span-continuous truss) [21]. In order to adapt to the longitudinal displacement of the bridge decks caused by temperature changes, huge expansion joints are mounted between the three segments at 1#, 4#, 7#, and 10# piers, with a maximum moving ability of 38 cm for 1#, 55 cm for 4# and 7#, and 34 cm for 10# [22]. Movable bearings are mounted at 1#, 3#, 4#, 6#, 7#, 9#, and 10#, while fixed bearings are on 0#, 2#, 5#, and 8#. Low alloy steel of 16Mnq is used as the main structure of the main girder and railway cross-section [23], with a CTE of $11.26 \times 10^{-6}/°C$. The overall layout of the NYRB is shown in Figure 3f. It should be noted that the bridge was closed for 27 months for comprehensive repair and maintenance work at the end of 2016.

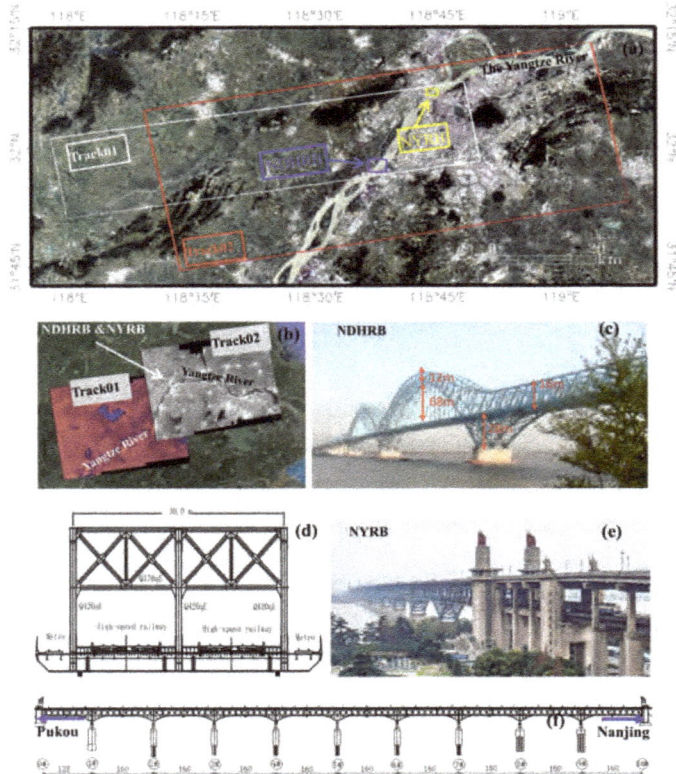

Figure 3. Sentinel SAR image coverage over the two bridges. (**a**) Location of the two bridges and burst coverage (white and red rectangles) of the two ascending SAR datasets used in this study; (**b**) Footprint of the two tracks; (**c**) Photo of the Nanjing-Dashengguan High-speed Railway Bridge (NDHRB), the heights of the structure are taken from Reference [24]; (**d**) Cross-section of the NDHRB; (**e**) Photo of the Nanjing-Yangtze River Bridge (NYRB); (**f**) Layout of the NYRB.

3.3. The Sentinel-1 Datasets

The two bridges are imaged in the overlap area of two ascending Sentinel-1 tracks: track 01 (the absolute orbit number of the first image, acquired on 25 April 2015, is 005639); and track 02 (the absolute orbit number of the first image, acquired on 2 April 2015, is 005437): see Figure 3b. Specifically, this occurs in a single burst of the third swath of track 01 (see the white rectangle in Figure 3a), and two bursts of the first swath of track 02 (see the red rectangle in Figure 3a). Seventy-five

IW mode SAR images, acquired between 25 April 2015 and 15 May 2018, are available for track 01. Track 02 has sixty-six IW mode SAR images, acquired between 8 April 2015 and 10 May 2018; the SAR image datasets of track 01 and track 02 are listed in Tables 1 and 2, and the ambient temperatures were acquired from the Pukou weather station, respectively. Due to the different swaths of the two tracks, the corresponding incidence angles are 45° for track 01, and 31° for track 02.

Table 1. Sentinel-1 SAR image dataset of Track 01 (the non-bold group is used for displacement modeling and the bold one for health evaluation).

No.	Date	T/°C	No.	Date	T/°C	No.	Date	T/°C	No.	Date	T/°C	No.	Date	T/°C
1	20150425	25.0	16	20160126	4.8	31	20160829	27.2	46	20170414	24.7	61	20171104	11.7
2	20150706	18.2	17	20160219	12.6	32	20161004	22.4	47	20170426	17.8	62	20171116	14.1
3	20150730	33.0	18	20160302	16.6	33	20161016	20.7	48	20170508	15.6	63	20171128	14.4
4	20150811	27.5	19	20160314	11.8	34	20161028	12.8	49	20170520	28.6	64	20171210	8.4
5	20150823	27.5	20	20160326	11.3	35	20161109	10.3	50	20170601	32.5	65	20171222	11.5
6	20150916	25.2	21	20160407	17.8	36	20161203	11.5	51	20170613	22.4	66	20180103	1.8
7	20150928	26.3	22	20160419	21.0	37	20161215	4.4	52	20170625	28.5	67	20180115	11.6
8	20151010	19.6	23	20160501	26.8	38	20161227	2.8	53	20170719	34.1	68	20180127	−1.2
9	20151022	22.6	24	20160513	18.4	39	20170108	5.7	54	20170731	33.8	69	20180220	4.2
10	20151103	15.8	25	20160525	27.7	40	20170201	3.7	55	20170812	26.2	70	20180304	16.4
11	20151115	17.1	26	20160606	24.9	41	20170213	12.6	56	20170824	32.1	71	20180328	24.6
12	20151127	4.5	27	20160630	30.7	42	20170225	11.0	57	20170905	26.7	72	20180409	23.3
13	20151209	10.3	28	20160724	36.3	43	20170309	16.1	58	20170917	27.0	73	20180421	22.5
14	20151221	7.0	29	20160805	29.7	44	20170321	13.2	59	20171011	15.1	**74**	**20180503**	**24.1**
15	20160114	3.6	30	20160817	33.0	45	20170402	19.6	60	20171023	15.1	**75**	**20180515**	**33.2**

Table 2. Sentinel-1 SAR image dataset of Track 02 (the non-bold group is used for displacement modeling and the bold one for health evaluation).

No.	Date	T/°C	No.	Date	T/°C	No.	Date	T/°C	No.	Date	T/°C	No.	Date	T/°C
1	20150408	10.8	16	20160601	20.2	31	20170304	16.4	46	20170912	26.6	61	20180311	18.6
2	20150502	17.9	17	20160719	30.3	32	20170316	11.9	47	20170924	21.4	62	20180323	18.5
3	20150701	27.8	18	20160812	33.7	33	20170328	19.5	48	20171006	19.1	63	20180404	11.9
4	20150725	28.5	19	20160929	18.3	34	20170409	11.8	49	20171018	15.9	64	20180416	16.9
5	20150818	28.4	20	20161011	18.2	35	20170421	20.4	50	20171030	12.5	65	**20180428**	**25.1**
6	20150911	25.1	21	20161023	16.2	36	20170503	21.1	51	20171111	13.9	66	**20180510**	**20.8**
7	20151005	20.4	22	20161104	16.9	37	20170515	20.2	52	20171123	11.2	67		
8	20151122	16.6	23	20161116	14.9	38	20170527	30.6	53	20171205	5.7	68		
9	20151216	4.3	24	20161128	9.5	39	20170608	28.4	54	20171217	1.9	69		
10	20160109	8.1	25	20161210	10.3	40	20170702	24.3	55	20171229	10.5	70		
11	20160202	2.5	26	20170103	11.2	41	20170714	33.8	56	20180110	2.8	71		
12	20160226	15.3	27	20170115	4.0	42	20170726	36.2	57	20180122	8.1	72		
13	20160321	15.0	28	20170127	6.5	43	20170807	32.2	58	20180203	-1.2	73		
14	20160414	24.2	29	20170208	1.0	44	20170819	30.1	59	20180215	6.6	74		
15	20160508	14.2	30	20170220	4.4	45	20170831	20.0	60	20180227	14.3	75		

4. Feasibility Study: SAR Measurement Sensitivity

This section presents a sensitivity analysis of the longitudinal deformations of the SAR-based measurements for both datasets. The aim of this study is to calculate the sensitivity parameter and evaluate the feasibility of the proposed PSI-based approach. For details on the sensitivity analysis of the line of sight (LOS) observation regarding the different sensors, see Reference [25].

Due to the Line-of-Sight (LOS) nature of the displacements measured using SAR interferometry, the structural displacement monitoring capability depends on the SAR geometry and the azimuth of the bridge's main axis. We assumed that the most important contribution of the temperature related movements was longitudinal [13]. Thus, considering only the displacements in the longitudinal direction, the relation between the LOS and the longitudinal deformation can be written as follows (see Figure 4):

$$d_L = \frac{d_{LOS}}{\sin\theta\cos(\alpha_{br} - \alpha_{rg})} \tag{5}$$

where d_{LOS} and d_L are the LOS and longitudinal deformation, respectively; θ is the incidence angle, and $\alpha = \alpha_{br} - \alpha_{rg}$ is the horizontal angle given by the difference of the SAR range azimuth, α_{rg}, and bridge longitudinal azimuth, α_{br}.

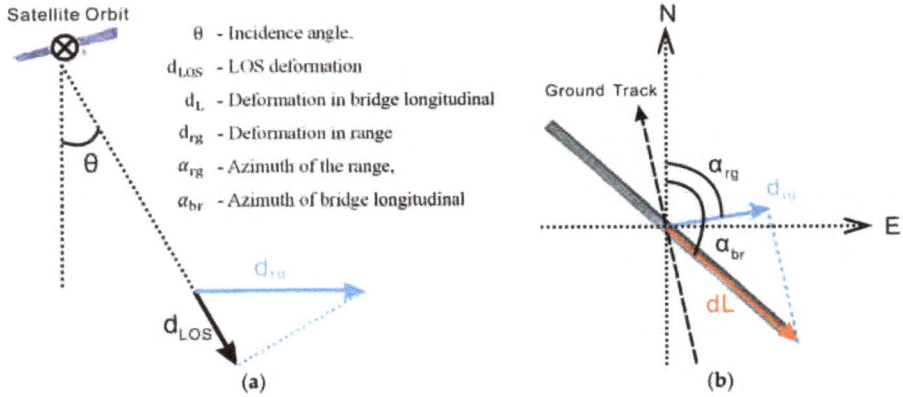

Figure 4. Relation between the line of sight (LOS) displacements and those in the bridge longitudinal direction. Scheme in the vertical plane (a) and in the horizontal one (b). It is worth noting that in this analysis it has been assumed that the bridge slope is almost zero.

Let us define $s = \sin \theta \cos \alpha$ as the sensitivity of the longitudinal displacements of bridge in the LOS. The larger the s is, the better the measurements are. Figure 5 illustrates the relationship between s, the radar incidence angle θ, and the horizontal angle α. When the SAR range direction is perpendicular to the bridge's main axis, the sensitivity goes to zero and the longitudinal displacements cannot be measured. The sensitivity of the NDHRB and the NYRB, calculated with the geometry of Track 01 and Track 02, is listed in Table 3.

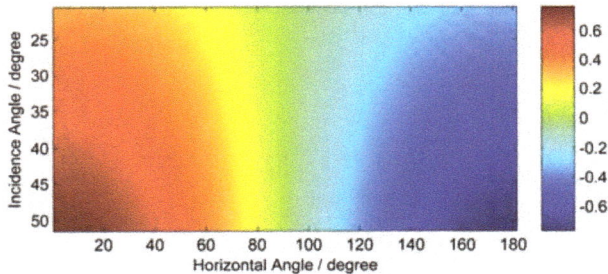

Figure 5. Relation of the sensitivity s as a function of the radar incidence angle θ and the horizontal angle α.

Table 3. Sensitivity to the longitudinal displacements of the NDHRB and the NYRB, computed with the geometry of Track 01 and Track 02.

	NDHRB		NYRB	
	Track 01	Track 02	Track 01	Track 02
θ/degree	45.0	31.0	45.0	31.0
α_{br}/degree	133.6	133.6	120.6	120.6
α_{rg}/degree	79.5	79.5	79.5	79.5
α/degree	54.1	54.1	41.1	41.1
s	0.41	0.30	0.53	0.39

5. PSI Processing

To evaluate bridge health using the method presented in Section 2, we divided each dataset into two groups (distinguished by bold and non-bold fonts in Tables 1 and 2): the first group (non-bold) was used for modeling the bridge displacement, while the second group (bold) was used for evaluating bridge health.

Software developed by CTTC was used for SAR data processing [26]. It consists of two main parts: the generation of differential interferograms; and the modeling and decomposition of phase components.

A redundant network was used for the phase component decomposition. All interferograms with temporal baselines of less than 132 days were generated (see Figure 6). A 3-arc SRTM DEM was used for topographic phase removing. In total, 585 interferograms were generated for Track 01 and 485 for Track 02. The maximum spatial baseline for Track 01 was 196 m (interferometric pair 20171011_20180103) and 263 m for Track 02 (interferometric pair 20150408_20150725). The minimum spatial baselines for the two tracks were 2 m (interferometric pair 20161227_20170213) and 1 m (interferometric pair 20170702_20170714), respectively. The SAR multi-looking was not applied to preserve the original resolution of the data.

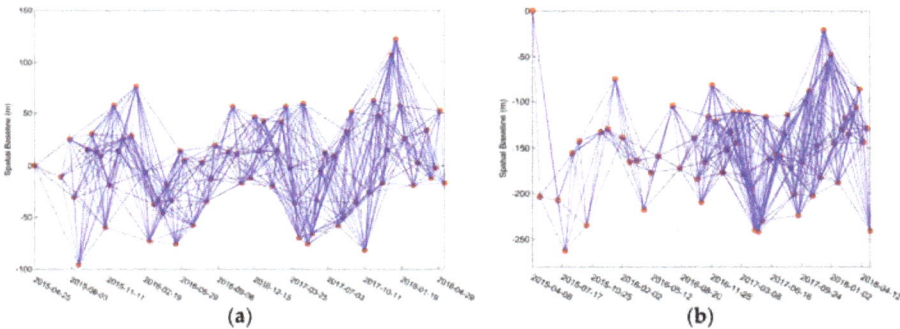

Figure 6. Spatial and temporal baselines of the Sentinel-1 datasets: Track 01 (**a**) and Track 02 (**b**).

The amplitude dispersion index (DA) was applied for pixel selection. The threshold was set to 0.2. Edges with $\gamma < 0.7$ were discarded, and a minimum number of 10 edges associated with a single point was set for the integration. Finally, the three-phase components (linear velocity, topographic error, and thermal dilation coefficient) were extracted numerically with the extended PSI model [14].

6. PSI Results

6.1. NDHRB

The topographic error is given by the difference between the Digital Terrain Model (DTM) used in the interferogram generation and the actual height of a given scatterer. Figure 7 shows the topographic error and its statistics using the two tracks: the two arcs of the bridge can be clearly identified. The distribution of the topographic error is similar in both cases: as seen below, both include a uniform distribution, which corresponds to the bridge arcs, and a normal distribution, which is related to the other part of the bridge.

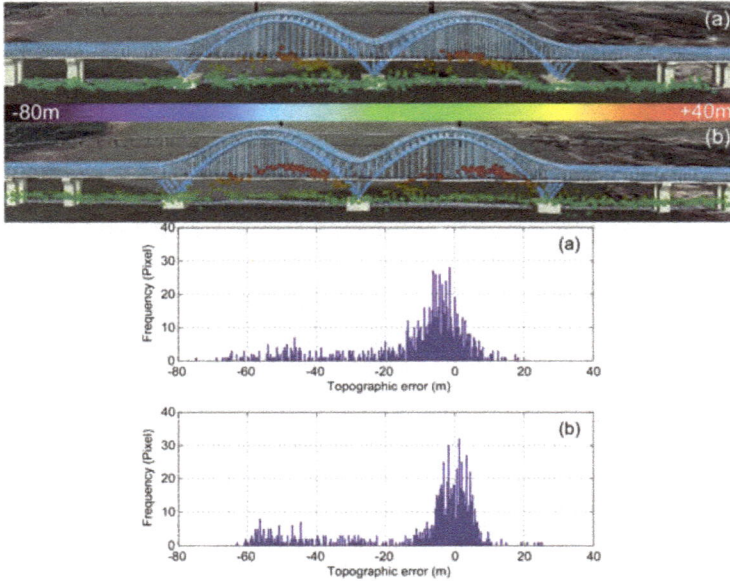

Figure 7. Estimated topographic errors and their statistics for the NDHRB. The figure marked with '(a)' is related to Track 01, and that with '(b)' refers to Track 02.

To study the displacements of the bridge deck, a mask was applied to the map of the thermal dilation parameter and linear velocity: only those points whose heights are between -10 m and 10 m were selected. This was followed by the projection of the LOS displacements into the longitudinal bridge direction. Figure 8 shows the estimated thermal dilation parameters in the LOS direction; Figure 9 presents their average cross values in the longitudinal direction for the two tracks, while Figure 10 shows the average cross values of the LOS linear velocities for the two tracks. The main results related to thermal dilation are summarized in Table 4. This is discussed in the following four sections.

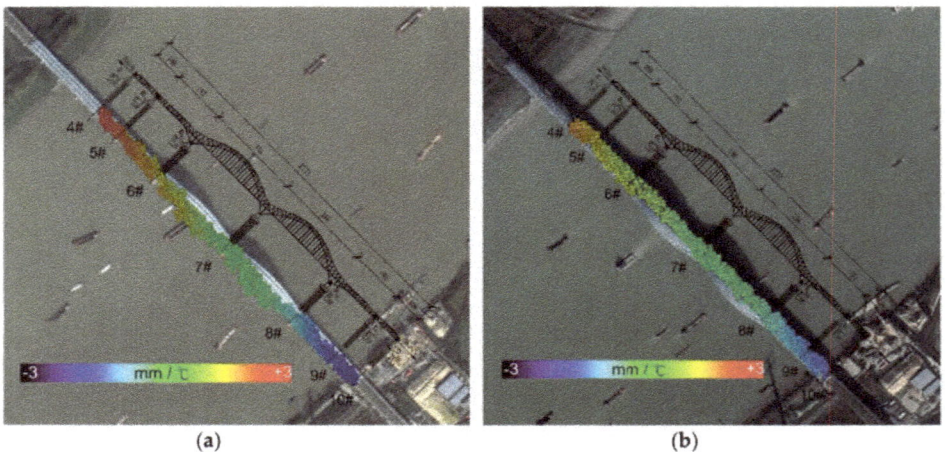

Figure 8. Estimated thermal dilation parameter in the LOS direction. The figure marked with '(a)' is related to Track 01, and that marked with '(b)' refers to Track 02.

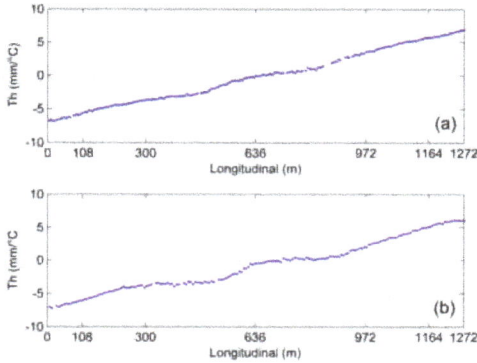

Figure 9. Average cross thermal dilation parameters in the longitudinal direction. The figure marked with '(**a**)' is related to Track 01, and that marked with '(**b**)' refers to Track 02.

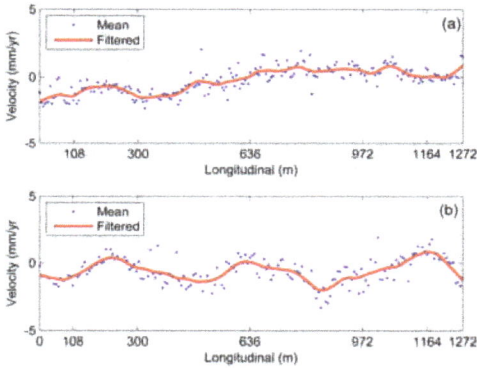

Figure 10. Average cross velocities in the longitudinal direction. The figure marked with '(**a**)' is related to Track 01, and that marked with '(**b**)' refers to Track 02.

Table 4. Thermal dilation parameters of the NDHRB.

	Track 01	Track 02
PS/pixels	903	942
$d_{LOS,Max}$ (mm/)	3.06	2.31
$d_{LOS,Min}$ (mm/°C)	−2.88	−2.00
$d_{LOS,total}$ (mm/°C)	5.94	4.31
$d_{L,total}$ (mm/°C)	14.33	14.24
CTE (/°C)	11.26×10^{-6}	11.19×10^{-6}

(1) A large number of Persistent Scatterer (PS) measurements were obtained on the deck of the NDHRB, for both Track 01 (903) and Track 02 (942). They are uniformly distributed, covering the entire bridge. Many PSs are from the steel truss girder and the bridge deck, including railway sleepers, tracks, and ballast. The change in the incidence angle of the radar has very little impact on the scattering characteristics of NDHRB.

(2) Similar thermal dilation characteristics were observed on the two tracks. Results show that the thermal dilation parameters on both sides of the bridge are almost equal but with opposite signs. The magnitude of the thermal dilation parameter increases with the distance from the bridge center, where the fixed bearing (7#) is located. The measured CTE (11.26×10^{-6} and 11.19×10^{-6} for Tracks 01 and 02, respectively) match the bridge properties mentioned in Section 1, and the results described

in Reference [13]. To validate the accuracy of the thermal dilations measured using PSI, the thermal dilations observed at the six movable bearings and the in-situ measurements [17] were compared (see Figure 11). Taking the in-situ measurements as Reference, the absolute mean error is 0.25 mm/°C.

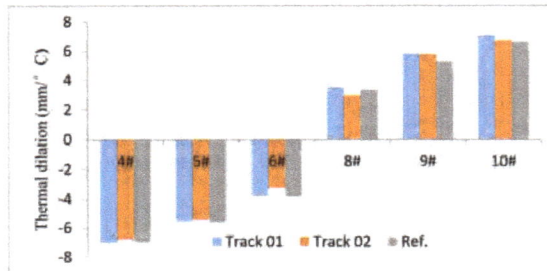

Figure 11. Thermal dilations measured by Track 01, Track 02, and the in-situsensors.

(3) The average cross values of the linear velocity calculated for the LOS displacement for both tracks are below 2 mm/year, and there is no clear correlation between the two tracks. Comparing this estimated linear deformation with the displacement caused by the thermal dilation: the relative thermal dilation parameters of the entire bridge, in the LOS, reached 5.94 mm/°C and 4.31 mm/°C (see Table 4), this means that a temperature variation of 30 °C causes at least 130 mm of LOS displacement; hence the estimated linear deformation is much smaller. The velocity values shown in Figure 10 could be due to residual non-modeled thermal dilation displacements, and can be neglected in bridge displacement modeling.

(4) Table 4 shows the thermal dilation parameters estimated using Track 01 and Track 02: the difference in the total longitudinal parameter ($d_{L,total}$) of the two tracks is 0.09 mm/°C, which corresponds to 2.7 mm with a temperature variation of 30 °C. These results depict the high sensitivity of the proposed approach.

(5) Considering the length of the NDHRB (1272 m), the CTE of the NDHRB can be estimated, corresponding to 11.26×10^{-6}/°C and 11.19×10^{-6}/°C for the two tracks: they agree well with each other (see Table 4). The differences between these values and 13.0×10^{-6}/°C, which is the CTE of Q420qE that dominates the expansion of the bridge, are 1.74×10^{-6}/°C and 1.81×10^{-6}/°C: their relative errors are 13% and 14%. The relatively smaller observed CTE values can be explained by the friction of movable bearings.

6.2. Results of NYRB

Due to the comprehensive maintenance of the NYRB, the interferometric coherence decreases dramatically for all SAR acquisitions; hence, only 45 images in Track 01 and 34 images in Track 02, from their first acquisitions, were used for PSI processing.

Figure 12 shows the LOS phase components of the NYRB estimated with Track 01 (upper) and Track 02 (lower). It is very clear that the PS density is quite different: Track 01 has far more measurements (640) than Track 02 (96). Hence, with more PS measurements covering the entire bridge, Track 01 is capable of providing valuable phase information, while Track 02 fails. The diversity of the PS density can be explained by the fact that the upper layer of the NYRB is a highway that is relatively flat for C-band radar signal, while the lower part of the bridge, constructed with metal truss, has strong backscattering. With the decrease in the radar incidence angle, the upper layer with less backscattering becomes the main scattering area, hence, the amount of PSs decreases dramatically.

Three segments of the NYRB are highlighted by the thermal dilation coefficients in Figure 12a, which corresponds with the architectural properties of the bridge (four fixed bearings mounted at 0#, 2#, 5#, and 8#, see the green arrows, and huge expansion joints mounted at 1#, 4#, 7#, and 10#, see the red arrows in Figure 12). On each segment, the thermal dilation is zero in the fixed pier, and the values increase towards each side up to the expansion joints, but with opposite signs. The thermal dilation parameters of the NYRB at the middle segment (480 m) are listed in Table 5. It can be seen that the difference in the estimated CTE and the structural property is very small ($0.80 \times 10^{-6}/°C$), corresponding to a relative error of 7.1%. The under-estimated value can be explained by the small internal stresses of the structure. The absolute values of the estimated topographic errors are all less than 10 m (see Figure 12b), which also correspond well with the flat characteristics of the NYRB. The estimated linear deformation rates shown in Figure 12c are mostly around 0 mm/year, which shows that the NYRB had no linear deformation during the monitoring period.

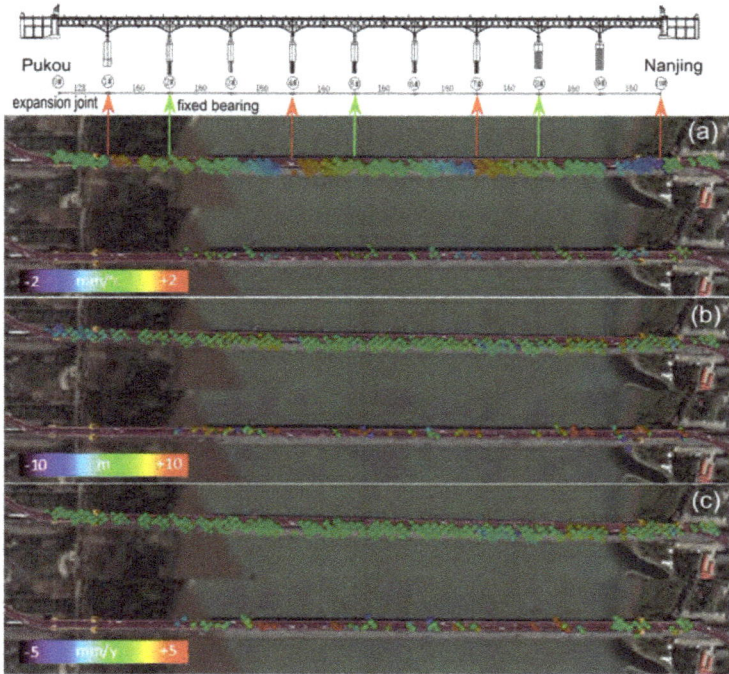

Figure 12. LOS phase components of the NYRB, estimated with Track 01 (upper) and Track 02 (lower); the reference point is marked with the red triangle, (**a**) thermal dilation coefficient, (**b**) topography error, and (**c**) linear velocity.

Table 5. Thermal dilation parameters of the NYRB at the middle segment.

	Track 01	Track 02
PS/pixels	640	96
$d_{LOS,Max}$ (mm/°C)	1.23	-
$d_{LOS,Min}$ (mm/°C)	1.43	-
$d_{LOS,total}$ (mm/°C)	2.66	-
$d_{L,total}$ (mm/°C)	5.02	-
CTE (/°C)	10.46×10^{-6}	-

7. Bridge Health Evaluation

We assumed that the bridges were in a healthy state during the SAR monitoring period in the first group. By considering the NDHRB as an example, the two datasets were used independently to evaluate the bridge health using the procedure presented in Section 2. The bridge states of the two acquisitions from each track in the second group—that is, 3 May and 15 May 2018 for Track 01, and 28 April and 22 May 2018 for Track 02—were evaluated. The longitudinal displacement measurements of the bridge obtained from unwrapped interferograms were compared with the modeled ones, and the differences were used for the hypothesis testing. Hence, the health of the bridges was evaluated on the final SAR imaging dates.

The accuracy of the structural displacement model was estimated by evaluating the difference between the modeled and observed displacement time series during the designed stable life period of the bridge. Figure 13 shows the measured displacement time series, the modeled displacement time series, and their difference for Track 01 over the NDHRB; the standard deviation of their difference is 5.9 mm, while the value for Track 02 is 6.2 mm. It should be noted that the linear term of the displacement residual in each acquisition was estimated and removed.

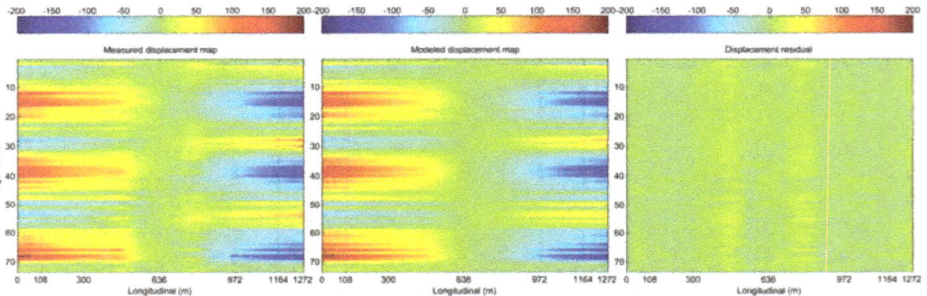

Figure 13. Accuracy evaluation of the bridge displacement model over Track 01. Measured displacement time series (**left**), modeled displacement time series (**middle**), and their difference (**right**).

Figure 14 shows the NDHRB health evaluation results using the proposed procedure in four SAR image acquisitions: 20180503, 20180515, 20180428, and 20180510. In each evaluation procedure, Figure 14 with '(a)' is the unwrapped interferogram, '(b)' is the measured average cross displacement, '(c)' is the modeled displacements, and '(d)' is the displacements difference between the measured and modeled value. The atmospheric effects are neglected assuming that the area is small enough to avoid significant contributions. Moreover, in order to ease the phase unwrapping, only the points located on the bridge deck are used. Considering two times the absolute mean error of the models (2×5.9 mm = 11.8 mm for Track 01, and 2×6.2 mm = 12.4 mm form Track 02), the up control line (UCL) and the lower control line (LCL) can be drawn at ±11.8 mm and 12.4 mm, respectively (see the red lines in '(d)' of Figure 14). It can be seen that all of the displacement differences are included in the control lines, while some large values are mainly caused by traffic along the bridge, e.g., pair 20180428_20180510. Therefore, in this case, there are no abnormal displacements of the entire bridge in the four periods observed.

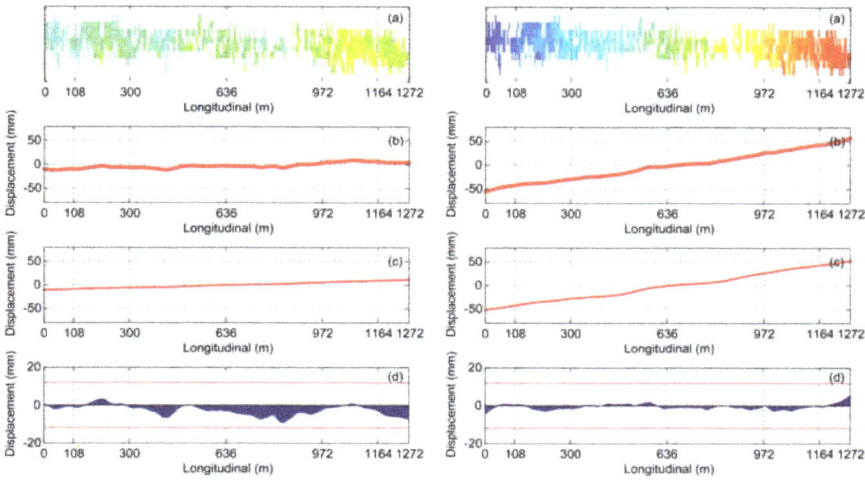

(I) Track 01. Interferometric pair 20180421_20180503 (**left**) and 20180503_20180515 (**right**).

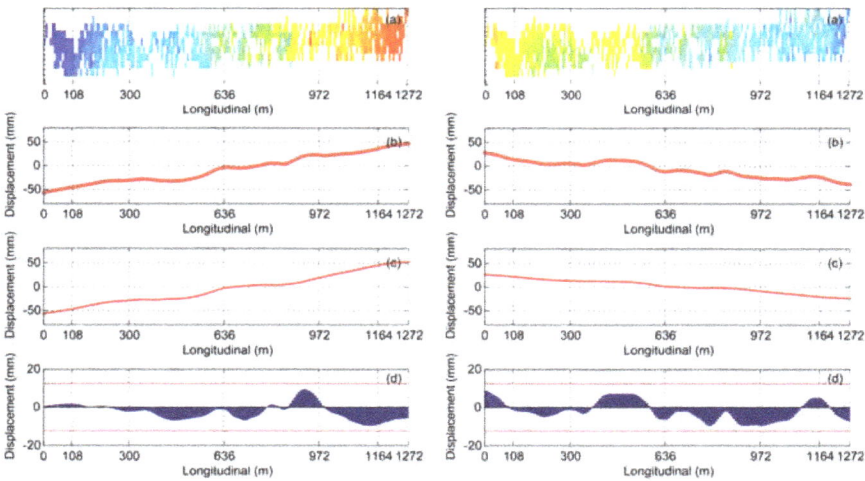

(II) Track 02. Interferometric pair 20180416_20180428 (**left**) and 20180428_20180510 (**right**).

Figure 14. NDHRB health evaluation using the proposed method for the two tracks. From top to bottom of each interferometric pair: longitudinal displacement interferometric measurements, averaged cross values, modeled value, and difference between the measurement and the modeled one.

8. Discussion

Displacement monitoring plays an important role in structural health evaluation. In this study, displacement monitoring and health evaluation of two bridges (the NDHRB and the NYRB) using the PSI technique and SAR interferometry were carried out.

We analyzed the sensitivity of Sentinel-1 space-borne SAR interferometry to measure structural displacements. The approach used can be replicated in different bridges to select the best track and frame before beginning the download of the images.

The results obtained have demonstrated the applicability of the proposed approach on two bridges with very different structural characteristics. The estimated sensitivity to anomalous displacements in the proposed health evaluation approach was around 1 cm. Such precision is good enough for a wide range of bridges. In this context, Figure 15 shows two temporal profiles of the movement of a point located at the middle of the bridge (192 m-span) and measured through an in-situ real aperture radar [27]. These time series show the vertical displacement of the point induced by a high-speed train (a) and a metro (b). It can be seen that all the induced displacements are below the centimeter.

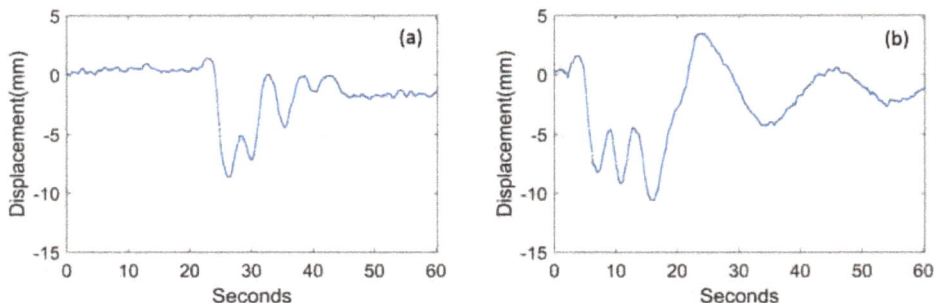

Figure 15. Vertical displacement time series induce by a high-speed train (**a**) and a metro (**b**) in the middle of the 192-m-span monitored using IBIS-S.

A total of 903 and 942 points were measured on the NDHRB using two Sentinel-1 tracks: such measurements made it possible to monitor the displacements of the entire bridge. Moreover, the use of two independent tracks in this bridge provided a cross-validation of the results obtained, which was useful for assessing the precision of the methods used. The number of PS measurements on the NYRB decreased dramatically as the radar incidence angle decreased from 45 degrees (in the third swath) to 31 degrees (in the first swath). In this case, it was only possible to obtain results with one trajectory.

Using two tracks of Sentinel-1 SAR images (75 and 76 images, respectively) was useful for assessing the results obtained from two independent datasets. However, the results obtained from both datasets could also be merged to provide higher temporal sampling for the SHM. During the period observed, the bridges underwent no actual displacements. Hence, the thermal dilation displacements were dominant. The total thermal dilation parameters of the NDHRB, estimated using the two datasets, were compared. A discrepancy of 0.09 mm/°C was found, which corresponds to a difference of 2.7 mm over the total length of the bridge, assuming a temperature variation of 30 °C. The total thermal dilation parameter was 14.28 mm/°C, which resulted in a CTE of 11.22×10^{-6}/°C: this corresponds well with material properties of the bridge. Similar results were obtained from the NYRB.

Finally, it is worth underlining that such an approach cannot be applied to all bridges due to different issues such as SAR geometry limitations, traffic along the bridge, or structural characteristics. However, depending on the location of the area of interest, the acquisition policy of Sentinel-1 constellation could help to minimize such limitations by offering the possibility of acquiring ascending and descending data, and, in some cases, of working with parallel adjacent tracks, as in the case of the two bridges analyzed. Moreover, the continuous acquisition mode providing an image every 6, 12, or 24 days, depending on the location, makes it possible to devise long-term monitoring plans.

9. Conclusions

In this paper, a procedure for continuously assessing the displacements of a bridge has been proposed. It has been successfully applied on two huge bridges—the NDHRB and the NYRB—by using medium-resolution Sentinel-1 SAR images. The results have been cross-validated by using two independent datasets obtaining millimeter-order differences. A bridge health evaluation method, based on longitudinal displacements covering the entire bridge deck, has been presented and validated

Remote Sens. **2018**, *10*, 1714

on the NDHRB. This method evaluates the abnormal displacements along the entire bridge deck, which is an advantage with respect to methods that only measure the displacements with respect to the piers. Its applicability has been illustrated over the two bridges, the NDHRB and the NYRB.

Author Contributions: Q.H. conceived, designed, and performed the experiments; Q.H., M.C. and Y.D. analyzed the data; O.M. and B.C. contributed the SAR analysis tools; Y.W. and J.J. contributed part of the data analysis; Q.H. and M.C. wrote the paper.

Funding: This study was supported by the Fundamental Research Funds for the Central Universities (2018B18814, 2018B699X14) and the Postgraduate Research & Practice Innovation Program of Jiangsu Province (KYCX18_0619). The CTTC activities have been partially funded by the Spanish Ministry of Economy and Competitiveness through the DEMOS project "Deformation monitoring using Sentinel-1 data" (Ref: CGL2017-83704-P).

Acknowledgments: We would like to thank Nuria Devanthery, Maria Cuevas-González, and Anna Barra, from the CTTC, for their support in the PSI data processing. The Sentinel-1A data were downloaded from the Sentinel-1 Scientific Data Hub.

Conflicts of Interest: The authors declare no conflict of interest.

References

1. Crosetto, M.; Monserrat, O.; Cuevas-Gonzalez, M.; Devanthery, N.; Crippa, B. Persistent Scatterer Interferometry: A review. *ISPRS J. Photogramm. Remote Sens.* **2016**, *115*, 78–89. [CrossRef]

2. Zhou, W.; Li, S.; Zhou, Z.; Chang, X. Remote Sensing of Deformation of a High Concrete-Faced Rockfill Dam Using InSAR: A Study of the Shuibuya Dam, China. *Remote Sens.* **2016**, *8*, 255. [CrossRef]

3. Zhou, W.; Li, S.; Zhou, Z.; Chang, X. InSAR Observation and Numerical Modeling of the Earth-Dam Displacement of Shuibuya Dam (China). *Remote Sens.* **2016**, *8*, 877. [CrossRef]

4. Ge, D.; Zhang, L.; Li, M.; Liu, B.; Wang, Y. Beijing subway tunnelings and high-speed railway subsidence monitoring with PSInSAR and TerraSAR-X data. In Proceedings of the 2016 IEEE International Geoscience and Remote Sensing Symposium (IGARSS), Beijing, China, 10–15 July 2016; pp. 6883–6886. [CrossRef]

5. Bakon, M.; Perissin, D.; Lazecky, M.; Papco, J. Infrastructure Non-linear Deformation Monitoring Via Satellite Radar Interferometry. *Procedia Technol.* **2014**, *16*, 294–300. [CrossRef]

6. Poreh, D.; Iodice, A.; Riccio, D.; Ruello, G. Railways' stability observed in Campania (Italy) by InSAR data. *Eur. J. Remote Sens.* **2016**, *49*, 417–431. [CrossRef]

7. Wang, C.; Zhang, Z.; Zhang, H.; Wu, Q.; Zhang, B.; Tang, Y. Seasonal deformation features on Qinghai-Tibet railway observed using time-series InSAR technique with high-resolution TerraSAR-X images. *Remote Sens. Lett.* **2017**, *8*, 1–10. [CrossRef]

8. Shamshiri, R.; Motagh, M.; Baes, M.; Sharifi, M.A. Deformation analysis of the Lake Urmia causeway (LUC) embankments in northwest Iran: Insights from multi-sensor interferometry synthetic aperture radar (InSAR) data and finite element modeling (FEM). *J. Geod.* **2014**, *88*, 1171–1185. [CrossRef]

9. Chang, L.; Dollevoet, R.P.B.J.; Hanssen, R.F. Nationwide Railway Monitoring Using Satellite SAR Interferometry. *IEEE J. STARS* **2017**, *10*, 596–604. [CrossRef]

10. Lazecky, M.; Hlavacova, I.; Bakon, M.; Sousa, J.J.; Perissin, D.; Patricio, G. Bridge Displacements Monitoring Using Space-Borne X-Band SAR Interferometry. *IEEE J. STARS* **2017**, *10*, 205–210. [CrossRef]

11. Zhao, J.; Wu, J.; Ding, X.; Wang, M. Elevation Extraction and Deformation Monitoring by Multitemporal InSAR of Lupu Bridge in Shanghai. *Remote Sens.* **2017**, *9*, 897. [CrossRef]

12. Wang, H.; Chang, L.; Markine, V. Structural Health Monitoring of Railway Transition Zones Using Satellite Radar Data. *Sensors* **2018**, *18*, 413. [CrossRef] [PubMed]

13. Huang, Q.; Crosetto, M.; Monserrat, O.; Crippa, B. Displacement monitoring and modelling of a high-speed railway bridge using C-band Sentinel-1 data. *ISPRS J. Photogramm. Remote Sens.* **2017**, *128*, 204–211. [CrossRef]

14. Monserrat, O.; Crosetto, M.; Cuevas, M.; Crippa, B. The Thermal Expansion Component of Persistent Scatterer Interferometry Observations. *IEEE Geosci. Remote Sens.* **2011**, *8*, 864–868. [CrossRef]

15. Li, S.; Li, H.; Liu, Y.; Lan, C.; Zhou, W.; Ou, J. SMC structural health monitoring benchmark problem using monitored data from an actual cable-stayed bridge. *Struct. Control Health Monit.* **2014**, *21*, 156–172. [CrossRef]

16. Watanabe, E.; Furuta, H.; Yamaguchi, T.; Kano, M. On longevity and monitoring technologies of bridges: A survey study by the Japanese Society of Steel Construction. *Struct. Infrastruct. E* **2014**, *10*, 471–491. [CrossRef]

17. Ding, Y.; Wang, G.; Sun, P.; Wu, L.; Yue, Q. Long-Term Structural Health Monitoring System for a High-Speed Railway Bridge Structure. *Sci. World J.* **2015**, *2015*, 250562. [CrossRef] [PubMed]

18. Huang, Q.; Crosetto, M.; Monserrat, O.; Crippa, B. Monitoring and evaluation of a long-span raiway bridge using Sentinel-1 data. *ISPRS Ann. Photogramm. Remote Sens. Spat. Inf. Sci.* **2017**, 457–463. [CrossRef]

19. Ferretti, A.; Prati, C.; Rocca, F. Permanent scatterers in SAR interferometry. *IEEE Trans. Geosci. Remote Sens.* **2001**, *39*, 8–20. [CrossRef]

20. Cigna, F.; Tapete, D.; Casagli, N. Semi-automated extraction of Deviation Indexes (DI) from satellite Persistent Scatterers time series: Tests on sedimentary volcanism and tectonically-induced motions. *Nonlinear Process. Geophys.* **2012**, *19*, 643–655. [CrossRef]

21. He, X.; Chen, Z.; Yu, Z.; Huang, F. Fatigue damage reliability analysis for Nanjing Yangtze river bridge using structural health monitoring data. *J. Cent. South Univ. Technol.* **2006**, *13*, 200–203. [CrossRef]

22. Zhu, G. Big expansion joints reconstruction for highway bridge of the Nanjing Yangtze river bridge. *Mod. Transp. Technol.* **2012**, *9*, 49–52.

23. He, X. Study on the Structural Health Monitoring of Nanjing Yangtze River Bridge and Its Key Technologies. Ph.D. Thesis, Central South University, Changsha Hunan, China, 2004.

24. Ding, Y.; Zhao, H.; Li, A. Temperature Effects on Strain Influence Lines and Dynamic Load Factors in a Steel-Truss Arch Railway Bridge Using Adaptive FIR Filtering. *J. Perform. Constr. Facil.* **2017**, *31*, 4017024. [CrossRef]

25. Chang, L.; Dollevoet, R.P.B.J.; Hanssen, R.F. Monitoring Line-Infrastructure with Multisensor SAR Interferometry: Products and Performance Assessment Metrics. *IEEE J. STARS* **2018**, *11*, 1593–1605. [CrossRef]

26. Crosetto, M.; Monserrat, O.; Cuevas, M.; Crippa, B. Spaceborne differential SAR interferometry: Data analysis tools for deformation measurement. *Remote Sens.* **2011**, *3*, 305–318. [CrossRef]

27. Luzi, G.; Monserrat, O.; Crosetto, M. The potential of coherent radar to support the monitoring of the health state of buildings. *Res. Nondestruct. Eval.* **2012**, *23*, 125–145. [CrossRef]

remote sensing

MDPI

Article

Multi-Sensor InSAR Analysis of Progressive Land Subsidence over the Coastal City of Urayasu, Japan

Yusupujiang Aimaiti *, Fumio Yamazaki and Wen Liu

Department of Urban Environment Systems, Chiba University, Chiba 263-8522, Japan;
fumio.yamazaki@faculty.chiba-u.jp (F.Y.); wen.liu@chiba-u.jp (W.L.)
* Correspondence: tuprak100@gmail.com; Tel.: +81-043-290-3528

Received: 7 July 2018; Accepted: 16 August 2018; Published: 18 August 2018

Abstract: In earthquake-prone areas, identifying patterns of ground deformation is important before they become latent risk factors. As one of the severely damaged areas due to the 2011 Tohoku earthquake in Japan, Urayasu City in Chiba Prefecture has been suffering from land subsidence as a part of its land was built by a massive land-fill project. To investigate the long-term land deformation patterns in Urayasu City, three sets of synthetic aperture radar (SAR) data acquired during 1993–2006 from European Remote Sensing satellites (ERS-1/-2 (C-band)), during 2006–2010 from the Phased Array L-band Synthetic Aperture Radar onboard the Advanced Land Observation Satellite (ALOS PALSAR (L-band)) and from 2014–2017 from the ALOS-2 PALSAR-2 (L-band) were processed by using multitemporal interferometric SAR (InSAR) techniques. Leveling survey data were also used to verify the accuracy of the InSAR-derived results. The results from the ERS-1/-2, ALOS PALSAR and ALOS-2 PALSAR-2 data processing showed continuing subsidence in several reclaimed areas of Urayasu City due to the integrated effects of numerous natural and anthropogenic processes. The maximum subsidence rate of the period from 1993 to 2006 was approximately 27 mm/year, while the periods from 2006 to 2010 and from 2014 to 2017 were approximately 30 and 18 mm/year, respectively. The quantitative validation results of the InSAR-derived deformation trend during the three observation periods are consistent with the leveling survey data measured from 1993 to 2017. Our results further demonstrate the advantages of InSAR measurements as an alternative to ground-based measurements for land subsidence monitoring in coastal reclaimed areas.

Keywords: ERS-1/-2; PALSAR; PALSAR-2; InSAR; land subsidence; reclaimed land; Urayasu City

1. Introduction

Land subsidence is one of the most serious environmental problems in many urban areas around the world [1]. In particular, coastal areas, which contain young and compressible deposits, are often vulnerable to subsidence caused by either anthropogenic or natural factors [2,3]. This phenomenon is evident in the coastal city of New Orleans, LA in the USA [4], Jakarta in Indonesia [5,6], Ho Chi Minh in Vietnam [7], Bangkok in Thailand [8], Shanghai and Shenzhen in China [9,10], Venice in Italy [11] and in the western Netherlands [12]. Continuous land subsidence causes remarkable economic losses in the form of building damages leading to high maintenance costs [13]. Thus, identifying land deformation trends is a crucial task to maintain the sustainability of coastal urban areas [14].

Over the past two decades, land subsidence monitoring has been significantly improved by the use of interferometric synthetic aperture radar (InSAR) techniques [15]. Although the traditional methods (i.e., global positioning system (GPS) and leveling) can also provide precise measurements, they cannot acquire dense ground displacement measurements with a large-scale coverage in a short time and at a low cost [16]. The advanced time-series InSAR techniques, such as persistent scatterers interferometry (PSI) and the small baseline subset (SBAS) technique, can achieve results in better spatial and temporal resolutions with higher precision [17–20]. Furthermore, the increase in the available synthetic aperture radar (SAR) satellites with different temporal and spatial resolutions has provided a great opportunity for researchers to perform long-term geohazard monitoring by combining observations from those satellites [21].

Urayasu City is located in the Tokyo Bay area, where more than 70% of the area was reclaimed from 1964 to 1980 [22,23]. The reclamation was performed using the sand and soil dredged from the seabed off the coast of Urayasu [24]. In addition, Urayasu City is located in an earthquake-prone area, which increases the risk of land subsidence due to the combined effects of seismicity and the natural consolidation of soil [25,26]. On 11 March 2011, a devastating earthquake of moment magnitude M_w 9.0 occurred off the coast of Tohoku, Japan, which caused severe damage to buildings and infrastructures and created large ground settlements of up to 60 cm in the reclaimed areas [27,28]. This catastrophic event has attracted a great deal of attention from researchers and organizations. The Geospatial Information Authority of Japan (GSI) carried out a leveling survey, comparing the results with light detection and ranging (LiDAR) survey data, and concluded that the surface subsidence was not caused only by the soil liquefaction but also by pro-earthquake consolidation [29]. Konagai et al. [30] mapped the soil subsidence using LiDAR data taken before and after the earthquake. Pasquali et al. [31] measured the land subsidence during 2006–2010 using both the Environment Satellite Advanced Sythetic Aperture Radar (ENVISAT ASAR) and the Advanced Land Observation Satellite Phased Array L-band Synthetic Aperture Radar ALOS PALSAR data. ElGharbawi and Tamura [32] estimated the liquefaction induced deformation using ALOS PALSAR images spanning from August 2006 to April 2011. Nigorikawa and Asaka [24] conducted a leveling survey from April 2011 to April 2013 and found accelerated land settlement only in the reclaimed land areas rather than in the natural alluvial low land and pointed out the settlement may still be ongoing. However, the previous studies mainly focused on the soil liquefaction-induced subsidence during the earthquake, and the long-term spatiotemporal evolution of land subsidence before and after the earthquake has not yet been clearly identified.

In this study, we used three different SAR datasets, the European Remote Sensing satellites (ERS-1/-2) and ALOS PALSAR & ALOS-2 PALSAR-2 to identify the trends of land subsidence dynamics in Urayasu City over a period of 24 years by using multitemporal InSAR techniques. Moreover, the InSAR results were compared with leveling survey data. The observed results may provide useful information for identifying and understanding the behavior of the slow subsidence phenomenon over a long-time period, which plays an important role in future risk mitigation strategies.

2. Study Area

Urayasu City is located in the Tokyo Bay area of Chiba Prefecture, from 139°56′22″E to 139°52′20″E and from 35°37′N to 35°40′23″N. The total area is 16.98 km², and the total population was 167,950 in February 2018 [33]. As shown in Figure 1, Urayasu City is divided into three areas, namely, Moto-Machi (old town), Naka-Machi (central town) and Shin-Machi (new town). Moto-Machi is a naturally formed Holocene lowland, and the other two areas were reclaimed from 1964 to 1980 [34,35]. Figure 1b and Table 1 shows the distribution and other detailed information of those reclaimed areas. The elevation in the old coastline area of Moto-Machi is approximately 0 to 2 m and gradually increases towards the coastal levee, becoming especially high in Sogo Park of Akemi district and the Tokyo Disney resort area (Figure 1c). The thickness of the alluvial soil layers varies from 20 m in the Moto-Machi area to 60–80 m in the Naka-Machi and Shin-Machi areas, which indicates the complexity of the soft soil distribution in those areas [22,23].

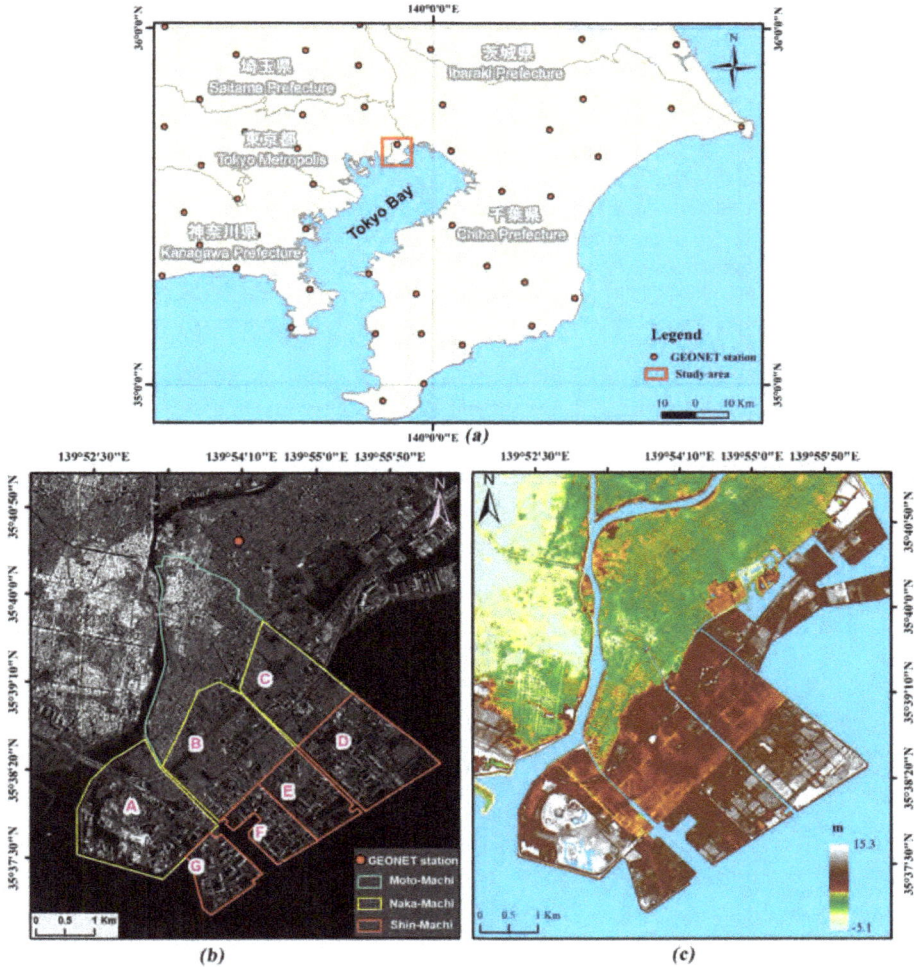

Figure 1. The map of the study area, Urayasu City, Japan. (**a**) The geographic location of Urayasu City; (**b**) the distribution and development history of the reclaimed areas, namely Moto-Machi (old town) outlined in green, Naka-Machi (central town) outlined in yellow and Shin-Machi (new town) outlined in red. A to G represent the reclaimed areas at different times. The background image is a Phased Array L-band Synthetic Aperture Radar onboard the Advanced Land Observation Satellite (ALOS-2 PALSAR-2) intensity image acquired on 4 December 2014; and (**c**) the topography of the study area [36].

Table 1. The detailed history of reclaimed areas and the districts.

Reclaimed Areas	Reclaimed Year	Districts
A	1975	Maihama
B	1968	Higashino, Tomioka, Imagawa, Benten and Tekkodori
C	1971	Kairaku, Mihama and Irifune
D	1978	Akemi and Hinode
E	1980	Takasu
F	1979	Minato
G	1979 and 1981	Chidori

3. Data Sets and Methods

3.1. Data Sets

In this study, the SAR data collected by the ERS-1/-2 and ALOS-1/-2 satellites were used to monitor the long-term deformation pattern of Urayasu City. The ERS-1/-2 data were provided by the European Space Agency (ESA) and the PALSAR & PALSAR-2 data by the Japan Aerospace Exploration Agency (JAXA). A total of 52 C-band ERS-1/-2 single look complex (SLC) scenes were acquired from the track/frame 489/2889 during the period from May 1993 to February 2006. Note that there is a data gap in 1994 and 1995 due to the limited acquisitions of ERS-1 data; 24 L-band ALOS PALSAR SLC data were acquired from the path/frame 58/2900 during the period from June 2006 to December 2010; 13 L-band ALOS-2 PALSAR-2 SLC data were acquired from the path/frame 18/2900 during the period from December 2014 to November 2017. The detailed acquisition parameters of these three SAR data are given in Table 2.

A 5-m high-resolution digital elevation model (DEM) provided by the GSI was used as a reference to remove the topographic phase in the multitemporal InSAR processing [36]. The daily GPS data observed by the GPS earth observation network system was used as reference point, and the leveling survey measurement data was used to validate the InSAR derived deformation. The daily GPS data was provided by the GSI of Japan. The leveling survey measurements have been conducted by the Chiba Prefecture on an annual basis, and the results are publicly available at their official website [37]. The archived leveling survey data before 2008 was obtained from the Chiba Prefectural Archives.

Table 2. Acquisition parameters of the ERS-1/-2, ALOS PALSAR and ALOS-2 PALSAR-2 data sets.

SAR [1] Sensor	ERS-1/-2 [2]	ALOS PALSAR	ALOS-2 PALSAR-2
Orbit direction	Descending	Descending	Descending
Operation mode	SAR/IM [3]	FBS/FBD [4]	Strip map (SM)1
Band (wavelength)	C-band (5.6 cm)	L-band (23 cm)	L-band (23 cm)
Resolution	20 m	10/20 m	3 m
Revisit cycle	35 days	46 days	14 days
Look angle	23°	34.3°	35.4°
Incidence angle	23.3°	38.7°	39.7°
Swath	100 km	70 km	50 km
Number of images	52	24	13
Temporal coverage	May 1993 to February 2006	June 2006 to December 2010	December 2014 to November 2017

[1] Synthetic aperture radar; [2] European Remote Sensing satellites; [3] IM: image mode; [4] FBS: fine beam single; FBD: fine beam double.

3.2. Methodology

The multitemporal InSAR methodologies involve the use of multiple SAR datasets to overcome the limitations of conventional InSAR (e.g., spatial and temporal decorrelations and atmospheric disturbance) and measure the land surface displacements with high precision [38–40]. In this study, the PSI and the SBAS were applied to the archived (i.e., ERS-1/-2 and ALOS PALSAR) and recent (i.e., ALOS-2 PALSAR-2) SAR data. The PSI method utilizes a time-series of radar images to identify high coherent points, the so-called persistent scatterers (PS) [17–19]; the SBAS method uses distributed scatterers from all available SAR images with corresponding small baselines in order to reduce the spatial decorrelation and obtain the time-series displacements [20]. The reason for using both techniques relies on the fact that the PSI applicability is limited to temporally uniform rates of displacement, while the SBAS has the ability to capture strong nonlinearities in the study area [41]. The PSI has a high sensitivity to slow displacements but suffers severe limitations in the capability to measure "fast" deformation phenomena, and the PS density is usually low in vegetated, forested and low-reflectivity areas (e.g. very smooth surfaces) [42], while the SBAS performs better in nonurban vegetated areas, and also in areas with high deformation rates [43,44].

As shown in Figure 2, the ERS-1/-2 and PALSAR data were processed using both the PSI and SBAS methods. Due to the limited number of PALSAR-2 acquisitions, we used only the SBAS method.

Remote Sens. **2018**, *10*, 1304

The SARscape®Modules (5.4) for ENVI (5.4) software suite (HARRIS Geospatial Solutions, Broomfield, CO, USA) was employed to perform the interferometric analyses. For the ERS-1/-2 data, we used the latest precise orbit products provided by the ESA to correct the orbit inaccuracies [45] and generate a total of 424 interferograms, including 36 for PSI processing and 388 for SBAS processing pairs (Figure 2a,b). The PSI pairs were generated with respect to the master image from 24 January 2000. The normal baselines range from 22 m to 557 m. A custom atmospheric filtering was performed with a low pass spatial filter with a 1.2 km × 1.2 km window and a temporal high pass filter at 365 days. The mean coherence threshold of 0.56 was used to identify the PS candidates. To obtain more accurate displacement measurements, the GPS base station was used as a reference point in the geocoding process (Figure 1b). The SBAS pairs were generated with respect to the multi-master images and by setting spatial and temporal threshold criteria. The threshold criteria for the absolute mean of the normal baselines was 210 m and that for the absolute mean of the temporal baselines was 937 days. Moreover, the image acquired on 2 August 1999 formed the largest number of interferometric pairs, when used as a master scene. For that reason, it was chosen as a reference (i.e., super master image). Therefore, all the slave scenes are co-registered to this reference geometry (Figure 2b). To increase the signal-to-noise ratio of the interferograms, a multi-looking factor of one in range and five in azimuth was used, producing a ground resolution of about 20 m. The topographic phases in both the PSI and SBAS interferograms were removed using the 5-m DEM data. After that, we visually checked the intermediate products (i.e., flattened and filtered (wrapped) interferograms and the unwrapped phases) to detect possible errors, which were caused by strong orbit inaccuracy, non-coherent pairs, atmospheric artefacts, residual topography etc., and 23 interferometric pairs were discarded from further processing. For refinement and re-flattening, we selected 45 reference points where the unwrapped phase value was close to zero and the flat areas were identified from the unwrapped interferograms and the topographic map. The linear inversion model was used to estimate the residual height and the displacement velocity for both the PSI and SBAS processing [20].

For the PALSAR data, we used both fine beam single (FBS) polarization and fine beam double (FBD) polarization images, with an HH polarization mode, and generated a total of 150 interferograms, including 21 for PSI processing and 129 for SBAS processing pairs (Figure 2c,d). The PSI pairs were generated with respect to the master image of 5 August 2009. The normal baselines range from 237 m to 3084 m. The same atmospheric filter which was used for the ERS-1/-2 PSI processing was also used to remove the atmospheric phase components. The mean coherence threshold of 0.75 was used to identify the PS candidates. The same GPS base station used for the ERS-1/-2 PSI processing was used as a reference point in the geocoding process. The SBAS pairs were generated with respect to the multi-master images and by setting spatial and temporal threshold criteria. The threshold criteria for the absolute mean of the normal baselines was 1084 m and that for the absolute mean of the temporal baselines was 453 days. The image acquired on 20 March 2009 was chosen as a super master image (Figure 2d). A multi-looking factor of one in range and five in azimuth was used, producing a ground resolution of about 15 m. The topographic phase in both the PSI and SBAS interferograms was removed using the same DEM used for the ERS data. Four interferometric pairs were removed due to the unwrapping errors. The same reference points used in ERS-1/-2 SBAS processing were also used for the refinement and re-flattening. The same linear inversion model was used for both the PSI and SBAS processing.

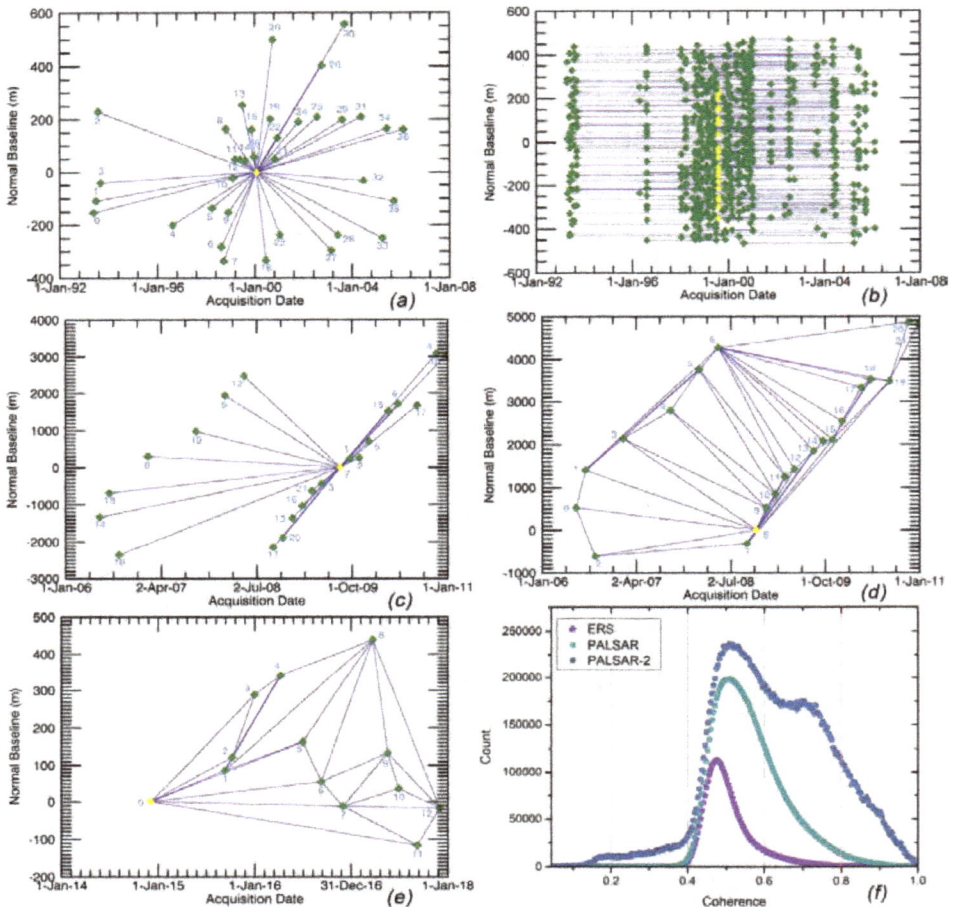

Figure 2. The temporal and spatial baseline distributions of the SAR interferograms from the ERS-1/-2, ALOS PALSAR and ALOS-2 PALSAR-2 data sets (a–e), where each acquisition is represented by a diamond associated to an ID number; the green diamonds represent the valid acquisitions and the yellow diamonds represent the selected master image of persistent scatterers interferometry (PSI) and super master image of the small baseline subset (SBAS). (**a**) Time–position plot of PSI interferograms generated by the ERS-1/-2 data, with 24 January 2000 as the master image; (**b**) time–baseline plot of SBAS interferograms generated by the ERS-1/-2 data, with 2 August 1999 as the super master image; (**c**) time–position plot of PSI interferograms generated by the ALOS PALSAR data, with 5 August 2009 as the master image; (**d**) time–position Delaunay 3D plot of SBAS interferograms generated by the ALOS PALSAR data, with March 20, 2009 as the super master image; (**e**) time–position Delaunay 3D plot of SBAS interferograms generated by the ALOS-2 PALSAR-2 data, with 4 December 2014 as the super master image; and (**f**) the histogram of the average coherence for the three satellite datasets. These connections in (**d,e**) are a subset of the whole main network and represent such interferograms that will be unwrapped in a 3D way.

The histogram of the average coherence for the PALSAR-2 data shows the relatively good coherence of PALSAR-2 when compared with the ERS-1/-2 and PALSAR data (Figure 2f). For the

PALSAR-2 data, we generated 78 interferograms for SBAS processing (Figure 2e), with respect to the multi-master images. The threshold criteria for the absolute mean of the normal baselines was 182 m and that for the absolute mean of the temporal baselines was 386 days. The image acquired on 4 December 2014 was chosen as a super master image. A multi-looking factor of six in range and seven in azimuth was used, producing a ground resolution of about 15 m. The topographic phase was removed using the same DEM used for the PALSAR data processing. To smooth the differential phase, the Goldstein filter was applied [46]. The minimum cost flow (MCF) network and Delaunay 3D method were employed to unwrap the differential interferograms [47,48] with an unwrapping coherence threshold of 0.35. The same reference points used in PALSAR SBAS processing was also used for the refinement and re-flattening. The linear inversion model was used in the processing. All the final displacement measurements were measured in the satellite line of sight (LOS) direction and were geocoded in the WGS84 reference ellipsoid with a 25-m ground resolution.

4. Results

4.1. Time-Series Analysis of the ERS-1/-2 Data from May 1993 to February 2006

The mean velocity (mm/year) maps of the final geocoded displacements generated from the ERS-1/-2 data are shown in Figure 3a (for PSI) and Figure 3b (for SBAS). The color cycle from green to purple indicates the positive to negative velocities in the LOS direction. The negative values indicate that the surface is moving away from the satellite (i.e., subsidence) while the positive values indicate the opposite direction of movement (i.e., uplift). As shown in Figure 3a,b, the major subsidence areas were highlighted by both InSAR measurements, which were located on the borders of the Naka-Machi and Shin-Machi areas. The results derived from the SBAS method show higher densities of the obtained points than those of the PSI. In the study area of over 860,256 pixels, 54,458 measurement points were obtained by the PSI method, and 89,251 points by the SBAS method. The presence of vegetation in Urayasu City—namely the palm trees in the streets and parks—might cause this difference. The histograms of the estimated displacement velocities by the PSI and SBAS for the study area are shown in Figure 4a,b, respectively. The average displacement rate and the standard deviation for the PSI were −1.0 and 4.9 mm/year, while those for the SBAS were −0.95 and 1.9 mm/year, respectively. In general, the ERS-1/-2 results show that approximately 85% of the PS points indicate displacement rates between −4 mm/year and 2 mm/year (Figure 4).

Figure 5 shows the measured displacement histories for eight representative points, which are shown in Figure 3. For both the PSI and SBAS measurements, the patterns of subsidence for each point show similar characteristics, such as an increase in subsidence rates. However, point P1 located in Moto-Machi shows very low subsidence rates (−0.1 and −0.9 mm/year for PSI and SBAS, respectively) compared to those in other areas. This suggests that the Moto-Machi area had relatively stable ground conditions during the ERS-1/-2 monitoring period. It is worth mentioning that the PSI's estimated displacement velocity is almost two times more than the SBAS results; this may be caused by the different reference points selected in the two methods. We also calculated the correlation coefficient between PSI and SBAS results over those selected points using the Pearson correlation coefficient [49,50]. Most of those points showed relatively good correlation, while the P1 and P3 showed low correlation. However, the points P1 (−0.1 mm/year vs. −0.8 mm/year) and P3 (−0.7 mm/year vs. −2.3 mm/year) both show a small displacement velocity. To provide a quantitative comparison of the estimated time series for those selected points, we calculated the velocity difference between the two methods. The smallest velocity difference was 0.7 mm/year (P1), while the largest velocity difference was 12.6 mm/year (P7). The average velocity difference for all points between the two methods was 4.6 mm/year.

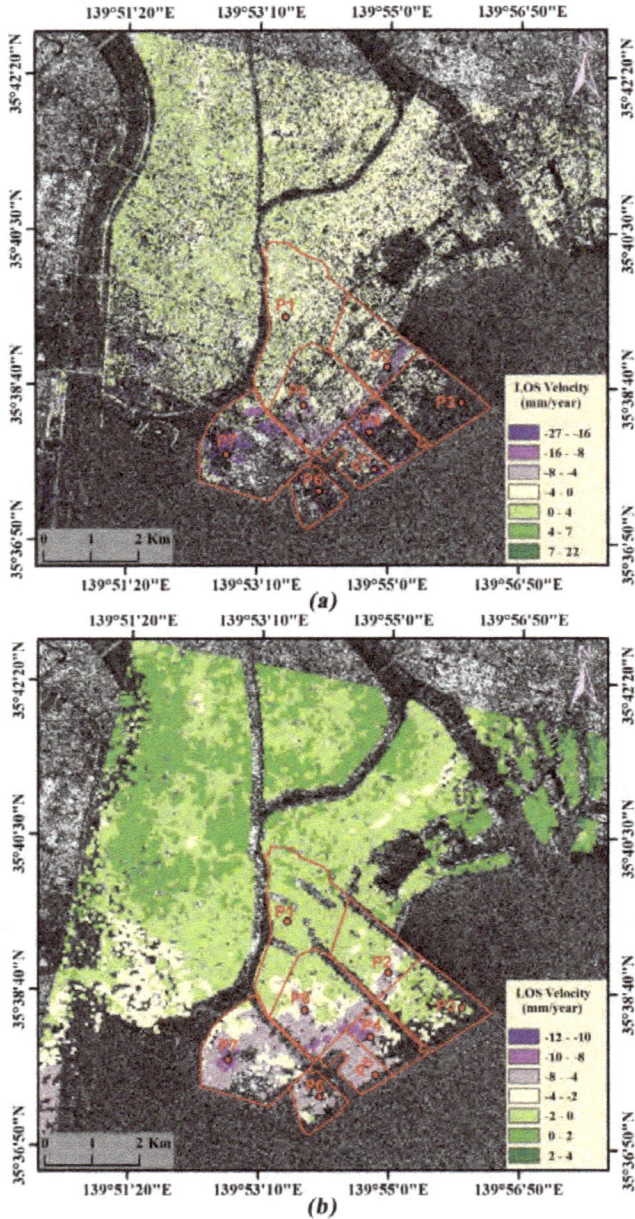

Figure 3. Line of sight (LOS) displacement velocity in Urayasu City from 1993 to 2006 for the ERS-1/-2 data: (**a**) Estimated mean displacement velocity using the PSI method; (**b**) estimated mean displacement velocity using the SBAS method. The background image is an ERS-2 intensity image acquired on 24 May 1999. The red points P1 to P8 are the selected points to show the time-series LOS displacements estimated by the PSI and SBAS measurements in (**a**,**b**), respectively.

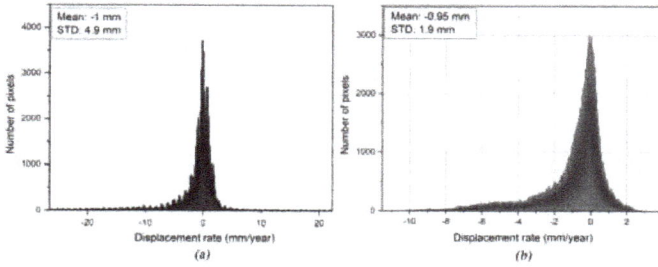

Figure 4. Histogram distribution for the ERS-1/-2-derived displacement rates from May 1993 to February 2006: (**a**) the corresponding histogram of the PSI measurements from the ERS-1/-2 data; and (**b**) the corresponding histogram of the SBAS measurements from the ERS-1/-2 data.

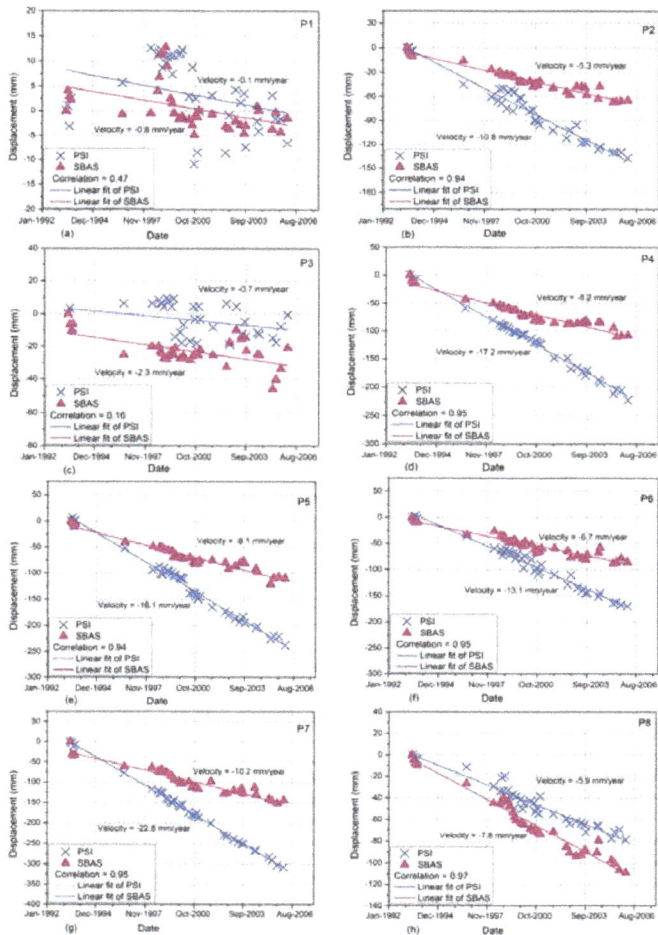

Figure 5. Time-series LOS displacement plots of the PSI and SBAS measurements from the ERS-1/-2 data (**a–h**) for the selected points P1 to P8, which are indicated by red points in Figure 3.

4.2. Time-Series Analysis of the PALSAR Data from June 2006 to December 2010

The mean velocity (mm/year) maps of the displacements for the period from June 2006 to December 2010 is shown in Figure 6a for PSI and Figure 6b for SBAS. The same color cycle from green to purple was used for those results. As shown in Figure 6a,b, the density of the measured points by the PSI is coarser than those by the SBAS, due to the existence of vegetation in the study area. In the study area of over 695,387 pixels, 50,441 measurement points were obtained by the PSI method, and 78,044 points by the SBAS method. The histograms of the estimated displacement velocity by the PSI and SBAS for the study area are shown in Figure 7a,b, respectively. The average displacement rate and the standard deviation for the PSI were -1.3 and 3.9 mm/year, whereas those for the SBAS were -1.7 and 3.3 mm/year, respectively. Overall, the PALSAR results show that approximately 85% of the PS points indicate displacement rates between -6 mm/year and 3 mm/year (Figure 7).

During the PALSAR monitoring period, most of the previously detected subsidence areas were also detected in this period, but the spatial distributions of subsidence are reduced (e.g., the areas such as points P2, P4, P5 and P8 located in Figure 6a,b). This indicates that most of those areas were experiencing continuous subsidence over the study period, but the magnitude was beginning to decrease. This is evident at the points P2 and P4 (Naka-Machi) and P8 (Shin-Machi) that showed a decrease in displacement velocity compared to the ERS-1/-2 monitoring period. In addition, the leveling data at the points U-8, U-10, U-11, U-13 and U-14 also reveal that the subsidence rate has begun to decrease from 2003 [37]. However, significant subsidence was identified in the coastal levee areas (i.e., the Maihama (A), Akemi and Hinode (D), Takasu (E), Minato (F) and Chidori (G) districts), which was not identified by the ERS-1/-2 data (Figure 6). In general, the PALSAR (L-band) has a longer wavelength than the ERS-1/-2 (C-band), which has less decorrelation over vegetated terrain and has better coherence [51]. Thus, the results of the PALSAR data offer a higher density of PS pixels. Therefore, we can assume that these areas may have been experiencing subsidence during the ERS-1/-2 monitoring period and may have been excluded from further processing due to the low coherence exhibited in these areas in the ERS-1/-2 data. Another reason for those differences is that the subsidence in the coastal levee may have started during the PALSAR monitoring period.

Figure 8a–h shows the measured displacement time-series for eight representative points, which are shown in Figure 6a,b (the same points in Figure 3a,b). From Figure 8a–h, we can see that the time-series LOS deformations derived by both the PSI and SBAS processing showed good agreement in the subsidence trend. The estimated deformation rates by the PSI and SBAS measurements on points P1, P2, P4, P7 and P8 showed a velocity difference of less than 3 mm/year, while the points P3, P5 and P6 showed the largest velocity difference of over 5 mm/year. The average velocity difference for all points between the two methods was 2.9 mm/year. In general, similar to the ERS-1/-2 monitoring period, the Moto-Machi area also showed very low subsidence rates in the PALSAR monitoring period. This may be related to the fact that, in most parts of the Moto-Machi area, the urban infrastructures and houses are built over the naturally formed Holocene lowland that has stable ground conditions over time.

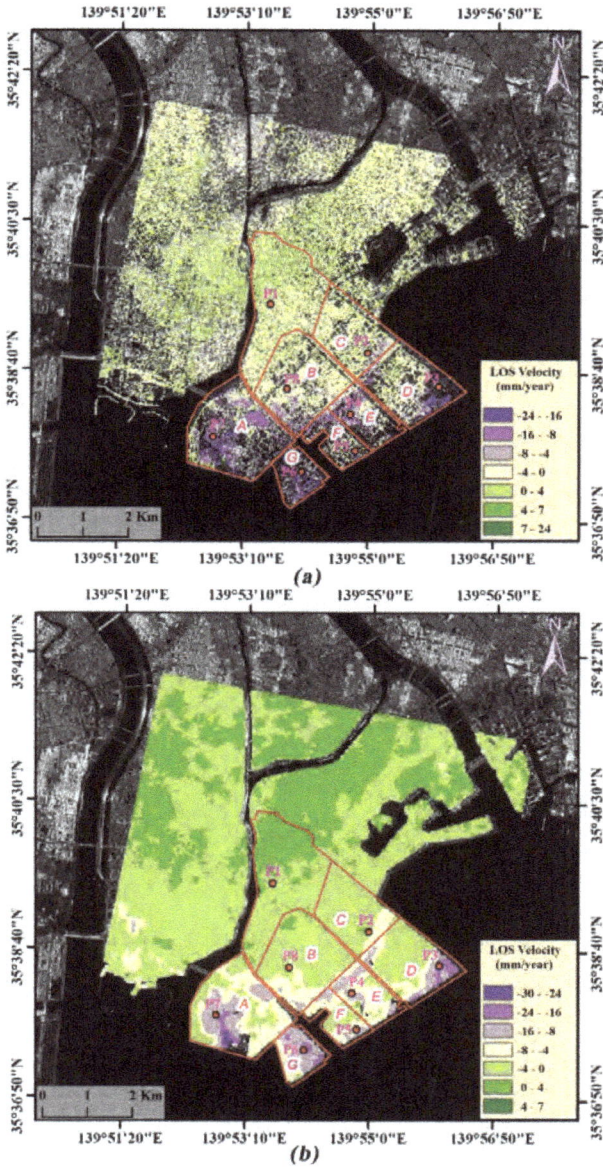

Figure 6. Mean LOS displacement velocity in Urayasu City from 2006 to 2010 for the PALSAR data: (**a**) estimated mean displacement velocity using the PSI method; (**b**) estimated mean displacement velocity using the SBAS method. The background image is a PALSAR-2 intensity image acquired on 04 December 2014. The red points P1 to P8 are the selected points to show the time-series LOS displacements estimated by the PSI and SBAS measurements in (**a**,**b**), respectively. A-G represent the reclaimed areas and districts which described in Table 1.

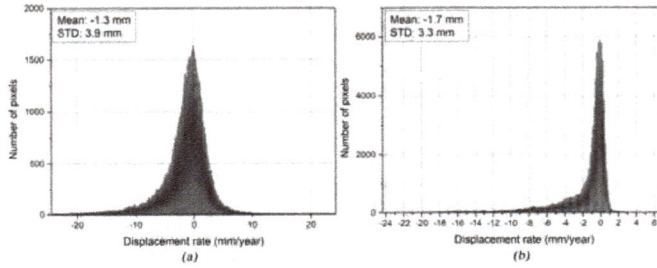

Figure 7. Histogram distribution for the PALSAR-derived results from June 2006 to December 2010. (**a**) The corresponding histogram of the PSI measurements from the PALSAR data; and (**b**) the corresponding histogram of the SBAS measurements from the PALSAR data.

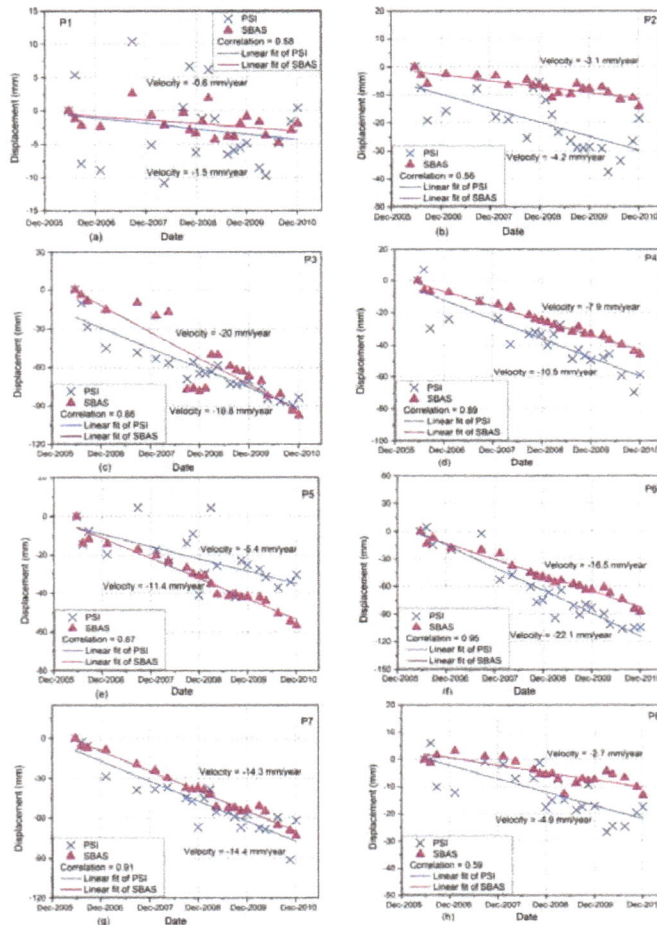

Figure 8. Time-series LOS displacement plots of the PSI and SBAS measurements (**a–h**) for points P1 to P8, which are indicated as red points in Figure 6a,b, respectively.

4.3. Time-Series Analysis of the PALSAR-2 Data from December 2014 to November 2017

The mean velocity (mm/year) maps of the displacements for the period from December 2014 to November 2017 are shown in Figure 9. The same color cycle from green to purple was used for the result. In the study area of over 690,336 pixels, 76,500 measurement points were obtained by the SBAS method. The histogram of the SBAS-derived displacement velocity for the study area is shown in Figure 10. The average displacement rate and the standard deviation are −0.5 mm/year and 1.9 mm/year, which are lower than those obtained with the ERS-1/-2 and PALSAR data. In general, the PALSAR-2 results show that approximately 85% of the PS points indicate displacement rates between −3 mm/year and 1 mm/year (Figure 10). To show the variations in the LOS displacement velocities at different locations over the three observation periods, six profiles across several locations in Urayasu City were selected (Figure 9). We can see from Figure 11 that these selected profiles show different displacement dispersion patterns, such as profiles P1–P1′ and P5–P5′ which show a dispersion of approximately −0.5 mm/year to −2.6 mm/year. Along profile P4–P4′, the subsidence rate increased from 0.1 to 21 mm/year within the distance of 0.6 km. The profiles in Figure 11b–d,f reveal that the PALSAR-estimated subsidence rate has a larger value than those from the ERS-1/-2 and PALSAR-2. Contrary to the ERS-1/-2 and PALSAR-estimated displacement velocity, the PALSAR-2 results show an uplift within the distance of 300 to 900 m in the profile P1–P1′ across the Moto-Machi area (Figure 11a). Moreover, both PALSAR and PALSAR-2 estimated displacement rates show a significantly decrease along P4–P4′ (Figure 11d).

During the PALSAR-2 monitoring period, because of the high spatial resolution and shorter revisiting time compared to the ERS-1/-2 and PALSAR data, a subsidence estimation with better spatial coverage and precision was achieved. Figure 9 shows that the three areas that have subsided during the previous monitoring periods have also showed land subsidence in this PALSAR-2 monitoring period (i.e., the border areas between Naka-Machi and Shin-Machi; the areas close to the levee of Hinode and Akemi (D); the Maihama area (A)). This may further imply that these areas were experiencing continuous subsidence during the entire monitoring period. Considering the existence of non-linear subsidence, the actual subsidence may not be a linear motion overtime, and the results by PSI and SBAS simply reflect the subsidence phenomena. However, the spatial extent and the magnitude of subsidence over Urayasu City is shrinking. The Moto-Machi area is in a relatively stable ground condition over the whole monitoring period, and the areas close to the borders of Moto-Machi and Naka-Machi began to stabilize over time. A further detailed discussion about the evolutions and the causes of land subsidence in Urayasu City are given in Section 5.4.

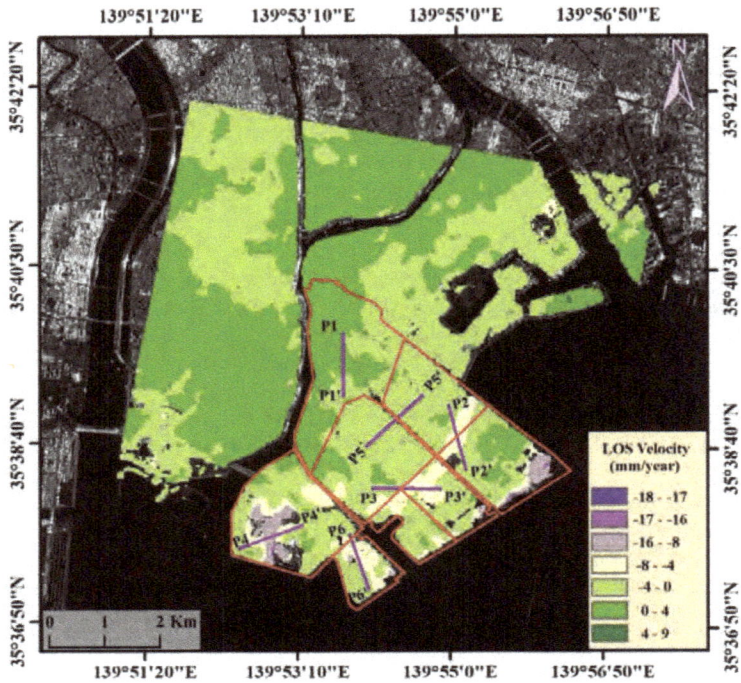

Figure 9. Mean LOS displacement velocity in Urayasu City from 2014 to 2017 for the PALSAR-2 data. The background image is a PALSAR-2 intensity image acquired on 4 December 2014. P1–P1' to P6–P6' are the selected profiles to show the displacement velocities at different sites.

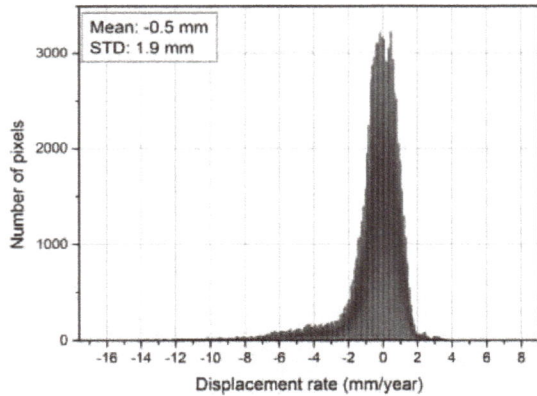

Figure 10. The corresponding histogram of the SBAS measurements from the PALSAR-2 data.

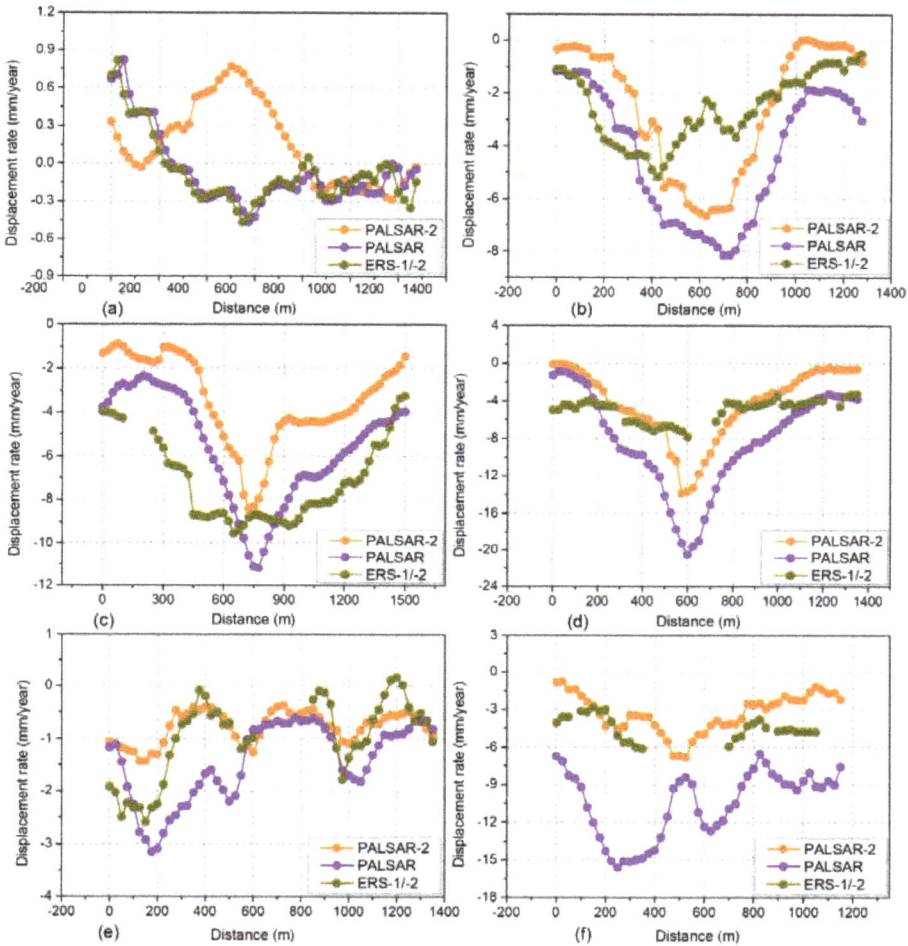

Figure 11. Mean LOS displacement velocities for the three observation periods (**a–f**) along the six profiles whose positions are indicated as purple lines in Figure 9.

5. Discussion

5.1. Comparison of the InSAR-Derived Results with the Leveling Data

To assess the accuracy of the InSAR-derived results over the three observation periods, a quantitative comparison of the time-series displacements with the leveling survey data provided by the Chiba Prefecture at 22 measurement points was performed. To locate each leveling point, we referenced the online version of the Chiba information map and the illustration figures of each leveling point provided [52]. For the InSAR measurement points, especially those in incoherent areas, the pixels that lay within 100 m of the corresponding leveling points were assigned, and the average velocity of these pixels was calculated. We selected the leveling data in the same overlapping periods as the three InSAR measurement periods. We assumed the horizontal deformation was negligible, and the LOS displacement velocity was converted into the vertical displacement velocity by dividing the cosine of the sensor incidence angle [53].

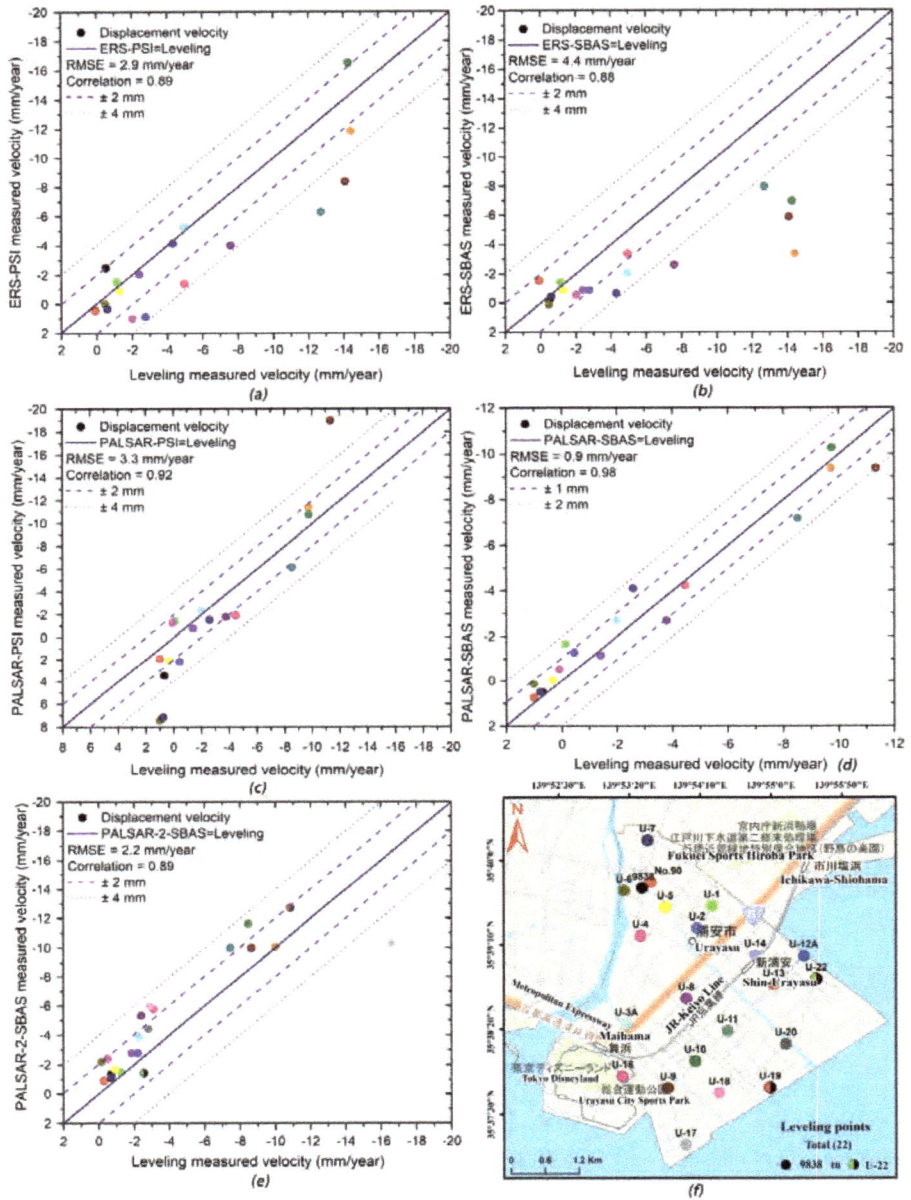

Figure 12. Comparison between InSAR-derived linear subsidence velocity and leveling measured linear subsidence velocity during the three InSAR observation periods: (**a**,**b**) ERS-1/-2 derived linear subsidence rate (May 1993 to February 2006) and leveling-derived linear subsidence rate (January 1993 to January 2006); (**c**,**d**) PALSAR-derived linear subsidence rate (June 2006 to December 2010) and leveling-derived linear subsidence rate (January 2006 to January 2011); (**e**) PALSAR-2-derived linear subsidence rate (December 2014 to December 2016) and leveling-derived linear subsidence rate (January 2015 to January 2017); and (**f**) spatial distribution of leveling points in Urayasu City.

Figure 12 shows the spatial distribution of the leveling points and the comparison between the leveling and InSAR-derived linear subsidence rate. Note that the number of leveling points are different among the different InSAR observation periods; 17 leveling points were used for the comparison of the ERS-1/-2 and PALSAR observation periods, while 21 leveling points were used for the PALSAR-2 observation period, which is due to five new leveling points being established after the 2011 Tohoku Earthquake and the leveling point U-12A being missing in 2016. We also used different plot scales (20 mm/year vs. 12 mm/year) and (2/4 mm vs. 1/2 mm for error lines), due to the smaller errors shown in PALSAR data using the SBAS method (Figure 12d). The comparison results show that the results from the ERS-1/-2 data using the SBAS method have the largest root mean square errors (RMSEs) of 4.4 mm/year, while the results from PALSAR and PALSAR-2 data using the SBAS method have the smallest RMSEs of 0.9 and 2.2 mm/year, respectively. For the ERS-1/-2 and PALSAR data, more than 12 out of the 17 measurement points showed a residual value of less than 4 mm/year (Figure 12a–d); for the PALSAR-2 data, and 14 out of the 21 measurement points showed a residual value of less than 2 mm/year (Figure 12e). As shown in Figure 12a,b,e, the results from the ERS-1/-2 and PALSAR-2 data using the PSI and SBAS method showed the largest discrepancies at several leveling points. This may have been caused by the low coherence of ERS-1/-2 datasets and the contribution of phase noise. The fewer PALSAR-2 image pairs and the sudden elevation changes in the ground, i.e., the leveling point U-17 subsided by the influence of construction work during 2015–2016 [37], may also affect the comparison result. Nevertheless, according to these comparisons, the InSAR-derived results agree relatively well with the result of the leveling measurements and suggest the reliability of the InSAR-measured subsidence rate.

5.2. Spatial and Temporal Patterns of Land Subsidence

To further reveal the land subsidence patterns in different districts over the three observation periods, we generated the spatial distribution map of difference of land subsidence rates (Figure 13) using the ArcGIS 10.3 (Esri, Redlands, CA, USA) spatial analyst tool. As the incidence angles of those sensors are different, before comparison, the LOS displacement velocity was converted into the vertical displacement velocity by dividing the cosine of the sensor incidence angle [53]. It can be seen from Figure 13a that the areas in the central town (i.e., Maihama (A), Tekkodori, Benten, Imagawa (B) and Irifune (C)) and new town (i.e., Takasu (E), Minato (F) and Chidori (G)) show slight to moderate subsidence with a 2–13 mm/year rate during the ERS-1/-2 observation period. From Figure 13b, we can see that the subsidence rate in some of the districts of the central town (e.g., Benten, Tekkodori and Imagawa (B)) has decreased up to 12 mm/year; while the areas in the new town showed increasing subsidence up to 28 mm/year, especially in Hinode (D) and Chidori (G). The comparison of PALSAR-2 and PALSAR estimated subsidence rate show that, the previous subsiding areas were experiencing a reduced subsiding rate, except some localized subsidence in the new town (Figure 13c). The comparison of PALSAR-2 and ERS-1/-2 estimated subsidence rate show that, the subsidence in both of the central town and new town has significantly decreased, except for areas in Maihama (A), Irifune (C), Hinode and Akemi (D) (Figure 13d). In general, most of those areas in the central town are residential and commercial amusement land, while the bew town are parks and industrial land. The subsidence in parks can only be caused by the natural soil consolidation, while in the residential, commercial and industrial areas, the subsidence may be caused from the integrated effect of numerous natural and anthropogenic processes.

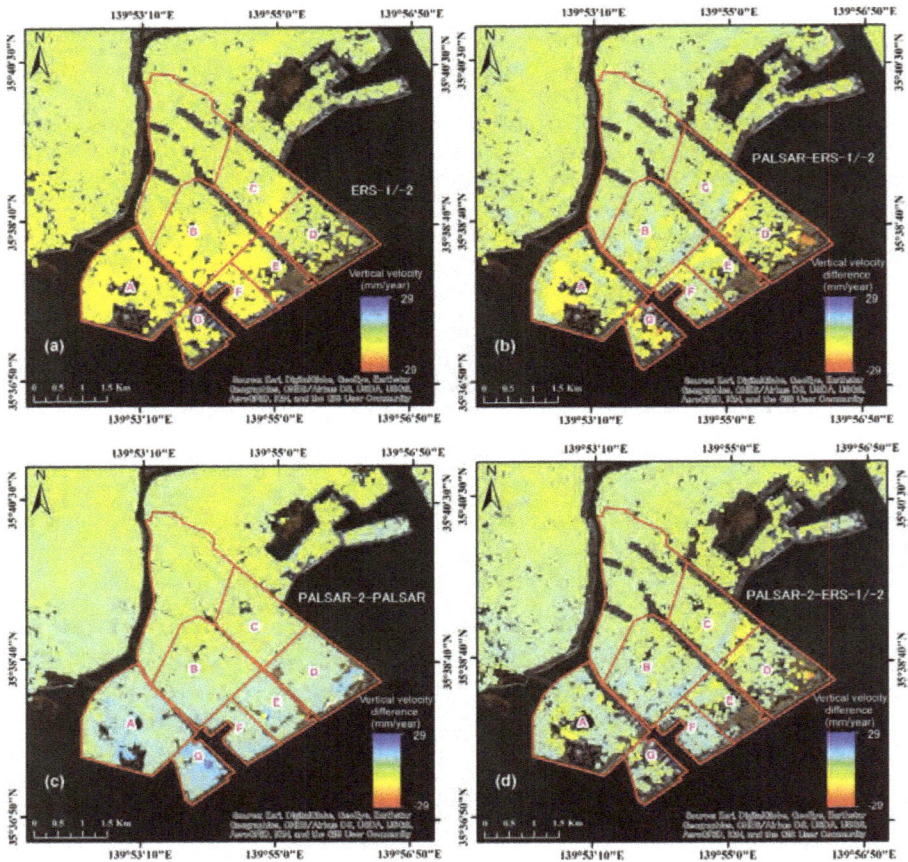

Figure 13. The spatial distribution map of difference of land subsidence rates during the three observation periods: (**a**) ERS-1/-2 derived subsidence rate using the SBAS method; (**b**) difference between ERS-1/-2 and PALSAR derived subsidence rates (subtracting ERS-1/-2 from PALSAR); (**c**) difference between PALSAR and PALSAR-2 derived subsidence rates (subtracting PALSAR from PALSAR-2); (**d**) difference between ERS-1/-2 and PALSAR-2 derived subsidence rates (subtracting ERS-1/-2 from PALSAR-2).

5.3. The Use of Different SAR Sensors in Land Subsidence Monitoring

The number of satellite data sources is currently increasing steadily. These datasets from the previous SAR sensors such as ESA archive (ERS-1/-2, ENVISAT) as well as the new generation of C, X and L-band SAR images provided by the RADARSAT-2, Sentinel-1A, ALOS-2, TerraSAR-X, Tandem-X and the COSMO-SkyMed constellation, etc. have enabled us to compute the time series of the occurred and on-going surface displacements from regional scale to individual buildings. In particular, the exploitation of the free and open access data archives collected by the Sentinel-1A system permit us to conduct continuous land deformation analysis over large areas.

In this study, three different SAR datasets, the ERS-1/-2, ALOS PALSAR and ALOS-2 PALSAR-2, were used to monitor the long-term land subsidence in Urayasu City. The C-band has a shorter wavelength and hence better displacement sensitivity, and the L-band has longer wavelength and lower frequency showing more extensive coverage over natural areas and less temporal decorrelation. The data acquired by these satellites cover long periods of time and enabled us to perform long-term deformation monitoring of the study area. However, those different sensors have different imaging parameters, e.g., spatial and temporal resolution, incidence angle, and wavelength, which show different characteristics in terms of their maximum detection gradient, degree of decorrelation, capability of noise rejection, etc. The different imaging parameters and the use of an uneven number of images among different sensors cause some difficulties in comparing their performance and the quality control of multi-sensor InSAR results. Moreover, the low resolution and the longer revisit time of ERS-1/-2 and PALSAR has prevented us from observing short-term land deformations caused by the anthropogenic activities. Furthermore, data gaps between the PALSAR and PALSAR-2 caused some difficulties in analyzing InSAR results.

5.4. Land Subsidence and Possible Causes

The origin of land subsidence in coastal areas can be summarized into two categories: either caused by natural causes (e.g., natural compaction/consolidation of soil or tectonic movements, such as earthquakes) or anthropogenic activities (e.g., oil, gas and ground water exploitation). In some cases, the pattern of land subsidence might be even more complicated when it is caused by the combined effects of multiple factors at different scales. In Urayasu City, since most of the areas are land-filled, the natural consolidation of soil is postulated to be the primary driver of land subsidence. To further analyze the relation between subsidence and soil geology, we compared the InSAR-derived subsidence areas with the geologic map showing the depth of the upper surface of the solid geological stratum in Urayasu City (Figure 14a) and found a remarkable spatial correlation between the geologic map of the soil properties and the subsiding areas. In most of the reclaimed zones, the upper layer of soil filled with hill sand and dredged sandy soil (FS) with a standard penetration test (SPT) N-value of 2–8; an alluvial sand layer (AS) with SPT N-value of 10–20 underlies the filled layer; a very soft alluvial clay (AC) is deposited under the AS layer with a low SPT N-value of 0–5; a diluvial (Pleistocene) dense sand layer (DS) with SPT N-value of 50 or greater is deposited blow the AC layer (Figure 14b) [23,24]. Along the line A–A', the thickness of AC layer increases significantly between the Naka-Machi and Shin-Machi area, and it continues towards the sea (Figure 14b). As the consolidation of soil occurs in soft clay deposits, the thick AC layers in Naka-Machi and Shin-Machi area are most probably responsible for the continuing subsidence in Urayasu City.

As shown in Figure 14a, the depth of the bottom of the alluvial layers increases from 20 m in Moto-Machi to about 40 m in Shin-Machi, with several narrow-buried valleys of up to 70 m in depth. The buried-valleys, which are about 60 m deep, exist directly below the Minato, Chidori, Tekkodori, Imagawa, Akemi and Irifune areas, causing complicated changes in the thickness of the soft ground in those areas, while the depth increases up to 80 m in Maihama where the largest subsidence occurred. This further suggests that the areas undergoing large subsidence correspond to those having thick layers of soft soil over a stiff basement. The Moto-Machi area, with soil deposits consisting of sandy soils with an alluvial origin, was quite stable over the observation period, while the Naka-Machi and Shin-Machi areas, with thick layers of fine-grained soft soil overlying a stiff basement, had significant land subsidence over the study period. However, considering the complexity of the land use and the anthropogenic activities in different districts of Urayasu City, the subsidence may not be solely caused by the natural consolidation, but also from the integrated effects of numerous natural and anthropogenic processes.

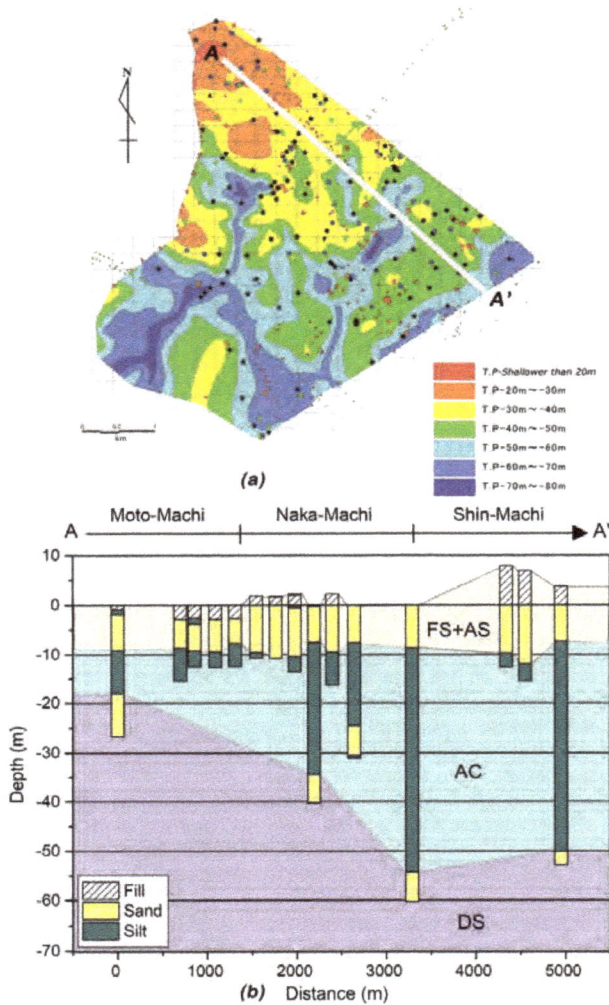

Figure 14. Depth of the upper surface of the solid geological stratum (**a**) in Urayasu City (adapted from the public report by the technical committee of Urayasu City [54]). The points refer to the locations of borehole sites; (**b**) soil cross sections along the A–A' line. FS + AS refer to filled sandy soil and alluvial sand layers, and AC and DS refer to the alluvial clay layer and diluvial dense sandy layer, respectively. The borehole investigation data were obtained from the Chiba Prefecture [55].

The additional load of buildings and structures is also considered to be one of the causes of land subsidence in urban areas [56]. In Urayasu City, since the establishment of Urayasu town in 1909 in the old town (Moto-Machi)—a naturally formed Holocene lowland—the natural soil consolidation might be gradually reduced and stopped. Besides this, the density of buildings in the old town are lower than the central town. Many houses, commercial buildings and public facilities were built in the central town during the first phase of the project, ending in 1975. Meanwhile, many high-rise buildings, universities, hotels and storehouses were built in the new town during the second phase of the project, ending in 1980 [22]. The additional load during and after the building construction, especially the high-rise buildings, could transfer a high loading to the ground and may eventually lead to substantial land subsidence. However, these buildings use a pile foundation to satisfy bearing capacity and deformation and may not show significant subsidence while the surrounding areas are subsiding. Figure 15 shows the InSAR-derived subsidence velocity (2006–2010) and the locations of high-rise buildings. Most of those buildings show stability, whereas their surroundings show land subsidence. However, further investigations are expected to determine the relationship between land subsidence and the building density/high rise buildings.

The Maihama district in the central town, where Tokyo Disneyland is located, showed significant subsidence throughout the whole InSAR observation period. However, in this area, the pattern of land subsidence may be even more complicated due to the continuous construction and renovation/redevelopment of the fantasy-land and other anthropogenic activities. The SAR images with a low resolution and longer revisiting time, and the linear inversion model used in the InSAR processing, may hinder the effective monitoring of short-term movements such as those induced by human activities and may cause some biased results. Therefore, more high-resolution SAR data with a short revisiting time and further investigation is required to understand the intricacies of the relationship between land subsidence, natural consolidation and load of buildings.

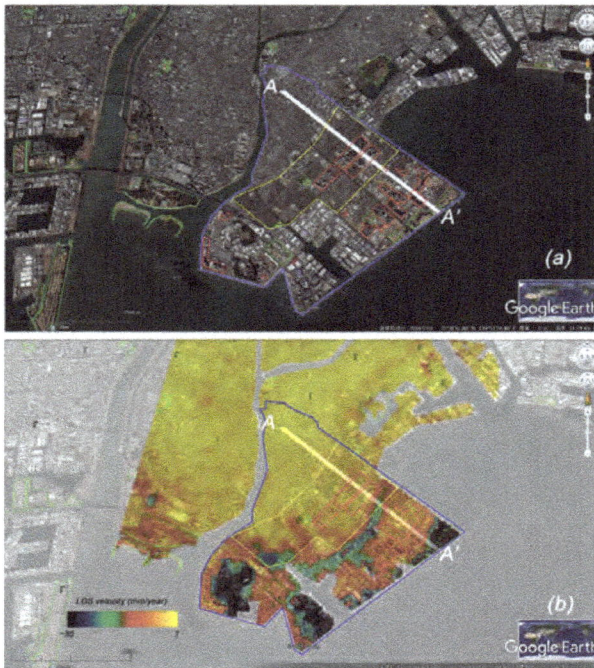

Figure 15. Subsidence rate map (2006–2010) generated with ALOS PALSAR data overlaid on a Google Earth image. The green polygons indicate the park area, red polygons indicate the location of high-rise buildings, the yellow polygon shows the highly populated residential area. The blue polygon indicates the border of Urayasu City and corresponds to the location of Figure 14a, and the A–A' line corresponds to the soil cross section in Figure 14a,b.

Ground water exploitation is one of the major causes of land subsidence in many coastal cities, such as in Jakarta [5], Bangkok [8] and Shanghai [57]. Nevertheless, this may not be the cause of land subsidence in Urayasu City; this is because the ground water exploitation was gradually reduced and stopped in 1993 [58], and the city receives water from a water purification plant which uses the main water sources of the Tone river and Edogawa river [59]. Moreover, since April 1992, Chiba Prefecture has been implementing restriction rules on groundwater use for the highly susceptible areas of land subsidence, including Urayasu City [60]. Thus, the ground water exploitation has insignificant impacts on land subsidence in Urayasu City.

As an earthquake-prone country, earthquakes happen frequently in Japan. Earthquakes have significant influences on coastal areas, especially on reclaimed land. In the 2011 Tohoku earthquake, houses and infrastructures were severely damaged due to soil liquefaction in Urayasu City [22]. In addition, long-term ground settlement was also observed after the earthquake, and the degree of subsidence was different in areas where reclaimed soils were improved or not [24]. In the areas where the soil was not improved, the subsidence may have been accelerated by the earthquake. The InSAR observation results derived from the PALSAR-2 data showed significant continuing land subsidence near the levee areas (mostly parks and vacant lands), which may have been accelerated by the effects of the earthquake. However, most areas showed a decrease of land subsidence, this may be related to the fact that the PALSAR-2 observations (December 2014 to November 2017) were collected almost 4 years after the Tohoku earthquake, and considering the soil aging effect and soil improvement, the land settlement in most of those areas caused by the natural soil consolidation and the earthquake might gradually decrease. It is worth mentioning that after the earthquake, the Urayasu government started to test several countermeasure methods, such as lowering the ground water level and grid wall soil improvement. Finally, Urayasu has adopted the grid wall soil improvements as a countermeasure to prevent future risks [61]. This project may also have played a positive role in alleviating the land subsidence in Urayasu City.

6. Conclusions

In this study, to monitor the long-term spatial patterns of land subsidence in Urayasu City, we used three sets of different SAR data and advanced InSAR techniques. The obtained InSAR results during the three observation periods from 1993–2010 and 2014 to 2017 show continuing subsidence occurring in several reclaimed areas of Urayasu City. The maximum subsidence rate from 1993 to 2006 was approximately 27 mm/year, from 2006 to 2010 it was 30 mm/year, and from 2014 to 2017 it was about 18 mm/year. The results were verified by comparing them with the leveling survey data. The comparison shows that the obtained InSAR results agree well with the leveling measurements, with a correlation value of over 0.8. The natural consolidation of soil in the reclaimed areas can be considered as a primary driver of land subsidence in Urayasu City, while the integrated effects of numerous natural and anthropogenic processes are also not negligible. Considering the soil aging effect, water-use restriction rules and soil improvement work performed by the government and land owners might also have played a positive role in alleviating the land subsidence and related disasters. However, further investigation is required to understand the intricacies of the relationship between the land subsidence and anthropogenic activities. The outcome of this research further proves the suitability and effectiveness of InSAR measurements in the land subsidence monitoring of coastal urban areas.

Author Contributions: Y.A. conceived the work, processed the SAR data and wrote the paper; F.Y. and W.L. supervised the data processing and revised the manuscript.

Funding: This research received no external funding.

Acknowledgments: We would like to thank JAXA for providing the ALOS-PALSAR and ALOS-2 PALSAR-2 data and ESA for providing the ERS-1/-2 data used in this study. We would also like to thank the editor and the three anonymous reviewers for their insightful comments that greatly helped to improve the quality of this manuscript.

Conflicts of Interest: The authors declare no conflict of interest.

References

1. Pradhan, B.; Abokharima, M.H.; Jebur, M.N.; Tehrany, M.S. Land subsidence susceptibility mapping at Kinta Valley (Malaysia) using the evidential belief function model in GIS. *Nat. Hazards* **2014**, *73*, 1019–1042. [CrossRef]
2. Chaussard, E.; Amelung, F.; Abidin, H.; Hong, S.-H. Sinking cities in Indonesia: ALOS PALSAR detects rapid subsidence due to groundwater and gas extraction. *Remote Sens. Environ.* **2013**, *128*, 150–161. [CrossRef]
3. Tessler, Z.D.; Vorosmarty, C.J.; Grossberg, M.; Gladkova, I.; Aizenman, H.; Syvitski, J.P.M.; Foufoula-Georgiou, E. Profiling risk and sustainability in coastal deltas of the world. *Science* **2015**, *349*, 638–643. [CrossRef] [PubMed]
4. Jones, C.E.; An, K.; Blom, R.G.; Kent, J.D.; Ivins, E.R.; Bekaert, D. Anthropogenic and geologic influences on subsidence in the vicinity of New Orleans, Louisiana. *J. Geophys. Res. Solid Earth* **2016**, *121*, 3867–3887. [CrossRef]
5. Abidin, H.Z.; Andreas, H.; Gumilar, I.; Fukuda, Y.; Pohan, Y.E.; Deguchi, T. Land subsidence of Jakarta (Indonesia) and its relation with urban development. *Nat. Hazards* **2011**, *59*, 1753–1771. [CrossRef]
6. Ng, A.H.M.; Ge, L.; Li, X.; Abidin, H.Z.; Andreas, H.; Zhang, K. Mapping land subsidence in Jakarta, Indonesia using persistent scatterer interferometry (PSI) technique with ALOS PALSAR. *Int. J. Appl. Earth Obs. Geoinf.* **2012**, *18*, 232–242. [CrossRef]
7. Ho Tong Minh, D.; Van Trung, L.; Le Toan, T. Mapping ground subsidence phenomena in Ho Chi Minh City through the radar interferometry technique using ALOS PALSAR data. *Remote Sens.* **2015**, *7*, 8543–8562. [CrossRef]
8. Aobpaet, A.; Cuenca, M.C.; Hooper, A.; Trisirisatayawong, I. InSAR time-series analysis of land subsidence in Bangkok, Thailand. *Int. J. Remote Sens.* **2013**, *34*, 2969–2982. [CrossRef]
9. Dong, S.; Samsonov, S.; Yin, H.; Ye, S.; Cao, Y. Time-series analysis of subsidence associated with rapid urbanization in Shanghai, China measured with SBAS InSAR method. *Environ. Earth Sci.* **2014**, *72*, 677–691. [CrossRef]
10. Xu, B.; Feng, G.; Li, Z.; Wang, Q.; Wang, C.; Xie, R. Coastal subsidence monitoring associated with land reclamation using the point target based SBAS-InSAR method: A case study of Shenzhen, China. *Remote Sens.* **2016**, *8*, 652. [CrossRef]
11. Tosi, L.; Teatini, P.; Strozzi, T. Natural versus anthropogenic subsidence of Venice. *Sci. Rep.* **2013**, *3*, 2710. [CrossRef] [PubMed]
12. Koster, K.; Erkens, G.; Zwanenburg, C. A new soil mechanics approach to quantify and predict land subsidence by peat compression. *Geophys. Res. Lett.* **2016**, *43*, 10792–10799. [CrossRef]
13. Raspini, F.; Loupasakis, C.; Rozos, D.; Adam, N.; Moretti, S. Ground subsidence phenomena in the Delta municipality region (Northern Greece): Geotechnical modeling and validation with Persistent Scatterer Interferometry. *Int. J. Appl. Earth Obs. Geoinf.* **2014**, *28*, 78–89. [CrossRef]
14. Normand, J.C.L.; Heggy, E. InSAR Assessment of Surface Deformations in Urban Coastal Terrains Associated with Groundwater Dynamics. *IEEE Trans. Geosci. Remote Sens.* **2015**, *53*, 6356–6371. [CrossRef]
15. Cianflone, G.; Tolomei, C.; Brunori, C.A.; Dominici, R. InSAR time series analysis of natural and anthropogenic coastal plain subsidence: The case of sibari (Southern Italy). *Remote Sens.* **2015**, *7*, 16004–16023. [CrossRef]
16. Hsieh, C.S.; Shih, T.Y.; Hu, J.C.; Tung, H.; Huang, M.H.; Angelier, J. Using differential SAR interferometry to map land subsidence: A case study in the Pingtung Plain of SW Taiwan. *Nat. Hazards* **2011**, *58*, 1311–1332. [CrossRef]
17. Ferretti, A.; Prati, C.; Rocca, F. Nonlinear Subsidence Rate Estimation Using Permanent Scatterers in Differential SAR Interferometry. *IEEE Trans. Geosci. Remote Sens.* **2000**, *38*, 2202–2212. [CrossRef]
18. Ferretti, A.; Prati, C.; Rocca, F. Permanent scatterers in SAR interferometry. *IEEE Trans. Geosci. Remote Sens.* **2001**, *39*, 8–20. [CrossRef]
19. Hooper, A.; Zebker, H.; Segall, P.; Kampes, B. A new method for measuring deformation on volcanoes and other natural terrains using InSAR persistent scatterers. *Geophys. Res. Lett.* **2004**, *31*, 1–5. [CrossRef]
20. Berardino, P.; Fornaro, G.; Lanari, R.; Sansosti, E. A new algorithm for surface deformation monitoring based on small baseline differential SAR interferograms. *IEEE Trans. Geosci. Remote Sens.* **2002**, *40*, 2375–2383. [CrossRef]

21. Armas, I.; Mendes, D.A.; Popa, R.G.; Gheorghe, M.; Popovici, D. Long-term ground deformation patterns of Bucharest using multi-temporal InSAR and multivariate dynamic analyses: A possible transpressional system? *Sci. Rep.* **2017**, *7*, 43762. [CrossRef] [PubMed]

22. Tokimatsu, K.; Tamura, S.; Suzuki, H.; Katsumata, K. Building damage associated with geotechnical problems in the 2011 Tohoku Pacific Earthquake. *Soils Found.* **2012**, *52*, 956–974. [CrossRef]

23. Yasuda, S.; Harada, K.; Ishikawa, K.; Kanemaru, Y. Characteristics of liquefaction in Tokyo Bay area by the 2011 Great East Japan Earthquake. *Soils Found.* **2012**, *52*, 793–810. [CrossRef]

24. Nigorikawa, N.; Asaka, Y. Leveling of long-term settlement of Holocene clay ground induced by the 2011 off the Pacific coast of Tohoku earthquake. *Soils Found.* **2015**, *55*, 1318–1325. [CrossRef]

25. Okada, N.; Ye, T.; Kajitani, Y.; Shi, P.; Tatano, H. The 2011 eastern Japan great earthquake disaster: Overview and comments. *Int. J. Disaster Risk Sci.* **2011**, *2*, 34–42. [CrossRef]

26. Zhou, L.; Guo, J.; Hu, J.; Li, J.; Xu, Y.; Pan, Y.; Shi, M. Wuhan surface subsidence analysis in 2015–2016 based on sentinel-1A data by SBAS-InSAR. *Remote Sens.* **2017**, *9*, 982. [CrossRef]

27. Bhattacharya, S.; Hyodo, M.; Goda, K.; Tazoh, T.; Taylor, C.A. Liquefaction of soil in the Tokyo Bay area from the 2011 Tohoku (Japan) earthquake. *Soil Dyn. Earthq. Eng.* **2011**, *31*, 1618–1628. [CrossRef]

28. Tokimatsu, K.; Katsumata, K. Liquefaction-induced damage to buildings in Urayasu city during the 2011 Tohoku Pacific earthquake. In Proceedings of the International Symposium on Engineering Lessons Learned from the 2011 Great East Japan Earthquake, Tokyo, Japan, 1–4 March 2012; pp. 665–674.

29. Imakiire, T.; Koarai, M. Wide-area land subsidence caused by "the 2011 off the Pacific Coast of Tohoku Earthquake". *Soils Found.* **2012**, *52*, 842–855. [CrossRef]

30. Konagai, K.; Kiyota, T.; Suyama, S.; Asakura, T.; Shibuya, K.; Eto, C. Maps of soil subsidence for Tokyo bay shore areas liquefied in the March 11th, 2011 off the Pacific Coast of Tohoku Earthquake. *Soil Dyn. Earthq. Eng.* **2013**, *53*, 240–253. [CrossRef]

31. Pasquali, P.; Cantone, A.; Riccardi, P.; De Filippi, M.; Ogushi, F.; Tamura, M.; Gagliano, S. Monitoring land subsidence in the tokyo region with sar interferometric stacking techniques. In *Engineering Geology for Society and Territory—Volume 5: Urban Geology, Sustainable Planning and Landscape Exploitation*; Springer: Berlin, Germany, 2015; pp. 995–999. ISBN 9783319090481.

32. ElGharbawi, T.; Tamura, M. Estimating deformation due to soil liquefaction in Urayasu city, Japan using permanent scatterers. *ISPRS J. Photogramm. Remote Sens.* **2015**, *109*, 152–164. [CrossRef]

33. Urayasu City. Population Statistics. Available online: http://www.city.urayasu.lg.jp/shisei/toukei/jinko/1002267.html (accessed on 20 March 2018). (In Japanese)

34. Urayasu City. The Project of Reclaimed Land. Available online: http://www.city.urayasu.lg.jp/shisei/profile/profile/1000020.html (accessed on 16 March 2018). (In Japanese)

35. Tokimatsu, K.; Suzuki, H.; Katsumata, K.; Tamura, S. Geotechnical Problems in the 2011 Tohoku Pacific Earthquakes. In Proceedings of the International Conference on Case Histories in Geotechnical Engineering, Chicago, IL, USA, 29 April–4 May 2013. Available online: http://scholarsmine.mst.edu/icchge/7icchge/session12/2 (accessed on 18 August 2018).

36. Geospatial Information Authority of Japan. Fundamental Geospatial Data Portal of GSI. Available online: https://fgd.gsi.go.jp/download/menu.php (accessed on 16 March 2018).

37. Chiba Prefecture. Chiba Prefecture Leveling Survey Results. Available online: http://www.pref.chiba.lg.jp/suiho/jibanchinka/torikumi/seikaomote.html (accessed on 16 March 2018). (In Japanese)

38. Hooper, A.J. A multi-temporal InSAR method incorporating both persistent scatterer and small baseline approaches. *Geophys. Res. Lett.* **2008**, *35*. [CrossRef]

39. Qu, F.; Lu, Z.; Zhang, Q.; Bawden, G.W.; Kim, J.W.; Zhao, C.; Qu, W. Mapping ground deformation over Houston-Galveston, Texas using multi-temporal InSAR. *Remote Sens. Environ.* **2015**, *169*, 290–306. [CrossRef]

40. Grzovic, M.; Ghulam, A. Evaluation of land subsidence from underground coal mining using TimeSAR (SBAS and PSI) in Springfield, Illinois, USA. *Nat. Hazards* **2015**, *79*, 1739–1751. [CrossRef]

41. Hooper, A.; Bekaert, D.; Spaans, K.; Arikan, M. Recent advances in SAR interferometry time series analysis for measuring crustal deformation. *Tectonophysics* **2012**, *514–517*, 1–13. [CrossRef]

42. Crosetto, M.; Monserrat, O.; Cuevas-González, M.; Devanthéry, N.; Crippa, B. Persistent Scatterer Interferometry: A review. *ISPRS J. Photogramm. Remote Sens.* **2016**, *115*, 78–89. [CrossRef]

43. Gourmelen, N.; Amelung, F.; Lanari, R. Interferometric synthetic aperture radar-GPS integration: Interseismic strain accumulation across the Hunter Mountain fault in the eastern California shear zone. *J. Geophys. Res. Solid Earth* **2010**, *115*. [CrossRef]

44. Chaussard, E.; Wdowinski, S.; Cabral-Cano, E.; Amelung, F. Land subsidence in central Mexico detected by ALOS InSAR time-series. *Remote Sens. Environ.* **2014**, *140*, 94–106. [CrossRef]

45. ESA PRARE Precise Orbit Product (ERS.ORB.POD). Available online: https://earth.esa.int/web/guest/-/prare-precise-orbit-product (accessed on 18 May 2018).

46. Goldstein, R.M.; Werner, C.L. Radar interferogram filtering for geophysical applications. *Geophys. Res. Lett.* **1998**, *25*, 4035–4038. [CrossRef]

47. Costantini, M. A novel phase unwrapping method based on network programming. *IEEE Trans. Geosci. Remote Sens.* **1998**, *36*, 813–821. [CrossRef]

48. Hooper, A.; Zebker, H.A. Phase unwrapping in three dimensions with application to InSAR time series. *J. Opt. Soc. Am. A* **2007**, *24*, 2737–2747. [CrossRef]

49. Lin, L.I.-K. A Concordance Correlation Coefficient to Evaluate Reproducibility. *Biometrics* **1989**, *45*, 255–268. [CrossRef] [PubMed]

50. Aimaiti, Y.; Yamazaki, F.; Liu, W.; Kasimu, A. Monitoring of Land-Surface Deformation in the Karamay Oilfield, Xinjiang, China, Using SAR Interferometry. *Appl. Sci.* **2017**, *7*, 772. [CrossRef]

51. Rosen, P.A.; Hensley, S.; Zebker, H.A.; Webb, F.H.; Fielding, E.J. Surface deformation and coherence measurements of Kilauea Volcano, Hawaii, from SIR-C radar interferometry. *J. Geophys. Res. Planets* **1996**, *101*, 23109–23125. [CrossRef]

52. Chiba Prefecture. Chiba Information Map. Available online: http://map.pref.chiba.lg.jp/pref-chiba/Portal (accessed on 16 March 2018).

53. Pepe, A.; Calò, F. A Review of Interferometric Synthetic Aperture RADAR (InSAR) Multi-Track Approaches for the Retrieval of Earth's Surface Displacements. *Appl. Sci.* **2017**, *7*, 1264. [CrossRef]

54. Regional Disaster Prevention Project of Urayasu City—Earthquake Disaster. Available online: https://www.city.urayasu.lg.jp/_res/projects/default_project/_page_/001/002/417/1-4.pdf (accessed on 19 June 2018). (In Japanese)

55. Chiba Prefecture. Chiba Prefecture Geological Environment Information Bank. Available online: https://www.pref.chiba.lg.jp/suiho/chishitsu.html (accessed on 7 July 2018). (In Japanese)

56. Chen, G.; Zhang, Y.; Zeng, R.; Yang, Z.; Chen, X.; Zhao, F.; Meng, X. Detection of land subsidence associated with land creation and rapid urbanization in the Chinese Loess Plateau using time series InSAR: A case study of Lanzhou New District. *Remote Sens.* **2018**, *10*, 270. [CrossRef]

57. Chai, J.-C.; Shen, S.-L.; Zhu, H.-H.; Zhang, X.-L. Land subsidence due to groundwater drawdown in Shanghai. *Géotechnique* **2004**, *54*, 143–147. [CrossRef]

58. Chiba Prefecture. Survey Results of Ground Water Use in Chiba Prefecture. Available online: https://www.pref.chiba.lg.jp/suiho/jibanchinka/torikumi/yousuiryou.html (accessed on 13 April 2018). (In Japanese)

59. Chiba Prefecture. Chiba Prefecture Water Source Information. Available online: https://www.pref.chiba.lg.jp/suidou/souki/2nd-page/suigen.html (accessed on 19 March 2018). (In Japanese)

60. Chiba Prefecture. Land Subsidence Survey Report on 2014. Available online: https://www.pref.chiba.lg.jp/suiho/press/2015/h26-jibanchinka.html (accessed on 19 March 2018). (In Japanese)

61. Ishii, I.; Towhata, I.; HiradateI, R.; Tsukuni, S.; Uchida, A.; Sawada, S.; Yamaguchi, T. Design of grid-wall soil improvement to mitigate soil liquefection damage in residential areas in Urayasu. *J. JSCE* **2017**, *5*, 27–44. [CrossRef]

remote sensing

MDPI

Article

A Methodology to Detect and Characterize Uplift Phenomena in Urban Areas Using Sentinel-1 Data

Roberta Bonì [1,*], Alberto Bosino [1], Claudia Meisina [1], Alessandro Novellino [2], Luke Bateson [2] and Harry McCormack [3]

[1]   Department of Earth and Environmental Sciences, University of Pavia, Via Ferrata 1, 27100 Pavia, Italy; alberto.bosino01@universitadipavia.it (A.B.); claudia.meisina@unipv.it (C.M.)
[2]   British Geological Survey, Natural Environment Research Council, Nicker Hill, Keyworth, Nottinghamshire NG12 5GG, UK; alessn@bgs.ac.uk (A.N.); lbateson@bgs.ac.uk (L.B.)
[3]   Compagnie Générale de Géophysique (CGG), NPA Satellite Mapping, Crockham Park, Edenbridge, Kent TN8 6SR, UK; harry.mccormack@cgg.com
*   Correspondence: roberta.boni01@universitadipavia.it; Tel.: +39-03-8298-5842

Received: 6 February 2018; Accepted: 12 April 2018; Published: 14 April 2018

Abstract: This paper presents a methodology to exploit the Persistent Scatterer Interferometry (PSI) time series acquired by Sentinel-1 sensors for the detection and characterization of uplift phenomena in urban areas. The methodology has been applied to the Tower Hamlets Council area of London (United Kingdom) using Sentinel-1 data covering the period 2015–2017. The test area is a representative high-urbanized site affected by geohazards due to natural processes such as compaction of recent deposits, and also anthropogenic causes due to groundwater management and engineering works. The methodology has allowed the detection and characterization of a 5 km^2 area recording average uplift rates of 7 mm/year and a maximum rate of 18 mm/year in the period May 2015–March 2017. Furthermore, the analysis of the Sentinel-1 time series highlights that starting from August 2016 uplift rates began to decrease. A comparison between the uplift rates and urban developments as well as geological, geotechnical, and hydrogeological factors suggests that the ground displacements occur in a particular geological context and are mainly attributed to the swelling of clayey soils. The detected uplift could be attributed to a transient effect of the groundwater rebound after completion of dewatering works for the recent underground constructions.

Keywords: Persistent Scatterer Interferometry (PSI); Sentinel-1; uplift; expansive soils; dewatering; London

1. Introduction

Ground displacements can be evidence of several processes of natural origin such as swelling/shrinkage of expansive soils, compaction of recent deposits, tectonic displacements associated to the occurrence of earthquakes or long-term tectonic movements and anthropogenic causes such as pumping-induced aquifer-system compaction [1]. In many instances, the movements are due to the interactions of multi-driving factors that act at various spatial and temporal scales [2]. Furthermore, ground motion can imply surface deformation with 3D displacement components, negative and positive vertical movements and/or horizontal (E-W) movements. Negative displacement corresponds to lowering of the earth surface named land subsidence meanwhile positive displacement is the uplift of the earth surface.

Uplift phenomena are less common and less studied than land subsidence. Positive movements (uplift) can occur as a result of various natural and human causes; for example, swelling of clay soils [3], fault effects [4], and water rebound in mining areas [5,6]. Uplift phenomena can lead to various environmental and engineering problems such as springs of polluted water [7] and damage to building foundations [8,9].

In the United Kingdom (UK), shrinking and swelling of clay lithologies represents one of the most damaging geohazards, costing the economy an estimated £3 billion over the past 10 years as reported by the Association of British Insurers [10]. Accordingly, the capability to detect and quantify the ground displacement of structures and infrastructure at regional and local scale would be a cost-effective tool that offers great value to insurance companies and government institutions.

Persistent Scatterer Interferometry (PSI) is a powerful remote sensing tool, capable of mapping displacements over wide areas at very high spatial resolutions. The technique is based on the processing of multiple interferograms derived from a large set of Synthetic Aperture Radar (SAR) images to obtain displacement time series, along the line of sight (LOS) of the satellite, of radar targets on the earth surface [11–13].

Several studies report on PSI applications for urban deformation monitoring such as the study of displacement time series of buildings, roads, railways, dams, and tunnels [14–22].

London is a megacity of the United Kingdom experiencing an increasing density of structures and infrastructure such as transport tunnels, requiring dewatering schemes to control groundwater during their construction [23,24]. Satellite-based data were previously used to characterize ground displacements in London. Aldiss et al. [25] used ERS and ENVISAT collected between March 1997 and December 2005 to carry out a geological interpretation of the subsidence and uplift trend. Cigna et al. [26] used ERS and ENVISAT covering the time intervals 1992–2000 and 2002–2010 to delineate the boundaries of the geohazards in London within the framework of the European Commission FP7-SPACE project PanGeo. Bateson et al. [27] used the satellite-based data covering the period 1997–2005 to validate the results of the modelled subsidence due to groundwater abstraction for the Merton area of south-west London. Boní et al. [28] exploited the ERS and ENVISAT covering the time intervals 1992–2000 and 2002–2010 to analyze the ground motion due to the groundwater level changes in London. More recently, high-resolution PSI data from the COSMO-SkyMed constellation has been used to study the effect of tunnel-induced subsidence damage assessment [29].

In this paper, new Sentinel-1 SAR data are used to measure the ongoing displacements in London covering the period from 2015 to 2017. The goals of this study are (1) the exploitation of new and freely available Sentinel-1 data to analyze uplift phenomena and (2) the development of a methodology for the geological interpretation of PSI results in urban areas. The developed methodology represents a refinement for uplift investigations using new Sentinel-1 data, of the methodology proposed in Boní et al. [30] for subsidence studies. The procedure is addressed to overcome limitations, such as the analysis of large data sets, by (i) improving the management and interpretation of dense time series guaranteed by Sentinel-1 (ii) and to provide an insight on the capability to monitor movements during engineering works given by the reduced revisit time (i.e., 6–12 days) of the latest spaceborne sensors.

Urban developments as well as geological, geotechnical, and hydrogeological factors have been compared with the average velocities and the displacement time series to identify the predisposing and triggering factors of the detected ground displacements. The results allow to detect and characterize uplift phenomena after the termination of engineering works such as dewatering process for structures and infrastructure network construction as in the case of Crossrail tunneling in London.

2. Study Area

With more than 60,000 boreholes sunk in Greater London alone, the geology of the London Basin has been widely described by Sumbler [31], Ellison et al. [32], and reviewed by Royse et al. [23]. The Basin consists largely of a broad gentle syncline of Mesozoic and Cenozoic units overlapping the Palaeozoic basement, the London Platform, of folded Silurian and foreland Devonian rocks at depths of ~300 m in central London. The main geological units of relevance are: The Upper Cretaceous Chalk Group, a fine grained and micro-porous limestone up to 400 m thick which mainly outcrops in the marginal areas of the basin (Figure 1). Unconformably overlying the Chalk is the oldest Paleogene deposit, the Thanet Sand Formation, a 30 m coarsening upwards succession of fine grained, sandy-silty terrigenous sediments [32]. Successively, the basin experienced the deposition of the Lambeth Group (LMBE) composed of 20/30 m illite, smectite and montmorillonite dominant clay intercalated with silty and sandy horizons and lenses. During the Eocene the deposition of the Thames Group, comprising the Harwich Formation and, then, London Clay Formation (LC), started in the basin. West of London, the youngest Eocene sediments in the London Basin are preserved: they are predominantly the sands and clay units of the Bracklesham Group.

The subsequent Quaternary deposits (superficial geology) are represented by the river terrace deposits of the Thames River, locally covered by alluvial deposits of the Thames River.

Due to the erosion of much of the overlying deposits, the LC is probably the most well-known of the units present in the London Basin with a significant influence on London's infrastructure: its widespread presence beneath much of central London, with a thickness between 90 m in the west and to 150 m in the east [33] and relatively homogeneous structure makes it a near perfect tunneling medium, thus, facilitating the development of the London Underground [23].

Nevertheless, spatially widespread illite/smectite clay minerals in the LC are particularly susceptible to seasonal processes of shrinking and swelling, potentially damaging buildings and infrastructure [34,35], and therefore represent a major concern for the insurance industry.

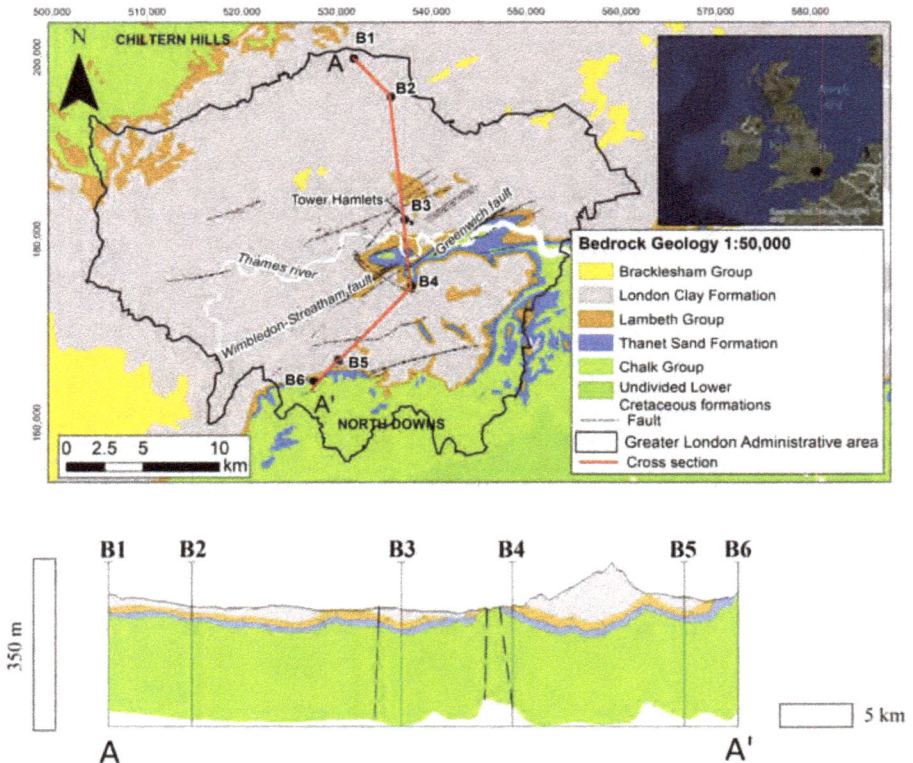

Figure 1. Location (sources: Esri, DeLorme, USGS, NPS) and geological map of London based upon the 1:50,000 bedrock geology, with the permission of the British Geological Survey. All rights reserved. British National Grid. Projection: Transverse Mercator. Datum: OSGB 1936. The geological cross section A-A' is also reported (modified from [28]).

The main aquifer of the Basin is represented by the Chalk Group. This is characterized by a primary and secondary porosity related to the matrix and the fracturing network, respectively [36–38]. The primary porosity of these rocks is about 35% [39] but the average conductivity is just about 0.001 m/day [40]. The basin is marked by several faults that act like a barrier or conduits for the groundwater flow [41,42]. The Chalk outcrops in the northern and southern part of the Basin where it is directly recharged by the rainfall infiltrations. In the central area of the Basin, the overlying Palaeogene formations confine the aquifer. Whereas the lithological variability of Palaeogene deposits leads to hydrogeological heterogeneity, these units do not form principle aquifers. Where sand-rich horizons are present in the Paleogene deposits (Thanet Formation and Lambeth Group), significant quantities of groundwater can be contained. The hydraulic continuity between the Chalk and Thanet Sands may be limited in places, and some continuity with the Lambeth Group depending on the clay and sand content [41]. Therefore, Chalk-Thanet Sands (deep aquifer) and Lambeth Group-Harwich Formation (intermediate aquifer) are regarded as separate aquifers for resource management [41]. The sand unit of Lambeth Group and Harwich Formation represents the intermediate aquifer that it encloses in the top and bottom by the clayey units of the London Clay and the Lambeth Group, respectively.

River Terrace Deposits (RTD) and Made Ground (MG) form a minor shallow aquifer, separated from deeper aquifers by clay layers.

The groundwater level (GWL) lowering across the London Basin mainly started in the mid-1850s with a progressive increase in abstraction from the Chalk aquifer. This abstraction became unsustainable leading to unconfined conditions in about 1940 and started to rise again after the mid-1970s [41,42]. In order to contrast this local groundwater variation, that can lead to serious effects on both building foundations and shallow structures such as the London Underground etc., the General Aquifer Research, Development and Investigation Team (GARDIT) strategy was developed by Thames Water, Environment Agency and London Underground and as a result, an observation borehole network within the Basin was established in order to control and manage the GWL in the London Basin [41,42]. In this area, more than 20 major tunneling projects started since the 1980s, the latest being: Pimlico and Wandsworth to Wimbledon cable tunnel (1992–1995), Jubilee Line Extension (1993–1999), Channel Tunnel Rail Link (2001–2006) and the National Grid power tunnels (2011 to present).

3. Data

A multidisciplinary approach has been implemented in order to carry out a comprehensive study of the investigated phenomena, by using satellite radar interferometric data and different information contained in geological, geotechnical, hydrogeological and buildings database.

3.1. PSI Data

The study has been performed using 79 SAR images acquired by Sentinel-1A/B satellites on the descending pass on track 201. The images were acquired from May 2015 to March 2017 with a nominal revisit cycle of 6/12 days. Processing was carried out using the GAMMA software—and, in particular, the Interferometric Point Target Analysis—IPTA package [43]. IPTA allows millimetric displacement measurements to be made using individual, highly reflective terrain-features that provide a persistent response throughout the multi-temporal dataset being analyzed. These 'persistent scatterers' (PS) generally correspond to parts of man-made structures such as buildings, bridges, pylons, etc., or hard, rocky terrain. A multi-master approach was used with displacement, elevation and thermal expansion coefficients estimated for each PS.

Results are in the satellite line-of-sight and therefore contain a combination of vertical and horizontal displacement components. The line-of-sight is defined by an incidence angle of 34 degrees from nadir and a look orientation of 282 degrees. The processing results show a total of 1,455,921 PS, over a processing area of around 1596 km^2, hence the target density amounts to 912 PS/km^2 (Figure 2).

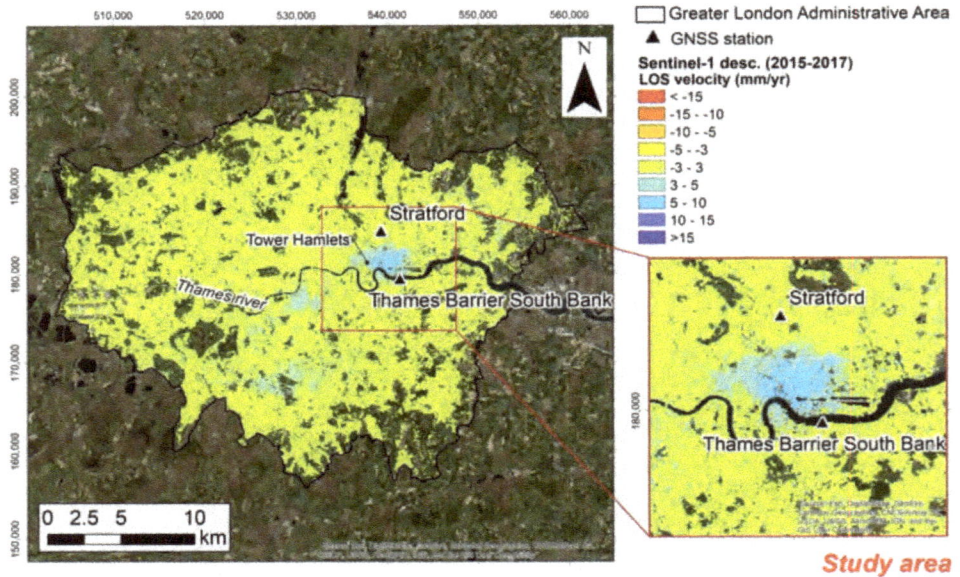

Figure 2. Average line of sight (LOS) velocity measured by the use of Sentinel-1 data during the period 2015 to 2017 across the study area. The location of the GNSS station is also reported.

The Sentinel-1 dataset covers a limited observation period, thus, the historical ascending ERS-1/2 and descending ENVISAT data, from 1992 to 2000 and from 2002 to 2010 [26,28], respectively, have been also analyzed for the characterization of the local scale ground instability (study area, see the location in Figure 2). PSI ground motion data acquired by ERS1/2 and ENVISAT data, are characterized by nominal repeat cycle of 35 days and were also processed using the GAMMA SAR and Interferometry software and, in particular, the IPTA algorithm [43]. This allows an investigation into the evolution of displacement from 1992 through to the current day.

3.2. Geological, Groundwater Level and Buildings Database

Different datasets have been considered to analyze the predisposing and triggering factors of ground motion in London. The exploited databases are described as follows.

- Groundhog Desktop from the British Geological Survey (BGS)

Groundhog Desktop is a free-to-use software tool for visualizing and interpreting a range of geological and environmental data such as boreholes, water levels, geo-technical and geo-chemical measurements, geological maps, conceptual models, and cross-sections [44]. The database contains more than 1,300,000 records of boreholes, shafts, and wells from all forms of drilling and site investigation work available in Great Britain. This database has been used to estimate the thickness of the clayey deposits.

- National Geotechnical Properties Database (NGPD) from BGS

The NGPD holds geotechnical information extracted from site investigation records provided by clients; consultants and contractors, and from field and, secondarily, from laboratory test results carried out by the BGS [45]. For each sample, the location, the lithology, the depth, and the geotechnical characteristics are documented. The Swelling Pressure of the clayey layer has been analyzed using this database.

- GeoSure from BGS

GeoSure identifies and classifies the susceptibility connected to areas of potential natural ground movement in Great Britain. The database provides information on the Volume Change Potential (VCP) and Volume Change Potential Range of the shrinking-swelling formations in three-dimensional space; at intervals down to 20 m in Greater London [46]. The VCP values are based on the Modify plasticity Index (Ip') proposed in the Building Research Establishment Digest 240 [46].

- Groundwater level data from Environment Agency (EA)

The EA maintains an extensive network of groundwater level observation boreholes within and outside the London Basin. Time series data for groundwater levels have been made available from the Environment Agency's WISKI database and plotted as water level in m AOD (above ordnance datum; UK sea level measurement).

- Building height map from Emu Analytics

The online database [47] shows the height of each buildings for England's largest urban areas, based upon the EA's LiDAR data up to 2015.

- Building age map from Consumer Data Research Centre

The database [48] shows, for the major cities of England and Wales, the age of constructions of residential structures present in it. The dwelling age data is supplied grouped in approximately ten-year age bands starting from 1900 with a count of the number of houses in each band plus a Pre-1900 band that groups all the previous buildings.

4. Methodology

In this section, the procedure to analyze uplift phenomena in urban areas by exploiting the new Sentinel-1 data is presented refining the one discussed in Bonì et al. (2016) [30] for subsidence studies. The procedure consists of three main phases (Figure 3). In the first phase, the displacement time series (TS) and the average velocity accuracy assessment is performed. In the second phase, different statistical tests are applied in order to find the spatio-temporal pattern of the principal components of movement, and the kinematic model of the targets. Finally, the third step consists of the mechanism recognition of the ground motion areas. Therefore, the integration of satellite data with geological, geotechnical, hydrogeological, urbanization and construction processes data is considered in order to investigate the causes of ground motion processes. The phases of the methodology are described in detail in the following subparagraphs.

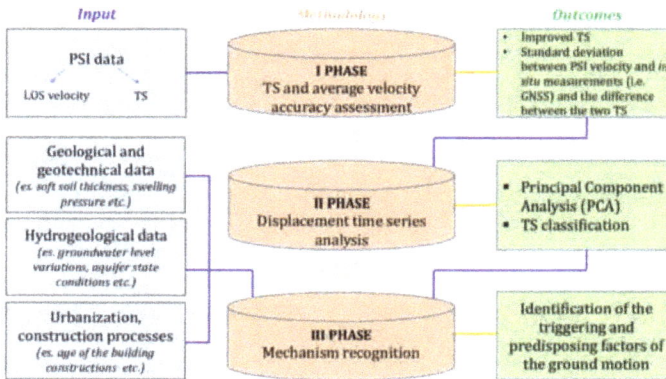

Figure 3. Flowchart of the methodological approach to detect and characterize uplift phenomena in urban areas.

4.1. Displacement Time Series and Average Velocity Accuracy Assessment

In order to properly exploit the TS and the average velocities, it is fundamental to take into account that these measurements can be sensitive to uncompensated orbital errors or uncompensated low frequency atmospheric effects [49]. These systematic errors can affect both TS and the average velocities and can be detected as regional trends in the whole dataset [50]. Furthermore, TS are also sensitive to the phase noise, therefore, a post-processing analysis of the PSI data is essential to avoid misinterpretation of unreal ground deformation. A TS check is performed by selecting the most coherent (>0.9) targets with an average LOS velocity in the range ±0.5 mm/year., where no significant movements are expected. The approach proposed by Notti et al. [51] was applied to correct uncertainty due to regional unreal trends, and anomalous displacement detected on certain dates (i.e., unreal movements at the same time as meteorological events, such as snowfall). Thus, the average TS of the selected targets is extracted and according to the wavelength of the sensor (C-band), a threshold ±5 mm of displacement along LOS is considered for detecting anomalous dates. The regression line of the average TS of the selected target is computed to investigate if the dataset is affected by regional trend or tilts.

Additionally, TS and the average velocity accuracy assessment is also performed using external data such as measurements acquired by levelling campaigns, inclinometers, or Global Navigation Satellite Systems (GNSS) stations. First, PSI time series and the average velocities are projected along the vertical direction, by dividing them for the cosine of the satellites incidence angle. Then, the average velocity is computed for the period that matches between the two independent techniques. Thus, the standard deviation between the PSI and GNSS average velocity (δ_{VEL}) is estimated. The same procedure is also applied for the displacement time series. In this case, the standard deviation of the difference between the PSI and GNSS time series (δ_{TS}) is computed. If the standard deviation of the PSI and GNSS average velocity and of the difference between the displacement time series is close to the sensitivity of the PSI technique the results can be assumed as consistent.

4.2. Displacement Time Series Analysis

PSI techniques allow the measurement of the average velocity and the displacement time series of a huge number of measurement points. The displacement time series are complex to interpret using manual analysis. Indeed, recent studies report some methodologies to overcome limits related to manual and visual analysis of TS [51–54] using automatic or semi-automatic time series analysis. These methodologies support the analysis of large PSI datasets in order to identify areas of interest using the time series trends. Here, two approaches have been implemented: (1) a statistical procedure to find the principal components (PC) of TS and (2) an automatic classification tool for TS based on statistical tests that analyze the variance.

PC analysis has been applied to satellite-based time series by implementing a matrix of PS location versus time [30]. The matrix contains in each column the LOS displacements for each SAR image, and in each row the displacement time series of the targets. The main outcomes are the correlation and covariance matrices, the eigenvalues and eigenvectors, the percent variance that each eigenvalue captures, and the PC score maps. In interpreting the principal components, PC scores and the eigenvectors related to each target are useful for knowing the distribution and the trends of the principal components. Furthermore, scree plots of the percentage of variance explained by each principal component is useful to find the number of significant PC of the dataset.

The second procedure is the TS analysis using the PS-Time program [54], which is a freely downloadable toolbox compiled in MATLAB [54]. The main outcome is the classification of TS in one of three predefined target trends such as uncorrelated (displacement fluctuates erratically over time), linear (linear and constant velocity) and non-linear (changes of the velocity over time and style of deformation that can be quadratic, bilinear, discontinuous with constant velocity and discontinuous with variable velocity) based on a sequence of statistical tests that discriminate different styles of ground deformation. Furthermore PS-time program permits the detection of the date (break) where abrupt changes in slope in non-linear TS are recorded.

Then, the outcomes of the two approaches are exploited to enhance the interpretation of the predisposing and triggering factors using cross-comparisons.

4.3. Mechanism Recognition

Geological interpretation of satellite data is performed using external data (such as geological, geotechnical, hydrogeological, urbanization and construction process data), by integrating them into a Geographical Information System (GIS). The recognition of the mechanisms is based on cross-comparison of the representative subsoil geological profiles, and the relative displacement time series with multidisciplinary information [14]. Furthermore, the analysis of the breaks and the detection of the deceleration and acceleration periods are fundamental to identify the predisposing and triggering factors.

5. Results

5.1. Displacement Time Series and Average Velocity Accuracy Assessment

The displacement time series were analyzed as described in 4.1. First, targets characterized by coherence higher than 0.9 and LOS velocity in the range ±0.5 mm/year were selected. By applying this filter to the whole dataset 14.5% of PS shows high coherence and low velocity rates. Then, the average displacement of the selected targets has been computed for each SAR scene, to extract the average TS (Figure 4). It is worth noting that Sentinel-1 data are affected by a higher noise than the long-term time series such as acquired by ERS-1/2 and ENVISAT satellites. The reason for this issue is mainly due to the much shorter time span covered by the Sentinel-1 data (2 years) [55].

However, the results show that anomalous LOS displacement at certain date and regional trends are not evident. Therefore, all SAR scenes have been exploited for the following analysis and post-processing corrections were not implemented.

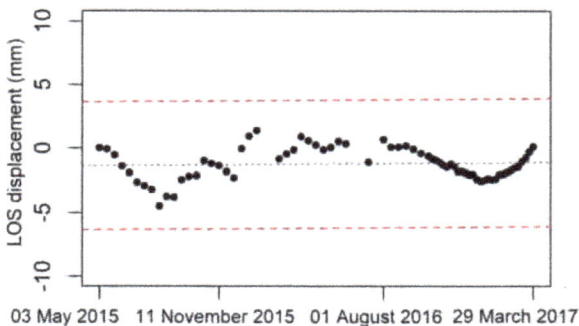

Figure 4. Average displacement time series (TS) of the targets characterized by coherence higher than 0.9 and LOS velocity in the range ±0.5 mm/year. The black dotted line represents the regression line of the average TS whereas the red ones are the upper and lower threshold line.

Furthermore, the measurements acquired by two GNSS stations located in London, and available through the British Isles Continuous GNSS Facility (http://www.bigf.ac.uk/), have been exploited (see the location in Figure 2) to obtain a comparison of the displacement time series and the average velocity using independent data. Firstly, LOS displacements TS and average velocities were projected along the vertical direction, by assuming that the displacement is essentially vertical [25,26]. This estimation was done dividing the LOS measurements by the cosine of the incidence angle.

The measurements acquired at the Stratford station (STRA) and Thames Barrier South Bank station (TBSB) were compared with the average velocity and TS of the nearest targets (around 12 and 30 m, respectively) characterized by a value of coherence higher than 0.9 (Figure 5). Also, in this case, TS acquired by the Sentinel-1 sensors are characterized by a higher noise than the measurements acquired by the two GNSS stations for the short total time span covered by this dataset [55]. The standard deviation of the velocity acquired by PSI data and the Stratford station (δ_{VEL}) reaches values of 1.35 mm/year. While the standard deviation of the difference between the PSI and GNSS time series (δ_{TS}) reaches values of 0.66 mm. The comparison between the PSI data and the measurements acquired at the Thames Barrier South Bank station shows values of 1.34 mm/year and 1.59 mm for δ_{VEL} and δ_{TS}; respectively. It is worth noting that the use of only two GNSS stations cannot adequately validate the PSI data at the scale of the basin. However, the results show that at local scale the PSI data are consistent with the displacements and velocities measured by the available GNSS stations. Indeed, the retrieved values for δ_{VEL} and δ_{TS} are comparable with the values (1 mm/year and 5 mm) obtained by previous authors using the C-band sensors [56]. Whereas, a higher accuracy was detected using X-band sensors (0.5–01 mm/year and 1 mm) by other authors [57,58].

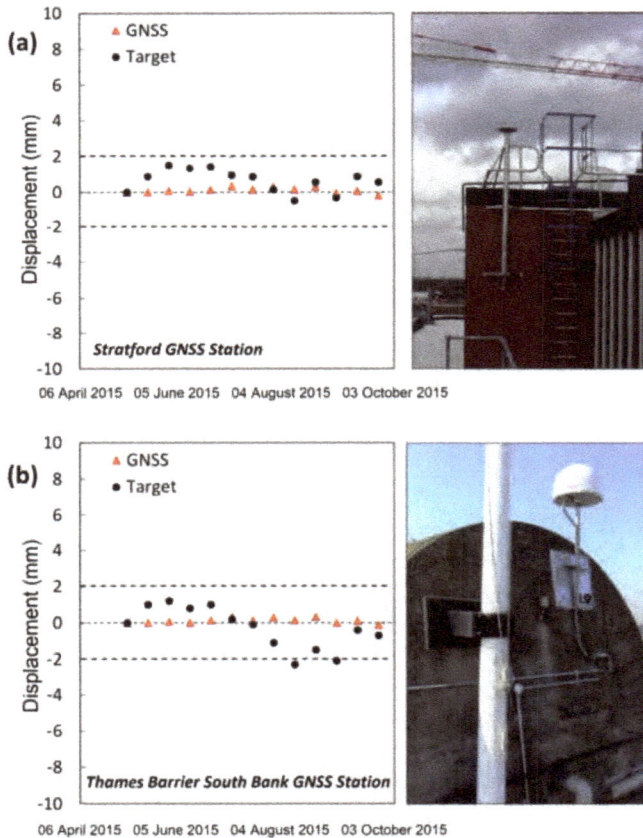

Figure 5. Comparison between the PSI and GNSS vertical displacement time series. See Figure 2 for the GNSS stations location. The antenna installation images are also reported (available data at http://www.bigf.ac.uk/files/network_maps/script_all_pcsn_30s.html).

5.2. Displacement Time Series Analysis

Following the time series accuracy assessment, TS were analyzed through different automatic statistical procedures; such as the Principal Component Analysis (PCA) and the TS classification in predefined target trends. The first procedure has been performed at large scale (the Greater London administrative area) using TS over the whole dataset, while the second one has been applied using TS at local scale over an area of 206 km^2 (study area, see the location in Figure 6). Indeed, even though the average velocity may be useful to detect physical processes characterized by linear trends, the same parameter seems not to be efficient in detecting non-linear and seasonal movements [59]. TS classification was implemented to investigate in detail the study area and was not performed at large scale for the computational load.

Principal Component Analysis approach has been applied using the procedure described in Section 4.2. Principal component analysis was performed using Sentinel-1 data to analyze the spatio-temporal deformation pattern during the period 2017–2017 as previously applied using different SAR data for land subsidence studies [30,53,59]. The results show that the first component of motion (PC1) explains 97.3% of the variance and 2.54% is explained by the second component of motion (Figure 6d). The other components are not significant (explained variance lower than 1%) and were not considered in the following analysis. Figure 6 shows the spatial distribution of the principal components score units. Positive scores for a PC have a TS trend similar to that of one of the eigenvector time functions, while those with negative scores show a TS trend opposite to that of the eigenvector time function [60]. Indeed, each PC is defined by a linear combination, whose coefficients, termed eigenvectors, are the magnitude of the contribution of each original variable to each PC [60].

The spatial pattern of the PC scores highlights that PC1 mainly affects the southwestern sector of the Greater London administrative area (Figure 6a); while the second component is mainly localized in the southwestern sector (Figure 6b) and in the uplift zone localized in the study area (see insets box in Figure 6b). A visual inspection of the principal components eigenvectors of the dataset covering Greater London administrative area (Figure 6c) highlights that the first component (PC1) corresponds to the long-term linear lowering of the earth's surface, while the second one corresponds to the long-term uplift ground motion with a non-linear trend. The uplift area is characterized by negative scores of the PC1 and high positive score of the second one. Therefore, the TS trend of the uplift phenomena is directly correlated with the PC2 trend. The PCA approach has allowed for the easy detection of not only the uplift but also the non-linear trend of this phenomena characterized by two breaks in the eigenvector time function (Figure 6c) by analyzing around 1,455,921 PS. Furthermore, the boundaries of the uplift zone have been defined using the approach proposed by Bonì et al., 2016 [53], by using a buffer area of 50 m around the PS characterized by PC2 score higher than the interquartile range (Figure 10).

TS classification has been applied to investigate in detail the time series trend in the uplifting area. The results of the automatic TS classification show that among the selected targets (around 260,000) in the study area, 74.5% are classified with a non-linear trend, whereas 1.6% are classified with a linear trend (Figure 7a). The remaining targets (23.9%) show uncorrelated time series. The non-linear time series are characterized by a strong non-linearity. Non-linear time series are bilinear trend and mainly, changes in trends in displacement time series were identified at August 2015 and May–August 2016 in the Tower Hamlets area. The results are consistent with the eigenvector of the second component of ground motion detected using the PCA approach. Indeed, by selecting the targets within the uplift zone (see the location in Figure 10) and by computing the average TS (Figure 7b), the break dates are evident. LOS velocity until August 2015 is −12.86 mm/year, whereas in the following period an uplift trend with LOS velocity of 8.31 mm/year is observed. Furthermore, starting in May–August 2016, a deceleration of the movements is detected, with LOS velocity of 1.23 mm/year.

The results of the analysis highlight that a remarkable ground uplift phenomenon has been detected in an area of 5 km^2 with deformation rates ranging from to 6 to 18 mm/year and the TS

Remote Sens. **2018**, *10*, 607

analysis give insight about acceleration and deceleration of the ground motion during the monitored period (2015–2017).

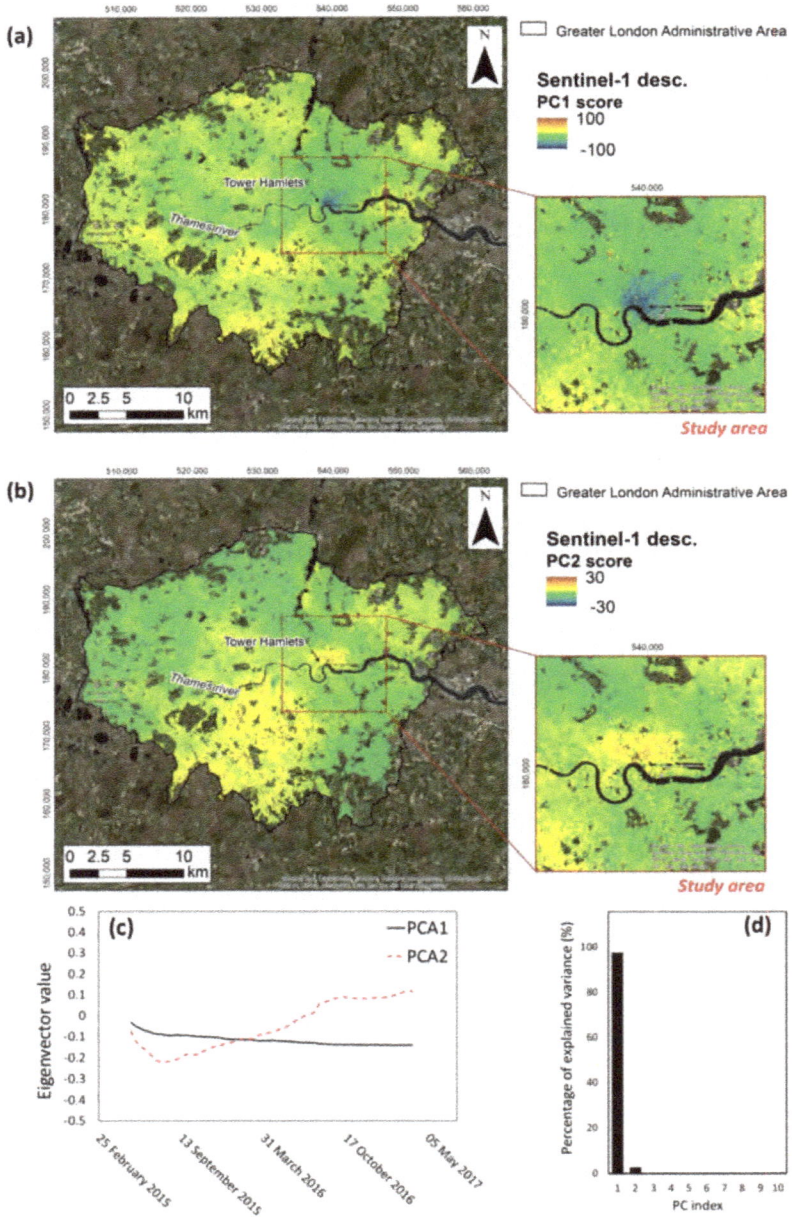

Figure 6. Principal component score maps of the first (**a**) and second (**b**) component of motion. Eigenvector value (**c**) of the principal component (PC) and the percentage of explained variance (**d**) are also reported.

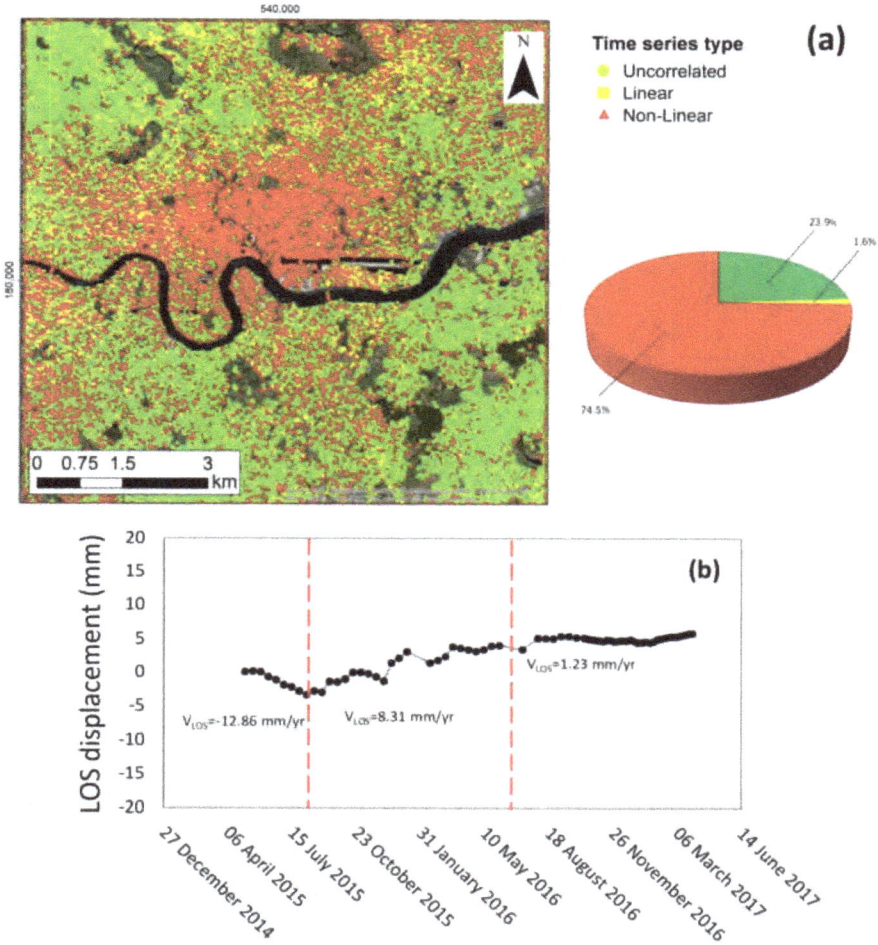

Figure 7. TS trends in the uplift zone (**a**). Average TS of the non-linear targets located in the uplift zone. The detected breaks (red dotted lines) are also reported (**b**). See the location of the uplift zone in Figure 10.

5.3. Mechanism Recognition

5.3.1. Analysis the Geological and Geotechnical Factors

The lithological information of 21 boreholes provided by BGS were exploited to estimate the thickness of the clayey soils in the study area (Figure 8a). First, the thickness of the LC was computed and then, the estimation of the clayey soils was estimated as the total thickness of the clayey soils considering the LC and LMBE (Figure 8d). Furthermore, the average LOS velocity in a 50 m buffer zone were estimated for each borehole. Considering the thickness of the LC, the comparison between the clayey soils thickness and the LOS velocity show a coefficient of determination (R^2) for the linear regression of 0.56 (Figure 8c). While by considering the clayey layers of the LC and the LMBE, a higher correlation is retrieved showing a coefficient of determination, R^2 for the linear regression of 0.75 (Figure 8d). Therefore, the uplift rates are higher when the clayey layers are thicker.

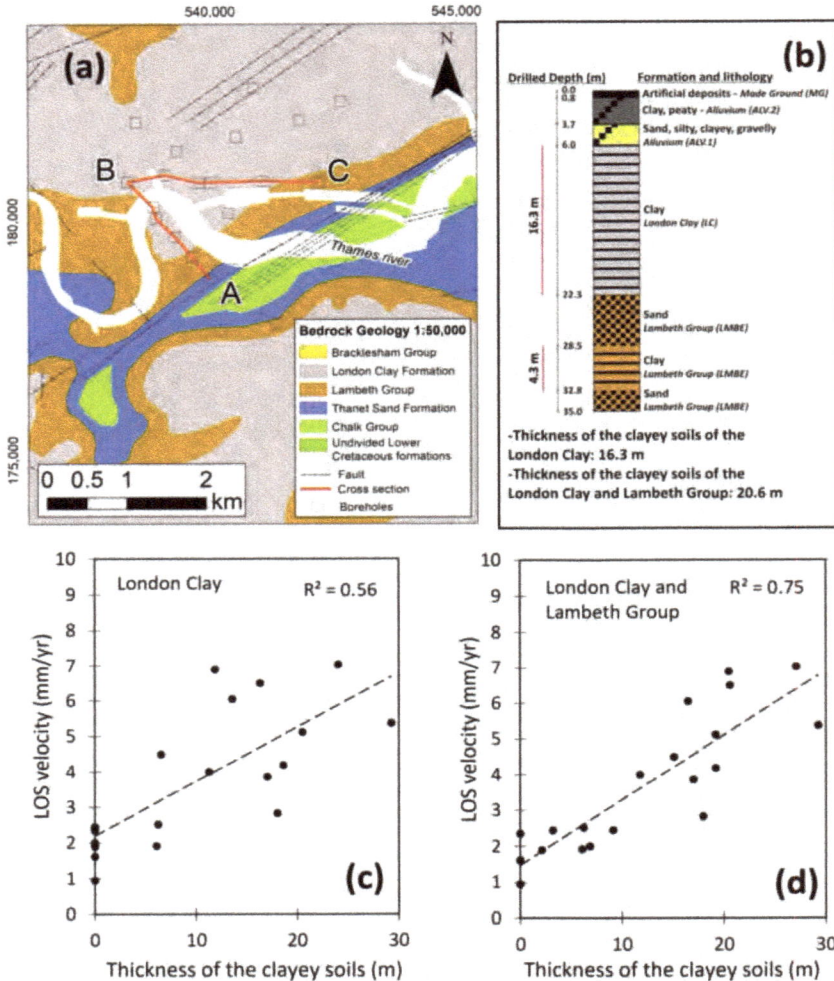

Figure 8. (**a**) Location of the boreholes used to estimate the thickness of the clayey soils. The cross sections (red line) are also reported; (**b**) Example of the procedure applied to estimate the thickness of the clayey soils of the London Clay and by computing the sum of the thickness of the clayey soils of the London Clay and Lambeth Group; (**c**) Comparison of the LOS velocity estimates in the period 2015–2017 with the thickness of clayey soils of the London clay using the boreholes information; (**d**) Comparison of the LOS velocity estimates in the period 2015–2017 with total thickness of clayey lithologies of the boreholes considering the London Clay and the Lambeth Group. Based upon the 1:50,000 bedrock geology, with the permission of the British Geological Survey. All rights reserved.

Then, the BGS geotechnical database [44] was analyzed to investigate the role of the geotechnical properties of the deposits located in the study area. A buffer of 50 m has been computed for each geotechnical borehole (Figure 9a) and the average LOS velocity has been extracted. Then, the LOS velocity in the buffers have been plotted versus the swelling pressure (SPRS) of these formations (Figure 9b). The values of swelling pressure represent the pressure exerted by a contained clay when absorbing water in a confined space and derive from consolidation tests performed using samples extracted at different depths for each borehole [44]. The analysis of the SPRS has been performed by diving the measures obtained for the alluvial of the Thames and River Terrace Deposits that represent the superficial deposits (samples carried out in the first 7 m) and for the deeper formations such as the LC and LMBE (samples carried out at depth higher than 7 m). It is worth noting that a direct and linear correlation between the swelling pressure and the average LOS velocity for the deeper layer is evident, while the superficial deposits do not show a linear correlation between the swelling pressure and LOS velocity which can be explained with the coarser granulometry, different mineralogy and thinner succession (<10 m on average) of the overlying material. Jones et al., 2017 [34] analyzed the swelling potential of the London Clay at different depths, and specifically: 0, −1, −2, −3, −4, −5, −10, −15 and −20 m across England and Wales. According to Jones et al., 2017 [34], in the study area are present potential swelling soils associated with a high Volume Change Potential (VCP). VCP is the relative change in volume of a soil to be expected with changes in soil moisture content and is manifested by shrinking and swelling of the ground [34]. The superficial deposits (up to −5 m deep) are characterized by VCP in the classes A, B, and C that correspond to non-plastic, low and medium plasticity classes, with Ip' up to 40% according the classification introduced by Jones et al., 2017 [34]. From −10 m the VCP assumes value referred to class D according to [34], that means formations with high plasticity up to 60%. Therefore, also the analysis of the plasticity index shows that the clayey layers between 10–15 m depth are characterized by high plasticity index and high swelling pressure and these layers could be responsible of the detected uplift.

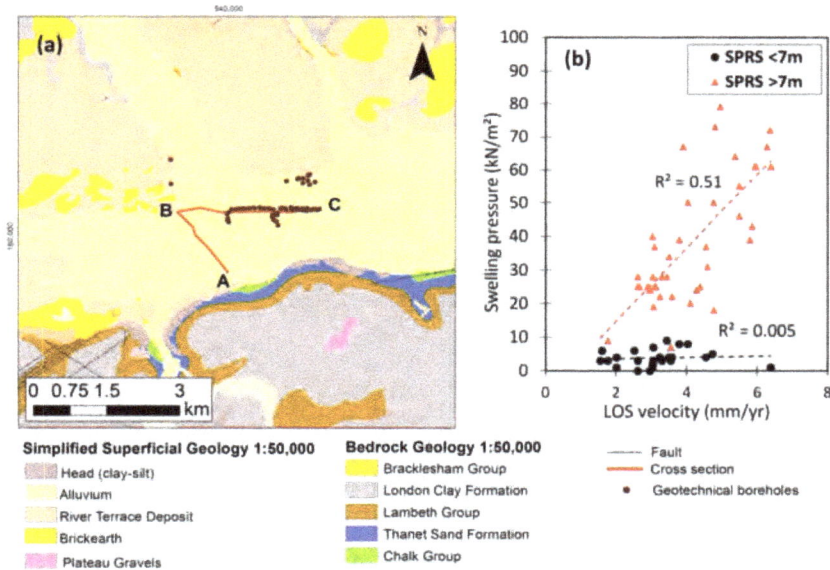

Figure 9. (a) Location of the geotechnical boreholes; (b) Comparison of the LOS velocity estimates in the period 2015–2017 with the swelling pressure of the superficial (depth lower than 7 m) and deeper layers (higher than 7 m). Simplified Superficial Geology of Greater London modified from the DiGMapGB50, the Digital Geological Map of Great Britain at the 1:50,000 scale and bedrock geology at the 1:50,000 scale, with the permission of the British Geological Survey. All rights reserved.

5.3.2. Comparison between the Groundwater Level Changes and Deformation Rates

After considering the clayey layer of the LC and partially contained in the LMBE as the predisposing factor of the uplift, the triggering factors of the occurrence of these movements in the period 2015–2017 were investigated. First, the groundwater level changes were analyzed to verify if the uplift is related to a groundwater rebound. Indeed, the uplift zone (Figure 10) is located in an area where the aquifer is confined by the LC and pore pressure changes could trigger ground motion. Therefore, historical ERS-1/2 and ENVISAT data were also exploited to compare the deformation rates and the piezometric level evolution in the last twenty-four years. Unfortunately, in the uplift area (Figure 10), piezometers time series provided by the Environment Agency are not available for the 1992–2016 period, but four exemplificative piezometers time series have been analyzed in proximity of the uplift area (Figure 10). It is worth noting that the use of the PSI data acquired by different sensors using different incidence angles clearly affects the capability to measure the vertical component of the displacement. More precisely, ERS-1/2 and ENVISAT with an incidence angle of 23° allows the estimation of 92% of vertical displacements, while Sentinel-1 satellites with a 34° incidence angle only detect 83% of vertical displacements. Therefore, LOS velocities have been projected along the vertical direction for each dataset in order to homogenize the datasets for the comparison with the groundwater level changes (Figure 10).

Groundwater level changes between January 2015 and January 2016 reported by the Environment Agency [41] record a rise of 1 m in the uplift zone (Figure 10). Furthermore, by cross-comparison of the available piezometric level measurements with the deformation rates, a direct correlation is evident (Figure 10) as previously reported [28].

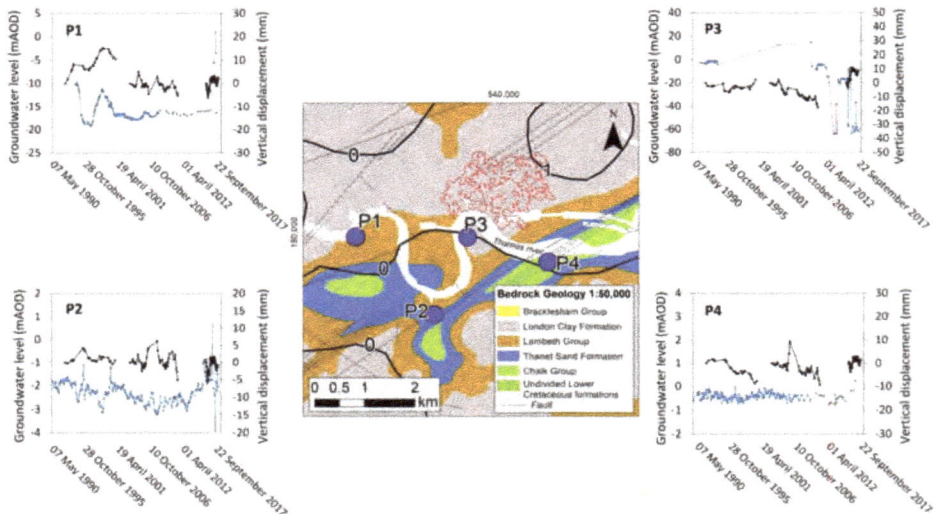

Figure 10. Cross-comparison between the deformation rates detected using ERS-1/2, ENVISAT and Sentinel-1 data and groundwater level changes. The black and the blue lines represent the vertical displacement and the groundwater level data, respectively. In the map; the black lines represent the groundwater level change (m) between January 2015 and January 2016 from [41] and the red line represents the uplift area. Based upon the 1:50,000 bedrock geology, with the permission of the British Geological Survey. Contains Environment Agency information © Environment Agency and/or database right 2017.

5.3.3. Comparison between Urbanization, Construction Processes and Uplift Rates

As already mentioned in Section 1, London represents a megacity where an increasing urban development has been reached. Therefore, the role of the urbanization for the ground displacement has been investigated. Building construction could be an accelerating factor for the consolidation processes [61], whereas in this case it was verified that the uplift could be due to local groundwater rebound after the termination of construction processes supposing dewatering operations. Thus, the age of the building constructions has been plotted versus the LOS velocity, with the results showing that newer building corresponds to areas with the highest terrain uplift (Figure 11). Furthermore, even if newer buildings have been constructed (2000–2009), the measured LOS velocity is low (higher than −3 and lower than 3 mm/year), whereas in the London Clay it is not present (see the cross-section A-B in Figure 11). While uplift rates higher than 7 mm/year are observed where the buildings have been built up in the period 2000–2009, the uplift ranges between 3 to 7 mm/year where the buildings have been constructed in the period 1993–1999.

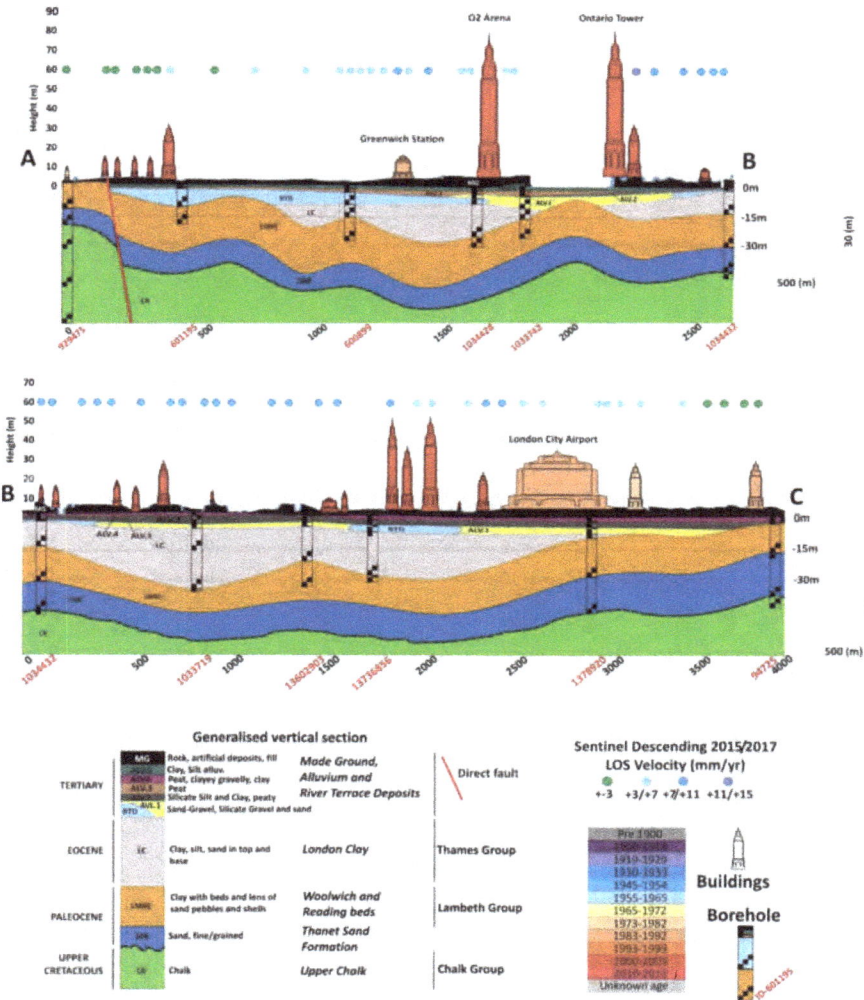

Figure 11. Cross section and buildings age. Based upon Groundhog Desktop data; with the permission of the British Geological Survey. See the location of the cross section in Figures 8 and 9.

Furthermore, the role of the engineering works for the Crossrail construction has also been assessed. Crossrail is a 42 km underground railway under London, running as far west as Reading in

Berkshire and as far east as Shenfield in Essex and includes the construction of a 21 km of twin-bore tunnel, 10 newly built, and 30 upgraded stations up to approximately 30 m below street level (Figure 12). During the construction, begun in 2012, controlled dewatering was performed at some stations in order to guarantee ground stability during the excavation process [62]. The dewatering operations started from 11 August 2008 for works at Canary Wharf Station (Figure 12) and ceased in August 2015, with a substantial reduction of pumping at Limmo (Figure 12). All dewatering operations were completed on 14 March 2016. Simultaneous dewatering was undertaken at multiple Crossrail sites. During the dewatering process the ground water level was monitored. More precisely, the dewatering was performed in the deep aquifer (Chalk-Thanet Sand) and in the shallow aquifer (Made Ground and River Terrace Deposits). In the intermediate aquifer (Lambeth Group-Harwich Formation), de-pressuration works were performed to reduce the water pressures, rather than removing the water itself. Therefore, the works connected with the Crossrail Project mainly affected the deep aquifer and did not affect the intermediate and shallow aquifers [62,63].

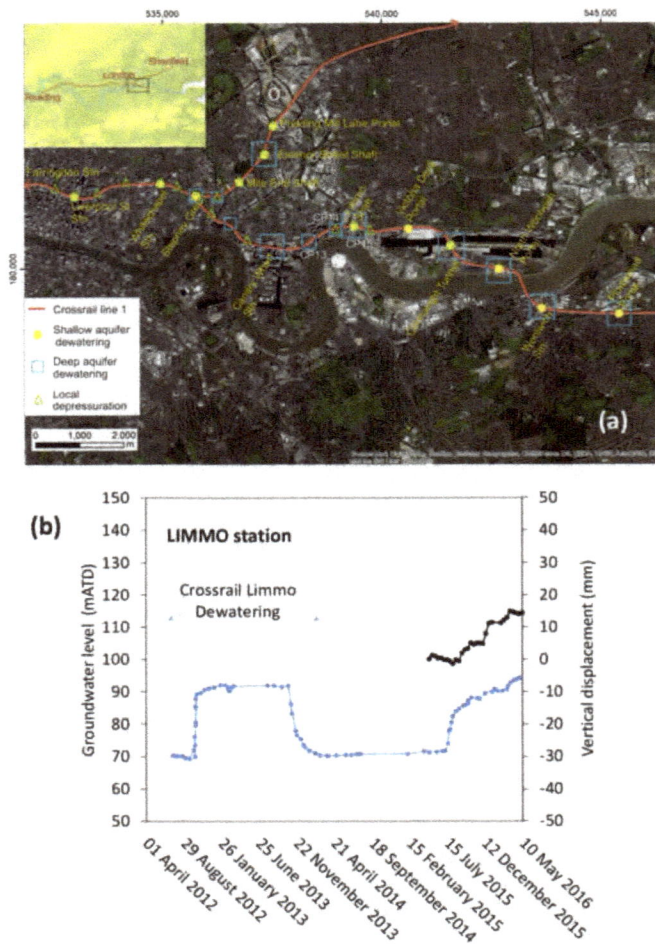

Figure 12. (**a**) Location of the Crossrail line 1 and the dewatering sites. (**b**) Cross-comparison of the average TS obtained using the Sentinel-1 data in a buffer zone of 50 m from Limmo shaft and the groundwater level changes measured at Limmo station. Dewatering periods are also reported.

The uplift area is located in the Crossrail worksite named the Limmo Shaft, where a 40 m deep shaft was required for the construction of 8.3 km of tunnel from the Limmo Peninsula to Farringdon (Figure 12a). In this area, the pumping started on the 4 November 2013 to support dewatering works CP13 and CP14 (Figure 12a) and was completed on the 14 March 2016 (Figure 12b). Dewatering of the major aquifer was performed, including the deep aquifer, in the Thanet Sand and Chalk strata. After the completion of the dewatering, groundwater recovery was observed from August 2015 to May 2016. The uplift rates detected using the Sentinel-1 data were compared with the groundwater level changes measured at Limmo station (Figure 12b). The results give insight about the direct correlation between the groundwater rebound after the completion of the dewatering procedure at Limmo. Indeed, the breaks in the TS date detected at August 2015 and May August 2016 represents the termination of the dewatering procedure and the end of the groundwater rebound, respectively. The groundwater rebound of around 20 m corresponds to about 15 mm of uplift. Moreover, a delay time of about 1 month has been detected between the beginning of the groundwater level recovery and the uplift trend.

6. Discussion

This work introduces a methodology for the geological interpretation of PSI data for the detection and characterization of uplift trends in an urban area. The systematic and reproducible procedure has been developed and tested using the new and freely available Sentinel-1 SAR data acquired over the London basin. The procedure improves the management and interpretation of dense time series as acquired by Sentinel-1 sensors and gives insight into the possibility of new applications of the recent sensors to monitor the movements during construction process and the effects after the termination of the engineering works.

From the geological point of view, the uplift zone insists on the depocenter of the syncline of the London Basin and it is bounded by faults in the southern area. The analysis of the geotechnical parameters has highlighted a high value of swelling pressure for the clayey soils of the London Clay and Lambeth Group (bedrock formations), and a low value for the alluvium soil and River Terrace Deposits present in the superficial deposits, confirming that bedrock formations, when wetted, have a great potential of volume change which can be responsible for the uplift. Indeed, in the study area, the piezometric level rose by 1 m in the period 2015–2016.

In the Tower Hamlets area (Figure 1), several large buildings built after the 2000s are also present. To construct their foundations the piezometric level was probably lowered below the swelling formations. Following the construction of the foundations the subsequent rise of the groundwater may have caused the swelling of the clayey formations. Considering that dewatering operations probably were implemented in the uplift area to proceed with construction works; the subsequent groundwater rebound after the completion of the works may have caused the swelling of the clayey formations. From the analysis of the geological, hydrogeological, structural, and anthropogenic factors it is possible to hypothesize that the uplift was caused by the rise of the piezometric level in this particular geological contest.

It is worth noting that the groundwater rebound due to the completion of the dewatering procedures for the Crossrail construction is correlated with the non-linear uplift trend detected using the Sentinel-1 data. The dewatering activities in the shallow aquifer were performed by local pumping or sump flows of the deposits enclosed within impermeable retaining walls and they did not generate effects on this aquifer outside the retaining walls [62]. Whereas, the dewatering operations of the deep aquifer generated temporary effects in the groundwater level [63]. Indeed, the maximum planned abstraction triggers; in January 2014, a drawdown cone measuring 5.9 km × 7 km in plan, with a maximum drawdown in the Chalk of about 35 m at the Canary Warf [62]. After the termination of Crossrail dewatering, the drawdown cone induced by Crossrail was dissipated and the groundwater level gradually recovered. Therefore, it is reasonable that the highest value of uplift occurs in an area where the recovery is fastest because the cone depression was greatest.

Therefore; the full groundwater level recovery that has been achieved at Limmo could be the cause of the uplift area, whereas in the Canary Wharf, the deep central cone still existed in 2016 because the dewatering was still ongoing. Thus, an uplift trend is also expected in the proximity of the Canary Wharf after the groundwater recovery in the following years.

7. Conclusions

In this work a methodology to detect and characterize uplift trends in urban areas using new satellite SAR sensors such as Sentinel-1 data is presented. The principal novelty of the proposed procedure is the full exploitation of the displacement time series and the average velocity obtained by the PSI technique. The methodology represents a systematic and reproducible approach to investigate uplift phenomena in urban areas. The proposed methodology consists of three phases:

1. Displacement time series and average velocity accuracy assessment: check of the systematic errors that could affect the whole dataset and local validation with external data such as GNSS measurements.
2. Displacement time series analysis: two approaches have been implemented, such as a statistical procedure to find the PC of TS and an automatic classification tool for TS based on statistical tests.
3. Mechanism recognition: cross-comparison between the uplift rates and external data (such as geological, geotechnical, hydrogeological, urbanization and construction processes data) in order to interpret the predisposing and triggering factors of the phenomena.

The methodology was tested in London, which is a representative high-urbanized area, using the Sentinel-1 data covering the period from 2015 to 2017. The results confirm its ability for the definition of the extension and rates of uplift phenomena and for the characterization of the phenomena. The results reveal that an area of about 5 km^2 localized between the O2 arena and the London City Airport has experienced an average uplift rates of 7 mm/year and a maximum value of 18 mm/year in the period from May 2015 to March 2017. The analysis of the displacement time series indicates a non-linear trend of the uplift area characterized by an acceleration during the period from August 2015 to May–August 2016, and decreasing uplift rates from 2017. The comparison between the spatio-temporal evolution of the movements and the predisposing and triggering factors gives insight about the correlation between the uplift trends and the thickness of the clayey soils within the LC and LMBE. Furthermore, the analysis of the geotechnical properties of the LC shows that the soils localized in the uplift area have high volume change potential between 10 to 15 m in depth. The groundwater level change of the uplift area is about of 1 m in the period 2015–2016. The uplift triggered by the groundwater rebound occurs in a peculiar geological context, such as the depocenter of the syncline of the London Basin, where the maximum thickness of clayey soils is reached, and the aquifer is characterized by confined conditions. Additionally, the role of recent construction has been investigated in detail. Indeed, uplift rates higher than 7 mm/year are observed where the buildings have been built up in the period 2000–2009 and the uplift ranges between 3 to 7 mm/year where the buildings have been constructed in the period 1993–1999.

Furthermore, the role of the engineering works for the Crossrail construction has also been assessed and the results reveal that the breaks in the TS coincides with the termination of the dewatering procedure and the end of the groundwater rebound, respectively.

In the era of big data with multiple SAR systems coming into service, our methodology has proven to be appropriate and efficient to leverage large volumes of heterogeneous data and is easily applicable in others urban areas. The findings in this work confirmed that these systems can support the knowledge of the effects after the termination of dewatering works such as for the Crossrail project in London.

The approach contributed to understanding the uplift phenomenon in London due to the interaction of anthropic activities and natural predisposing factors. Overall this work contributes to an increased understanding of the ground motion response after dewatering operations and the

Remote Sens. **2018**, *10*, 607

analysis could play a significant role on evaluating the induced displacements by future construction processes in other similar geological settings.

Acknowledgments: Groundwater level data for the London Basin were provided by the Environment Agency (License No. NR45879; non-commercial use), and the ERS and ENVISAT PS data via the EC FP7 PanGeo project. Sentinel-1 data were provided by the European Space Agency (ESA). This paper is published with permission of the Executive Director of BGS; NERC. The authors would like to thank the editor and three anonymous referees for their valuable comments and suggestions to improve the quality of the paper.

Author Contributions: Roberta Bonì wrote the paper and she conceived and designed the methodology to detect and characterize uplift phenomena in urban areas using Sentinel-1 data in the framework of the project entitled "Advanced detection, interpretation and modelling of ground motion areas (A-GMA)" at the University of Pavia. Alberto Bosino analyzed the PSI and geological data in the framework of the Erasmus + Traineeship program at the British Geological Survey (BGS) in March–May 2017. Claudia Meisina provided guidance and support throughout the research process to develop the methodology and she provided support for the analysis of the geotechnical data. Alessandro Novellino and Luke Bateson supervised Alberto Bosino during his visit at BGS and supported the analysis of the PSI and geological data. Alessandro Novellino provided guidance for the collection and analysis of the geological; groundwater level and buildings database. Claudia Meisina and Roberta Bonì supervised Alberto Bosino in the framework of his master thesis. Harry McCormack processed and analyzed ERS-1/2, ENVISAT and Sentinel-1 data for the London Basin. All authors co-wrote and reviewed the manuscript. The authors would like to acknowledge Lee Jones and Ricky Terrington from BGS for their valuable suggestions to improve the quality of the paper. Alessandro Novellino and Luke Bateson publish with the permission of the Executive Director of BGS.

Conflicts of Interest: The authors declare no conflict of interest.

References

1. Galloway, D.L.; Jones, D.R.; Ingebritsen, S.E. *Land Subsidence in the United States: U.S. Geological Survey*; Circular 1182; U.S. Geological Survey: Reston, VA, USA, 1999; p. 177.

2. Chaussard, E.; Milillo, P.; Bürgmann, R.; Perissin, D.; Fielding, E.J.; Baker, B. Remote Sensing of Ground Deformation for Monitoring Groundwater Management Practices: Application to the Santa Clara Valley During the 2012–2015 California Drought. *J. Geophys. Res. Solid Earth* **2017**, *122*, 8566–8582. [CrossRef]

3. Deffontaines, B.; Kaveh, F.; Fruneau, B.; Arnaud, A.; Duro, J. Monitoring Swelling Soils in Eastern Paris (France) through DinSAR and PSI Interferometry: A Synthesis. In *Engineering Geology for Society and Territory*; Lollino, G., Manconi, A., Clague, J., Shan, W., Chiarle, M., Eds.; Springer International Publishing: Cham, Switzerland, 2015; Volume 5, pp. 195–202, ISBN 978-3-319-09048-1.

4. Amelung, F.; Gallowey, D.L.; Bell, J.; Zebker, H.; Laczniak, R.J. Sensing the ups and downs of Las Vegas: InSAR reveals structural control of land subsidence and aquifer-system deformation. *Geology* **1999**, *27*, 483–486. [CrossRef]

5. Bateson, L.; Cigna, F.; Boon, D.; Sowter, A. The application of the Intermittent SBAS (ISBAS) InSAR method to the South Wales Coalfield, UK. *Int. J. Appl. Earth Obs. Geoinf.* **2015**, *34*, 249–257. [CrossRef]

6. Gee, D.; Bateson, L.; Sowter, A.; Grebby, S.; Novellino, A.; Cigna, F.; Marsh, S.; Banton, C.; Wyatt, L. Ground motion in areas of abandoned mining: Application of the intermittent SBAS (ISBAS) to the Northumberland and Durham coalfield, UK. *Geosciences* **2017**, *7*, 85. [CrossRef]

7. Johnston, D.; Potter, H.; Jones, C.; Rolley, S.; Watson, I.; Pritchard, J. Environmental Agency Report. In *Abandoned Mines and the Water Environment*; Environment Agency: Bristol, UK, 2008.

8. Brake, B.T.; Hanssen, R.F.; Van der Ploeg, M.J.; De Rooij, G.H. Satellite-based radar interferometry to estimate large-scale soil water depletion from clay shrinkage: Possibilities and limitations. *Vadose Zone J.* **2013**, *12*. [CrossRef]

9. Kurka, M.; Gutjahr, K.H. Observation of Expansive Clay Movement with DInSAR. G. In *Engineering Geology for Society and Territory*; Lollino, G., Manconi, A., Clague, J., Shan, W., Chiarle, M., Eds.; Springer International Publishing: Cham, Switzerland, 2015; Volume 5, pp. 151–154, ISBN 978-3-319-09048-1.

10. Association of British Insurers. Subsidence—Dealing with the Problem. [Cited 3 August 2006]. 2006. Available online: http://www.abi.org.uk (accessed on 12 November 2017).

11. Ferretti, A.; Prati, C.; Rocca, F. Permanent scatterers in SAR interferometry. *IEEE Trans. Geosci. Remote Sens.* **2001**, *39*, 8–20. [CrossRef]

12. Fornaro, G.; Reale, D.; Serafino, F. Four-dimensional SAR imaging for height estimation and monitoring of single and double scatterers. *IEEE Trans. Geosci. Remote Sens.* **2009**, *47*, 212–237. [CrossRef]

13. Crosetto, M.; Monserrat, O.; Cuevas-González, M.; Devanthéry, N.; Crippa, B. Persistent scatterer interferometry: A review. *ISPRS J. Photogramm. Remote Sens.* **2016**, *115*, 78–89. [CrossRef]

14. Peduto, D.; Huber, M.; Speranza, G.; van Ruijven, J.; Cascini, L. DInSAR data assimilation for settlement prediction: Case study of a railway embankment in The Netherlands. *Can. Geotech. J.* **2017**, *54*, 502–517. [CrossRef]

15. Confuorto, P.; Di Martire, D.; Centolanza, G.; Iglesias, R.; Mallorqui, J.J.; Novellino, A.; Plank, S.; Ramondini, M.; Thuro, K.; Calcaterra, D. Post-failure evolution analysis of a rainfall-triggered landslide by multi-temporal interferometry SAR approaches integrated with geotechnical analysis. *Remote Sens. Environ.* **2017**, *188*, 51–72. [CrossRef]

16. Meisina, C.; Zucca, F.; Fossati, D.; Ceriani, M.; Allievi, J. Ground deformation monitoring by using the permanent scatterers technique: The example of the Oltrepo Pavese (Lombardia, Italy). *Eng. Geol.* **2006**, *88*, 240–259. [CrossRef]

17. Huang, Q.; Crosetto, M.; Monserrat, O.; Crippa, B. Displacement monitoring and modelling of a high-speed railway bridge using C-band Sentinel-1 data. *ISPRS J. Photogramm. Remote Sens.* **2017**, *128*, 204–211. [CrossRef]

18. North, M.; Farewell, T.; Hallett, S.; Bertelle, A. Monitoring the Response of Roads and Railways to Seasonal Soil Movement with Persistent Scatterers Interferometry over Six UK Sites. *Remote Sens.* **2017**, *9*, 922. [CrossRef]

19. Lan, H.; Li, L.; Liu, H.; Yang, Z. Complex Urban Infrastructure Deformation Monitoring Using High Resolution PSI. *IEEE J. Sel. Top. Appl. Earth Obs. Remote Sens.* **2012**, *5*, 643–651. [CrossRef]

20. Tomás, R.; Cano, M.; Garcia-Barba, J.; Vicente, F.; Herrera, G.; Lopez-Sanchez, J.M.; Mallorquí, J.J. Monitoring an earthfill dam using differential SAR interferometry: La Pedrera dam, Alicante, Spain. *Eng. Geol.* **2013**, *157*, 21–32. [CrossRef]

21. Milillo, P.; Perissin, D.; Salzer, J.T.; Lundgren, P.; Lacava, G.; Milillo, G.; Serio, C. Monitoring dam structural health from space: Insights from novel InSAR techniques and multi-parametric modeling applied to the Pertusillo dam Basilicata, Italy. *Int. J. Appl. Earth Obs. Geoinf.* **2016**, *52*, 221–229. [CrossRef]

22. Strozzi, T.; Delaloye, R.; Poffet, D.; Hansmann, J.; Loew, S. Surface subsidence and uplift above a headrace tunnel in metamorphic basement rocks of the Swiss Alps as detected by satellite SAR interferometry. *Remote Sens. Environ.* **2011**, *115*, 1353–1360. [CrossRef]

23. Royse, K.R.; de Freitas, M.; Burgess, W.G.; Cosgrove, J.; Ghail, R.C.; Gibbard, P.; King, C.; Lawrence, U.; Mortimore, R.N.; Owenj, H.; et al. Geology of London, UK. *Proc. Geol. Assoc.* **2012**, *123*, 22–45. [CrossRef]

24. Bricker, S.H.; Banks, V.J.; Galik, G.; Tapete, D.; Jones, R. Accounting for groundwater in future city visions. *Land Use Policy* **2017**, *69*, 618–630. [CrossRef]

25. Aldiss, D.; Burke, H.; Chacksfield, B.; Bingley, R.; Teferle, N.; Williams, S.; Blackman, D.; Burren, R.; Press, N. Geological interpretation of current subsidence and uplift in the London area; UK; as shown by high precision satellite-based surveying. *Proc. Geol. Assoc.* **2014**, *125*, 1–13. [CrossRef]

26. Cigna, F.; Jordan, H.; Bateson, L.; McCormack, H.; Roberts, C. Natural and anthropogenic geohazards in greater London observed from geological and ERS-1/2 and ENVISAT persistent scatterers ground motion data: Results from the EC FP7-SPACE PanGeo Project. *Pure Appl. Geophys.* **2015**, *172*, 2965–2995. [CrossRef]

27. Bateson, L.B.; Barkwith, A.K.A.P.; Hughes, A.G.; Aldiss, D.T. *Terrafirma: London H-3 Modelled Product: Comparison of PS Data with the Results of a Groundwater Abstraction Related Subsidence Model*; British Geological Survey Commissioned Report; OR/09/032; British Geological Survey: Keyworth, Nottingham, UK, 2009; 47p.

28. Bonì, R.; Cigna, F.; Bricker, S.; Meisina, C.; McCormack, H. Characterisation of hydraulic head changes and aquifer properties in the London Basin using Persistent Scatterer Interferometry ground motion data. *J. Hydrol.* **2016**, *540*, 835–849. [CrossRef]

29. Milillo, P.; Giardina, G.; DeJong, M.J.; Perissin, D.; Milillo, G. Multi-Temporal InSAR Structural Damage Assessment: The London Crossrail Case Study. *Remote Sens.* **2018**, *10*, 287. [CrossRef]

30. Bonì, R.; Pilla, G.; Meisina, C. Methodology for detection and interpretation of ground motion areas with the A-DInSAR time series analysis. *Remote Sens.* **2016**, *8*, 686.

31. Sumbler, M.G. *British Regional Geology: London and the Thames Valley*, 4th ed.; HMSO for the British Geological Survey: London, UK, 1996.

32. Ellison, R.A.; Woods, M.A.; Allen, D.J.; Forster, A.; Pharaoh, T.C.; King, C. *Geology of London: Special Memoir for 1:50,000 Geological Sheets 256 (North London)*; 257 (Romford); 270 (South London); and 271 (Dartford) (England and Wales); British Geological Survey: London, UK, 2004.

33. Sellwood, B.W.; Sladen, C.P. Mesozoic & Tertiary argillaceous units: Distribution and composition. *Q. J. Eng. Geol. Hydrogeol.* **1981**, *14*, 263–275.

34. Jones, L.D.; Terrington, R. *Methods for Modelling the 3D Volume Change Potential of UK Clay Soils*; British Geological Survey Research Report; RR/17/008; British Geological Survey: London, UK, 2017; 57p.

35. Jones, L.D.; Terrington, R. Modelling volume change potential in the London Clay. *Q. J. Eng. Geol. Hydrogeol.* **2011**, *44*, 109–122. [CrossRef]

36. Barker, J.A. Transport in fractured rock. In *Applied Groundwater Hydrology*; Downing, R.A., Wilkinson, W.B., Eds.; Claredon Press: Oxford, UK, 1991; pp. 199–216.

37. Price, M. Fluid flow in the Chalk of England. In *Fluid Flow in Sedimentary Basins and Aquifers*; Goff, J.C., Williams, B.P.J., Eds.; Geological Society London Special Publications: London, UK, 1987; Volume 34, pp. 141–156.

38. Price, M.; Downing, R.A.; Edmunds, W.M. The Chalk as an aquifer. In *The Hydrogeology of the Chalk of North-West Europe*; Downing, R.A., Price, M., Jones, G.P., Eds.; Clarendon Press: Oxford, UK, 1993; pp. 14–34.

39. Bloomfield, J.P.; Brewerton, L.J.; Allen, D.J. Regional trends in matrix porosity and dry density of the chalk of England. *Q. J. Eng. Geol.* **1995**, *28*, S-131–S-142. [CrossRef]

40. Allen, D.J.; Brewerton, L.J.; Coleby, L.M.; Gibbs, B.R.; Lewis, M.A.; MacDonald, A.M.; Wagstaff, S.J.; Williams, A.T. *The Physical Properties of Major Aquifers in England and Wales*; British Geological Survey Technical Report WD/97/34; British Geological Survey: London, UK, 1997.

41. Environment Agency. *Management of the London Basin Chalk Aquifer*; Status Report 2016; Environment Agency Report; Environment Agency: Bristol, UK, 2016.

42. Environment Agency. *Management of the London Basin Chalk Aquifer*; Status Report 2017; Environment Agency Report; Environment Agency: Bristol, UK, 2017. Available online: https://www.gov.uk/government/publications/london-basin-chalk-aquifer-annual-status-report (accessed on 13 April 2018).

43. Werner, C.; Wegmüller, U.; Wiesmann, A.; Strozzi, T. Interferometric point target analysis with JERS-1 L-band SAR data. In Proceedings of the IEEE International Geoscience and Remote Sensing Symposium 2003, IGARSS 2003, Toulouse, France, 21–25 July 2003; Volume 7, pp. 4359–4361.

44. Self, S.J.; Entwisle, D.C.; Northmore, K.J. *The Structure and Operation of the BGS National Geotechnical Properties Database Version 2*; British Geological Survey Internal Report IR/12/056; British Geological Survey: London, UK, 2012.

45. Jones, L. *User Guide for the 3D Shrink-Swell (GeoSure Extra) Dataset*; British Geological Survey Open Report; OR/16/043; British Geological Survey: London, UK, 2016; 14p.

46. Building Research Establishment (BRE). *Low-Rise Buildings on Shrinkable Clay Soils*; BRE Digest; Construction Research Communications: London, UK, 1993; Volume 240–242.

47. Building Heights in England. Available online: http://buildingheights.emu-analytics.net (accessed on 12 November 2017).

48. Consumer Data Research Centre. Dwellings: Modal Age. Available online: https://maps.cdrc.ac.uk/ (accessed on 12 November 2017).

49. Ding, X.L.; Li, Z.W.; Zhu, J.J.; Feng, G.C.; Long, J.P. Atmospheric effects on InSAR measurements and their mitigation. *Sensors* **2008**, *8*, 5426–5448. [CrossRef] [PubMed]

50. Crosetto, M.; Monserrat, O.; Iglesias, R.; Crippa, B. Persistent Scatterer Interferometry. *Photogram. Eng. Remote Sens.* **2010**, *76*, 1061–1069. [CrossRef]

51. Notti, D.; Calò, F.; Cigna, F.; Manunta, M.; Herrera, G.; Berti, M.; Meisina, C.; Tapete, D.; Zucca, F. A user-oriented methodology for DInSAR time series analysis and interpretation: Landslides and subsidence case studies. *Pure Appl. Geophys.* **2015**, *172*, 3081–3105. [CrossRef]

52. Cigna, F.; Tapete, D.; Casagli, N. Semi-automated extraction of Deviation Indexes (DI) from satellite Persistent Scatterers time series: Tests on sedimentary volcanism and tectonically-induced motions. *Nonlinear Processes Geophys.* **2012**, *19*, 643–655. [CrossRef]

53. Chaussard, E.; Bürgmann, R.; Shirzaei, M.; Fielding, E.J.; Baker, B. Predictability of hydraulic head changes and characterization of aquifer-system and fault properties from InSAR-derived ground deformation. *J. Geophys. Res. Solid Earth* **2014**, *119*, 6572–6590. [CrossRef]

54. Berti, M.; Corsini, A.; Franceschini, S.; Iannacone, J.P. Automated classification of Persistent Scatterers Interferometry time series. *Nat. Hazards Earth Syst. Sci.* **2013**, *13*, 1945–1958. [CrossRef]

55. Wegmüller, U.; Werner, C.; Wiesmann, A.; Strozzi, T.; Kourkouli, P.; Frey, O. Time-series analysis of Sentinel-1 interferometric wide swath data: Techniques and challenges. In Proceedings of the 2016 IEEE International Geoscience and Remote Sensing Symposium (IGARSS), Beijing, China, 10–15 July 2016; pp. 3898–3901.

56. Casu, F.; Manzo, M.; Lanari, R. A quantitative assessment of the SBAS algorithm performance for surface deformation retrieval from DInSAR data. *Remote Sens. Environ.* **2006**, *102*, 195–210. [CrossRef]

57. Fornaro, G.; Reale, D.; Verde, S. Bridge Thermal Dilation Monitoring With Millimeter Sensitivity via Multidimensional SAR Imaging. *IEEE Geosci. Remote Sens. Lett.* **2013**, *10*, 677–681. [CrossRef]

58. Nicodemo, G.; Peduto, D.; Ferlisi, S.; Maccabiani, J. Investigating building settlements via very high resolution SAR sensors. In *2017 Life-cycle of Engineering Systems: Emphasis on Sustainable Civil Infrastructure*; Bakker, J., Frangopol, D.M., van Breugel, K., Eds.; Taylor & Francis Group: London, UK, 2016; pp. 2256–2263.

59. Bonì, R.; Meisina, C.; Cigna, F.; Herrera, G.; Notti, D.; Bricker, S.; McCormack, H.; Tomás, R.; Béjar-Pizarro, M.; Mulas, J.; et al. Exploitation of satellite A-DInSAR time series for detection, characterization and modelling of land subsidence. *Geosciences* **2017**, *7*, 25. [CrossRef]

60. Richman, M.B. Rotation of principal components. *J. Climatol.* **1986**, *6*, 293–335. [CrossRef]

61. Solari, L.; Ciampalini, A.; Raspini, F.; Bianchini, S.; Moretti, S. PSInSAR Analysis in the Pisa Urban Area (Italy): A Case Study of Subsidence Related to Stratigraphical Factors and Urbanization. *Remote Sens.* **2016**, *8*, 120. [CrossRef]

62. Crossrail, Crossrail Project Dewatering Works—Close-out Report. 2016. Available online: https://learninglegacy.crossrail.co.uk/wp-content/uploads/2017/01/7A-026_1-Dewatering-Close-out-report.pdf (accessed on 12 November 2017).

63. Black, M. Crossrail project: Managing geotechnical risk on London's Elizabeth line. In Proceedings of the Institution of Civil Engineers-Civil Engineering; Thomas Telford Ltd.: London, UK, 2017; Volume 170, pp. 23–30.

remote sensing

MDPI

Article

Analysis of Secular Ground Motions in Istanbul from a Long-Term InSAR Time-Series (1992–2017)

Gokhan Aslan [1,2,*], Ziyadin Cakır [3], Semih Ergintav [4], Cécile Lasserre [1,5] and François Renard [1,6]

[1] Université Grenoble-Alpes, CNRS, IRD, IFSTTAR, ISTerre, 38000 Grenoble, France;
cecile.lasserre@univ-grenoble-alpes.fr (C.L.); francois.renard@geo.uio.no (F.R.)

[2] Eurasia Institute of Earth Sciences, Istanbul Technical University, 34469 Istanbul, Turkey

[3] Department of Geological Engineering, Istanbul Technical University, 34469 Istanbul, Turkey;
ziyadin.cakir@itu.edu.tr

[4] Kandilli Observatory and Earthquake Research Institute (KOERI), Bogazici University, 34684 Istanbul,
Turkey; semih.ergintav@boun.edu.tr

[5] Université de Lyon, UCBL, ENSL, CNRS, LGL-TPE, 69622 Villeurbanne, France

[6] Physics of Geological Processes (PGP), The NJORD Centre, Department of Geosciences, UiO,
NO-0316 Oslo, Norway

* Correspondence: gokhan.aslan@univ-grenoble-alpes.fr; Tel.: +33-789-786233

Received: 31 January 2018; Accepted: 1 March 2018; Published: 6 March 2018

Abstract: The identification and measurement of ground deformations in urban areas is of great importance for determining the vulnerable parts of the cities that are prone to geohazards, which is a crucial element of both sustainable urban planning and hazard mitigation. Interferometric synthetic aperture radar (InSAR) time series analysis is a very powerful tool for the operational mapping of ground deformation related to urban subsidence and landslide phenomena. With an analysis spanning almost 25 years of satellite radar observations, we compute an InSAR time series of data from multiple satellites (European Remote Sensing satellites ERS-1 and ERS-2, Envisat, Sentinel-1A, and its twin sensor Sentinel-1B) in order to investigate the spatial extent and rate of ground deformation in the megacity of Istanbul. By combining the various multi-track InSAR datasets (291 images in total) and analysing persistent scatterers (PS-InSAR), we present mean velocity maps of ground surface displacement in selected areas of Istanbul. We identify several sites along the terrestrial and coastal regions of Istanbul that underwent vertical ground subsidence at varying rates, from 5 ± 1.2 mm/yr to 15 ± 2.1 mm/yr. The results reveal that the most distinctive subsidence patterns are associated with both anthropogenic factors and relatively weak lithologies along the Haramirede valley in particular, where the observed subsidence is up to 10 ± 2 mm/yr. We show that subsidence has been occurring along the Ayamama river stream at a rate of up to 10 ± 1.8 mm/yr since 1992, and has also been slowing down over time following the restoration of the river and stream system. We also identify subsidence at a rate of 8 ± 1.2 mm/yr along the coastal region of Istanbul, which we associate with land reclamation, as well as a very localised subsidence at a rate of 15 ± 2.3 mm/yr starting in 2016 around one of the highest skyscrapers of Istanbul, which was built in 2010.

Keywords: time series analysis; InSAR; PS; landslide; subsidence; land reclamation; urbanization; risk; Istanbul; Turkey

1. Introduction

Very rapid social and economic transformation in recent decades caused a huge rural-to-urban migration all over the world, which has fueled urban growth. This massive population shift has brought many complex challenges together with regard to sustainable development and natural disaster management. Istanbul is the largest city in Turkey, with a population of approximately

14 million inhabitants, and one of the most rapidly growing cities in Europe [1]. According to the Istanbul Transportation Master Plan (ITMP), when taking into account the consequence of Turkey's economic growth in the last two decades and the large amount of immigration, projections indicate that the population will overcome 20 million inhabitants in 2023 [2]. This rapid population growth poses major threats to the city's development when considering its vulnerability to natural disasters such as earthquakes, landslides, and floods, due to heavy and unplanned urbanization practices. Besides, the short distance (~20 km) of the main active branch of the North Anatolian Fault to the city poses a major threat to Istanbul [3].

Spaceborne interferometric synthetic aperture radar (InSAR) is a powerful remote sensing tool that enables observations of Earth's surface day and night under all weather conditions with high precision. Over the past decades, the method has been widely exploited in order to measure and monitor ground deformation induced by earthquakes [4–7] volcanoes [8], the withdrawal of ground water or other fluids [9,10], soil consolidation [11,12], mining [13], landslides [14,15], permafrost melting [16], ground subsidence [17,18], land reclamation [19], and sinkholes [20]. Previous studies have shown the capacity of InSAR methods to measure and map land subsidence due to various anthropogenic factors, including ground water extraction in megacities such as Tianjin [21,22], Shanghai [23], Mexico City [24], and lithological factors in Bandung basin, Indonesia [25]. Among InSAR techniques, persistent scatterers InSAR (PS-InSAR) was developed to tackle limitations related to temporal and geometrical decorrelation and atmospheric effects [26–31]. It enables monitoring the temporal evolution of the ground motion by exploiting multiple SAR images acquired over the same area. It uses the radar return signal reflected from persistent scatterers (PS, pointwise phase-stable targets) such as rooftops, large rock outcrops, bridges, or motorways, where the spatial density of such PS is high [27]. PS-InSAR analyses provide a time series of PS displacements and average surface velocities by searching a motion model that is relative to a reference point, and assumed to be motionless.

In the megacity of Istanbul, PS-InSAR time series analyses have allowed the monitoring of ground motions induced by anthropogenic activities [32], lithological features [33], and tectonic activities [7,34]. However, there has been little discussion about the long-term temporal evolution of the ground motion and its possible causes. This study presents a PS-InSAR analysis of the secular ground motion in the urbanized metropolitan area of Istanbul. The processed InSAR data (Figure 1) span nearly 25 years, from 1992 to 2017, with multi-sensor images acquired along ascending and descending orbits. Most of the surface motion anomalies that we identify are associated with ground subsidence that has been induced by various factors, including natural compaction, and anthropogenic activities. These subsidence anomalies are carefully measured and analyzed from the perspective of urbanization and the assessment of geohazards for the city of Istanbul. The causes of subsidence in cities are diverse, and include factors such as lithology (i.e., rock type), variations in soil moisture content, groundwater exploitation, and overburden loads associated with human activity. In Istanbul, considering the proximity of several segments of the active North Anatolian Fault (NAF) in the Sea of Marmara, which have remained unbroken since 1776, the characterisation of subsidence susceptibility for Istanbul is crucial with regard to hazard mitigation and urban planning, as it can identify the vulnerable parts of the region that are prone to possible future earthquake damage. Thus, the main goal of the present study was to use a long-term PS-InSAR time series to: (1) quantify subsidence phenomena and discuss associated causes such as lithology-controlled natural compaction and anthropogenic activities, and (2) monitor the temporal evolution of the subsidence.

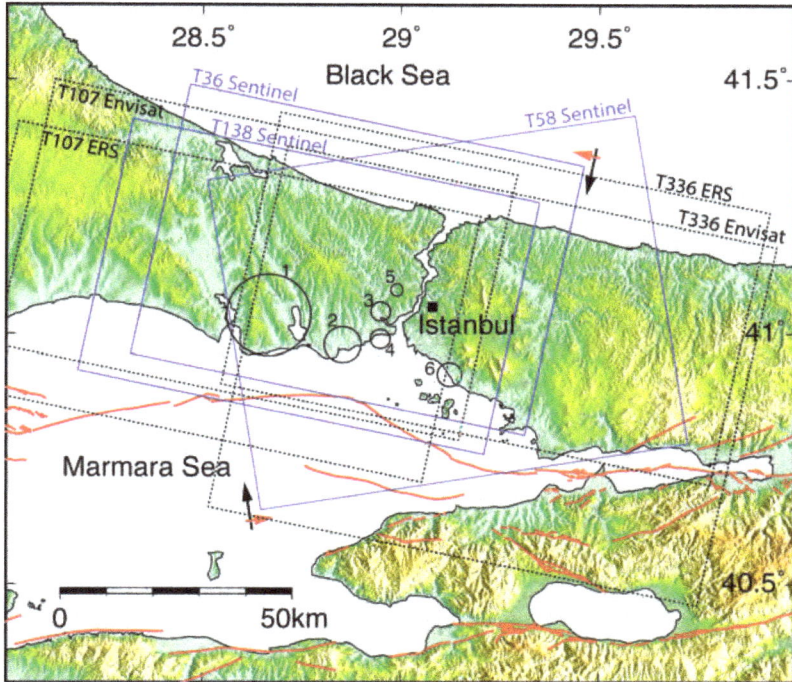

Figure 1. Study area and satellite synthetic aperture radar data coverage used in the present study. The shaded topography is given by the Shuttle Radar Topography Mission (SRTM) along the North Anatolian Fault (NAF) in the Sea of Marmara, and major faults are drawn in red [35]. Rectangles labeled with sensor and track numbers indicate the coverage of the SAR images that were used in the present study. The red and black arrows indicate satellite's line-of-sight look and flight directions, respectively. Circles with numbers show the study regions in the paper (details are given in the text).

2. Background of the Study Areas

The present study primarily focusses on six areas where an InSAR time series enabled the detection of anomalous ground displacement: the region of Haramidere, where a series of active natural landslides had been previously mapped (circle 1 in Figures 1 and 2), and the Ayamama floodplain (circle 2 in Figure 1), a geological setting that is discussed in Section 2.1 (Figure 2), as well as several local subsidence areas that are related to anthropogenic activities (circles 3 to 6 in Figure 1).

2.1. Geological Setting of Study Areas 1 and 2

The Haramidere and the Ayamama streams are located near the boundary between the Istranca Methamorphic Unit (Paleozoic–Mesozoic) and Istanbul Unit (Paleozoic). These units are covered by the Eocene sedimentary sequence of the Thrace basin [36]. The Paleozoic metamorphic basement of the study area consists in the east in a thick turbiditic sandstone–shale sequence of the Carboniferous age, while the Eocene cover is made of limestone, marl, and claystone units, which are transgressive on Çatalca metamorphic units in the west (Figure 2).

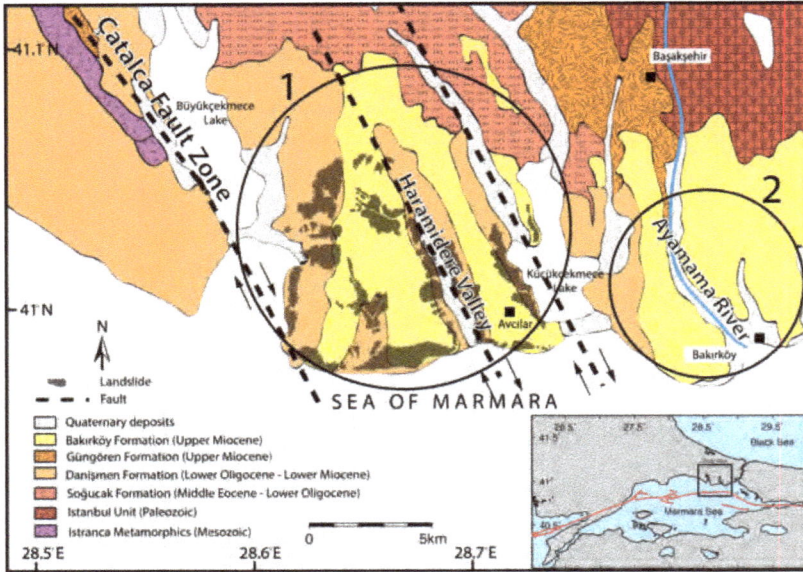

Figure 2. Simplified geological and structural map of study areas 1 and 2 (circles 1 and 2 as in Figure 1). Numerous active landslides (dark green patches) were mapped between the Küçükçekmece and Büyükçekmece lakes [simplified from Ergintav et al., Duman et al. and Ozgul et al. [37–39]. The inset map shows a figure area north of the Sea of Marmara, and the main segments of the NAF in red [35].

2.2. Study Area 1: Haramidere Valley and Avcilar Neighborhood

The Haramidere valley is located between Buyukcekmece Lake and Kucukcekmece Lake in the Avcilar Peninsula. It is located about 15 km north of the NAF, which cuts across the Sea of Marmara (Figure 2). Although the Avcilar district is located at a distance of 120 km west of the epicenter of the 17 August 1999 Izmit earthquake, it was the only area in the Istanbul metropolitan region that suffered extensive damage [40]. During this earthquake, 27 buildings collapsed, 2076 other buildings were heavily damaged, and 273 casualties were reported in Avcilar [41,42]. The maximum ground acceleration that was measured on soft sediments was 0.25 g, which is six times higher than the peak ground acceleration recorded on the bedrock in the center of Istanbul [40]. This difference is the result of the amplification of seismic waves in surficial layers with soft lithology [40,42–46]. Despite a low background seismicity, which suggested no active faulting in the region [47], destructive and widespread damage during the 1999 Izmit earthquake drew considerable attention on this area, requiring the reassessment of active faulting. In order to investigate any relationship between the faults and damage observed in Avcilar and the vicinity, fault-mapping was refined from field studies and seismic reflection analyses [37,48–50]. These studies showed that the presence of secondary faults might be an important driving factor for the localised seismic amplification [7,51].

Landslides have been identified as another important geohazard in the suburb of Avcilar for many years [38,52,53]. Recent events in that area often result from the reactivation of old landslides due to the overloading of the existing landslides by new constructions. The investigation of the old landslides in more detail is very important in order to anticipate measures that could be taken to avoid future possible damage in the urbanized center of the Avcilar peninsula. A total of 391 landslides were mapped (Figure 2) in the region [38]. Approximately half of all of the landslides were distributed between Buyukcekmece and the Haramidere valley, which are important local landforms in the region

(Figure 2). Duman et al. [52,53] used field geological observations to argue that the parameters that control landslide initiation are shallow groundwater levels, lithology, and liquefaction.

2.3. Study Area 2: Ayamama River Valley

The Ayamama valley is located in the western part of Istanbul, east of the Haramidere valley (Figure 2). The river flows north–south from the eastern part of the Basaksehir district, and towards the Sea of Marmara in the Bakirkoy district [54]. It runs through the heavily urbanised and highly populated area of the European side of Istanbul. The lower parts of its basin show various land use patterns and a high density of population [55]. The Ayamama stream is known to produce seasonal flooding. According to the municipality of Istanbul, sedimentation and illegal urbanisation in the riverbeds have decreased floodplain capacity, which has subsequently caused periodic floods and overflows. One of the worst flooding events that affected the region was on 9 September 2009, which caused 31 casualties and 50 injuries. In the city's development plan, the Ayamama river basin and its surrounding zone had been set aside for recreational areas. However, after an amendment was made to develop it into residential area in 1997, industrial and residential land use rapidly increased along the axis of the stream [56].

2.4. Study Areas 3 to 6: Subsidence Events Associated with Land Reclamation and Urbanisation

Istanbul lies on both sides of the Bosphorus Strait (İstanbul Boğazı), and has been subjected to heavy and unplanned urbanisation. This rapid urbanisation gave rise to land reclamation along the coastal areas of the Marmara Sea in order to provide new recreational areas to compensate for the destruction of green areas [57]. Nearly 2.6 square kilometers of land have been gained from the Istanbul coast by filling up the sea since 2000 [58]. We have focussed on the two recreational areas in the Yenikapi neighborhood and the Maltepe district for the investigation of the ground deformation related to land reclamation by the PS-InSAR method (sites 4 and 6). Another two local subsidence phenomena have been observed along the banks of the Golden Horn (Haliç, site 3), and around a skyscraper located in the Levent neighborhood of Istanbul (site 5).

3. Datasets and Methodology

3.1. Datasets

The SAR data used in the present study consisted of 291 C-band (5–6 GHz, ~6 cm wavelength) images acquired with various sensors, including 51 (ERS 1/2) images on two overlapping ascending tracks spanning from 1992 to 2001, 52 ENVISAT images on two overlapping ascending tracks spanning from 2003 to 2011, and 188 Sentinel 1A/B TOPS (Terrain Observation with Progressive Scans in azimuth) images in one ascending and two descending orbits acquired between 2014–2017 (Figure 3, see Table 1 for details). The Istanbul metropolitan area is entirely covered by all of the tracks. These multiorbit/sensor datasets enable us to examine the consistency between different sets of observations by inter-comparisons.

Table 1. Characteristics of each processed track.

Track	C-Band/Satellite	Geometry	Time Interval	Incidence Angle at Swath Center	Interferograms Used	Density [2] (PS/km^2)
T107	ERS 1/2	Descending	1992–2001	~23°	28	93
T336	ERS 1/2	Descending	1992–2001	~23°	21	212
T107	ENVISAT	Descending	2003–2011	~23°	25	190
T336	ENVISAT	Descending	2003–2011	~23°	25	176.8
T58	SENTINEL [1]	Ascending	2015–2017	~39.2°	56	250.4
T36	SENTINEL [1]	Descending	2014–2017	~33.7°	65	289.6
T138	SENTINEL [1]	Descending	2014–2017	~43.9°	65	252.7

[1] SENTINEL 1 A/B TOPS; [2] Density of permanent scatterers (PS) in the urban areas.

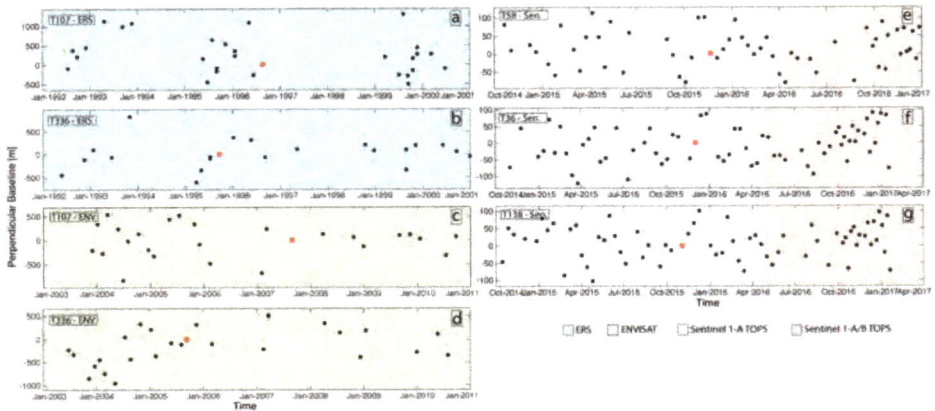

Figure 3. Baseline versus time plots for the seven tracks used in this study. The red dots indicate the master image used as a reference for each track. For the Sentinel data, the period when the two satellites 1A/1B were operational is indicated in orange (before this period, only satellite 1A was operational).

3.2. Methodology

Single look complex images and interferograms of the Envisat/ASAR and ERS archives were generated using the ROI_PAC software [59] and the Delft Object-oriented Radar Interferometric Software (DORIS), respectively [60]. All of the interferograms of the SENTINEL 1A/B satellites datasets were generated using the latest version of the software "Generic Mapping Tools Synthetic Aperture Radar (GMTSAR)" [61] using the Shuttle Radar Topography Mission (SRTM) 3-arcsecond digital elevation model for correcting topographic contributions to the radar phase. All of the interferograms were generated based on a single master network in order to use them in the PS-InSAR analysis. The choice of the master images for each track (red dots on Figure 3) were made so as to minimize the spatial and temporal baselines. The stacks of interferograms were processed with the PS-InSAR approach using the software package STAMPS (permanent scatterers InSAR analysis), which allows the identification of radar benchmarks by selecting pixels on the basis of their noise characteristics [62,63]. For the selection of permanent scatterers, STAMPS takes into account the statistical relationship between amplitude and phase stability, which is quantified by the ratio between the standard deviation and the mean of the amplitude through time for each pixel (amplitude dispersion index). In the present study, we selected a threshold value of 0.42 for this amplitude dispersion index [27], which minimises the random amplitude variability, and eliminates highly decorrelated pixels in some areas covered with vegetation, agricultural fields, or snow.

4. Results

InSAR-Derived Land Subsidence Maps

Figure 4 shows the mean line-of-sight (LOS) velocity fields that were calculated from a PS-InSAR time series on the metropolitan city of Istanbul. A main deformation feature, which is common to all of the tracks, is the NNW–SSE elongated area in the southwest region of Istanbul, along the Haramidere valley on the Avcilar peninsula (Figure 4, circle 1 in a, site 1, as described in Sections 2.1 and 2.2). The associated motion is away from the satellite for both the descending and ascending tracks, suggesting a dominant subsidence signal, peaking at up to 10 mm/year in the line-of-sight. We identified another likely subsiding area along the Ayamama stream valley. This anomaly is elongated in a WNW–ESE direction, and is located along the western side of the Kucukcekmece Lake (Figure 4a circle 2, named as site 2). The area labeled with circle 3 in Figure 4a (named as site 3) covers the primary inlet of the Bosphorus, and is called Golden Horn urban waterway. This Golden Horn's bank is also likely subject to subsidence. Other potential subsidence anomalies caused by the settlement of the reclamation in the coastal areas were identified along the northern coast of the Marmara Sea (circles 4 and 6 in Figure 4a corresponding to sites 4 and 6). Lastly, we point to a local subsidence signal at a rather fast rate that was observed around a skyscraper (Figure 4a, circle and site 5) located in the Levent neighbourhood of Istanbul. The spatial and temporal variations of these different deformation patterns are discussed along with their possible underlying causes in the following sections.

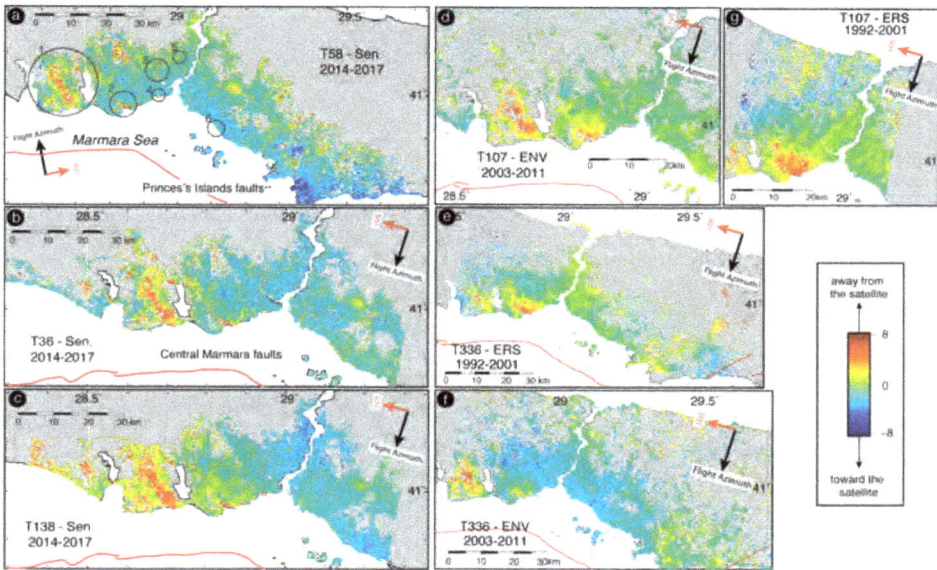

Figure 4. Averaged line-of-sight velocity maps of the Istanbul metropolitan area from an interferometric synthetic aperture radar (InSAR) time series analysis, with varying time spans depending on the sensor. Negative velocities (cold colors) represent the displacement of the ground toward the satellite, and positive velocities (warm colors) indicate the displacement away from the satellite. Red lines in the Sea of Marmara indicate the submarine branches of the NAF. Average line-of-sight velocity (**a**) for Sentinel ascending track 58. The solid black circles labeled from 1 to 6 indicate the locations of the subsidence anomalies that are discussed in the present study. 1: Haramidere River, 2: Ayamama Stream Valley, 3: Golden Horn, 4: Yenikapi reclamation area, 5: Skyscrapers in Levent neighbourhood, 6: Maltepe reclamation area; (**b**) for SENTINEL 1A/B descending track 36; (**c**) for Sentinel descending track 138; (**d**) for Envisat descending track 107; (**e**) for ERS descending track 336; (**f**) for ENVISAT descending track 336 and (**g**) for ERS descending track 107.

5. Discussion

5.1. Self-Consistency between InSAR Measurements

We first quantitatively assessed the consistency of InSAR mean velocity measurements calculated from different datasets, across different time periods, for the widest areas of subsidence in the Haramidere and Avcilar valleys. An inter-comparison of the line-of-sight displacement rates in these areas was performed using a two-step procedure. In the first step, the mean velocity fields were downsampled by a factor of 10 in order to minimise errors within the geolocalisations (spatial mismatch between measurements). In a second step, we selected all of the pixels in the localised deforming areas of Haramidere Valley and Avcilar neighbourhood that were common to a pair of overlapping tracks, and compared the distribution of their subsidence rates for each pair of tracks (Figure 5). The consistency of velocities was good overall (mean correlation = 0.62). The observed differences could have originated from various factors, such as: (1) different incidence angles depending on the satellite, as they were not taken into account in the inter-comparison (Table 1), (2) different temporal coverages, (3) geolocalization uncertainty, (4) InSAR processing errors, and (5) seasonal effects [64].

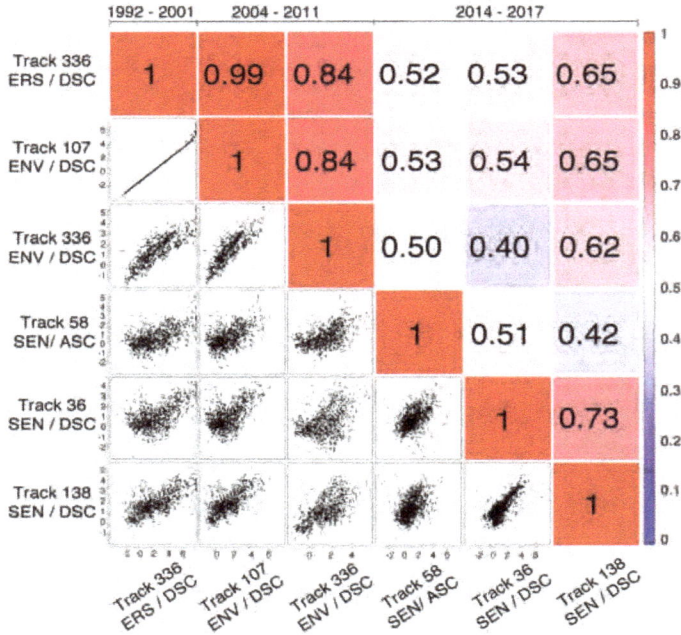

Figure 5. Quantitative comparison of the line-of-sight displacement rates between all of the tracks used in the present study. The upper-right triangle matrix shows pairwise correlations of different tracks with correlation values and color intensities (blue and red indicate low and high correlation, respectively). In the lower-left triangle, the black dots denote the points that can be extracted on both tracks. SEN and ENV in the panel represent SENTINEL and ENVISAT, respectively.

5.2. Site 1: Haramidere Valley

Close-up views of the elongated pattern of subsidence in the Haramidere valley are shown on 6 for all tracks. In order to quantify the vertical, subsidence component of the motion, we decomposed the mean PS-InSAR line-of-sight velocity fields into east–west and vertical components using the method described by Samieie-Esfahany et al. [65]. We only used the velocity fields calculated from the Sentinel 1A/B images (two tracks, 58 [Asc] and 138 [Dsc]) that covered the same time interval to calculate this decomposition. We assumed that there was no north–south displacement, due to the low sensitivity of the LOS data to this component of displacement. Doing so, we reduced the number of unknown variables for each permanent scatterer point to two displacement components, d_{down}, the vertical displacement (positive downward), and d_{west}, the horizontal displacement in the east–west direction (positive toward west). In a first step, we resampled the mean line-of-sight velocities for the ascending and descending tracks onto a 0.0005° × 0.0005° regular grid (approximately 10 m spacing). We used the nearest neighbour procedure in the resampling of the persistent scatterer pixels that were within 30 m of the center of each grid nodal point. In a second step, we selected all of the pixels that exist in both the ascending and descending tracks. For further interpretation of the displacements, we referenced the two tracks using reference points located in an area that was assumed to be stable (circle in the NE part of Figure 6). In a third step, the decomposition of the line-of-sight velocity fields into east–west and vertical components was calculated, taking into account the local incidence angle of the satellite view (Figure 7).

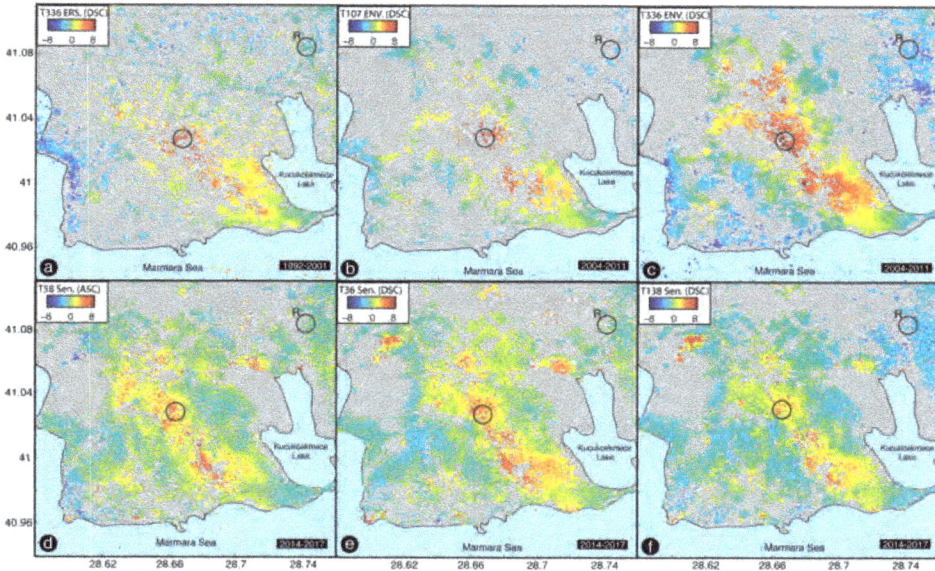

Figure 6. Zoom views of displacement rates in the Haramidere Valley and its surrounding area. The mean velocity value of the persistent scatterers (PS)-InSAR points within the solid black circle in the center of the maps has been used to illustrate the temporal evolution of the subsidence associated with weak lithology (Figure 8). It is referenced to the mean value of the PS-InSAR points within the circle labeled R, which is considered a stable area. The DSC and ASC labels are for the descending and ascending orbits, respectively.

The vertical displacement field shows that subsidence occurs on both banks of the Haramidere valley, and follows an elongated area in the valley in a northwest–southeast direction (Figure 7c). The maximum subsidence is centered on the valley (Figure 7f). This region has a long, slow-moving landslides history, and is located in an area with shallow water level, poor soil conditions, and weak lithology, which are all parameters that are considered as favoring landslides [17,40,66]. The subsidence that we observed coincided overall with previously mapped active landslide zones. However, the contours of these mapped landslides did not match precisely with the areas that had the highest subsidence in our vertical velocity map. The horizontal component in the east–west direction, which was derived by decomposition, showed a horizontal movement in the opposite direction on both banks of the valley, toward the valley center (Figure 7b,e). The sign change in the east–west velocity was also centered on the valley axis, as the maximum subsidence. We concluded that the observed signal was consistent, rather, with the subsidence of the soil/surface on the valley slope (Figure 7d), moving downslope due to landslide or soil creep gravity. This downslope movement had both horizontal and vertical components, which were well captured by the InSAR data.

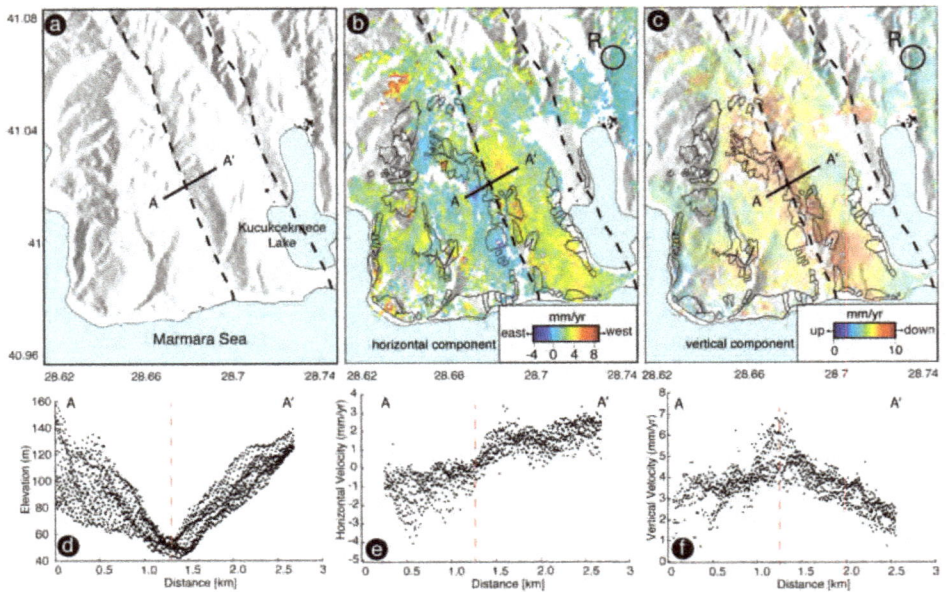

Figure 7. Decomposition of horizontal and vertical components of ground displacement using only S-1 datasets. (**a**) The shaded topography was given by the Shuttle Radar Topography Mission (SRTM) along the Avcilar region. Fault lines were simplified from Ergintav et al. [37]; (**b**) Vertical component. Patches with thick dark boundaries correspond to the landslides that were identified geological maps, as shown on Figure 2 [simplified from Duman et al. and Ozgul et al. [38,39]; (**c**) Horizontal component in the east–west direction. (**d**)Valley-perpendicular elevation profile extracted from (**a**); (**e,f**) are horizontal and vertical velocity profiles extracted from (**b,c**) respectively.

In order to analyse the subsidence temporal evolution (Figure 8), we selected permanent scatterer points located in an area that was previously detected as undergoing landslide activity, and where the subsidence rate was among the highest observed in the present study. For the sake of consistency between the datasets acquired from different viewing geometries, these selected PS points were from an area where the horizontal velocity was considered negligible (see the circle in center of Figure 6), so that the line-of-sight velocities were converted into vertical velocities by a simple geometrical equation. The date of the first SAR image used here was taken as the reference time of the time series. As seen in Figure 8, the three datasets used for the Haramidere valley and Avcilar area had different starting dates and temporal coverage. For comparison, we set one reference time as 26 May 1992 for the three datasets, and the time series that were mapped from the Envisat and Sentinel datasets were shifted with a constant, which was calculated by assuming that the site was undergoing subsidence with a constant rate (Figure 8).

Although the subsidence rates that were calculated from the ERS and Envisat datasets were consistent with each other and matched well, the subsidence rates that were obtained from the Sentinel datasets were slightly smaller. Two reasons might lie behind this difference. The first reason is related to the relatively short duration of the SENTINEL 1 A/B time series, which could alter the accuracy of the rate estimate. The second reason may be the retardation of the settlement due to a long-term decay of the soil consolidation rate related to ground water extractions. Such an exponential decay of ground subsidence was proposed to explain an InSAR time series on the Great Salt Lake in Utah [67].

Ground subsidence in the Avcilar peninsula has been previously reported at a mean rate of 6 mm/yr using an InSAR analysis of ERS-1 and ERS-2 satellite images taken between 1992–1999 [17], and at a rate of 10 mm/yr using ERS 1/2 and ENVISAT satellite images taken between 1992–2010 [64]. These authors concluded that the spatial coverage of the land subsidence in this area, which was overall consistent with ours, was associated with partially saturated and unconsolidated shallow layers of soil formation with a relatively weak lithological profile. The results of the present study thus support the observation of land subsidence in Avcilar, at similar to higher rates. Such types of lithology-controlled subsidence have also been observed in the Bandung basin on the island of Java in Indonesia, using an ALOS-1 dataset, although at much higher rates (up to 12 cm/yr), along the boundaries between consolidated rocks and unconsolidated sediments [25].

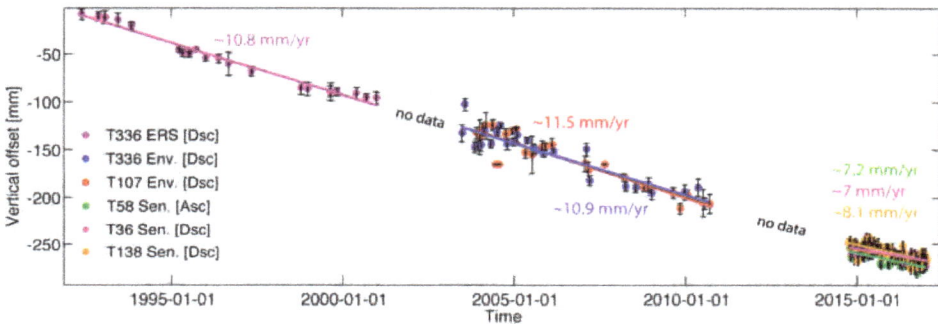

Figure 8. Time series of the vertical displacement at the selected PS points circled in center of Figure 6 (referenced to points in area labeled R in Figure 6).

5.3. Site 2: Ayamama River

Another subsidence zone lies along the dense settlements of the Ayamama stream banks in the western part of Istanbul (site 2, Figure 4a). Here, the river has the typical morphology of a delta, and the streambed is mostly composed of young alluvial deposits, varying in thickness in the range 3–10 m [68]. The subsidence rates we measured were about 6 mm/yr at maximum in the line-of-sight, corresponding to a maximum vertical subsidence rate of about 10 mm/yr (Figure 9). The area affected by subsidence shrunk gradually (Figure 9) during the observation period, which we interpreted as resulting from the flood prevention and remediation project of the river and stream system, which started in 2008. The subsidence pattern along the Ayamama stream was firstly reported by Walter et al. [33] with a more limited dataset, who suggested that the ongoing land subsidence in the region is related to the sediment consolidation process, and that there might be direct or indirect consequences of the destructive flood events, such as the one in 2009. This is thus consistent with our interpretation of the recovery of the region following the restoration of the river and stream system.

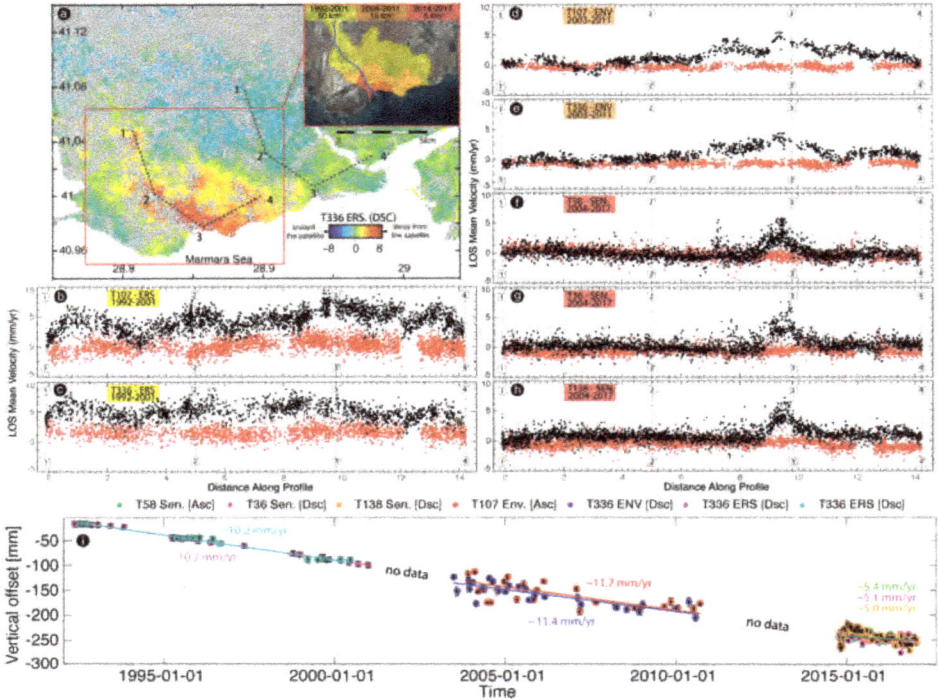

Figure 9. Spatiotemporal characteristics of subsidence along the Ayamama river valley study area. (a) Mean displacement rates in line-of-sight (LOS) obtained from a track 336 ERS dataset. The black dashed lines indicate two profiles, one in an area with an active subsidence (profile labeled 1–4), and the other one with the same length in an area considered as stable and used as a reference (profile labeled 1′–4′). The inset map indicates the temporal and spatial pattern of subsidence for the region, from 1992 to 2017; (b–h) Rates along the profiles 1–4 (black) and 1′–4′ (red) taken for each track; (i) Temporal evolution of the coastal subsidence of selected points around point 3 in Figure 9a.

5.4. Sites 4, 6, and 3

We also identified other sites along the terrestrial and coastal region of Istanbul undergoing ground subsidence. The subsidence that we related to land reclamation in two recreational areas in the Yenikapi neighbourhood (Figure 10a, site 4 in Figure 4a) and the Maltepe district (Figure 10d, site 6 in Figure 4a), which were constructed in 2014, was observed at a vertical rate of 10 mm/yr. This subsidence was presumably related to the compaction induced by the primary consolidation process of the alluvial clay beneath the reclamation zone. The subsidence rates that were measured in both reclamation areas were likely dependent on the physical characteristic and thickness of the underlying alluvial deposit and used matrix of the filling material [68,69]. Another local subsidence area related to a similar phenomena was observed along the shores of the Golden Horn (Haliç) (Figure 10b, site 3 in Figure 4a). A significant part of this subsidence was located on reclamation areas that have been transformed into parks and recreational facilities along both banks of the Golden Horn. Besides that, the shorelines in this area have also undergone significant urban changes within the frame of the renovation project of all of the waterfronts in Istanbul. During the renovation, sediments made of loose clay deposits were removed from the shallowest parts of the Golden Horn, which might have triggered subsidence in the nearby waterfront areas. The subsidence pattern along the Golden

Horn that we observed was consistent with the coastal subsidence that was previously described in the Figures 5 and 7 of [31], which, used a dataset of high-resolution TerraSAR-X SAR images covering the period 2010–2012. In their analyses of subsidence evolution over the urbanised region of Istanbul, these authors similarly concluded that the patterns of settlement along the Golden Horn shores were caused by anthropogenic factors arising from unsustainable urban development.

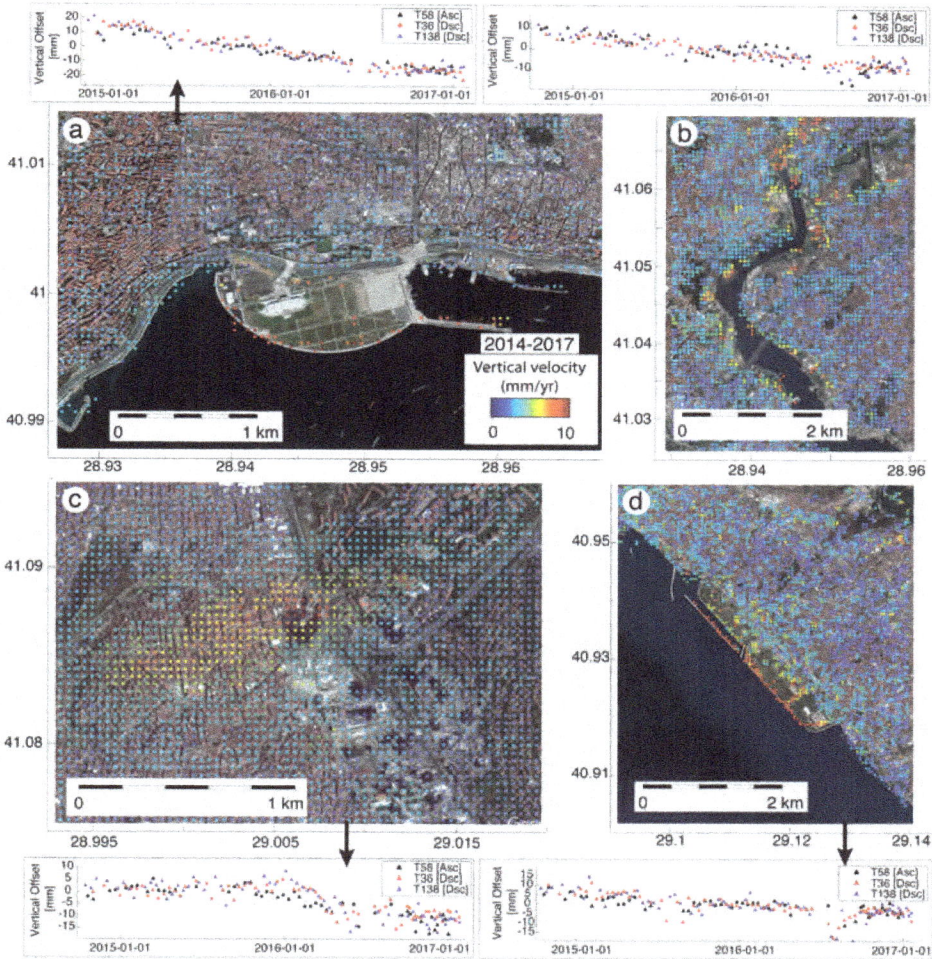

Figure 10. Vertical velocities obtained by the decomposition of mean velocity fields of Sentinel 1 data (T58 ascending track and T138 descending track) superimposed on a Google Earth image of Istanbul, and the relevant time series of the vertical displacement. Black, red, and blue triangles represent the ascending T58, descending T36, and descending T138 tracks, respectively. (**a**) Yenikapi coastal and land reclamation area (circle 4 in Figure 4a). The color scale represents the vertical displacement of the surface; (**b**) Golden Horn area (circle 3 in Figure 4a); (**c**) Highly urbanised area of Istanbul, with subsiding persistent scatterer points clustered around the highest skyscraper of Istanbul (circle 5 in Figure 4a); (**d**) Maltepe reclamation zone (circle 6 in Figure 4a).

5.5. Site 5

High skyscrapers might suffer rapid settlement and declination, and cause local ground surface subsidence due to the consolidation of the underlying soft soil deposits. In this study, we detected a very localised subsidence pattern around a skyscraper built in 2010, and other high-rise buildings in the Levent neighbourhood of Istanbul starting from 2016, with an average subsidence rate of about 15 mm/yr (Figure 10c). A time series analysis of the point targets surrounding the skyscraper showed a very rapid increase of subsidence rate during the first four months of 2016, which might be related to the groundwater lowering around the foundation of the building, which may cause a downward movement of the surface surrounding the building.

6. Conclusions

Istanbul has been subject to intense industrialisation and population increase, especially since the 1960s, which is causing very rapid urbanisation and heavy land-use changes. We identified several sites along the terrestrial and coastal region of Istanbul that have been undergoing vertical ground subsidence at rates ranging from 5 mm/yr to 15 mm/yr. In the present study, a PS-InSAR time series analysis was performed using 291 C-band SAR images in order to characterise these subsidence phenomena by combining multi-track/sensor InSAR datasets and provide insights into the potential hazards induced by local soil conditions and human activities. Using the PS-InSAR technique, enough time-coherent pixels were obtained over six different sites in Istanbul. The most extended, clearly visible subsidence signals were detected over the western part of Istanbul consistently in all of the frames. The spatiotemporal variability of the ground displacement was measured over the last 25 years in Avcilar, which suffered extensive damage during the 1999 Izmit earthquake and along the Haramidere valley, where a long-lasting landslide history has been reported. The time series analysis in this region revealed that the Haramidere valley banks are undergoing ground vertical subsidence at a rate of 10 ± 2 mm/yr. Another subsidence area has been reported along the Ayamama river banks, which is made of shallow alluvial deposits due to the local substratum consolidation process, at a rate of 10 ± 1.8 mm/yr, and the surface area affected by subsidence has been shrinking gradually following the restoration of the river and stream system, which started in 2008. The inter-comparison of PS-InSAR measurements from different satellite sensors for the western part of Istanbul showed that the correlation between the mean velocity fields were high between 1992–2011, and decreased with time due to spatiotemporal changes in the ground deformation. Sentinel-1A/B InSAR measurements acquired between 2014–2017 showed that reclaimed lands in both the European (Yenikapi reclamation area) and Asian (Maltepe reclamation area) coastlines of Istanbul underwent significant subsidence of up to 8 ± 1.3 mm/yr as a result of the primary consolidation process of the alluvial clay beneath the filling material. Lastly, a very localised subsidence pattern was detected around a skyscraper, with average subsidence rate of 15 ± 1.2 mm/yr.

On the whole, we can conclude that during the interseismic period, human-driven changes produced a more significant control on the coastal subsidence in Istanbul than natural factors. In future studies, such high-resolution SAR data over the very dense urban areas of Istanbul could help continuously monitor the urbanised areas suffering from land subsidence. With this purpose, future plans include the processing of high-resolution X-band sensors COSMO-Skymed and TerraSAR-X datasets in order to better quantify the surface settlement and reveal the underlying causes of these settlements, and thus provide more complete data to public organisations that are in charge of sustainable urban policies and hazard mitigation.

Acknowledgments: SAR (ERS, Envisat and Sentinel) data sets were obtained through the GSNL Marmara Region Permanent Supersite. Processing of Sentinel 1A/B images is performed at TUBITAK ULAKBIM, High Performance and Grid Computing Center (TRUBA resources). This study received funding from the Norwegian Research Council, project HADES, grant 250661 to F.R., TUBITAK project 113Y102 and AFAD project UDAP-G-16-02 to Z.C. and Boğaziçi University Research Fund Grant Number 12200 to S.E. This is part of the Ph.D. dissertation of Gokhan Aslan who is supported by the French Embassy in Turkey (Bourse Etudes scholarship program 889075G),

Remote Sens. **2018**, *10*, 408

University Grenoble Alpes IDEX project scholarship, Université Grenoble Alpes LabeX OSUG@2020 project of C.L. and CMIRA scholarship program provided by the Rhone-Alpes Region.

Author Contributions: G.A. carried out SAR data processing, data analysis and manuscript writing and revision. Z.C. supervised the SAR data processing, contributed to manuscript revision. S.E., C.L. and F.R. contributed to results analysis and manuscript revisions.

Conflicts of Interest: The authors declare no conflict of interest.

References

1. Van Leeuwen, K.; Sjerps, R. Istanbul: The challenges of integrated water resources management in Europe's megacity. *Environ. Dev. Sustain.* **2016**, *18*, 1–17. [CrossRef]
2. European Bank for Reconstruction and Development. Available online: http://www.ebrd.com/english/pages/project/eia/42163nts.pdf (accessed on 15 October 2017).
3. Helgert, T.; Heidbach, O. Slip-rate variability and distributed deformation in the Marmara Sea fault system. *Nat. Geosci.* **2010**, *3*, 132–136. [CrossRef]
4. Massonnet, D.; Feigl, K.; Rossi, M.; Adrangna, F. Radar Interferometric mapping of deformation in the year after The Landers Earthquake. *Nature* **1994**, *369*, 227–230. [CrossRef]
5. Zebker, H.A.; Rosen, P.A.; Goldstein, R.M.; Gabriel, A.; Werner, C.L. On the derivation of coseismic displacement fields using differential radar interferometry: The Landers earthquake. *J. Geophys. Res. Solid Earth* **1994**, *99*, 19617–19634. [CrossRef]
6. Çakir, Z.; Chabalier, J.-B.; Armijo, R.; Meyer, B.; Barka, A.; Peltzer, G. Coseismic and early postseismic slip associated with the 1999 Izmit earthquake (Turkey), from SAR interferometry and tectonic field observations. *Geophys. J. Int.* **2003**, *155*, 93–110. [CrossRef]
7. Diao, F.; Walter, T.; Solaro, G.; Wang, R.; Bonano, M.; Manzo, M.; Ergintav, S.; Zheng, Y.; Xiong, X.; Lanari, R. Fault locking near Istanbul: Indication of earthquake potential from InSAR and GPS observations. *Geophys. J. Int.* **2006**, *205*, 490–498. [CrossRef]
8. Biggs, J.; Robertson, E.; Cashman, K. The lateral extent of volcanic interactions during unrest and eruption. *Nat. Geosci.* **2016**, *9*, 308–311. [CrossRef]
9. Tomás, R.; Herrera, G.; Cooksley, G.; Mulas, J. Persistent scatterer interferometry subsidence data exploitation using spatial tools: The vega media of the Segura river basin case study. *J. Hydrol.* **2011**, *400*, 411–428. [CrossRef]
10. Ruiz-Constán, A.; Ruiz-Armenteros, A.M.; Lamas-Fernández, F.; Martos-Rosillo, S.; Delgado, J.M.; Bekaert, D.P.S.; Sousa, J.J.; Gil, A.J.; Caro, C.M.; Hanssen, R.F.; et al. Multi-temporal InSAR evidence of ground subsidence induced by groundwater withdrawal: The Montellano aquifer (SW Spain). *Environ. Earth Sci.* **2016**, *75*, 1–16. [CrossRef]
11. Kim, S.W.; Wdowinski, S.; Dixon, T.H.; Amelung, F.; Won, J.S.; Kim, J.W. InSAR-based mapping of the surface subsidence in Mokpo City, Korea, using JERS-1 and ENVISAT SAR data. *Earth Planets Space* **2008**, *60*, 453–461. [CrossRef]
12. Kim, S.W.; Wdowinski, S.; Dixon, T.H.; Amelung, F.; Woo, J.; Won, J.S. Measurements and predictions of subsidence induced by soil consolidation using persistent scatterer InSAR and a hyperbolic model. *Geophys. Res. Lett.* **2010**, *37*, L05304. [CrossRef]
13. Abdikan, S.; Arıkan, M.; Şanlı, F.B.; Çakir, Z. Monitoring of coal mining subsidence in peri-urban area of Zongundak city (NW Turkey) with persistent scatterer interferometry using ALOS-PALSAR. *Environ. Earth Sci.* **2014**, *71*, 4081–4089. [CrossRef]
14. Hu, X.; Lu, Z.; Pierson, T.C.; Kramer, R.; George, D.L. Combining InSAR and GPS to determine transient movement and thickness of a seasonally active low-gradient translational landslide. *Geophys. Res. Lett.* **2008**, *45*, 1453–1462. [CrossRef]
15. Colesanti, C.; Wasowski, J. Satellite SAR interferometry for wide-area slope hazard detection and site-specific monitoring of slow landslides. In Proceedings of the Ninth International Symposium on Landslides, Rio de Janeiro, Brazil, 28 June–2 July 2004.
16. Liu, G.; Jia, H.; Nie, Y.; Li, T.; Zhang, R.; Yu, B.; Li, Z. Detecting subsidence in coastal areas by ultrashort-baseline TCPInSAR on the time series of high-resolution TerraSAR-X images. *IEEE Trans. Geosci. Remote Sens.* **2014**, *52*, 1911–1923. [CrossRef]

17. Akarvardar, S.; Feigl, K.L.; Ergintav, S. Ground deformation in an area later damaged by an earthquake: Monitoring the Avcilar district of Istanbul, Turkey, by satellite radar interferometry. *Geophys. J. Int.* **2009**, *178*, 976–988. [CrossRef]
18. Ferretti, A.; Colombo, D.; Fumagalli, A.; Novali, F.; Rucci, A. InSAR data for monitoring land subsidence: Time to think big. *Proc. Int. Assoc. Hydrol. Sci.* **2015**, *372*, 331–334. [CrossRef]
19. Xu, B.; Feng, G.; Li, Z.; Wang, Q.; Wang, C.; Xie, R. Coastal subsidence monitoring associated with land reclamation using the point target based SBAS-INSAR method: A case study of Shenzhen, China. *Remote Sens.* **2016**, *8*, 652. [CrossRef]
20. Intrieri, E.; Gigli, G.; Nocentini, M.; Lombardi, L.; Mugnai, F.; Fidolini, F.; Casagli, N. Sinkhole monitoring and early warning: An experimental and successful GB-InSAR application. *Geomorphology* **2015**, *241*, 304–314. [CrossRef]
21. Luo, Q.; Perissin, D.; Zhang, Y.; Jia, Y. L- and X-band multi-temporal InSAR analysis of Tianjin subsidence. *Remote Sens.* **2014**, *6*, 7933–7951. [CrossRef]
22. Zhang, K.; Ge, L.; Li, X.; Ng, A.-M. Monitoring ground surface deformation over the North China Plain using coherent ALOS PALSAR differential interferograms. *J. Geod.* **2013**, *87*, 253–265. [CrossRef]
23. Chen, J.; Wu, J.; Zhang, L.; Zou, J.; Liu, G.; Zhang, R.; Yu, B. Deformation trend extraction based on multi-temporal InSAR in Shanghai. *Remote Sens.* **2013**, *5*, 1774–1786. [CrossRef]
24. Osmanoglu, B.; Dixon, T.H.; Wdowinski, S.; Cabral-Cano, E.; Jiang, Y. Mexico city subsidence observed with persistent scatterer InSAR. *Int. J. Appl. Earth Obs. Geoinf.* **2011**, *13*, 1–12. [CrossRef]
25. Khakim, M.Y.N.; Tsuji, T.; Matsuoka, T. Lithology-controlled subsidence and seasonal aquifer response in the Bandung basin, Indonesia, observed by synthetic aperture radar interferometry. *Int. J. Appl. Earth Obs. Geoinf.* **2014**, *32*, 199–207. [CrossRef]
26. Ferretti, A.; Prati, C.; Rocca, F. Nonlinear subsidence rate estimation using permanent scatterers in differential SAR interferometry. *IEEE Trans. Geosci. Remote Sens.* **2000**, *38*, 2202–2212. [CrossRef]
27. Ferretti, A.; Prati, C.; Rocca, F. Permanent scatterers in SAR interferometry. *IEEE Trans. Geosci. Remote Sens.* **2001**, *39*, 8–20. [CrossRef]
28. Ferretti, A.; Novali, F.; Bürgmann, R.; Hilley, G.; Prati, C. InSAR permanent scatterer analysis reveals ups and downs in San Francisco Bay area. *Eos Trans. Am. Geophys. Union* **2004**, *85*, 317–324. [CrossRef]
29. Colesanti, C.; Ferretti, A.; Novali, F.; Prati, C.; Rocca, F. SAR monitoring of progressive and seasonal ground deformation using the permanent scatterers technique. *IEEE Trans. Geosci. Remote Sens.* **2003**, *41*, 1685–1701. [CrossRef]
30. Colesanti, C.; Ferretti, A.; Prati, C.; Rocca, F. Monitoring landslides and tectonic motions with the permanent scatterer technique. *Eng. Geol. Spec. Issue Remote Sens. Monit. Landslides* **2003**, *68*, 3–14. [CrossRef]
31. Colesanti, C.; Ferretti, A.; Locatelli, R.; Novali, F.; Savio, G. Permanent scatterers: Precision assessment and multi-platform analysis. In Proceedings of the Geoscience and Remote Sensing Symposium, Tolouse, France, 21–25 July 2003.
32. Calo, F.; Abdikan, S.; Gorum, T.; Pepe, A.; Kilic, H.; Sanli, F.B. The space-borne SBAS-DInSAR technique as a supporting tool for sustainable urban policies: The case of Istanbul Megacity, Turkey. *Remote Sens.* **2015**, *7*, 16519–16536. [CrossRef]
33. Walter, T.R.; Manzo, M.; Manconi, A.; Solaro, G.; Lanari, R.; Motagh, M.; Woith, H.; Parolai, S.; Shirazei, M.; Zschau, J.; et al. Satellite monitoring of hazards: A focus on Istanbul, Turkey. *Eos Trans. Am. Geophys. Union* **2010**, *91*, 313–324. [CrossRef]
34. Diao, F.; Walter, T.R.; Minati, F.; Wang, R.; Costantini, M.; Ergintav, S.; Xiong, X.; Prats-Iraola, P. Secondary Fault Activity of the North Anatolian Fault near Avcilar, Southwest of Istanbul: Evidence from SAR Interferometry Observations. *Remote Sens.* **2016**, *8*, 846. [CrossRef]
35. Emre, Ö.; Duman, T.Y.; Özalp, S.; Elmaci, S.; Olgun, S.; Saroglu, F. Active Fault Map of Turkey with and Explanatory Text. In *General Directorate of Mineral Research and Exploration*; Special Publication Series-30; General Directorate of Mineral Research and Expansion (MTA): Ankara, Turkey, 2003.
36. Okay, A.I.; Satır, M.; Tuysuz, O.; Akyuz, S.; Chen, F. The tectonics of the Strandja Massif: Late Variscan and mid-Mesozoic deformation and metamorphism in the north Aegean. *Int. J. Earth Sci.* **2001**, *90*, 217–233. [CrossRef]

37. Ergintav, S.; Demirbag, E.; Ediger, V.; Saatcilar, R.; Inan, S.; Cankurtaranlar, A.; Dikbas, A.; Bas, M. Structural framework of onshore and offshore Avcilar, Istanbul under the influence of the North Anatolian Fault. *Geophys. J. Int.* **2001**, *185*, 93–105. [CrossRef]

38. Duman, T.Y.; ve Arama Enstitüsü, M.T. *İstanbul Metropolü Batısındaki (Küçükçekmece-Silivri-Çatalca Yöresi) Kentsel Gelişme Alanlarının Yerbilim Verileri*; Özel Yayın Serisi 3; Maden Tetkik ve Arama Genel Müdürlüğü (MTA): Ankara, Turkey, 2004.

39. Özgül, N.; Üner, K.; Akmeşe, İ.; Bilgin, İ.; Kokuz, R.; Özcan, İ.; Tekin, M. *İstanbul il Alanının Genel Jeoloji Özellikleri*; İstanbul Büyükşehir Belediyesi Deprem ve Zemin İnceleme Müdürlüğü: Istanbul, Turkey, 2005; 79p.

40. Tezcan, S.S.; Kaya, E.; Bal, E.; Ozdemir, Z. Seismic amplification at Avcilar, Istanbul. *Eng. Struct.* **2002**, *24*, 661–667. [CrossRef]

41. Ozmen, B. *Damage Investigation of 17 August 1999 Izmit Earthquake*; Turkish Earthquake Foundation Press: Istanbul, Turkey, 2000; p. 140.

42. Sen, S. A fault zone cause of large amplification and damage in Avcilar (west of Istanbul) during 1999 Izmit earthquake. *Nat. Hazards* **2007**, *43*, 351–363. [CrossRef]

43. Meremonte, M.; Ozel, O.; Cranswick, E.; Erdik, M.; Safak, E.; Overturf, D.; Frankel, A.; Holzer, T. *Damage and Site Response in Avcılar, West of Istanbul*; Aykut, B., Ed.; The 1999 Izmit and Duzce Earthquake, Preliminary Results; Istanbul Technical University Press: Istanbul, Turkey, 2000; pp. 265–267.

44. Cranswick, E.; Ozel, O.; Meremonte, M.; Erdik, M.; Safak, E.; Mueller, C.; Overturf, D.; Frankel, A. Earthquake damage, site response and building response in Avcılar, west of Istanbul, Turkey. *J. Hous. Sci. Appl.* **2000**, *24*, 85–96.

45. Ozel, O.; Cranswick, E.; Meremonte, M.; Erdik, M.; Safak, E. Site effects in Avcilar, west of Istanbul, Turkey, from strong and weak motion data. *Bull. Seismol. Soc. Am.* **2002**, *92*, 499–508. [CrossRef]

46. Ergin, M.; Özalaybey, S.; Aktar, M.; Yalcin, M.N. Site amplification at Avcılar, Istanbul. *Tectonophysics* **2004**, *391*, 335–346. [CrossRef]

47. Eyidogan, H. Analysis of Micro-Seismicity of Istanbul Greater City Area and Active Faults. In *Proceedings of the International Workshop on "Comparative Studies of the North Anatolian Fault (Northwest Turkey) and the San Andreas Fault (Southern California)"*; ITU: Istanbul, Turkey, 2006; p. 54.

48. Sen, S. Cekmece Golleri Arasındaki Bolgesinin Jeolojisi ve Sedimenter Ozellikleri. Unpublished MSc Thesis, İstanbul Üniversitesi Fen Bilimleri Enstitüsü, Istanbul, Turkey, 1994.

49. Şen, Ş.; Koral, H.; Önalan, M. Sedimentary and tectonic evidence for the relationship between the Istranca Massif, the Paleozoic of Istanbul and overlying Tertiary sequence. In Proceedings of the International Symposium on the Petroleum Geology and Petroleum Potential of the Black Sea Area, Sile, Istanbul, Turkey, 22–24 September 1996; pp. 237–244.

50. Kandilli Observatory Earth Research Institution (KOERI). *The Earthquake Activities of the Marmara Region between 1990 and 2000*; Bogazici Univ Kandilli Observatory and Earthquake Research Institute: Istanbul, Turkey, 2000.

51. Dalgıç, S. Factors affecting the greater damage in the Avcılar area of Istanbul during the 17 August 1999 Izmit earthquake. *Bull. Eng. Geol. Environ.* **2004**, *63*, 221–232. [CrossRef]

52. Duman, T.Y.; Can, T.; Gokceoglu, C.; Nefeslioglu, H.A. Landslide susceptibility mapping of Cekmece area (Istanbul, Turkey) by conditional probability. *Hydrol. Earth Syst. Sci. Discuss.* **2005**, *2*, 155–208. [CrossRef]

53. Duman, T.Y.; Can, T.; Ulusay, R.; Keçer, M.; Emre, Ö.; Ateş, S.; Gedik, I. A geohazard reconnaissance study based on geoscientific information for development needs of the western region of Istanbul (Turkey). *Environ. Geol.* **2005**, *4*, 871–888. [CrossRef]

54. Einfalt, T.; Keskin, F. *Analysis of the Istanbul Flood 2009, BALWOIS 2010 Scientific Conference on Water Observation and Information System for Decision Support*; Ministry of Environment and Physical Planning of Republic of Macedonia: Ohrid, Republic of Macedonia, 2010.

55. Komuscu, A.I.; Celik, S. Analysis of the Marmara flood in Turkey, 7-1- September 2009: An assessment from hydrometerological perspective. *Nat. Hazards* **2013**, *66*, 781–808. [CrossRef]

56. Ozcan, O.; Musaoglu, N. Vulnerability analysis of floods in urban areas using remote sensing and GIS. In Proceedings of the 30th EARSeL Symposium: Remote Sensing for Science, Education and Culture, Paris, France, 31 May–4 June 2010.

57. Burak, S.; Kucukakca, E. Impact of Land Reclamation on the Coastal Areas in Istanbul. In Proceedings of the EGU General Assembly Conference, Vienna, Austria, 12–17 April 2015; Volume 17.

58. TMMOB, Hurriyet Daily News. Available online: http://www.hurriyetdailynews.com/six-square-kilometers-of-istanbuls-land-reclaimed-from-the-sea-117884 (accessed on 28 November 2017).

59. Rosen, P.A.; Hensley, S.; Joughin, I.R.; Li, F.K.; Madsen, S.N.; Rodriguez, E.; Goldstein, R.M. Synthetic aperture radar interferometry. *Proc. IEEE* **2000**, *88*, 333–382. [CrossRef]

60. Kampes, B.; Stefania, U. Doris: The Delft Object-Oriented Radar Interferometric Software. In Proceedings of the 2nd International Symposium on Operationalization of Remote Sensing, Enschede, The Netherlands, 16–20 August 1999; p. 16.

61. Sandwell, D.; Mellors, R.; Tong, X.; Wei, M.; Wessel, P. Open Radar Interferometry Software for Mapping Surface Deformation. *Eos Trans. AGU* **2011**, *92*, 234. [CrossRef]

62. Hooper, A. A multi-temporal InSAR method incorporating both persistent scatterer and small baseline approaches. *Geophys. Res. Lett.* **2008**, *35*, L16302. [CrossRef]

63. Hooper, A.; Bekaert, D.; Spaans, K.; Arıkan, M. Recent advances in SAR interferometry time series analysis for measuring crustal deformation. *Tectonophysics* **2012**, *514–517*, 1–13. [CrossRef]

64. Ge, L.; Ng, A.H.; Li, X.; Abidin, H.Z.; Gumilar, I. Land subsidence characteristics of Bandung Basin as revealed by ENVISAT ASAR and ALOS PALSAR interferometry. *Remote Sens. Environ.* **2014**, *154*, 46–60. [CrossRef]

65. Samieie-Esfahany, S.; Hanssen, R.; van Thienen-Visser, K.; Muntendam-Bos, A. On the effect of horizontal deformation on InSAR subsidence estimates. In Proceedings of the Fringe 2009 Workshop, Frascati, Italy, 30 November–4 December 2009; Volume 30.

66. Dogan, U.; Oz, D.; Ergintav, S. Kinematics of landslide estimated by repeated GPS measurements in the Avcilar region of Istanbul, Turkey. *Stud. Geophys. Geodyn.* **2012**, *57*, 217–232. [CrossRef]

67. Hu, X.; Oommen, T.; Lu, Z.; Wang, T.; Kim, J.W. Consolidation settlement of Salt Lake County tailings impoundment revealed by time-series InSAR observations from multiple radar satellites. *Remote Sens. Environ.* **2017**, *199–209*. [CrossRef]

68. Kilic, H.; Ozener, P.T.; Yildirim, M.; Ozaydin, K.; Adatepe, S. Evaluation of Kucukcekmece region with respect to soil amplification. In Proceedings of the International Conference on Soil Mechanics and Geotechnical Engineering, Osaka, Japan, 12–16 September 2005.

69. Plant, G.W.; Covil, C.S.; Hughes, R.A. *Site Preparation for the New Hong Kong International Airport—Thomas Telford*; Institution of Civil Engineers Publishing: London, UK, 1998.

remote sensing

MDPI

Article

Spatio-Temporal Characterization of a Reclamation Settlement in the Shanghai Coastal Area with Time Series Analyses of X-, C-, and L-Band SAR Datasets

Mengshi Yang [1,2], Tianliang Yang [3,4], Lu Zhang [1,5], Jinxin Lin [3,4], Xiaoqiong Qin [1] and Mingsheng Liao [1,5,*]

[1]   State Key Laboratory of Information Engineering in Surveying, Mapping and Remote Sensing, Wuhan University, Wuhan 430079, China; yangms@whu.edu.cn (M.Y.); luzhang@whu.edu.cn (L.Z.); qinxiaoqiong@whu.edu.cn (X.Q.)
[2]   Department of Geoscience and Remote Sensing, Delft University of Technology, 2628CN Delft, The Netherlands
[3]   Key Laboratory of Land Subsidence Monitoring and Prevention, Ministry of Lanf and Resources, Shanghai 200072, China; sigs_ytl@163.com (T.Y.); ljxsupper@126.com (J.L.)
[4]   Shanghai Institute of Geological Survey, Shanghai 200072, China
[5]   Collaborative Innovation Center for Geospatial Technology, Wuhan 430079, China
*   Correspondence: liao@whu.edu.cn

Received: 19 January 2018; Accepted: 21 February 2018; Published: 22 February 2018

Abstract: Large-scale reclamation projects during the past decades have been recognized as one of the driving factors behind land subsidence in coastal areas. However, the pattern of temporal evolution in reclamation settlements has rarely been analyzed. In this work, we study the spatio-temporal evolution pattern of Linggang New City (LNC) in Shanghai, China, using space-borne synthetic aperture radar interferometry (InSAR) methods. Three data stacks including 11 X-band TerraSAR-X, 20 L-band ALOS PALSAR, and 35 C-band ENVISAT ASAR images were used to retrieve time series deformation from 2007 to 2010 in the LNC. An InSAR analysis from the three data stacks displays strong agreement in mean deformation rates, with coefficients of determination of about 0.9 and standard deviations for inter-stack differences of less than 4 mm/y. Meanwhile, validations with leveling data indicate that all the three data stacks achieved millimeter-level accuracies. The spatial distribution and temporal evolution of deformation in the LNC as indicated by these InSAR analysis results relates to historical reclamation activities, geological features, and soil mechanisms. This research shows that ground deformation in the LNC after reclamation projects experienced three distinct phases: primary consolidation, a slight rebound, and plateau periods.

Keywords: reclamation settlements; Lingang New City; time series InSAR analysis; terraSAR-X; ENVISAT ASAR; ALOS PALSAR

1. Introduction

Reclamation is a potential solution to the growing demand for new land for living and development in cities. Over the past decades, the Netherlands, China, the UK, Japan, Singapore, and other coastal countries have extensively exploited sea enclosures and reclamation for coastal city development [1]. Because reclamation usually consists of dumping un-compacted filling materials over unconsolidated marine sediments, land subsidence has always been a significant issue in such areas, and could lead to damage to structures and poses a threat to public safety and the environment [2]. Therefore, it is crucial to monitor land subsidence in reclaimed areas to facilitate an understanding of the evolutionary processes so that proper measures can be taken to plan construction efficiently and mitigate loss.

Remote Sens. **2018**, *10*, 329; doi:10.3390/rs10020329   185   www.mdpi.com/journal/remotesensing

Ground motion can be measured by geodetic techniques such as GPS or the leveling of specific targets. However, this task is labor-intensive and time-consuming, and more importantly, it can only provide discrete information at a limited number of points. In contrast to these traditional techniques, repeat-pass space-borne synthetic aperture radar (SAR) interferometry (InSAR) permits large-scale ground subsidence field retrieval without any prior knowledge, especially with the advent of time series InSAR analysis [3]. These advanced InSAR methods [4–6] are capable of estimating time series deformation through an analysis of phase signals over selected Coherent Points (CPs) and have become a mature tool for land subsidence monitoring in urban areas, the reliability of which has been verified by many studies [7,8]

Lingang New City (LNC), Shanghai, China, is the second-largest reclaimed area in the world. Previous research [9,10] on subsidence monitoring in LNC focused on the data processing algorithm for SAR data, rather than on the interpretation of the deformation observed through InSAR results in relation to reclamation activities and geological features. Jiang et al [11] presented InSAR-derived results with analysis of the reclamation activities in Hong Kong and Macao [12], but the reclamation-induced subsidence in LNC is more complicated. The LNC was built in multi-phase reclamation projects; therefore, the evolutionary process in the LNC is not clearly identified. To fill this gap, we collected multi-band SAR data stacks and employed time series InSAR analyses to explore the spatio-temporal patterns of land subsidence in LNC. The SAR data stacks include 11 X-band TerraSAR-X, 20 L-band ALOS PALSAR, and 35 C-band ENVISAT ASAR images during the period from 2007 to 2010. Using these data, we deduced the evolutionary process in the reclaimed area through an analysis of the observed deformation, reclamation activities, and geological features in LNC. Meanwhile, this research demonstrates that the results achieved from multi-source datasets acquired by SAR satellites with different system configurations can provide a reference for selecting appropriate SAR datasets for monitoring subsidence of the reclaimed area and in similar areas.

This paper is structured as follows. Section 2 presents the basic information of the study area. Data and the methodology of retrieving deformation observations with time series SAR data are discussed in Section 3. The InSAR-derived subsidence rate maps and time-series deformation map of the CPs are shown in Section 4. Section 5 compares the results achieved from different SAR datasets, validates the InSAR results with 13 leveling benchmarks, and discusses the correspondence between the observed deformations and reclamation activities, geological features, and soil mechanisms. Some conclusions are drawn in Section 6.

2. Study Area

The study area of Lingang New City (LNC) is geographically located at the south-eastern corner of Shanghai and close to the intersection of Yangtze River and Hangzhou Bay (approximately longitudes from 121.75°E to 122°E and latitudes from 30.84°N to 31.05°N). The LNC is 37 km from Pudong International Airport and 80 km from Hongqiao Airport, as shown in Figure 1. The total planned area of LNC is 311.6 km², including 133.3 km² (about 42 percent of total land area) created through reclamation projects [13]. The planning area of LNC is nearly one-third of Hong Kong's land area (1104.43 km²) and around nine times as much as Macao's land area (32.8 km²). Since 1973, five phases of the reclamation project have been carried out in the coastal areas of Shanghai, with five dams named by the year they were built: 1973, 1979, 1985, 1994 and the newest dam, built in 2002 [13]. Accordingly, the LNC area consists of various zones formed over different periods.

Three Landsat TM/ETM+ images acquired on 3 November 1999, 16 December 2003, and 6 May 2009 show the changes of this area in Figure 2. The coastal land shown in Figure 2a was formed before the new reclamation project. A new reclamation project was launched in 2003, and construction is scheduled to be completed in 2020 [13]. Figure 2b shows the LNC at the beginning of the project in 2003. The new land outside the coastline in 1999 was completed in 2009, and manmade features including a seawall, watercourse, and artificial lake can be easily identified in the 2009 image (see Figure 2c). The LNC was planned as a development area with multiple roles as an industrial area, shipping hub, and economic center.

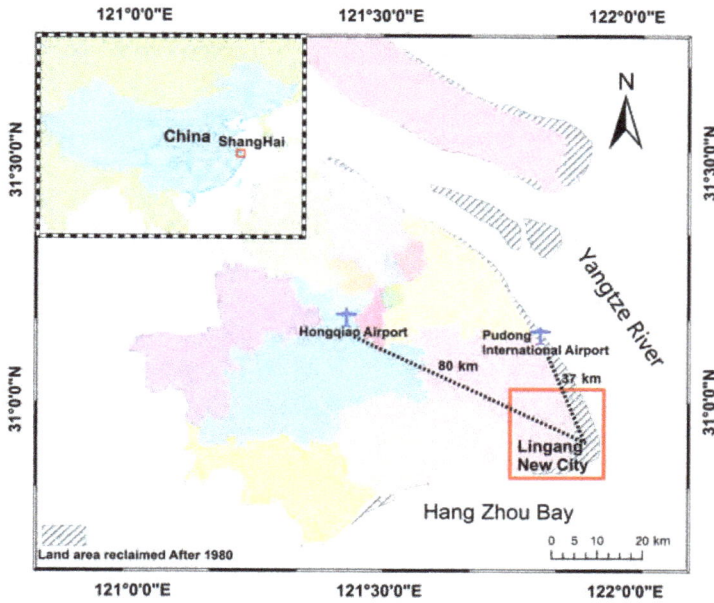

Figure 1. Geographic locations of LNC.

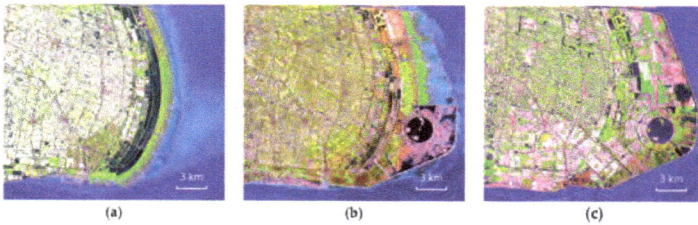

Figure 2. Landsat TM/ETM+ images over LNC in (**a**) 3 November 1999 and, (**b**) 16 December 2003, and (**c**) 6 May 2009.

LNC also contributes to the urban ecological system. According to the land use plan for the LNC in 2003 [13], less than 15% of the land is used for urban construction, around 55% of the land is agricultural, and around 30% is undeveloped. Therefore, LNC has wetlands and rich vegetation, which mitigates weather patterns and helps maintain ecological diversity.

3. Data and Methodology

3.1. Data

We collected 66 satellite SAR images in three stacks, including 11 Stripmap (SM) mode images acquired by TerraSAR-X, 20 Fine Beam Single-polarization (FBS) and Fine Beam Dual-polarization (FBD) mode images acquired by ALOS PALSAR, and 35 Image Mode Single Look Complex (IMS) images obtained by ENVISAT ASAR. Figure 3a shows the coverage of the three data stacks, and Table 1 summarizes the detailed parameters of these datasets. They are not fully overlapped in time. The three data stacks are obtained by the ascending mode, which cannot extract the east-west and vertical (up-down) components of deformations by a combination of three ascending and descending stacks. However, data stacks with slightly different imaging geometry provide a chance to cross-check against estimated results to reveal the deformation behavior in LNC. Moreover, previous results of LNC are based on C-band ENVISAT ASAR [9,10] from February 2007 to May 2010 and X-band COSMO-SkyMed [10] from December 2013 to March 2016. Here, we attached more attention to the results from X-band TerraSAR-X and especially L-band ALOS PALSAR.

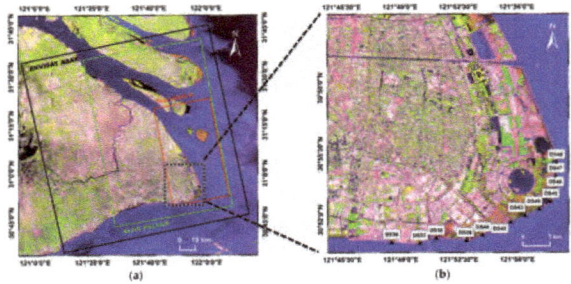

Figure 3. (**a**) The coverage of TerraSAR-X, ALOS PALSAR, and ENVISAT ASAR data; (**b**) The detailed map of LNC and the locations of leveling benchmarks. Black triangles indicate the locations of benchmarks. The background figure is a Landsat TM/ETM+ image acquired on 1 November 2010.

Table 1. Basic parameters of the three SAR data stacks.

Satellite/Parameter	TERRASAR-X	ALOS PALSAR	ENVISAT ASAR
Band (wavelength in cm)	X (3.1)	L (23.6)	C (5.6)
Acquisition dates	20091225~ 20101223	20070107~ 20100718	20070206~ 20100910
Number of images	11	20	35
Acquisition mode	SM	FBS&FBD	IMS
Pass direction	Ascending	Ascending	Ascending
Incident angle (°)	26.5	36.8	22.1
Heading (°)	349.24	347.21	346.80
Spatial coverage of full scene (range in km × azimuth in km)	30 × 50	70 × 56	100 × 100
Slant range spacing (m)	0.9	4.7 (FBS)/9.4(FBD)	7.8
Azimuth spacing (m)	2.0	3.1	4.0
Nominal critical baseline (m)	4000	9800	930
Track and frame	T5F167	T441F610	T497F603–621

The SAR data are paired to form interferograms with high coherence following small baseline rules. The maximum perpendicular baselines and temporal baselines were set to 300 m and 120 days for TerraSAR-X, 400 m; 210 days for Envisat ASAR; and 2800 m and 300 days for ALOS PALSAR. In total, 139 pairs were selected to form interferograms, including 19 from TerraSAR-X, 46 from ALOS PALSAR, and 74 from ENVISAT ASAR data stacks. The temporal distribution of all the interferograms is illustrated in Figure 4. The distribution maps of perpendicular baselines and temporal baselines of the three stacks are given as the supplementary material.

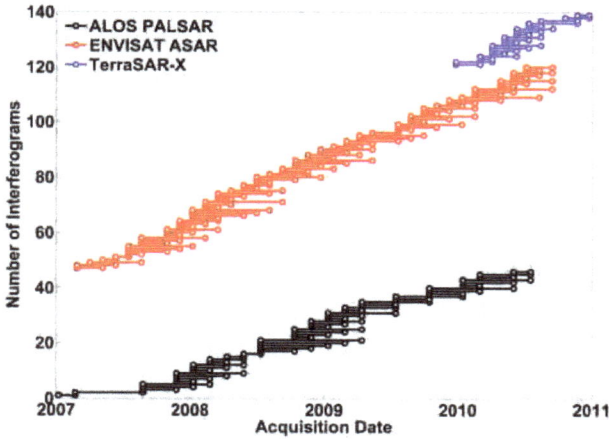

Figure 4. Temporal distributions of interferograms generated from the three data stacks.

Another dataset used in this study is the in-situ leveling measurements. There are thirteen leveling benchmarks, distributed along the seawall in the LNC area, as shown in the Figure 3b. Two leveling measurement campaigns were carried out at all these 13 benchmarks on 9th January 2009 and 9th January 2010. The leveling data was acquired by the Shanghai Institute of Geological Survey (SIGS) following the specification of second-order leveling (±2 mm), and was processed with rigorous adjustment based on the theory of least squares estimation. The reference of leveling measurements is Chinese national height datum 1985. The height differences between two campaigns (one year interval) of 13 benchmarks were converted to the yearly velocities so that they could be used for validating InSAR results.

3.2. The SBAS-InSAR Technique

In this study, the SBAS-InSAR method [5,14] was employed to derive subsidence measurements in the LNC from the three satellite SAR data stacks. The steps in time series SAR data processing include differential interferograms generation, coherent point selection, network formation, phase unwrapping, and temporal-spatial filtering.

Differential interferograms are generated by separately subtracting the flat earth and topographic phase components from the original interferograms of the three data stacks. The phase observation at pixel x in the ith differential interferogram gives:

$$\psi_{int,x,i} = W\left\{ \phi_{def,x,i} + \phi_{atm,x,i} + \Delta\phi_{\theta,x,i} + \Delta\phi_{orb,x,i} + \phi_{n,x,i} \right\} \tag{1}$$

where W is the wrapping operator that truncates phase into the interval $[-\pi, \pi]$ by modulo 2π, and $\phi_{def,x,i}$ is the phase component corresponding to the target movement along the satellite line-of-sight (LOS) direction. The value $\phi_{atm,x,i}$ is the phase signal induced by the difference in atmospheric path delay between two observations, $\Delta\phi_{\theta,x,i}$ is the residual phase of look angle error, $\Delta\phi_{orb,x,i}$ is the phase of residual inaccuracy orbit, and $\phi_{n,x,i}$ is the noise term.

CPs that maintain stable and strong backscattering signals over long time series observations were extracted from each data stack for differential interferometric phase analysis. In the LNC, there are a lot of vegetation-covered areas and built structures; thus, the phase noise is used as the indicator for CPs selection. It can identify reliable CPs in vegetated areas and areas with built structures. Phase noise γ_x is defined as StaMPS [14]:

$$\gamma_x = \frac{1}{N}\left|\sum_{i=1}^{N}\exp\left\{j\left(\psi_{\text{int},x,i} - \widetilde{\psi}_{\text{int},x,i} - \Delta\hat{\phi}_{\theta,x,i}^{nc}\right)\right\}\right| \tag{2}$$

where N is the number of images, $\widetilde{\psi}_{\text{int},x,i}$ is the wrapped estimated of the spatially correlated parts of $\psi_{\text{int},x,i}$, and $\Delta\phi_{\theta,x,i}^{nc}$ is the uncorrelated spatial component of $\Delta\phi_{\theta,x,i}$. Supposing that $\phi_{def,x,i}$, $\phi_{atm,x,i}$, and $\Delta\phi_{orb,x,i}^{nc}$ are spatially correlated over a certain distance, and $\Delta\phi_{\theta,x,i}$ and $\phi_{n,x,i}$ are uncorrelated over the same distance, then the uncorrelated spatial components of $\phi_{def,x,i}$, $\phi_{atm,x,i}$, and $\Delta\phi_{orb,x,i}$ are small and are ignored. The phase noise of each pixel can be estimated by γ_x. Then, the CP candidates exceeding the noise threshold (in this case, we used 0.2π) are eliminated. Afterwards, for each stack, a triangulated irregular network (TIN) was built over the final CPs to establish the spatial network for phase unwrapping. Phase observations of the final group of CPs are unwrapped in both spatial and temporal dimensions to retrieve the deformation rates, as well as time series displacements. The detailed estimation algorithm of SBAS can be found in [5,14].

3.3. Vertical Displacement Estimation

Displacements by SBAS are given with respect to a reference pixel P, and relative to one reference acquisition time T. To compare results from different stacks, the three results must refer to a common pixel at the same reference time. A reference pixel is a common, stable pixel appearing in all three sets of results. In this work, the reference point is in the west corner of the Shanghai Ocean University, located in the stable area as identified in the Shanghai geological environment bulletin based on ground measurements including the leveling measurements [15,16]. A reference time is a common date of the three data stacks. However, three satellites have different revisiting intervals (11 days for TerraSAR-X, 35 days for ENVISAT ASAR, 46 days for ALOS PALSAR) and TerraSAR-X data covers only from 2009 to 2010, hence there are no data acquired on the same date. A compromise method is selecting acquisition dates adjacent to each other to achieve a common reference time. According to the acquisition date of SAR data, we have chosen 11-July-2010 for TerraSAR-X, 18-July-2010 for ALOS PALSAR, and 5-July-2010 for ENVISAT ASAR. Thus, the three sets of results refer to a generic pixel and the same reference date, the displacements of pixel P in T is zero, and all measurements are relative to this spatial-temporal reference point.

Furthermore, InSAR-derived displacements by SBAS are in LOS that must be converted to geographic space. Displacement in the LOS direction is a projection of the real deformation in east-west, north-south, and up-down directions. As SAR satellites are in a near-polar orbit with an angle between the flight direction and the north direction close to 10 degrees, InSAR observations have limited sensitivity to north-south displacement. Moreover, previous research indicated negligible west-east deformation in LNC [9,10]. Hence, we convert all the LOS measurements into a vertical direction according to incidence angles:

$$d_v = \frac{d_{LOS}}{\cos\theta} \tag{3}$$

where d_v is the vertical deformation, d_{LOS} represents the displacements in the LOS direction, and θ is the incidence angle. Polynomial model fitting then estimates the subsidence rates with the extracted vertical deformation. For deformation decomposition, the optimal solution is combing the ascending and descending orbit for extracting deformation components in the east-west, north-south, and up-down directions. However, we do not have a descending data stack for this area during the same observation period. Another possibility is that we can compare the results from data stacks with slightly different imaging geometry to validate our assumption that vertical deformation is dominant. Such conversion also enables the validation of results against in-situ leveling measurements.

3.4. Results Validation

We evaluate the accuracy of InSAR-derived deformation measurements in two ways: an inner-precision check by consistency analysis among results and validation with independent ground measurements.

To perform consistency analysis among the results from the three SAR data stacks, we selected CPs with high coherence identified within the LNC area for comparative analyses. The threshold of coherence was set as 0.8 for initial candidate CP selection. In a given cross-stack CP pair (a, b), i.e., *a* and *b* are CPs chosen from two data stacks of A and B individually. If *b* is the nearest neighbor CP from B in object space for *a*, and vice versa, then, *a* and *b* will be identified as a common CP subject, if the Euclidean distance between them is less than 50 m. For each pair of data stacks, a linear fitting model of the mean deformation rate derived from one stack with respect to one derived from the other is established by least square estimation for the group of common CPs. The determination coefficient R^2, mean absolute difference, and standard deviation of absolute differences of CP pairs indicate the consistency of the results.

The absolute accuracy of deformation measurements derived from the three SAR data stacks is evaluated with respect to independent ground measurements. The leveling measurements are obtained at the benchmarks along the seawall in LNC following the specifications for second-order leveling (± 2 mm) and were processed with rigorous adjustment based on the theory of least squares estimation. Indeed, two epochs of leveling measurements for 13 benchmarks in LNC are insufficient for evaluating all the InSAR results for multi epochs. However, these measurements do provide a reference independent of SAR satellite observations. Moreover, the locations of benchmarks are along the seawall, whose stability must be continuously monitored.

In consideration of the spatial and temporal mismatch between InSAR results and leveling measurements, two preprocessing steps are taken to enable the absolute accuracy assessment.

First, since there are only two phases of leveling measurements in January 2009 and January 2010, it is preferable to use InSAR-derived mean deformation rates over the same period for comparison. For ASAR and PALSAR data stacks, the appropriate subsets of derived time series deformations that are temporally closest to the leveling measurements are used to estimate the desired mean deformation rates. However, for the TerraSAR-X data stack, such a selection is not possible because there is almost no overlap in time between X-band observations and leveling measurements. Therefore, the deformation rates derived from the full X-band data stack are used in the comparison by assuming there is a linear deformation. Thus, the deformation rates of 2009 and 2010 are consistent.

Second, it is impossible to make direct one-to-one comparisons between CPs and leveling benchmarks because they are not located at exactly the same geographic position. As an alternative, a more practical average-to-one approach [17] is adopted to make the comparison. To be specific, for each leveling benchmark b_i, all CPs with a distance to b_i that is less than a predefined threshold T_d are chosen to form a group. Then, the arithmetic average of mean deformation rates at all CPs within this group is calculated as a representative to be used in comparison against the annual mean deformation rate measured by leveling. In this study, the value of distance threshold T_d was empirically set at 100 m. The mean values of differences and the corresponding standard deviations of InSAR results and leveling measurements are the indicators of absolute precision.

4. Results

4.1. Subsidence Rate Map

Using the method described in Section 3, we obtained the subsidence rates at selected CPs within the LNC from the three data stacks, as shown in Figure 5. Negative values of deformation rates indicate subsidence, while positive values represent uplift motions. In general, moderate (between -20 and -5 mm/y) to strong (<-20 mm/y) subsidence was observed in the newly reclaimed land close to the coastline. In contrast, most areas in the western part of the LNC appeared quite stable (between −5 and 5 mm/y), while gentle uplifts (between 5 and 10 mm/y) were clustered in the middle zone.

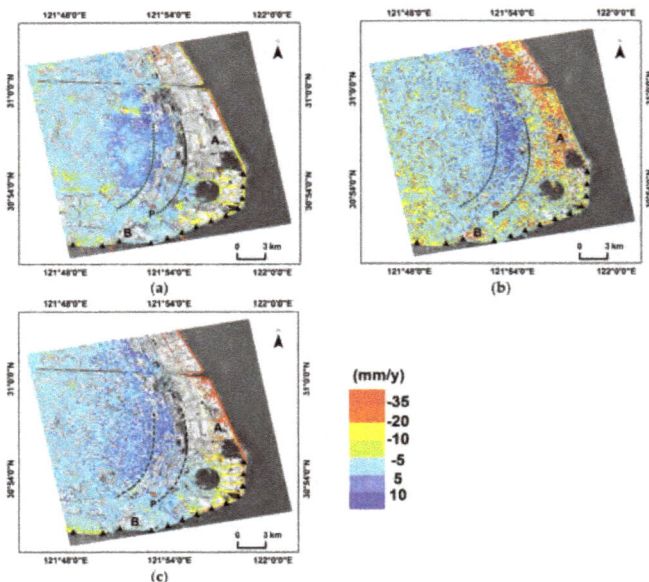

Figure 5. Motion rates of LNC derived by time series analyses using: (**a**) X-band TerraSAR-X; (**b**) L-band ALOS PALSAR; (**c**) C-band ENVISAT ASAR data stacks. Red Star indicates the reference point. Black triangles indicate the locations of leveling benchmarks. The background is a mean amplitude map of 11 TerraSAR-X images.

Similar spatial variability also exists over the newly reclaimed land, with subsidence in the northern part being more serious than in the southern part. According to a construction order [13], we divided the study area into three zones, i.e., zone1 formed before 1973, zone 2 formed between 1973 and 1994, and zone 3 formed after 2002. Locations of the three zones are illustrated in Figure 6. The southern part in zone 3 is the center of the LNC, and over the past decade, has developed more rapidly than the northern part in zone 3 (see Figure 2). Many factories, office blocks, residential buildings, and public infrastructures have been built in the southern part. Consequently, most CPs in the southern part were identified from new buildings and roads constructed on relatively stable piled foundations. Therefore, subsidence in the southern part is significantly slower than in the northern part, where there is subsiding soil. The deformation rates derived from the three different stacks show similar inhomogeneous spatial patterns as described. Furthermore, the distributions of selected CPs are not uniform. The CP density in the offshore area is lower than that in the inland area. Such a non-uniform spatial pattern can be largely attributed to the different stages of consolidation in the reclaimed zones. The relationship between observed deformation and reclamation projects is discussed in Section 5.3.

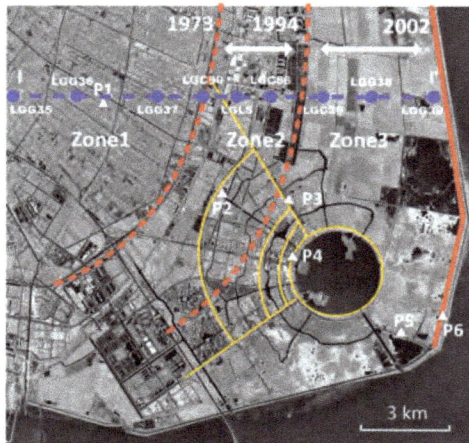

Figure 6. Partition diagram of LNC, zone1 formed before 1973, zone 2 formed between 1973 and 1994, and zone 3 formed after 2002.

SBAS analyses were implemented on CPs that maintained coherence over time; the number and density of CPs will primarily determine the spatial sampling frequency. In general, a higher density of CPs is more likely to afford details of single objects. In this study, the total number and mean density (unit: CPs/km^2) of CP points detected in the LNC area are 48872 and 128 for TerraSAR-X, 60683 and 112 for PALSAR, and 53783 and 70 for ASAR.

The highest mean density of CPs was detected in the urban area by the X-band TerraSAR-X data stack, followed by L-band and C-band data stacks. Two factors are involved. First, TerraSAR-X data has a higher spatial resolution than ASAR and PALSAR data, with which more details of ground features can be observed and thus more CPs can be detected at the same target. Second, the high density of detected CPs may benefit from the relatively short observation period, as well as from the frequent TerraSAR-X acquisitions.

As shown in Figure 5, in the southern part of zone 3 where a residential urban area is located, all the three data stacks manifested similar capabilities of detecting numerous CPs with stable and strong backscattering signals. Nevertheless, the north part of zone 3, marked as A in Figure 5, in the rural area covered with vegetation, an entirely different pattern appeared. Very few CPs were detected from X- and C-band data stacks, whereas quite a few CPs were still identified from the L-band data stack. Such an advantage of using L-band data can be attributed to its capabilities of deeper penetration into vegetation and better resistance to temporal de-correlation related to the longer wavelength.

To examine more closely the performance of the PALSAR data stack, we focused on two marked areas, A and B, shown in Figure 5. In the marked area A, few points are identified by X-band (Figure 5a) and C-band (Figure 5c) data stacks, and most of the points were detected along the road. However, the result of the L-band (Figure 5b) data stack provides detailed deformation rates except for the two small ponds in this area. In the marked area B, results based on X-band (Figure 5a) and C-band (Figure 5c) data stacks failed to detect the deformation in the marked area B because no CP was selected. The result of the L-band (Figure 5b) data stack shows there is a severe subsidence. A quantitative comparison among the three stacks and absolute precision validation will be elaborated in Sections 5.1 and 5.2.

4.2. Time Series Deformation of Selected CPs

To evaluate the spatial distribution and temporal evolution of deformation, six CPs were selected in different zones: P1 in zone 1; P2 in zone 2; and P3, P4, P5, and P6 in zone 3. The locations of the selected CPs are shown in Figure 6. P1 is in the stable area shown in Figure 5, P2 is in the middle area with slight uplifts, and P3, P4, P5, and P6 are in the newly reclaimed land that is close to the coastline. There are three extra points selected in zone 3 because of the spatial variability of subsidence within this area, as denoted in Figure 5.

Figure 7 shows the time series deformation of six CPs: (a) P1, (b) P2, (c) P3, (d) P4, (e) P5, and (f) P6 from ASAR, PALSAR, and TerraSAR-X data stacks. Time series deformation measured at CPs reflects the detailed changes of the CPs in the observation period relative to the measurement in the selected reference time. The data acquisition times of ASAR and PALSAR were from 2007 to 2010 and TerraSAR data was collected from 2009 to 2010. Thus, the overlapping time in July 2010 was chosen as the reference time. As indicated in Figure 7, P1 is quite stable as it underwent only a slight fluctuation around zero from 2007 to 2010; whereas P2 shows a trend of gentle uplift before September 2009 and afterward zero-centered subtle fluctuation. The other four CPs (P3, P4, P5, and P6) experienced linear subsidence at various velocities from 2007 to 2010, with the fastest and slowest subsidence appearing at P6, located on the seawall, and P3 at a more inland location. However, such a subsidence trend at all four points slowed down in 2009 and then transformed into a relatively mild pattern of subsidence, which might indicate that the newly reclaimed land has entered the stage of long-term slow compression. In Sections 5.3–5.5 we will further discuss the spatial distribution and temporal evolution of deformation in LNC in joint analyses of reclamation activities, local geological data, and the soil mechanisms.

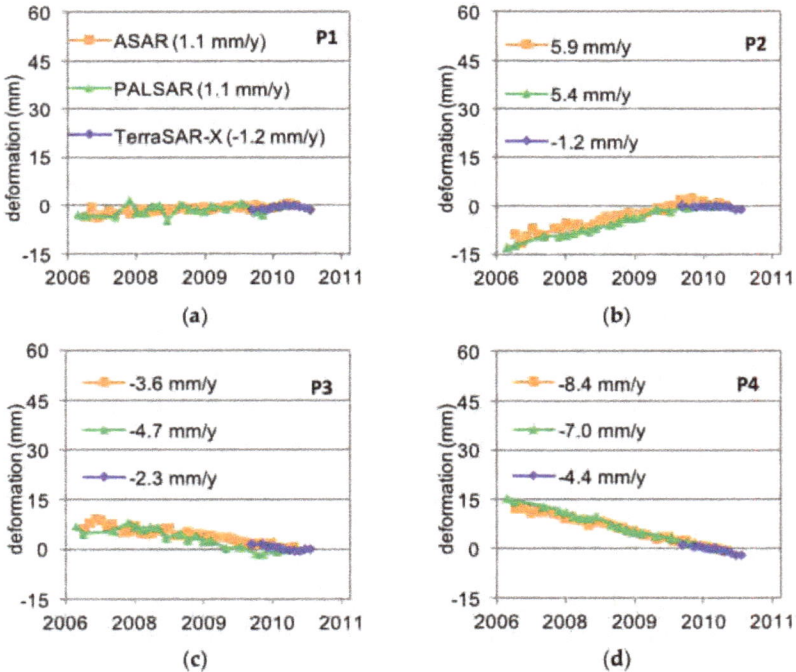

Figure 7. *Cont.*

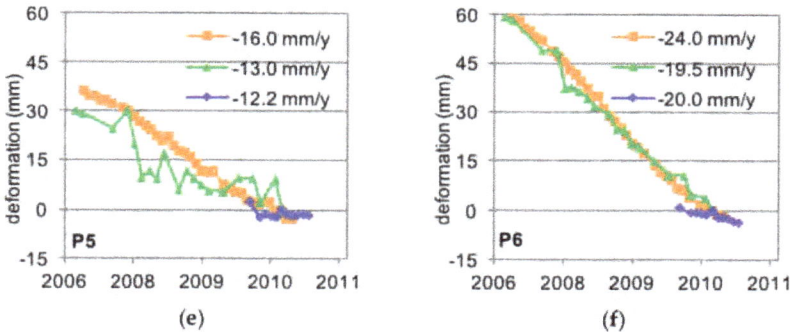

Figure 7. Time series displacements at six typical CPs: (**a**) P1, (**b**) P2, (**c**) P3, (**d**) P4, (**e**) P5, and (**f**) P6 derived from ASAR, PALSAR, and TerraSAR-X data stacks.

5. Discussion

5.1. Consistency Analysis among InSAR-Derived Results

Consistency analysis among the results from the three SAR data stacks was carried out. We estimated a linear fitting model of the mean deformation rate derived from one stack with respect to a model derived from the other two stacks by least square estimation using the entire group of common CPs. Scatter plots for these models are shown in Figure 8, with the corresponding mathematical expressions and related statistics summarized in Table 2. When making comparisons between TerraSAR-X and the other two data stacks, the time interval used for mean deformation rate estimation was set as from late 2009 to middle 2010.

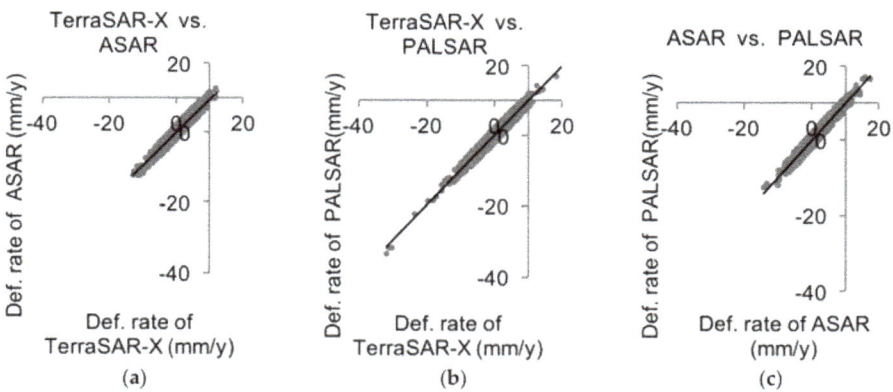

Figure 8. Comparison of mean deformation rates among the three data stacks: (**a**) TerraSAR-X vs. PALSAR, (**b**) TerraSAR-X vs. ASAR, (**c**) ASAR vs. PALSAR.

Table 2. Results of cross validation among the three SAR data stacks.

Satellite/Parameter	TERRASAR-X VS. PALSAR	TERRASAR-X VS. ASAR	ASAR VS. PALSAR
Number of common CPs	1863	2987	1866
Mathematical expression of linear model	Y = 0.9x	Y = 0.9x	Y = 0.9x
Coefficient of determination R^2	0.9	0.9	0.9
Mean absolute difference (mm/y)	0.9	0.7	0.2
Standard deviation of absolute differences (mm/y)	3.6	3.4	3.9

From Figure 8 and Table 2, it can be seen that the three coefficients of determination are around 0.9, which indicates that highly similar spatial patterns in the mean deformation rate were detected by the three SAR data stacks. The mean and standard deviation of absolute differences for the three pairs were less than 1 mm/y and 4 mm/y, respectively, showing an overall good agreement among the deformation measurements derived from the three SAR data stacks. Several CPs in areas with relatively fast subsidence (mean deformation rate < −20 mm/y) were identified by both X-band and L-band data stacks simultaneously. These were excluded from the groups of identical CPs for linear fitting between the C-band result and that of X-band or L-band. This phenomenon might be caused by the weaker capability of the ASAR data stack to detect fast deformation, as compared to the TerraSAR-X and PALSAR observations used in this study.

This comparison was based on the extracted vertical displacement. Thus, the good agreement among these InSAR-derived results also validates our claim that the east-west component of displacements, with respect to the vertical deformation, is negligible.

5.2. Validation with Leveling

The absolute accuracy of deformation measurements derived from the three SAR data stacks was evaluated with respect to independent ground measurements. The results of the absolute accuracy assessment at 13 leveling benchmarks are plotted in Figure 9. According to the leveling measurements, large subsidence (<−20 mm/y) occurred along the eastern section of the seawall in the LNC (from DS43 to DS48), while in the southern section (from DS36 to DS42), only slight to moderate subsidence was detected. Such a spatial variability was also revealed by InSAR results from the three data stacks, showing overall good agreement with leveling measurements. The arithmetic mean values of differences are 2.7, 1.5, and 0.2 (mm/y) for TerraSAR-X, PALSAR, and ASAR, respectively, and the corresponding standard deviations are 4.7, 5.3, and 4 (mm/y). Therefore, we can conclude that all the three data stacks achieved millimeter-level accuracy. The relatively larger differences for X-band data are likely due to temporal mismatch, while the best performance by C-band data might be attributed to its more frequent observations during the period.

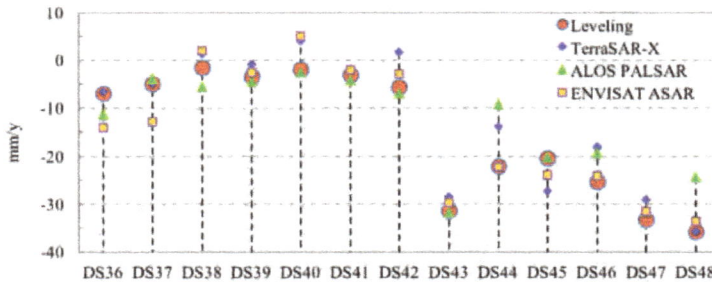

Figure 9. Validation of InSAR-derived mean deformation rates at leveling benchmarks.

5.3. Analysis of Observed Subsidence and Reclamation Evolution

We also explored the relationship between InSAR-observed ground subsidence and reclamation evolution. As shown in Figure 6, we divided the study area into three zones considering that the construction order for LNC Zone 1 was established before 1973, in zone 2 the construction order was set between 1973 and 1994, and in zone 3, it was formulated from 2002.

As shown in Figure 5, many CPs were detected in zone 1 and zone 2. Since the reclamation projects in these two zones have been completed for more than a decade, most CPs in these two zones exhibited relatively good stability (between −5 and 5 mm/y). Compared to the stable older zones, an overwhelming subsiding tendency with relatively high deformation rates (<−10 mm/y) was detected in zone 3, reclaimed after 2002. The fillings used for reclamation in the LNC are characterized by a high void ratio, large compressibility, and low strength. This zone was still experiencing a consolidation procedure, consistent with the established Terzaghi theory which assumes that the primary consolidation could last for a few years [18]. This situation was also verified by the time series deformation of CPs in the three zones. As indicated in Figure 7, P1 in zone 1 was quite stable from 2007 to 2010, whereas P2 in zone 2 shows a slight uplift. P3, P4, P5, and P6 illustrate the spatial variability of subsidence within zone 3.

To quantitatively evaluate the corresponding relationships between the reclamation phases and observed subsidence, the mean deformation rates over the three zones derived from the three SAR data stacks were calculated and summarized in Table 3. The mean deformation rates of points of the three zones describe the main trends within each zone, and the standard deviations of zones illustrate variation in the velocities of deformation in each zone and probable error. Therefore, the standard deviation increase from zone 1 to zone 3 indicates that the spatial variation was the largest in zone 3. Moreover, the standard deviation of ALOS PALSAR is larger than the other two, which indicates that the phase noise in the PALSAR data is larger. All the three mean values across zone 1 are around ±1 mm/y, which again confirms its relative stability. In the meantime, the moderate positive value and large negative value of the mean deformation rate for zone 2 and zone 3 indicate, individually, a gentle uplift and rapid subsidence during the study period. Therefore, it is evident that these three zones were in different consolidation periods, corresponding to the reclamation activities at different times.

Table 3. Deformation rates of the three zones (unit: mm/y).

Mean Deformation Rate (Standard Deviation)		SAR SENSOR		
		TERRASAR-X	ALOS PALSAR	ENVISAT ASAR
zone	1	−0.5 (3.9)	−0.8 (9.9)	1.3 (5.0)
	2	2.4 (4.3)	4.8 (11.9)	6.1 (6.4)
	3	−7.9 (9.1)	−12.1 (19.5)	−11.3 (20.3)

According to the spatial distribution and temporal evolution of deformation, we infer three stages of surface changes in LNC after the reclamation project. For newly reclaimed areas such as zone 3, the consolidation will last for a few years. This primary consolidation is the first stage of changes after reclamation. Then, it enters the second stage, which is a slight rebound after long-term compression. The second stage elaborates the changes in the area, which build for a while during a reclamation project; zone 2 has been established for more than ten years. Finally, subsidence will stay at a steady level after long-term changes. Zone 1 exhibits this stability, which has developed over a period of more than 30 years.

5.4. Geological Features

In this section, we further analyze the observed settlements with the geological features in the LNC. The different components and lithology of the underlying soil are important factors influencing the subsidence of LNC [15,16,19,20]. Soft soil in LNC can be divided into seven engineering geological layers (①–⑦) based on their geological ages, soil behaviors, and physical and mechanical properties. Table 4 gives detailed information about each layer.

Table 4. The division of engineering geologic layers in LNC.

Geological Age		Layer Number and Lithology	Deposit Type	Distribution Area	Foundation Conditions
Holocene Q_h	Q_h^3	①₁ Dredger fill	Artificial	Whole area	Not as foundation
		①₃ Dredger fill	Reclaiming project	The eastern and southern part of LNC	Prone to liquefaction
		②₁ Silty clay	Supralittoral	The western part of LNC	Compression layer
		②₃ Sandy silt	Mesolittoral	Whole area	Prone to quicksand
	Q_h^2	④ Muddy clay	Littoral-shallow sea	Whole area	Compression layer
	Q_h^1	⑤₁₋₁ Clay	Supralittoral	Whole area	Compression layer
		⑤₁₋₂ Silty clay	Supralittoral	Widely distributed	Compression layer
		⑤₂ Sandy silt	Swampy	Sporadically distributed	Poor holding layer for foundation
		⑤₃ Silty clay with silt	Swampy	Paleo-rivers area	
		⑤₄ Silty clay	Swampy	Paleo-rivers area	
Late Pleistocene Q_{p3}	Q_{p3}^2	⑥ Clay	Plain-lake	Mainly in LNC area	Good holding layer for construction piles
		⑦₁ Silt with silty	Estuary Marina	Whole area	
		⑦₂ Silt with silty	Estuary Marina	Whole area	

The dredger fill layer ①₃ is mainly distributed in the eastern and southern part of the LNC. Dredger fill is a kind of unconsolidated soil with a high water content, large void ratio, and high compressibility. Therefore, land subsidence is prone to occur during the initial period of a reclamation project. Moreover, the thickness of dredger fill is different as the LNC was built in a multi-phase reclamation project. In the areas built before 1994 (zone 1 and zone 2), the thickness of the dredger fill is around 2–3 m. The fill is in normal consolidation phase. The fill in zone 3, the area reclaimed after 1994, is unconsolidated with an increasing thickness of 3–6 m from the inland to the coast. Therefore, most of the subsidence occurred in the eastern and southern part of LNC.

Silty clay ②₁ is mainly distributed in the west and north part of the LNC at a depth of 0.5–1.5 m with a thickness of 0.5–2 m. A silty clay layer can be used as the natural foundation of small construction due to its low water content, compressibility, and void ratio. This layer shows a relatively stable engineering characteristic compared to ①₃. This agrees with our results that the western and northern parts of the LNC are more stable than the southern and eastern parts of the LNC. Sandy silt ②₃ is distributed across the whole area, which is prone to quicksand. The thickness of the shallow sand layer is around 6 to 14 m and increases from the northwestern to southeast. Therefore, the different consolidation stages of ①₃, and the distribution of ①₃ and ②₁, as well as the thickness changing of shallow sand layer ②₃ together, contribute to the spatial variability of subsidence in the LNC.

The soft soil layer ④ and clay layer ⑤ were distributed in the whole LNC area. Muddy clay ④ is a soft soil typical of Shanghai with a high water content and compressibility. It is easily deformed during construction. Clay layers ⑤ are not a good foundation for buildings except the secondary hard soil layer⑤₄. The clay layer ⑤₁ and ⑤₂ are the compression layers of the larger construction projects. Attention must be paid to the compaction of the soft soil layer ④ and clay layer ⑤ in civil engineering. The widespread hard soil layer ⑥ and sand bed ⑦ originated in the late Pleistocene and are a suitable foundation layer for structures, having nothing to do with the displacements in the LNC.

Figure 10 shows a geological section map of a profile line I-I' as indicated in Figure 6, from 45 m below ground to several meters above ground. Gray represents layer ①₁, faint yellow represents silty clay ②₁ and ⑤₄, cyan represents ①₃, orange with dark dots represents ②₃, blue represents ④, orange with oblique line represents ⑤₁₋₁, faint yellow represents ⑤₁₋₂, green represents ⑥, beige with dotted line represents ⑦₁, and yellow represents ⑦₂. Nine drill holes of the profile line are in different zones: LGG35, LGG36, and LGG37 in zone 1; LGC90, LGL5, and LGC86 in zone 2; and LGC39, LGG38, and LGG39 in zone 3. Hence, the changes in the geological section of nine drill holes describe the underlying soil features of the area formed in different periods.

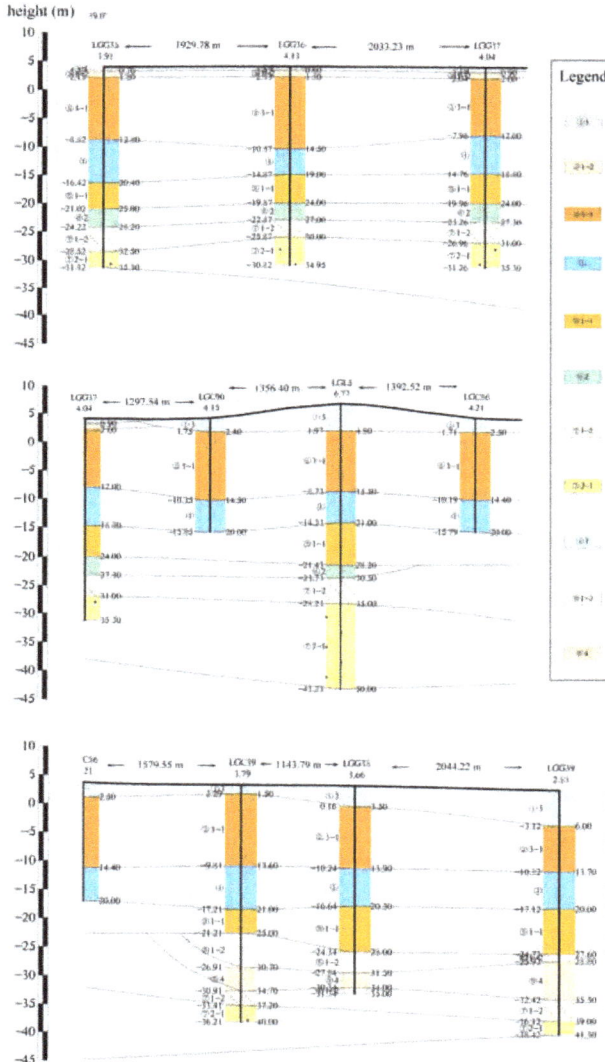

Figure 10. Engineering geologic layers of profile line I-I'. The location of profile line I-I' is indicated in Figure 6.

As illustrated in Figure 10, the geological section of nine drill holes is roughly the same. Specifically, the thickness of each layer of LGG35, LGG36, and LGG37 is relatively the same, which indicates a stable geological condition; while the thickness of holes in zone 2 and zone 3 increases relative to the holes in zone 1. Furthermore, the thickness changes are mainly in the compression layers: topsoil of loose and newly deposited soil layers ①₃, Sandy silt ②₃, soft soil layers ④ and clay layer ⑤, and especially ⑤₁. The consolidation of underlying soil causes the subsidence in the LNC. As described, the different consolidation stages of dredger fill①₃, the distribution of the dredger fill layer ①₃, relatively stable silty clay layer ②₁, the thickness changing of the shallow sand layer ②₃, soft soil layer ④ and the clay layer ⑤ together, contribute to the spatial variation of subsidence in the LNC.

5.5. Compression Mechanism of Hydraulic Fill

In this section, we discuss the observed deformation with the compression mechanism of hydraulic fill. The hydraulic fill in the LNC is mainly the soft alluvial soil [19,20]; therefore, the three stages are explained by the compression mechanism of soft soil. There are two critical compressibility indexes for soft soil: the compression index and resilience index. Previous studies [18,21,22] give the relationship between soil matric suction and the compression index/resilience index of the Shanghai area, which can be used for analyzing the spatial distribution and temporal evolution of deformation in the LNC. Figure 11 gives the effects of soil matrix suction on the compression index and resilience index. The compression index declines with soil matric suction growth, slowly arriving at a plateau. While the resilience index first decreases with an increase in matric suction, it goes up as the suction continuously increases.

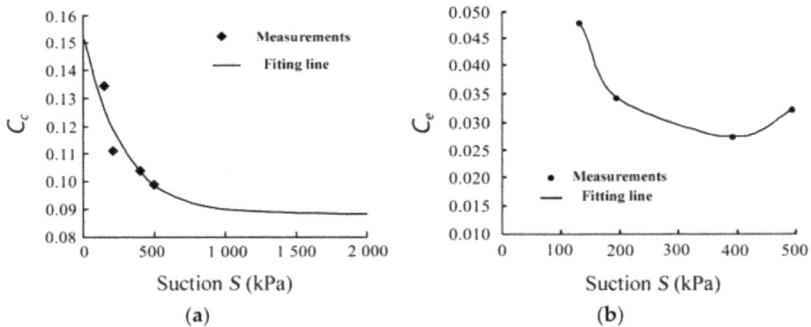

Figure 11. The compressibility of Shanghai soft soil [21,22]: (**a**) relationship between compression index C_c and matric suction S, (**b**) effects of matric suction S on the resilience index C_e.

We explain the temporal evolution of subsidence in the LNC with the compression mechanism of hydraulic fill. We argue that the external forces in a newly reclaimed area are constant, but at the beginning, the consolidation pressure increases as the matric suction increases. Thus, the void ratio decreases gradually, which shows the continuous subsidence in the area during the first stage. Meanwhile, the compression index first decreases and then reaches stability gradually, with an increase in matric suction. However, the rebound index first decreases with an increase of matric suction. When the matric suction is higher than a certain amount, it increases with the rise of matric suction. The slight rebound in the area after a period of reclamation project reflects the growth in the rebound index. Stability occurs when the rebound index and compression index reach equilibrium. Thus, the third stage reflects the balance of rebound index and compression index, and this might take the time to reach, in this case, more than 30 years.

6. Conclusions

In this paper, time series InSAR analyses are carried out with X-, C-, and L-band SAR data stacks to retrieve the subsidence velocity and time series of surface deformation at the LNC in the Shanghai metropolis. A cross-comparison of the results of the three data stacks and validation with leveling measurements demonstrates the performance of these data in land subsidence monitoring. Furthermore, an evaluation of spatial distribution and temporal evolution of deformation in the LNC was conducted by joint analyses of deformation measurements, historical reclamation activities, local geological data, and soil mechanisms. The main conclusions are summarized as follows:

(1) Cross-comparisons of the three results suggest that PALSAR works at a longer wavelength, which makes it much less affected by undesired temporal decorrelation and has advantages when mapping newly reclaimed areas. Cross-validation shows a good agreement among mean deformation rates measured at CPs shared by the three data stacks, with the coefficients of determination around 0.9 and the standard deviations of inter-stack differences less than 4 mm/y. The good agreement validates our argument of the negligibility of the east-west component of displacement with respect to the vertical deformation. Validations with leveling data collected at benchmarks along the seawall indicate that all the three data stacks achieved millimeter-level accuracy. The mean values of differences were 2.7, 1.5, and 0.2 (mm/y) for TerraSAR-X, PALSAR, and ASAR, respectively, and the corresponding standard deviations were 4.7, 5.3, and 4 (mm/y).

(2) The results from the three data stacks show a similar spatial variability of land subsidence across the LNC area. Specifically, overall good stability was observed within the area built before 1973 in the west of LNC, while moderate to large subsidence occurred within the coastal area built after 2002 in the east, and gentle uplift existed within the area built in 1994. A quantitative evaluation of observed subsidence with the historical reclamation activities indicates the spatial variability of land subsidence in the LNC, related to multi-phase reclamation and urban construction projects.

(3) The analysis of local geological data indicates that the consolidation of underlying soil causes the subsidence in LNC. Specifically, the different consolidation stages of dredger fill①$_3$, the distribution of dredger fill layer ①$_3$, relatively stable silty clay layer ②$_1$, the thickness changing of the shallow sand layer ②$_3$, soft soil layer ④ and the clay layer ⑤ together, contribute to the spatial variation of subsidence in the LNC.

(4) Three stages of evolution in the reclaimed area were derived from the observed subsidence in the LNC, verified by the soft soil mechanisms. The first stage is the primary consolidation stage, which can last for a few years. The second stage is the slight rebound stage after long-term compression. The final stage is a state of stable equilibrium.

Acknowledgments: This work is financially supported by the National Natural Science Foundation of China (No. 41571435; No. 61331016). The authors would like to thank DLR, ESA, and JAXA for providing the data stacks of TerraSAR-X (TSX-Archive-2012 AO project COA1755), ENVISAT ASAR (Dragon-3 project, ID 10569), and ALOS PALSAR (ALOS-RA4 project, PI No. 1247 and 1440), respectively, and the Shanghai Institute of Geological Survey (SIGS) for providing leveling data and geological data. The Landsat TM/ETM+ images and SRTM DEM were provided by NASA and USGS. We thank Stephen C. McClure for providing assistance in language editing.

Author Contributions: All four authors contributed to this work. Mengshi Yang implemented the methodology and finished the manuscript. Tianliang Yang, Xiaoqiong Qin and Jinxin Lin made contributions to the data collection and results interpretation. Mingsheng Liao and Lu Zhang designed the research program, supervised the research, and provided valuable suggestions for the revision.

Conflicts of Interest: The authors declare no conflict of interest

References

1. McLeod, E.; Poulter, B.; Hinkel, J.; Reyes, E.; Salm, R. Sea-level rise impact models and environmental conservation: A review of models and their applications. *Ocean Coast Manag.* **2010**, *53*, 507–517. [CrossRef]
2. Haas, J.; Ban, Y. Urban growth and environmental impacts in Jing-Jin-Ji, the Yangtze, River Delta and the Pearl River Delta. *Int. J. Appl. Earth Obs. Geoinf.* **2014**, *30*, 42–55. [CrossRef]

3.   Crosetto, M.; Monserrat, O.; Cuevas-González, M.; Devanthéry, N.; Crippa, B. Persistent Scatterer Interferometry: A review. *ISPRS.J. Photogramm.* **2016**, *115*, 78–89. [CrossRef]
4.   Ferretti, A.; Prati, C.; Rocca, F. Permanent scatterers in SAR interferometry. *IEEE Trans. Geosci. Remote Sens.* **2001**, *39*, 8–20. [CrossRef]
5.   Berardino, P.; Fornaro, G.; Lanari, R.; Sansosti, E. A new algorithm for surface deformation monitoring based on small baseline differential SAR interferograms. *IEEE Trans. Geosci. Remote Sens.* **2002**, *40*, 2375–2383. [CrossRef]
6.   Van Leijen, F.J. Persistent Scatterer Interferometry Based on Geodetic Estimation Theory. 2014. Available online: https://www.ncgeo.nl/index.php?option=com_k2&view=item&id=2669:persistent-scatterer-interferometry-based-on-geodetic-estimation-theory&Itemid=350&lang=en (accessed on 22 February 2018).
7.   Casu, F.; Manzo, M.; Lanari, R. A quantitative assessment of the SBAS algorithm performance for surface deformation retrieval from DInSAR data. *Remote Sens. Environ.* **2006**, *102*, 195–210. [CrossRef]
8.   Raucoules, D.; Bourgine, B.; de Michele, M.; Le Cozannet, G.; Closset, L.; Bremmer, C.; Veldkamp, H.; Tragheim, D.; Bateson, L.; Crosetto, M.; et al. Validation and intercomparison of Persistent Scatterers Interferometry: PSIC4 project results. *J. Appl. Geophys.* **2009**, *68*, 335–347. [CrossRef]
9.   Zhao, Q.; Pepe, A.; Gao, W.; Lu, Z.; Bonano, M.; He, M.L.; Wang, J.; Tang, X. A DInSAR Investigation of the ground settlement time evolution of ocean-reclaimed lands in Shanghai. *IEEE J. Sel. Top. Appl.* **2015**, *8*, 1763–1781. [CrossRef]
10.   Pepe, A.; Bonano, M.; Zhao, Q.; Yang, T.; Wang, H. The use of C-/X-Band Time-Gapped SAR data and geotechnical models for the study of Shanghai's ocean-reclaimed lands through the SBAS-DInSAR technique. *Remote Sens.* **2016**, *8*, 911. [CrossRef]
11.   Jiang, L.; Lin, H. Integrated analysis of SAR interferometric and geological data for investigating long-term reclamation settlement of Chek Lap Kok Airport, Hong Kong. *Eng. Geol.* **2010**, *110*, 77–92. [CrossRef]
12.   Jiang, L.; Lin, H.; Cheng, S. Monitoring and assessing reclamation settlement in coastal areas with advanced InSAR techniques: Macao city (China) case study. *Int. J. Remote Sens.* **2011**, *32*, 3565–3588. [CrossRef]
13.   Shanghai Urban Planning and Design Research Institute. The Master Plan of Lingang New City. *Shanghai Urban Plan. Rev.* **2009**, *4*, 11–26. (In Chinese)
14.   Hooper, A. A multi-temporal InSAR method incorporating both persistent scatterer and small baseline approaches. *Geophys. Res. Lett.* **2008**, *35*. [CrossRef]
15.   Shanghai Municipal Bureau of Planning, and Land Resources. Shanghai Geological Environmental Bulletin (2009). Available online: http://www.shgtj.gov.cn/dzkc/dzhjbg/201007/t20100728_407604.html (accessed on 22 February 2018). (In Chinese)
16.   Shanghai Municipal Bureau of Planning, and Land Resources. Shanghai Geological Environmental Bulletin (2010). Available online: http://www.shgtj.gov.cn/dzkc/dzhjbg/201307/t20130718_600136.html (accessed on 22 February 2018). (In Chinese)
17.   Fuhrmann, T.; Caro Cuenca, M.; Knöpfler, A.; Van Leijen, F.J.; Mayer, M.; Westerhaus, M.; Hanssen, R.F.; Heck, B. Estimation of small surface displacements in the Upper Rhine Graben area from a combined analysis of PS-InSAR, levelling and GNSS data. *Geophys. J. Int.* **2015**, *203*, 614–631. [CrossRef]
18.   Terzaghi, K.; Peck, R.B.; Mesri, G. *Soil Mechanics in Engineering Practice*; John Wiley & Sons: Hoboken, NJ, USA, 1996; pp. 1–592.
19.   Shen, S. Geological environmental character of Lin-Gang new city and its influences to the construction. *J. Shanghai Geol.* **2008**, *105*, 24–28. (In Chinese)
20.   Shi, Y.; Yan, X.; Zhou, N. Land subsidence induced by recent alluvia deposits in Yangtze River delta area, a case study of Shanghai Lingang New City. *J. Eng. Geol.* **2007**, *15*, 391–402. (In Chinese)
21.   Xie, N.; Sun, J. The rheological properties of soft soils in Shanghai. *J. Tongji Univ. Nat. Sci.* **1996**, *24*, 233–237. (In Chinese)
22.   Ye, W.; Zhu, Y.; Chen, B.; Ye, B. Compressibility of Shanghai unsaturated soft soil. *J. Tongji Univ. Nat. Sci.* **2011**, *39*, 1458–1462. (In Chinese)

remote sensing

MDPI

Article

Wuhan Surface Subsidence Analysis in 2015–2016 Based on Sentinel-1A Data by SBAS-InSAR

Lv Zhou [1,2,3], Jiming Guo [1,4,*], Jiyuan Hu [1], Jiangwei Li [5], Yongfeng Xu [6], Yuanjin Pan [7] and Miao Shi [5]

[1] School of Geodesy and Geomatics, Wuhan University, Wuhan 430079, China; zhoulv_whu@163.com (L.Z.); plgk@whu.edu.cn (J.H.)
[2] Guangxi Key Laboratory of Spatial Information and Geomatics, Guilin University of Technology, Guilin 541004, China
[3] Key Laboratory for Digital Land and Resources of Jiangxi Province, East China University of Technology, Nanchang 330013, China
[4] Key Laboratory of Precise Engineering and Industry Surveying of National Administration of Surveying, Mapping and Geoinformation, Wuhan University, Wuhan 430079, China
[5] Wuhan Geomatics Institute, Wuhan 430022, China; 13995650278@163.com (J.L.); shimiao@whu.edu.cn (M.S.)
[6] ChangJiang Wuhan Waterway Bureau, Wuhan 430014, China; xyf01041411@126.com
[7] State Key Laboratory of Information Engineering in Surveying, Mapping and Remote Sensing, Wuhan 430079, China; pan_yuanjin@163.com
* Correspondence: jmguo@sgg.whu.edu.cn; Tel.: +86-133-0862-0798

Received: 18 July 2017; Accepted: 18 September 2017; Published: 22 September 2017

Abstract: The Terrain Observation with Progressive Scans (TOPS) acquisition mode of Sentinel-1A provides a wide coverage per acquisition and features a repeat cycle of 12 days, making this acquisition mode attractive for surface subsidence monitoring. A few studies have analyzed wide-coverage surface subsidence of Wuhan based on Sentinel-1A data. In this study, we investigated wide-area surface subsidence characteristics in Wuhan using 15 Sentinel-1A TOPS Synthetic Aperture Radar (SAR) images acquired from 11 April 2015 to 29 April 2016 with the Small Baseline Subset Interferometric SAR (SBAS InSAR) technique. The Sentinel-1A SBAS InSAR results were validated by 110 leveling points at an accuracy of 6 mm/year. Based on the verified SBAS InSAR results, prominent uneven subsidence patterns were identified in Wuhan. Specifically, annual average subsidence rates ranged from −82 mm/year to 18 mm/year in Wuhan, and maximum subsidence rate was detected in Houhu areas. Surface subsidence time series presented nonlinear subsidence with pronounced seasonal variations. Comparative analysis of surface subsidence and influencing factors (i.e., urban construction, precipitation, industrial development, carbonate karstification and water level changes in Yangtze River) indicated a relatively high spatial correlation between locations of subsidence bowl and those of engineering construction and industrial areas. Seasonal variations in subsidence were correlated with water level changes and precipitation. Surface subsidence in Wuhan was mainly attributed to anthropogenic activities, compressibility of soil layer, carbonate karstification, and groundwater overexploitation. Finally, the spatial-temporal characteristics of wide-area surface subsidence and the relationship between surface subsidence and influencing factors in Wuhan were determined.

Keywords: SBAS-InSAR; surface subsidence; Sentinel-1A; Wuhan; engineering construction; carbonate karstification; water level changes

1. Introduction

Surface subsidence is one of the main engineering geological problems worldwide, and it is caused by consolidation and compression of underground unconsolidated strata because of non-human-related (e.g., earthquake and natural consolidation of soil) or human-related (e.g., groundwater extraction and underground construction) activities [1–3]. Surface subsidence is also one of the major regional geological disasters that cause serious damage to buildings, infrastructures, roads, and bridges and affect human safety in cities [4–6].

In recent years, Interferometric Synthetic Aperture Radar (InSAR) technique has been widely used to investigate surface subsidence [7–9]. Compared with traditional monitoring methods, such as Global Navigation Satellite System (GNSS) [10,11], leveling [12,13], geological and geophysical investigation methods [14], InSAR can detect and monitor regional-scale surface subsidence at low costs with centimeter-to-millimeter accuracy [9,15]. However, problems accompany InSAR due to scatterer changes with time [16,17]. This phenomenon leads to signal decorrelation (e.g., temporal and spatial decorrelation) and reduces monitoring accuracy of InSAR [18]. To overcome limitations of this technique, time-series InSAR (TS-InSAR) has been proposed [19–21]. TS-InSAR extracts deformation information by simultaneously processing multi-SAR images obtained on different dates. Ferretti et al. [22] proposed a TS-InSAR method, which was referred to as Persistent Scatterer InSAR (PS InSAR), that can overcome decorrelation and atmospheric delay problems by identifying and analyzing point-like stable reflectors (PSs). Small Baseline Subset InSAR (SBAS InSAR) [23] is another TS-InSAR method that effectively mitigates decorrelation phenomena by analyzing distributed scatterers (DSs) with high coherence based on an appropriate combination of interferograms [24]. To combine advantages of PS InSAR and SBAS InSAR, Multi-temporal InSAR (MTInSAR) [25], which can simultaneously detect PSs and DSs, was presented to retrieve surface subsidence. Aside from the above-mentioned TS-InSAR methods, other approaches, e.g., the Temporally Coherent Point InSAR (TCPInSAR) [26], the Quasi-PS (QPS) [27], and the Intermittent SBAS (ISBAS) [28], were proposed for different monitoring situations. These TS-InSAR methods have been widely applied in urban surface subsidence monitoring in Beijing [2,29], Mexico [30], Hanoi [31], Shanghai [32,33], and Jinan [34].

Wuhan, as a central city in Central China, has suffered from serious surface subsidence over the past decades because of rapid urban development [35]. Surface collapse in Wuhan was first recorded in 1931 [36]. This surface collapse caused the Yangtze River dike to burst and to flood the Baishazhou area. Since 1978, varying scales and degrees of surface collapses intermittently occurred in various areas (e.g., Hanyang Steel Mill, Wuchang Lujia Street Middle School, Fenghuo Village, and Qingling (QL) Township) in Wuhan [37]. In recent years, owing to groundwater overexploitation, metro construction, and karst collapse, multiple subsidence areas (e.g., Houhu (HH) and Jianghan (JH) subsidence areas) have formed in Wuhan [38], and settlement range and magnitude of these subsidence areas gradually expanded. At present more than 300 benchmarks were arranged in Wuhan by Wuhan Geomatics Institute to monitor surface subsidence. However, given the low spatial resolution and high cost of leveling, difficulty arises from obtaining regional subsidence information and distribution of subsidence bowl. Consequently, effective data (e.g., SAR data) and methods are needed to monitor surface subsidence distribution and state in Wuhan for disaster prevention and sustainable development. Several studies have monitored and analyzed subsidence in Wuhan using TS-InSAR methods with SAR data. Bai et al. [35] retrieved surface subsidence in Wuhan from October 2009 to August 2010 using TerraSAR-X images based on MTInSAR. Results indicated that subsidence rates ranged from −63.7 mm/year to 17.5 mm/year in the study area. Costantini et al. [38] investigated spatial-temporal characteristics of subsidence in Hankou District, Wuhan, from June 2013 to June 2014 with COSMO-SkyMed SAR images. However, the above-mentioned studies adopted high-cost SAR data to extract deformation information in Wuhan before 2014 and only investigated major urban areas in Wuhan, neglecting large non-central regions where various surface subsidence patterns also exist. Prior to this study, wide-coverage surface subsidence information in Wuhan was still little known. Fortunately, Sentinel-1A data are easily and freely accessible. The Interferometric Wide swath

(IW) products of Sentinel-1A, which images three sub-swaths by adopting the Terrain Observation with Progressive Scans (TOPS) SAR technique, offer SAR images at 5 m × 20 m (range × azimuth) spatial resolution around a 250 km-wide area [39–41]. Owing to large swath width, surface subsidence monitoring of large areas can be easily realized using Sentinel-1A SAR images. Therefore, in this study, Sentinel-1A IW single-look complex (SLC) products are used to capture wide-area subsidence information in Wuhan, and the relationship between surface subsidence and influencing factors (e.g., urban construction, industrial development, carbonate karstification, and water level changes) were analyzed in detailed based on leveling data, daily water level changes data on Yangtze River, and distribution data of carbonate rock belt and industrial areas, etc.

In this study, to analyze spatial-temporal characteristics of wide-area surface subsidence in Wuhan, we first adopted the SBAS InSAR method to derive surface subsidence rate and time series using leveling data and 15 free Sentinel-1A TOPS images acquired between 11 April 2015 and 29 April 2016. Subsequently, 110 leveling benchmarks were used to verify the SBAS InSAR-derived results. Water level changes in Yangtze River, precipitation, groundwater, urban construction, and geological conditions were considered influencing factors, and mechanisms underlying their effects on surface subsidence were studied. Finally, the correlation between surface subsidence and these influencing factors were analyzed in detail.

2. Study Area and Data

2.1. Study Area

Wuhan is the central city of Central China. This city is located in the east of Jianghan Plain and southern slope of Ta-pieh Mountains (Figure 1a). Geographical coordinates of Wuhan include 113°41′–115°05′E and 29°58′–31°22′N. Yangtze and Han rivers pass through the central urban area of Wuhan and divide it into three main regions, i.e., Hankou (HK), Wuchang (WC), and Hanyang (HY). Many lakes (e.g., East Lake, Tangxun Lake, and Liangzi Lake) and rivers are distributed in Wuhan, in which water areas account for 25.79% of total area of the city. The study area is outlined by red rectangle in Figure 1a, covering most of Wuhan City. Figure 1b shows SAR mean intensity map of the Sentinel-1A, the map covers the study area. Length of mean intensity map in east–west and south–north directions measures approximately 46.62 and 34.22 km, respectively.

The general terrain in Wuhan is low in south and high in the north. The northern part features low mountain and hilly regions with elevation ranging from 100 m to 500 m. The middle part is mainly the relatively flat middle reaches of the Yangtze River Plain. The southern part is surrounded by hills and mounds, and its average elevation reaches approximately 55 m. Six independent carbonate rock belts with the trend of WNW to ESE are distributed in Wuhan, and its distribution area reaches 1100 km^2 [37]. Since 1931, at least 15 karst surface collapse disasters have been recorded in Wuhan [36]. Wuhan also serves as an important industrial base, science and education base, and comprehensive transportation hub in China due to its geographical location. In recent years, in line with economic development, numerous buildings and subways were constructed and have gradually led to multiple surface inhomogeneous subsidence areas (e.g., Hankou subsidence area) in Wuhan.

2.2. Data

In this paper, 15 ascending Sentinel-1A TOPS SAR images (C-band) acquired from 11 April 2015 to 29 April 2016 covering Wuhan were selected to estimate vertical average surface subsidence velocity and subsidence time series. Specific parameters of Sentinel-1A data are illustrated in Table 1. The three arc-second Shuttle Radar Topography Mission (SRTM) DEM provided by the National Aeronautics and Space Administration (NASA) was adopted to remove topographic phases. Precise Orbit Determination (POD) data released by the European Space Agency (ESA) were used to the orbital refinement and phase re-flattening. To validate SBAS InSAR-derived results, 110 benchmarks (location distribution of benchmarks will be illustrated in Section 5.1) provided by the Wuhan Geomatics

Institute were used, and the leveling data were acquired from 2013 to 2016. Daily water level changes in Yangtze River collected from 11 April 2015 to 29 April 2016 were provided by Changjiang Wuhan Waterway Bureau. These information and precipitation data from meteorological station of Wuhan were utilized to investigate the relationship between surface subsidence and water level changes.

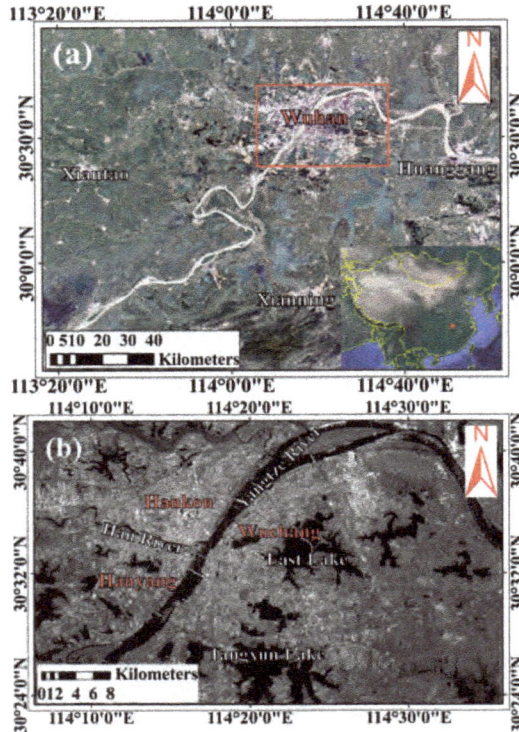

Figure 1. Location of study area and Sentinel-1A TOPS SAR data coverage: (**a**) the study area is outlined by a red rectangle; and (**b**) a SAR mean intensity image of Sentinel-1A covering the study area.

Table 1. Specific parameters of Sentinel-1A TOPS data.

Parameters	Description
Product type	Sentinel1 SLC IW
Track number	113
Central incidence angle on the test site (degree)	41.9
Azimuth angle (degree)	90
Slant range resolution (m)	2.3
Azimuth resolution (m)	13.9
Orbit direction	Ascending
Polarization	VV

3. Methodology

3.1. Fundamental Principle of SBAS-InSAR Technique

First, the SBAS-InSAR technique generates an appropriate combination of differential interferograms produced by SAR data pairs based on baseline threshold values. Then, this technique estimates deformation information of every single differential interferogram and regards them as observed values. Finally, SBAS-InSAR retrieves deformation rate and time series based on observed values acquired in the previous step [23,42].

Considering a set of $N + 1$ SAR acquisitions acquired at ordered times $(t_0, \cdots, t_N)$ covering the study area, M interferograms are generated based on baseline threshold values. M satisfies the following inequality:

$$\frac{N+1}{2} \leq M \leq N\left(\frac{N+1}{2}\right) \tag{1}$$

Assuming interferogram j is generated by combining SAR acquisitions at times t_A and t_B $(t_B > t_A)$, after removing flat-earth and topographic phases, the interferometric phase in pixel of azimuth and range coordinates (x, r) can be expressed as follows [43]:

$$\begin{aligned} \delta\phi_j(x,r) &= \phi(t_B, x, r) - \phi(t_A, x, r) \\ &\approx \phi_{def,j}(x,r) + \phi_{topo,j}(x,r) + \phi_{atm,j}(x,r) + \phi_{noise,j}(x,r) \end{aligned} \tag{2}$$

where $\phi(t_B, x, r)$ and $\phi(t_A, x, r)$ represent phase values of SAR images at t_B and t_A, respectively. $\phi_{def,j}(x,r)$ refers to deformation phase between times t_B and t_A. $\phi_{topo,j}(x,r)$ corresponds to residual phase due to inaccuracies in reference DEM. $\phi_{atm,j}(x,r)$ depicts atmospheric phase error. $\phi_{noise,j}(x,r)$ denotes random noise phases (e.g., orbital errors, thermal noise, and spatial decorrelation).

To achieve deformation information, components $\phi_{topo,j}(x,r)$, $\phi_{atm,j}(x,r)$, and $\phi_{noise,j}(x,r)$ should be separated from $\delta\phi_j(x,r)$. After removing the above-mentioned components, a system of M equations in N unknowns can be obtained from Equation (2). The matrix form of the system can be expressed as follows:

$$A\phi = \delta\phi \tag{3}$$

where A corresponds to an $M \times N$ coefficient matrix, $\forall j = 1, \cdots M$. M and N represent the numbers of interferograms and SAR acquisitions, respectively. $\phi^T = [\phi(t_1), \cdots, \phi(t_N)]$ denotes the vector of unknown phase values related to high-coherence pixels. $\delta\phi^T = [\delta\phi_1, \cdots, \delta\phi_N]$ represents the vector of unwrapped phase values associated with differential interferograms.

To retrieve deformation rates of high-coherence pixels, Equation (3) can be organized as follows:

$$Bv = \delta\phi \tag{4}$$

where B represents an $M \times N$ coefficient matrix, and v^T can be expressed as follows:

$$v^T = \left[v_1 = \frac{\phi_1}{t_1 - t_0}, \cdots, v_N = \frac{\phi_N - \phi_{N-1}}{t_N - t_{N-1}}\right] \tag{5}$$

Deformation rate can be achieved from Equation (4) by least squares (LS) or singular value decomposition (SVD) method [44]. Finally, the corresponding deformation time series can be derived according to the time span between SAR acquisitions.

3.2. Data Processing

In this study, to obtain surface subsidence rate and time series in the Wuhan region, we adopted the SBAS-InSAR technique to process 15 Sentinel-1A TOPS SAR images over the study area. The main steps are as follows:

3.2.1. Generation of Multiple Differential Interferograms

The image acquired on 20 October 2015, was selected as super master image for the interferometric combinations, and all slave images were co-registered and resampled to the super master image. Interferometric pairs were selected based on the spatial baseline shorter than 300 m and temporal baseline less than 200 days. The longest temporal baseline used in the analysis was 155 days.

Meanwhile, interferometric pairs with low coherence and poor unwrapping were removed. Finally, a combination with 92 differential interferograms was generated (Figure 2).

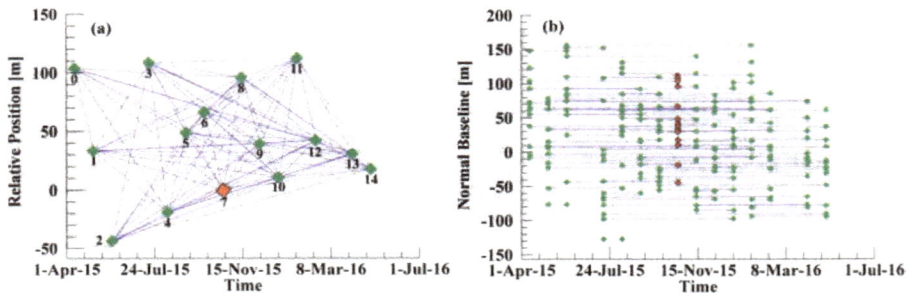

Figure 2. (a) Time–position of Sentinel-1A image interferometric pairs; and (b) time–baseline of Sentinel-1A image interferometric pairs. The red diamond denotes the super master image. Blue lines represent interferometric pairs. Green diamonds denote slave images.

3.2.2. Removal of Flat-Earth and Topographic Phases

POD data released by the ESA were adopted to remove the flat-earth phase. Topographic phase was eliminated using the three arc-second SRTM DEM provided by the NASA.

3.2.3. Orbital Refinement and Phase Re-Flattening

In this study, after adaptive filtering [45] and phase unwrapping (using Delaunay minimum cost flow), based on leveling data provided by the Wuhan Geomatics Institute, 24 stable points distributed in the study area were selected to execute orbital refinement and phase re-flattening for all interferometric pairs.

3.2.4. Subsidence Rate and Time Series Retrieval

No isolated interferogram clusters existed during combination of differential interferograms. Hence, subsidence rates were estimated using the LS method [25]. To retrieve subsidence time series, the estimated linear subsidence was subtracted first from raw subsidence time series. The remaining components comprised atmospheric phase, random noise phases, and nonlinear subsidence. Atmospheric and noise phases components are characterized by high spatial correlation but show a notably low temporal correlation. Subsequently, nonlinear subsidence component was separated through spatial and temporal bandpass filtering [46]. Finally, subsidence time series were retrieved by adding up linear subsidence and nonlinear subsidence components.

4. Results

4.1. Subsidence Rate Map

Figure 3 shows the subsidence rate map derived using SBAS InSAR; the map is superimposed on the Google Earth image of Wuhan acquired in 2016. Based on collected leveling data, a stable point located at 30°30′25″N and 114°22′44″E (red triangle in Figure 3) was selected as reference point, and subsidence rates in the study area were considered based on the reference point. Positive values of rate map indicate that the surface is uplifting in the vertical direction, whereas negative values denote surface subsidence in the vertical direction. Total number of PS points extracted from Sentinel-1A TOPS dataset by SBAS InSAR technique reached 2,101,453. Average density of PS points in the study area totaled 1317 PS points/km^2. SBAS InSAR-derived results in Figure 3 reveal that annual average subsidence rates in Wuhan range from −82 mm/year to 18 mm/year, and the largest subsidence rate was detected in HH region, Hangkou District (HK).

As shown in Figure 3, prominent uneven subsidence patterns were identified in Wuhan. The major subsidence areas include Region S1 (i.e., HK), Region S2 (mainly includes Qingshan (QS) and Yangluo (YL) Districts), Region S3 (i.e., WC), and Region S4 (mainly include QL Township, HY and Hongshan (HS) Districts). Regions S1 and S3 are central urban areas of Wuhan. The main large-scale industrial areas of Wuhan, e.g., Wuhan Iron and Steel Company (WISC), Sinopec Wuhan Company (SWC), and Huaneng Yangluo Power Plant (HYPP), are located in Region S2. Among the above-mentioned four regions, Region S1 features the most number of subsidence bowls, and most of the serious subsidence bowls are also distributed in Regions S1. Region S4 features the smallest subsidence bowls, but its subsidence range is gradually expanding. Additionally, these subsidence areas are mainly distributed along the Yangtze River. In addition to Regions S1–S4, other areas in study area present small surface subsidence, and most subsidence rates are less than 10 mm/year. More detailed subsidence analysis and explanation for Regions S1–S4 will be discussed in Section 5.

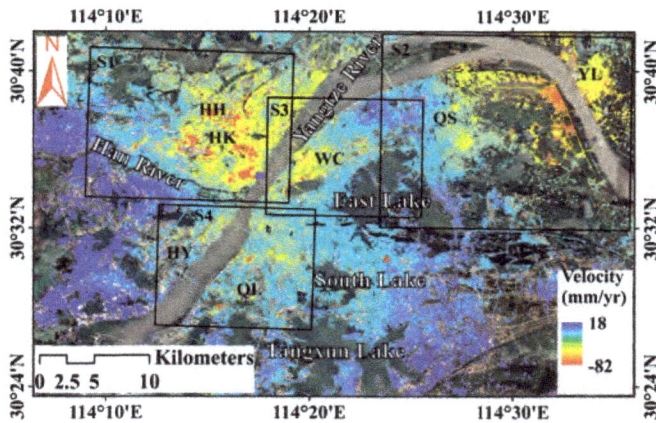

Figure 3. Vertical deformation rates derived by SBAS InSAR for the whole study area during the period from 11 April 2015 to 29 April 2016. The background is a Google Earth image acquired in 2016. Red triangle and blue square denote locations of reference point and Hankou hydrological station, respectively. Regions S1–S4 marked with black rectangles are major subsidence areas in Wuhan. These areas will be further analyzed in the discussion section. HK, HH, WC, QS, YL, HY, and QL are the abbreviations of Hankou, Houhu, Wuchang, Qingshan, Yangluo, Hanyang, and Qingling, respectively.

4.2. Subsidence Time Series

Figure 4 shows surface cumulative subsidence time series in Wuhan from 11 April 2015 to 29 April 2016. The Sentinel-1A TOPS SAR image acquired on 11 April 2015 was considered reference time of the time series in SBAS InSAR method. Maximum cumulative subsidence in this period is −86 mm, and it was noted at HH in HK, as marked by the black rectangle in Figure 4n. Eastern and western parts of Wuhan are relatively stable, and most of the cumulative subsidence in these areas range from −10 mm to 10 mm. Magnitude of subsidence in major subsidence areas (i.e., Regions S1–S4) gradually increased with time. Size of these subsidence areas also gradually expanded. However, changes in subsidence magnitude and range were small between May 2015 and July 2015. Wuhan experienced heavy rainfall in May, June, and July 2015, which possibly recharged groundwater effectively during the indicated months. Consequently, surface subsidence in Wuhan was possibly affected by rainfall and shows nonlinear subsidence with pronounced seasonal variations.

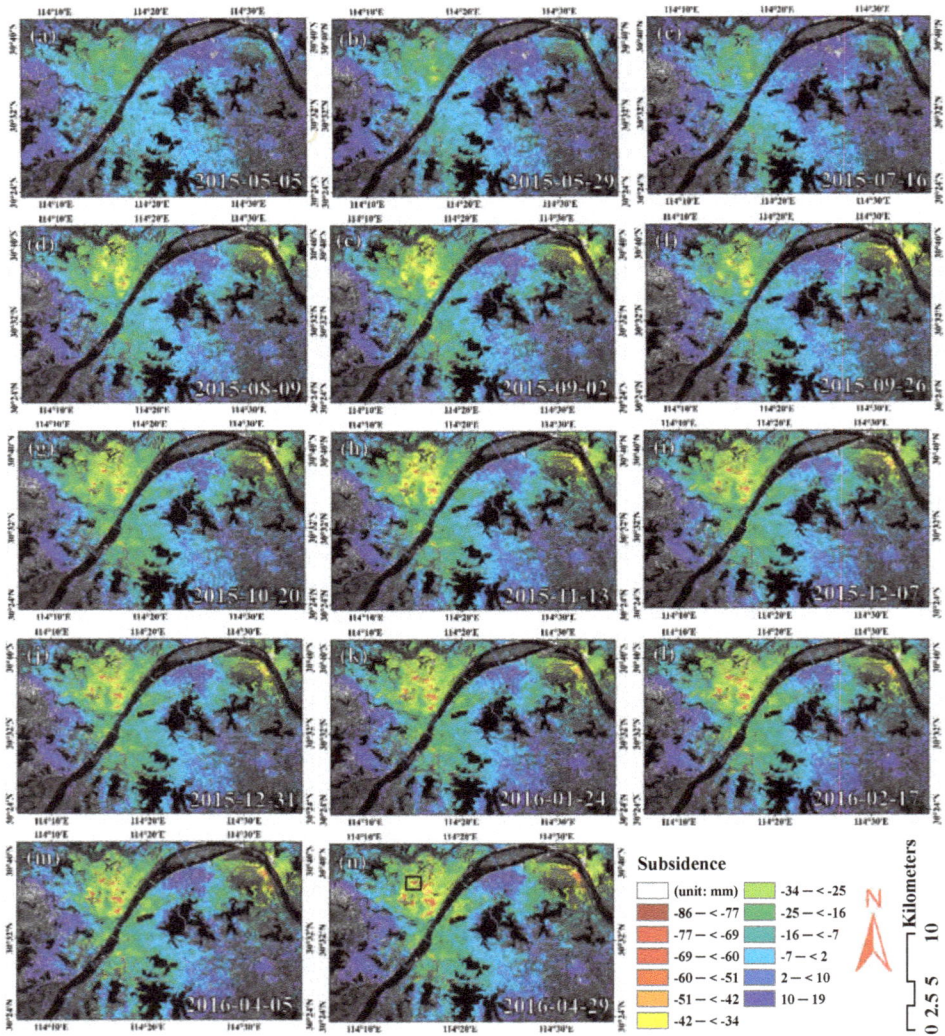

Figure 4. Cumulative subsidence (in the vertical direction) time series from 2015 to 2016. The image acquired on 11 April 2015 was not shown because it was selected as a reference image. The background is Sentinel-1A TOPS mean intensity image of Wuhan. (**a–n**) Cumulative subsidence in 14 stages in Wuhan. (**a**) 11 April 2015–5 May 2015; (**b**) 11 April 2015–29 May 2015; (**c**) 11 April 2015–16 July 2015; (**d**) 11 April 2015–9 August 2015; (**e**) 11 April 2015–2 September 2015; (**f**) 11 April 2015–26 September 2015; (**g**) 11 April 2015–20 October 2015; (**h**) 11 April 2015–13 November 2015; (**i**) 11 April 2015– 7 December 2015; (**j**) 11 April 2015–31 December 2015; (**k**) 11 April 2015–24 January 2016; (**l**) 11 April 2015–17 February 2016; (**m**) 11 April 2015–5 April 2016; (**n**) 11 April 2015–29 April 2016. HH region in HK is marked by black rectangle in (**n**).

4.3. Internal Precision Checking

To assess internal precision of subsidence rates extracted from Sentinel-1A TOPS data by SBAS InSAR technique, standard deviations of subsidence rates were statistically analyzed. Figure 5 shows distributions of standard deviations of subsidence rates. The standard deviations were obtained by computing the deviations of the linear fitting of velocities. If a PS point shows a strong nonlinear motion, it resulted in a large residual with respect to the linear model, i.e., in a high standard deviation value. Standard deviation of SBAS InSAR-derived results using Sentinel-1A TOPS data is 3 mm/year. Maximum standard deviation of PS point reaches 11 mm/year, and standard deviations of subsidence rates of 86.27% PS points are less than 6 mm/year. According to the above analysis, surface subsidence derived by SBAS InSAR technique using Sentinel-1A TOPS data features high reliability and precision.

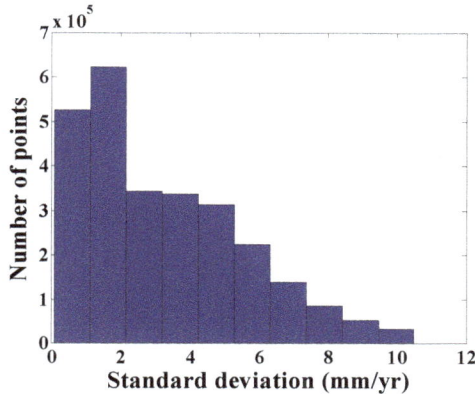

Figure 5. Distributions of standard deviations of subsidence rates.

5. Discussion

5.1. Validation with Leveling

To further quantitatively verify accuracy of surface subsidence monitoring using Sentinel-1A TOPS data by SBAS InSAR, a comparative analysis of differences between SBAS InSAR- and leveling-derived results was performed. A total of 110 benchmarks (i.e., BM1–BM110) were used in validation. Figure 6 shows location distribution of benchmarks for monitoring surface subsidence. Before validation, overlapping period data of Sentinel-1A TOPS and leveling data were selected. Given the limited opportunities for a benchmark and corresponding PS point located in the same place, Kriging method was adopted using ArcGIS 10.2 software and its geostatistical analyst extension to extract subsidence of the PS point corresponding to the nearest benchmark.

The linear regression of average subsidence rate was performed between SBAS InSAR- and leveling-derived results, as illustrated in Figure 7a. SBAS InSAR and leveling rates present a relatively high correlation, with a correlation coefficient value of 0.72. Figure 7b shows differences between SBAS InSAR- and leveling-derived results. Differences mainly ranged from −10 mm/year to 10 mm/year. Table 2 lists statistical results of differences. Mean error, maximum (MAX) and minimum (MIN) errors totaled −1, 11, and −13 mm/year, respectively. Root-mean-square error (RMSE) of differences averaged 6 mm/year. Therefore, validation results suggest that SBAS InSAR-derived results agree well with results obtained by leveling data, and SBAS InSAR technique can successfully extract surface subsidence information in Wuhan using Sentinel-1A TOPS data with an accuracy of 6 mm/year. This accuracy is similar to the subsidence monitoring accuracies derived by Luo et al. [47] and Zhang et al. [48].

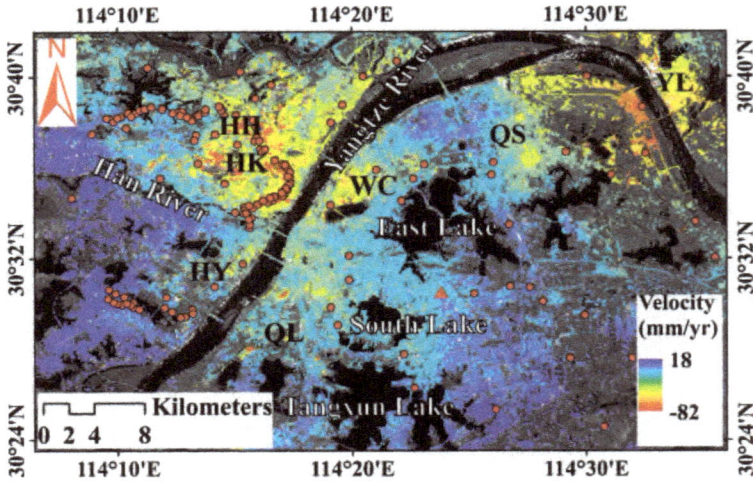

Figure 6. Distribution of benchmarks for surface subsidence monitoring over Wuhan. The red circle and triangle denote leveling point and reference point, respectively.

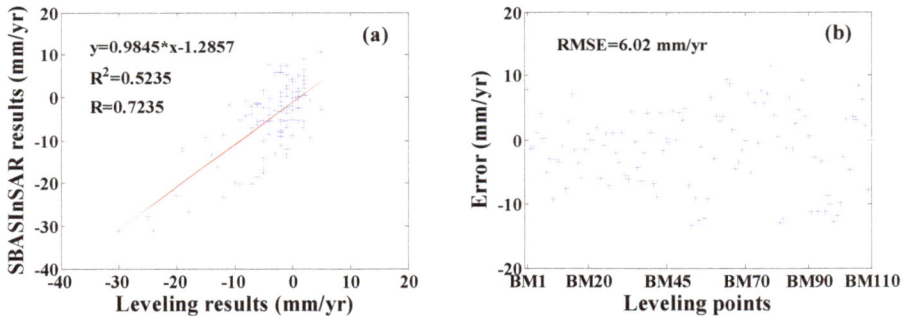

Figure 7. (**a**) Regression analysis between surface vertical deformation rates derived by SBAS InSAR and leveling; and (**b**) differences between SBAS InSAR- and leveling-derived results.

Table 2. Comparison of average subsidence rates between SBAS InSAR- and leveling-derived results. Mean error is the mean of the differences between SBAS InSAR- and leveling-derived rates. RMSE represents the standard deviation of the differences between SBAS InSAR- and leveling-derived rates (unit: mm/year).

Method	Mean Error	RMSE	MAX Error	MIN Error
Kriging	−1	6	11	−13

5.2. Surface Subsidence Associated with Urban Construction and Precipitation

In recent years, as the largest land and water transportation hub city in central China, Wuhan has witnessed significant changes, especially in urban construction. At present, Wuhan is in a critical period of urban development, and over 10,000 construction sites simultaneously operate throughout the city. HK (Region S1) and WC (Region S3) (Figure 3) are economic and cultural centers of Wuhan, respectively, and most urban construction activities are concentrated in these areas. Remarkable uneven subsidence patterns were detected in these regions (Figure 3). Therefore, the relationship between surface subsidence and urban construction corresponding to Regions S1 and S3 was analyzed in detail. Figures 8a and 9a show zoomed subsidence rate maps of Regions S1 and S3, respectively.

As shown in Figure 8a, four significant subsidence areas were detected in Regions S1, i.e., HH, Jianghan District (JH), Sun Yat-Sen Street (SYSS), and Wuhan Central Business District (CBD). Their surface subsidence distribution is similar to results extracted by Costantini et al. [38]. The most serious subsidence area is HH, where maximum subsidence rate exceeds −80 mm/year. Maximum subsidence rates in JH and SYSS reach −51 and −65 mm/year, respectively. Additionally, many large-scale buildings (e.g., Wuhan Center, a 438 m-tall skyscraper under construction) were constructed in the CBD area during the study period. As a result, density of PS points in CBD area was sparse, and detected maximum subsidence rate totaled −59 mm/year. HH, JH, and SYSS subsidence areas are mainly distributed along Metro Line 6, which is still under construction, and these subsidence areas gradually connect into a continuous area. To validate the uneven subsidence, fieldwork was carried out on HH, SYSS, and CBD areas in October 2016. Figure 8c–e illustrates significant effects caused by uneven subsidence observed in two buildings and a road.

In Region S1, extensive urban construction activities, including those for Metro Line 6 and Wuhan World Trade Center, were simultaneously under development during the study period (Figure 8a). During excavation of deep foundation pit and underground structure, groundwater level was often higher than construction surface. To ensure smooth excavation and avoid underwater operation, the foundation pit should be discharged of water. The entire Metro Line 6 and various buildings in CBD area were simultaneously under construction. Hence, a large volume of groundwater was extracted, and groundwater level gradually declined. This phenomenon reduced uplift pressure of groundwater in soil layer for the aboveground structures, compressed the soil, and finally led to surface subsidence. Region S1 is mainly located in the alluvial plain formed by joint actions of Yangtze River, Han River, and lakes, so that compressibility of alluvial deposit in this area is high. The above analysis indicates that extensive urban construction plays a dominant role in surface subsidence in Region S1. Surface subsidence is also affected by compressibility of soil layer.

Thus, to analyze temporal evolution of surface subsidence and the relationship between surface subsidence and precipitation, subsidence time series of four PS points (i.e., A, B, C, and D shown in Figure 8a) and monthly average precipitation during the study period were compared and analyzed. Figure 8b illustrates obtained results. Subsidence time series of selected PS points showed a nonlinear decline with seasonal variability. Accumulated subsidence of selected point is different depending on their spatial location. Accumulated subsidence of A located in HH area (approximately −80 mm) is significantly larger than that of other points, and it is related to poor soil-bearing capacity of the silt layer in HH area [49]. Subsidence of selected points from May 2015 to July 2015 was visibly small. As illustrated by the red line in Figure 8b, monthly average precipitations from May to July were the highest during the study period in Wuhan. These precipitations effectively replenished groundwater and slowed down surface subsidence rate. After July, precipitation declined noticeably, and subsidence increased as the temperature and domestic water consumption in Wuhan increased significantly, and groundwater was increasingly extracted, leading to increase in surface subsidence rates after July 2015. Comparative analysis suggests the following: surface subsidence in Region S1 experiences pronounced seasonal variations; surface subsidence is correlated not only to precipitation but also to other factors; and precipitation is a dominant factor influencing seasonal component of surface subsidence.

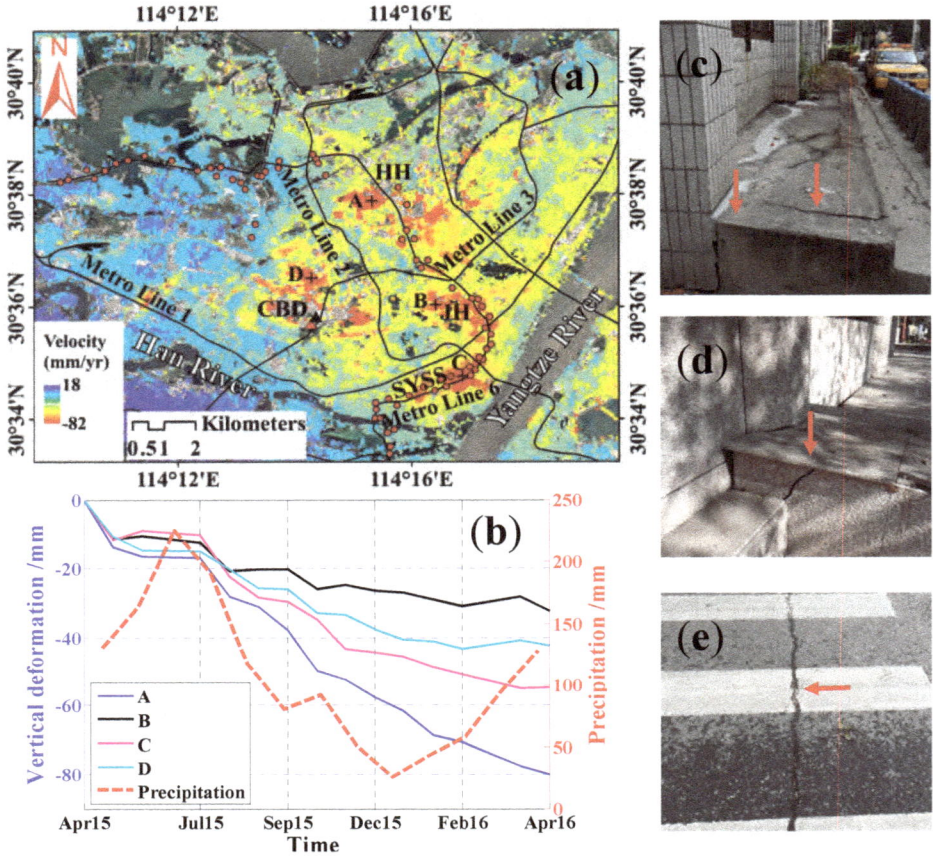

Figure 8. Surface vertical deformation rate map derived by SBAS InSAR superimposed on Google Earth image covering (**a**) Region S1. Black lines denote Wuhan Metro Lines, including metro lines in operation and under construction. Red circles represent leveling points distributed along Metro Line 6, which is under construction. Black and red triangles denote Wuhan Center and Wuhan World Trade Center, respectively. (**b**) Surface subsidence (in the vertical direction) time series with respect to PS points (marked by black crosses in Figure 8a) labeled as A, B, C, and D in Figure 8a versus average monthly precipitation of Wuhan area. (**c–e**) Structural damage caused by surface subsidence.

Figure 9 shows zoomed subsidence rate map of Region S3. In comparison with Region S1, subsidence rates in this region are much lower, and maximum subsidence rate just exceeds −40 mm/year. Two significant subsidence areas, namely, Xudong (XD) and Hubei University (HBU) subsidence areas (red ellipses in Figure 9), were identified in Region S3. The Metro Line 8 under construction runs through XD area along a northwest–southeast direction. The HBU area is located in the northern region of Sha Lake, where multiple super-tall residential communities are distributed along the lake. Distribution of settlement areas indicates that surface subsidence in Region S3 correlates with urban construction.

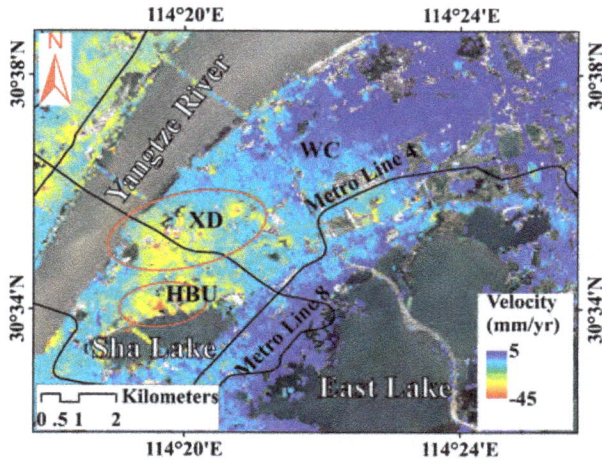

Figure 9. Surface vertical deformation rate map derived by SBAS InSAR superimposed on Google Earth image covering Region S3. Black lines denote the Wuhan Metro Lines, including the metro lines in operation and under construction. Red ellipses represent the two subsidence areas, i.e., XD and HBU subsidence bowls.

5.3. Effects of Surface Subsidence coupled with Industrial Development

Wuhan is one of the important old industrial bases in China. Many important industries are distributed in Wuhan. Multiple large-scale industrial plants (e.g., WISC, SWC, Yangluo Cement Plant (YCP), and HYPP) are located in Region S2. As shown in Figure 10a, severe subsidence areas are mainly yellow or red areas, and the most severe subsidence area is situated in SWC (white polygons in Figure 10a). The SWC subsidence area features the most serious subsidence rate, with maximum exceeding −46 mm/year. The subsidence bowl and the industrial area in WISC are consistent to some extent, although the distribution range of the subsidence bowl is smaller than that of the industrial area.The spatial distribution of the subsidence bowl overlaps the industrial regions in SWC, YCP and HYPP. In addition to industrial districts, a subsidence area has formed east of the YL urban area, and its highest subsidence rate reaches up to −32 mm/year.

Fieldwork was implemented in severe subsidence areas of Region S2 in October 2016. Figure 10b shows an approximately 2 cm-width crack on the step of a building located in YCP subsidence area. Figure 10c illustrates a 1.5 cm-wide crack on the roadside step caused by differential settlement situated in YL urban area.

Three profiles in representative areas were selected for subsidence analysis (A-A′, B-B′, and C-C′ in Figure 10). From A-A′ in Figure 11a, two remarkable subsidence bowls formed in the WISC area, and the largest subsidence along profiles A-A′ reaches up to −23 mm. In this area, the main factor inducing surface subsidence is groundwater overdraft caused by industrial development. By combining Figures 10a and 11b, we can conclude that a large subsidence bowl has formed in SWC area. The profile length reaches approximately 2.6 km, and maximum subsidence along profile B-B′ totals −36 mm. Figure 11c illustrates subsidence variation along a profile through the YL urban area in approximately east–west direction. Maximum subsidence measures −31 mm, which is close to point C′.

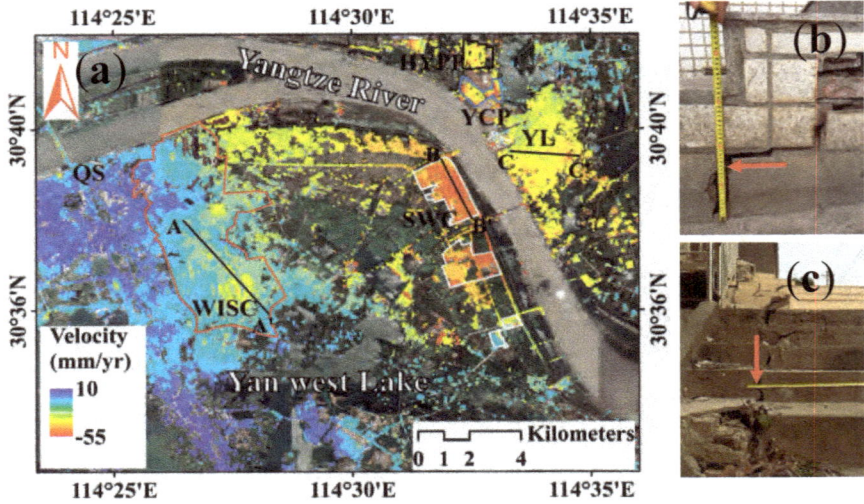

Figure 10. Surface vertical deformation rate map superimposed on Google Earth image covering (a) Region S2. Red, white, blue and black polygons denote the industrial areas of WISC, SWC, YCP, and HYPP, respectively. Black lines represent subsidence profiles that will be further analyzed in the Section 5.3. (**b,c**) Structural damage due to surface subsidence.

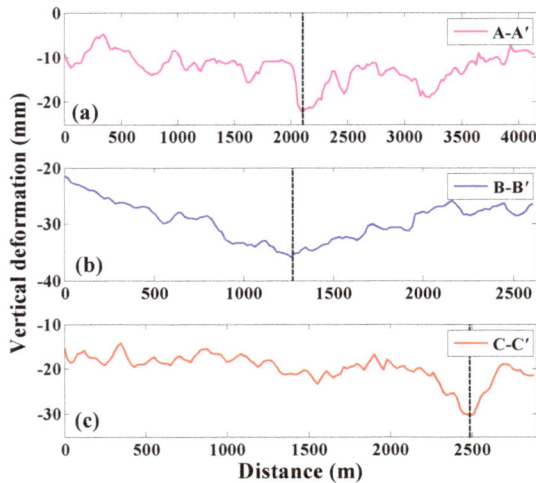

Figure 11. (**a–c**) Variation in accumulated subsidence (in the vertical direction) during the period from 11 April 2015 to 29 April 2016 along profiles A-A′, B-B′, and C-C′ (in Figure 10). Black dotted lines mark locations of the largest accumulated subsidence of profiles.

Overall, as multiple large-scale industrial plants are distributed in Region S2, high amounts of groundwater are pumped for industrial production, reducing pore water pressure in the aquifer of the overlaying soil and leading to uneven surface subsidence. A relatively high spatial correlation also exists between locations of surface subsidence bowl and that of industrial area.

5.4. Correlation between Surface Subsidence and Carbonate Karstification

Covered karst is widely distributed in Wuhan, and the area covered by quaternary loose sediments on the karst can easily induce surface collapse [50]. The covered karst area of the first terrace on both sides of Yangtze River are extremely prone to surface subsidence. Karst surface collapse is mainly distributed in HS, HY, and WC Districts. From 1977 to 2014, 29 karst surface collapses occurred in various locations in Wuhan, with 22 occurring in Region S4 (red squares in Figure 12) [51]. The Baishazhou Carbonate Rock belt (the area enclosed by red lines in Figure 12) is located in Region S4, where karst cave encountering rate of borehole and linear karst rate of borehole total 46.0% and 6.0%, respectively, and silty-fine sand is directly located above carbonate formations [52]. Cohesion of holocene silty-fine sand is small or equal to zero. Hence, silty-fine sand can easily enter the underlying karst channel under the influence of gravity and seepage force, thereby resulting in surface subsidence or karst surface collapse.

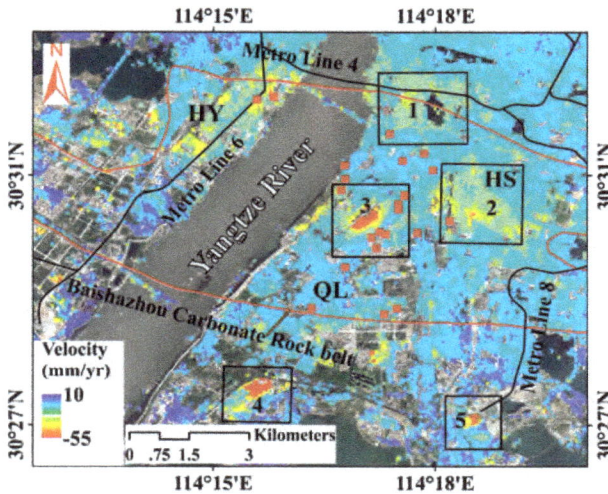

Figure 12. Surface vertical deformation rates map derived by SBAS InSAR superimposed on Google Earth image covering Region S4. Black lines denote the Wuhan Metro Lines, including metro lines in operation and under construction. The area enclosed by red line represents the Baishazhou Carbonate Rock belt. The red square denotes karst surface collapse. Areas 1, 2, 3, 4, and 5 marked with black rectangles are significant subsidence areas in Region S4.

As illustrated in Figure 12, uneven settlement is notable in Region S4, and five significant subsidence bowls were detected, namely, areas 1, 2, 3, 4, and 5, as labeled in Figure 12. Maximum subsidence rates in subsidence bowls 1, 2, 3, 4, and 5 reach up to −20, −29, −48, −53, and −35 mm/year, respectively. Subsidence rates in HY located west of the Yangtze River are small compared with that of the east, and subsidence areas are mainly distributed along Metro Line 6, which is under construction. Most subsidence bowls and surface collapses are located in the Baishazhou Carbonate Rock belt in the eastern part of Yangtze River (Figure 12). Groundwater level in this area is closely related to water level in Yangtze River, whereas active water level variations in Yangtze River can exacerbate groundwater cycle and promote development of underground karst caves. This situation can easily lead to surface subsidence and then surface collapse. The above analysis suggests that surface subsidence in Region S4 is seriously affected by carbonate karstification. Both water level variations of Yangtze River and urban engineering construction influence surface subsidence in this area.

5.5. Comparison between Surface Subsidence Changes and Water Level Changes

To analyze the relationship between surface subsidence and water level changes in Yangtze River, six PS points distributed along the sides of Yangtze River (the five-pointed star in Figure 13) and data on daily water level changes observed by Hankou hydrological station (black square in Figure 13) were selected to compare and analyze their changes time series.

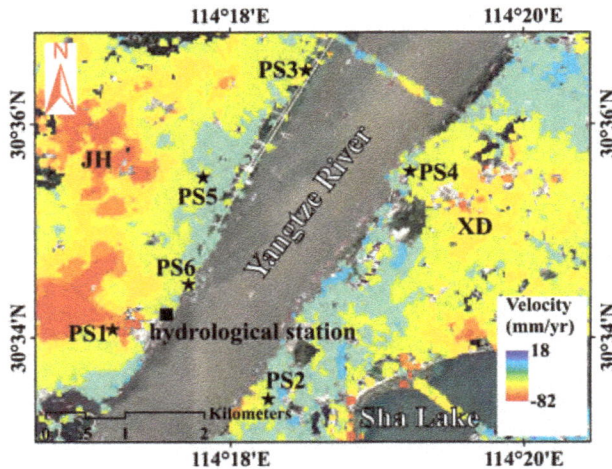

Figure 13. Surface vertical deformation rate map superimposed on Google Earth image covering both sides of the Yangtze River. The black square and five-pointed star denote the Hankou hydrological station and PS point, respectively. JH and XD are the abbreviations of Jianghan and Xudong, respectively.

As illustrated in Figure 14a, water level changes in Yangtze River exhibit obvious seasonal variability during the study period. Water level was typically high from May to July but low from January to March. On the other side, surface subsidence time series of the six PS points also showed a nonlinear decline with seasonal oscillation. Water level was high from May to July, and subsidence was low because of high water level in Yangtze River during this period. These conditions can effectively replenish groundwater and mitigate surface subsidence. From July 2015 to March 2016, water level dropped by 9 m, and average accumulated subsidence of six PS points reached approximately −19 mm.

To quantitatively analyze the relationship between surface subsidence and water level changes, we adopted the grey relational analysis (GRA) [53,54] to analyze proximity between subsidence and water level changes. Proximity is described by the grey relational grade (GRG), which is regarded a measure of similarities of discrete time-series data [55,56]. Temporal samplings of Sentinel-1A TOPS and water level data differ. Hence, we first interpolated subsidence time-series data to the same temporal sampling as water level data. Subsequently, we adopted GRA to calculate the GRG between subsidence and water level time series and the GRG between detrended subsidence and water level time series (results are listed in Table 3). The GRG indicates the magnitude of correlation between subsidence and water level time series. The closer the GRG value is to 1, the better the correlation [57,58]. Table 3 lists the standard deviation of subsidence with respect to the six PS points (i.e., PS1–PS6) and the standard deviations of the six selected PS points are approximately 1 mm. All GRGs between subsidence and water level time series are greater than 0.8, it is concluded that subsidence of PS points in Figure 14a is relatively highly related to the decline of water level.

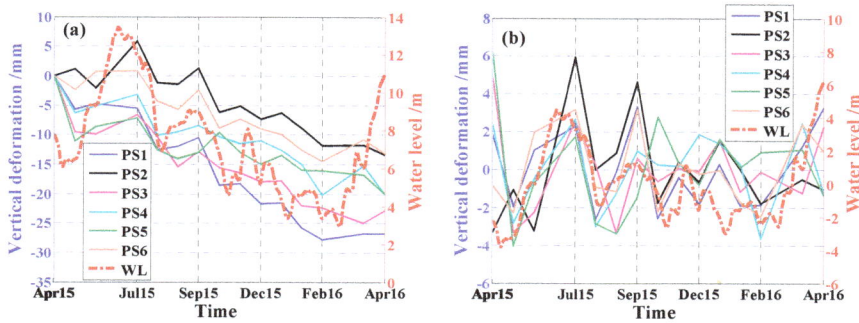

Figure 14. (**a**) Surface vertical deformation time series with respect to PS points labeled as PS1–PS6 in Figure 13 versus water level (WL) changes (red-dotted line) in Yangtze River; (**b**) Detrended surface vertical deformation time series relevant to the above PS points versus detrended water level changes (red dotted line) in Yangtze River.

Table 3. The GRG between the subsidence and water level of Yangtze River before and after the linear trend is removed and the standard deviations of displacement with respect to the six PS points.

PS Points	GRG		Standard Deviation
	Time Series	Detrended Time Series	Displacement (mm)
PS1	0.82	0.92	1.2
PS2	0.91	0.97	0.7
PS3	0.87	0.96	0.7
PS4	0.88	0.94	1.1
PS5	0.92	0.96	1.1
PS6	0.88	0.96	1.4

To further analyze the relationship between surface subsidence changes and water level changes, linear trends in surface subsidence changes and water level changes were removed by the least squares (LS) method. After removing the linear trend, the settlement changes of these PS points exhibit a certain periodic characteristics, as shown in Figure 14b. The peaks between detrended water level changes and subsidence changes show a similar pattern, but subsidence reaches its peak later than the water level. Furthermore, the GRGs between detrended subsidence and water level time series are close to 1, as shown in Table 3, indicating changes of subsidence may be closely related to that of water level. Therefore, we infer that seasonal signal is present in subsidence changes time series. In addition, as shown in Figure 14b, the fluctuation magnitude of displacement curves is about 3–4 mm, which is larger than the standard deviation of subsidence of the six selected PS points. This is because the curves shown in Figure 14b include nonlinear variations and error (i.e., the incompletely removed systematic error). While the standard deviation of subsidence were derived from the systematic error.

Overall, surface subsidence along sides of the Yangtze River correlated with water level changes. Specifically, seasonal component of subsidence time series is probably influenced by water level changes. Because of insufficient data, we cannot investigate the seasonal variations in subsidence time series in detail. In addition, longer time series will help to confirm the correlation between surface subsidence and water level changes. Therefore, we will study the periodic characteristics of signals using more data in the future research.

6. Conclusions

This paper presented spatial-temporal distribution of wide-area surface subsidence in Wuhan as derived by 15 Sentinel-1A TOPS SAR images using SBAS-InSAR technique. Cross-validation was conducted between InSAR- and leveling-derived results. The relationship between surface subsidence patterns and anthropogenic activities (e.g., metro construction, large-scale building construction and industrial development) and natural factors (e.g., precipitation, carbonate karstification and water level changes) were analyzed in detail, based on Sentinel-1A data, leveling data, daily water level changes data on Yangtze River, and distribution data of carbonate rock belt and industrial areas, etc. The main conclusions are as follows:

(1)   Surface subsidence in Wuhan is remarkably uneven, and four significant subsidence areas were detected in Wuhan. These surface subsidence areas are mainly distributed in the central urban areas of Wuhan (i.e., HK and WC Districts), industrial areas in QS and YL Districts, and Baishazhou Carbonate Rock belt. Annual average subsidence rates in Wuhan range from −82 mm/year to 18 mm/year. The most serious subsidence bowl was identified in HH area with a maximum rate exceeding −80 mm/year. Additionally, surface subsidence time series shows nonlinear subsidence with pronounced seasonal variations.

(2)   Internal precision checking indicated that standard deviation of Sentinel-1A SBAS InSAR results in the study area is 3 mm/year, and 86.27% of the PS point standard deviations are within 6 mm/year, implying high reliability and precision of surface subsidence derived by SBAS InSAR technique using Sentinel-1A data. Results obtained by SBAS InSAR and leveling showed good agreement. Specifically, RMSE and mean error reach 6 and −1 mm/year, respectively.

(3)   Surface subsidence in Wuhan is seriously affected by urban construction and industrial development, and spatial distribution of subsidence bowl is relatively highly correlated to that of engineering construction and industrial areas. In addition, carbonate karstification in Wuhan also plays a significant impact factor in surface subsidence. Seasonal variations in surface subsidence are correlated to water level changes and precipitation. The GRGs between detrended subsidence and water level time series are close to 1, indicating changes of subsidence may be closely related to that of water level. However, anthropogenic activities pose more notable influence on surface subsidence in Wuhan than natural factors.

Acknowledgments: This work was supported by the National Nature Science Foundation of China (Grant Nos. 41474004, 41461089 and 41604019); the Key Laboratory for Digital Land and Resources of Jiangxi Province, East China University of Technology (No. DLLJ201711); and the Open Fund of Guangxi Key Laboratory of Spatial Information and Geomatics (Grant No. 15-140-07-32). The authors wish to thank the ESA for arranging the Sentinel-1A data, NASA for providing the SRTM3 DEM data, ESA for releasing the POD data, the Wuhan Geomatics Institute for providing leveling data and the Changjiang Wuhan Waterway Bureau for providing the daily water level changes of Yangtze River.

Author Contributions: All authors contributed to the manuscript and discussed the results. Jiming Guo and Lv Zhou developed the idea that led to this paper. Lv Zhou performed Sentinel-1A data processing and analyses, and contributed to the manuscript of the paper. Jiming Guo provided critical comments and contributed to the final revision of the paper. Jiangwei Li performed the leveling data processing and analyzed the leveling-derived results. Yongfeng Xu performed the water level changes data processing and provided critical comments. Jiyuan Hu and Yuanjin Pan interpreted the SBAS InSAR-derived results and revised the manuscript. During the revision of manuscript, Miao Shi helped to calculate the GRG between subsidence and water level time series and analyze the relationship between surface subsidence and water level changes, and revised the manuscript. In addition, all authors contributed to the final revision of manuscript.

Conflicts of Interest: The authors declare no conflict of interest.

References

1.   Ge, L.; Ng, A.; Li, X.; Abidin, H.; Gumilar, I. Land subsidence characteristics of Bandung Basin as revealed by ENVISAT ASAR and ALOS PALSAR interferometry. *Remote Sens. Environ.* **2014**, *154*, 46–60. [CrossRef]

2.   Guo, J.M.; Zhou, L.; Yao, C.; Hu, J. Surface Subsidence Analysis by Multi-Temporal InSAR and GRACE: A Case Study in Beijing. *Sensors* **2016**, *16*, 1495. [CrossRef] [PubMed]

3.   Du, Z.; Ge, L.; Li, X.; Ng, A. Subsidence Monitoring over the Southern Coalfield, Australia Using both L-Band and C-Band SAR Time Series Analysis. *Remote Sens.* **2016**, *8*, 543. [CrossRef]

4. Qu, F.; Zhang, Q.; Lu, Z.; Zhao, C.; Yang, C.; Zhang, J. Land subsidence and ground fissures in Xi'an, China 2005–2012 revealed by multi-band InSAR time-series analysis. *Remote Sens. Environ.* **2014**, *155*, 366–376. [CrossRef]
5. Yang, H.L.; Peng, J.H. Monitoring Urban Subsidence with Multi-master Radar Interferometry Based on Coherent Targets. *J. Indian Soc. Remote Sens.* **2015**, *43*, 529–538. [CrossRef]
6. Duan, G.Y.; Gong, H.L.; Liu, H.; Zhang, Y.; Chen, B.; Lei, K. Monitoring and Analysis of Land Subsidence Along Beijing-Tianjin Inter-City Railway. *J. Indian Soc. Remote Sens.* **2016**, *44*, 915–931.
7. Amelung, F.; Galloway, D.L.; Bell, J.W.; Zebker, H.A.; Laczniak, R.J. Sensing the ups and downs of Las Vegas: InSAR reveals structural control of land subsidence and aquifer-system deformation. *Geology* **1999**, *27*, 483–486. [CrossRef]
8. Motagh, M.; Walter, T.R.; Sharifi, M.A.; Fielding, E.; Schenk, A.; Anderssohn, J.; Zschau, J. Land subsidence in Iran caused by widespread water reservoir overexploitation. *Geophys. Res. Lett.* **2008**, *35*, L16403. [CrossRef]
9. Chaussard, E.; Amelung, F.; Abidin, H.; Hong, S.H. Sinking cities in Indonesia: ALOS PALSAR detects rapid subsidence due to groundwater and gas extraction. *Remote Sens. Environ.* **2013**, *128*, 150–161. [CrossRef]
10. Poland, M.; Bürgmann, R.; Dzurisin, D.; Lisowski, M.; Marsterlark, T.; Owen, S.; Fink, J. Constraints on the mechanism of long-term, steady subsidence at Medicine Lake volcano, northern California, from GPS, leveling, and InSAR. *J. Volcanol. Geotherm. Res.* **2006**, *150*, 55–78. [CrossRef]
11. Baldi, P.; Casula, G.; Cenni, N.; Loddo, F.; Pesci, A. GPS-based monitoring of land subsidence in the Po Plain (Northern Italy). *Earth Planet. Sci. Lett.* **2009**, *288*, 204–212. [CrossRef]
12. Carminati, E.; Martinelli, G. Subsidence rates in the Po Plain, northern Italy: the relative impact of natural and anthropogenic causation. *Eng. Geol.* **2002**, *66*, 241–255. [CrossRef]
13. Psimoulis, P.; Ghilardi, M.; Fouache, E.; Stiros, S. Subsidence and evolution of the Thessaloniki plain, Greece, based on historical leveling and GPS data. *Eng. Geol.* **2007**, *90*, 55–70. [CrossRef]
14. Anell, I.; Thybo, H.; Artemieva, I.M. Cenozoic uplift and subsidence in the North Atlantic region: Geological evidence revisited. *Tectonophysics* **2009**, *474*, 78–105. [CrossRef]
15. Gabriel, A.K.; Goldstein, R.M.; Zebker, H.A. Mapping small elevation changes over large areas: Differential radar interferometry. *J. Geophys. Res.* **1989**, *94*, 9183–9191. [CrossRef]
16. Zebker, H.A.; Villasenor, J. Decorrelation in Interferometric Radar Echoes. *IEEE Trans. Geosci. Remote Sens.* **1992**, *30*, 950–959. [CrossRef]
17. Biswas, K.; Chakravarty, D.; Mitra, P.; Misra, A. Spatial-Correlation Based Persistent Scatterer Interferometric Study for Ground Deformation. *J. Indian Soc. Remote Sens.* **2017**, *45*, 1–14. [CrossRef]
18. Hanssen, R.F. *Radar Interferometry Data Interpretation and Error Analysis*, 1st ed.; Springer: Berlin, Germany, 2001.
19. Ferretti, A.; Savio, G.; Barzaghi, R.; Borghi, A.; Musazzi, S.; Novali, F.; Prati, C.; Rocca, F. Submillimeter accuracy of InSAR time series: Experimental validation. *IEEE Trans. Geosci. Remote Sens.* **2007**, *45*, 1142–1153. [CrossRef]
20. Vajedian, S.; Motagh, M.; Nilfouroushan, F. StaMPS improvement for deformation analysis in mountainous regions: Implications for the Damavand volcano and Mosha fault in Alborz. *Remote Sens.* **2015**, *7*, 8323–8347. [CrossRef]
21. Zhang, L.; Ding, X.; Lu, Z. Modeling PSInSAR time series without phase unwrapping. *IEEE Trans. Geosci. Remote Sens.* **2011**, *49*, 547–556. [CrossRef]
22. Ferretti, A.; Prati, C.; Rocca, F. Permanent scatterers in SAR interferometry. *IEEE Trans. Geosci. Remote Sens.* **2001**, *39*, 8–20. [CrossRef]
23. Beradino, P.; Fornaro, G.; Lanari, R.; Sansosti, E. A new algorithm for Surface deformation monitoring based on small baseline differential SAR interferograms. *IEEE Trans. Geosci. Remote Sens.* **2002**, *40*, 2375–2383. [CrossRef]
24. Lanari, R.; Mora, O.; Manunta, M.; Mallorquí, J.J.; Beradino, P.; Sansosti, E. A small-baseline approach for investigating deformations on full-resolution differential SAR interferograms. *IEEE Trans. Geosci. Remote Sens.* **2004**, *42*, 1377–1386. [CrossRef]
25. Hooper, A. A multi-temporal InSAR method incorporating both persistent scatterer and small baseline approaches. *Geophys. Res. Lett.* **2008**, *30*, L16302. [CrossRef]
26. Zhang, L.; Ding, X.; Lu, Z. Ground settlement monitoring based on temporarily coherent points between two SAR acquisitions. *ISPRS J. Photogramm.* **2011**, *66*, 146–152. [CrossRef]

27. Perissin, D.; Wang, T. Repeat-pass SAR interferometry with partially coherent targets. *IEEE Trans. Geosci. Remote Sens.* **2011**, *50*, 271–280. [CrossRef]
28. Bateson, L.; Cigna, F.; Boon, D.; Sowter, A. The application of the Intermittent SBAS (ISBAS) InSAR method to the South Wales Coalfield, UK. *Int. J. Appl. Earth Obs. Geoinf.* **2015**, *34*, 249–257. [CrossRef]
29. Chen, B.; Gong, H.; Li, X.; Lei, K.; Zhu, L.; Gao, M.; Zhou, C. Characterization and causes of land subsidence in Beijing, China. *Int. J. Remote Sens.* **2017**, *38*, 808–826. [CrossRef]
30. Castellazzi, P.; Arroyo-Domínguez, N.; Martel, R.; Calderhead, A.; Normand, J.; Gárfias, J.; Rivera, A. Land subsidence in major cities of Central Mexico: Interpreting InSAR-derived land subsidence mapping with hydrogeological data. *Int. J. Appl. Earth Obs. Geoinf.* **2016**, *47*, 102–111. [CrossRef]
31. Le, T.; Chang, C.P.; Nguyen, X.; Yhokha, A. TerraSAR-X Data for High-Precision Land Subsidence Monitoring: A Case Study in the Historical Centre of Hanoi, Vietnam. *Remote Sens.* **2016**, *8*, 338. [CrossRef]
32. Dong, S.; Samsonov, S.; Yin, H.; Ye, S.; Cao, Y. Time-series analysis of subsidence associated with rapid urbanization in Shanghai, China measured with SBAS InSAR method. *Environ. Earth Sci.* **2014**, *72*, 677–691. [CrossRef]
33. Wu, J.; Hu, F. Monitoring Ground Subsidence along the Shanghai Maglev Zone Using TerraSAR-X Images. *IEEE Geosci. Remote Sens.* **2017**, *14*, 117–121. [CrossRef]
34. Liu, X.; Cao, Q.; Xiong, Z.; Yin, H.; Xiao, G. Application of small baseline subsets D-InSAR technique to estimate time series land deformation of Jinan area, China. *J. Appl. Remote Sens.* **2016**, *10*, 026014. [CrossRef]
35. Bai, L.; Jiang, L.; Wang, H.; Wang, H.; Sun, Q. Spatiotemporal Characterization of Land Subsidence and Uplift (2009–2010) over Wuhan in Central China Revealed by TerraSAR-X InSAR Analysis. *Remote Sens.* **2016**, *8*, 350. [CrossRef]
36. Fan, S. A Discussion on Karst Collapse in Wuhan (Hubei). *Resour. Environ. Eng.* **2006**, *20*, 608–616.
37. Luo, X. Division of "Six Belts and Five Types" of carbonate region and control of karst geological disaster in Wuhan. *J. Hydraul. Eng.* **2014**, *45*, 171–179.
38. Costantini, M.; Bai, J.; Malvarosa, F.; Minati, F.; Vecchioli, F.; Wang, R.; Hu, Q.; Xiao, J.; Li, J. Ground deformations and building stability monitoring by COSMO-SkyMed PSP SAR interferometry: Results and validation with field measurements and surveys. In Proceedings of the IGARSS 2016—IEEE International Geoscience and Remote Sensing Symposium, Beijing, China, 10–15 July 2016.
39. Velotto, D.; Bentes, C.; Tings, B.; Lehner, S. First Comparison of Sentinel-1 and TerraSAR-X Data in the Framework of Maritime Targets Detection: South Italy Case. *IEEE J. Ocean. Eng.* **2016**, *41*, 993–1006. [CrossRef]
40. Sowter, A.; Amat, M.B.C.; Cigna, F.; Marsh, S.; Athab, A.; Alshammari, L. Mexico City land subsidence in 2014–2015 with Sentinel-1 IW TOPS: Results using the Intermittent SBAS (ISBAS) technique. *Int. J. Appl. Earth Obs. Geoinf.* **2016**, *52*, 230–242. [CrossRef]
41. Nannini, M.; Prats-Iraola, P.; Zan, F.D.; Geudtner, D. TOPS Time Series Performance Assessment with TerraSAR-X Data. *IEEE J. Sel. Top. Appl. Earth Obs. Remote Sens.* **2016**, *9*, 3832–3848. [CrossRef]
42. Hu, B.; Wang, H.; Sun, Y.; Hou, J.; Liang, J. Land-term land subsidence monitoring of Beijing (China) using the Small Baseline Subset (SBAS) technique. *Remote Sens.* **2014**, *6*, 3648–3661. [CrossRef]
43. Hooper, A.; Segall, P.; Zebker, H. Persistent scatterer interferometric synthetic aperture radar for crustal deformation analysis, with application to Volcán Alcedo, Galápagos. *J. Geophys. Res. Atmos.* **2007**, *112*, B07407. [CrossRef]
44. Casu, F.; Manzo, M.; Lanari, R. A quantitative assessment of the SBAS algorithm performance for surface deformation retrieval from DInSAR data. *Remote Sens. Environ.* **2006**, *102*, 195–210. [CrossRef]
45. Goldstein, R.M.; Werner, C.L. Radar interferogram filtering for geophysical applications. *Geophys. Res. Lett.* **1998**, *25*, 4035–4038. [CrossRef]
46. Ferretti, A.; Prati, C.; Rocca, F. Nonlinear subsidence rate estimation using permanent scatterers in differential SAR interferometry. *IEEE Trans. Geosci. Remote Sens.* **2000**, *38*, 2202–2212. [CrossRef]
47. Luo, Q.L.; Perissin, D.; Zhang, Y.; Jia, Y. L- and X-Band Multi-Temporal InSAR Analysis of Tianjin Subsidence. *Remote Sens.* **2014**, *6*, 7933–7951. [CrossRef]
48. Zhang, Y.; Wu, H.; Kang, Y.; Zhu, C. Ground Subsidence in the Beijing-Tianjin-Hebei Region from 1992 to 2014 Revealed by Multiple SAR Stacks. *Remote Sens.* **2016**, *8*, 675. [CrossRef]
49. Wu, C. Study and analysis of settlement in the estate of Houhu, Wuhan. *Technol. Innov. Appl.* **2014**, *22*, 216.

50. Zhong, Y.; Zhang, M.K.; Pan, L.; Zhao, S.K.; Hao, Y.H. Risk Assessment for Urban Karst Collapse in Wuchang District of Wuhan Based on GIS. *J. Tianjin Norm. Univ. (Nat. Sci. Ed.)* **2015**, *35*, 48–53.
51. Xu, G. Mechanism Study and Hazard Assessment of Cover Karst Sinkholes in Wuhan City, China. Ph.D. Thesis, China University of Geosciences, Wuhan, China, 2016.
52. Luo, X.J. Features of the shallow karst development and control of karst collapse in Wuhan. *Carsologica China* **2013**, *32*, 419–432.
53. Deng, J. The GRA in Cause-Effect Space of Resources. *J. Grey Syst.* **2009**, *21*, 113–119.
54. Deng, J. To Analyze the Connotation and Extension (C & E) of Grey Theory. *J. Grey Syst.* **2012**, *24*, 293–298.
55. Ip, W.C.; Hu, B.Q.; Wong, H.; Xia, J. Applications of grey relational method to river environment quality evaluation in China. *J. Hydrol.* **2009**, *379*, 284–290. [CrossRef]
56. Wei, G.W. Gray Relational Analysis Method For Intuitionistic Fuzzy Multiple Attribute Decision Making. *Expert Syst. Appl.* **2011**, *38*, 11671–11677. [CrossRef]
57. Du, J.C.; Kuo, M.F. Grey relational-regression analysis for hot mix asphalt design. *Constr. Build. Mater.* **2011**, *25*, 2627–2634. [CrossRef]
58. Mu, R.; Zhang, J.T. Research of hierarchy synthetic evaluation based on grey relational analysis. *Syst. Eng. Theor. Prac.* **2008**, *28*, 125–130.

remote sensing

MDPI

Article

SAR Tomography as an Add-On to PSI: Detection of Coherent Scatterers in the Presence of Phase Instabilities

Muhammad Adnan Siddique [1,*], Urs Wegmüller [2], Irena Hajnsek [1,3] and Othmar Frey [1,2*]

[1] Chair of Earth Observation and Remote Sensing, ETH Zurich, 8093 Zurich, Switzerland; Irena.Hajnsek@dlr.de

[2] GAMMA Remote Sensing AG, CH-3073 Gümligen, Switzerland; wegmuller@gamma-rs.ch

[3] Microwaves and Radar Institute, German Aerospace Center (DLR), 82234 Wessling, Germany

* Correspondence: siddique@ifu.baug.ethz.ch (M.A.S.); frey@ifu.baug.ethz.ch (O.F.)

Received: 28 April 2018; Accepted: 20 June 2018; Published: 25 June 2018

Abstract: The estimation of deformation parameters using persistent scatterer interferometry (PSI) is limited to single dominant coherent scatterers. As such, it rejects layovers wherein multiple scatterers are interfering in the same range-azimuth resolution cell. Differential synthetic aperture radar (SAR) tomography can improve deformation sampling as it has the ability to resolve layovers by separating the interfering scatterers. In this way, both PSI and tomography inevitably require a means to detect coherent scatterers, i.e., to perform hypothesis testing to decide whether a given candidate scatterer is coherent. This paper reports the application of a detection strategy in the context of "tomography as an add-on to PSI". As the performance of a detector is typically linked to the statistical description of the underlying mathematical model, we investigate how the statistics of the phase instabilities in the PSI analysis are carried forward to the subsequent tomographic analysis. While phase instabilities in PSI are generally modeled as an additive noise term in the interferometric phase model, their impact in SAR tomography manifests as a multiplicative disturbance. The detection strategy proposed in this paper allows extending the same quality considerations as used in the prior PSI processing (in terms of the dispersion of the residual phase) to the subsequent tomographic analysis. In particular, the hypothesis testing for the detection of coherent scatterers is implemented such that the expected probability of false alarm is consistent between PSI and tomography. The investigation is supported with empirical analyses on an interferometric data stack comprising 50 TerraSAR-X acquisitions in stripmap mode, over the city of Barcelona, Spain, from 2007–2012.

Keywords: synthetic aperture radar (SAR); SAR tomography; deformation monitoring; persistent scatterer interferometry (PSI); urban deformation monitoring; radar interferometry; displacement mapping; spaceborne SAR; differential interferometry; differential tomography

1. Introduction

Persistent scatterer interferometry (PSI) [1–7] is nowadays an operational geodetic technique for the monitoring of surface deformation with spaceborne synthetic aperture radar (SAR) data stacks. These stacks typically comprise several repeat-pass SAR acquisitions, spanning from months to years. PSI techniques attempt to extract the interferometric phase components correlated with the scatterer motion. The quality of the deformation estimates is tied to the precision of the interferometric phases. Temporal and geometric decorrelation, as well as uncompensated platform motion and atmosphere-induced optical path delay variations, are among the factors that cause random instabilities in phase. For these reasons, a quality control is necessary during the processing as well as when reporting the final results.

The single dominant scatterers that exhibit long-term phase stability are generally termed as *persistent scatterers* (PS). PSI processing approaches often use a classifier to identify a priori a set of PS candidates, e.g., the permanent scatterers [1] approach uses the *dispersion index* as a proxy for phase stability. The PSI approaches based on the interferometric point target analysis (IPTA) framework, as in [3,8], employ low *spectral diversity* [3,9–11] as a proxy for phase stability in addition to the stability of the backscattering amplitude. Low dispersion index and low spectral diversity are indicative of good phase quality. The observed differential interferometric phases are fit to a phase model and the unknown parameters, such as the deformation velocity and the residual topography, are thereby estimated. The dispersion of the residue of the fit is a means to characterize the quality of the estimates. It is often used to compute the multi-interferogram complex coherence (MICC) [1,12,13] which can in turn be used as a test statistic to perform statistical detection i.e., to decide among the hypotheses whether a given PS candidate is a phase coherent single scatterer or if it comprises noise only. The statistics of the noise impact the probability of false alarm in the detection process.

An inherent limitation associated with PSI techniques is the fact that a phase-only model cannot consider multiple coherent scatterers with different complex reflectivity interfering in the same range-azimuth resolution cell. The cumulative phase response in this case is mismatched to the interferometric phase model, which is essentially based on the assumption of a single scatterer. Consequently, it may lead to erroneous estimation of the deformation parameters. Therefore, PSI processing approaches typically reject the cells that contain backscattering contributions from multiple scatterers, as for the case of layovers.

The aforementioned limitation can be alleviated by SAR tomography [14–17], which exploits both the amplitude and the phase of the received signal, thereby permitting a higher order analysis [18]. It allows 3-D reconstruction of the scene reflectivity—a feature that renders it possible to resolve the layover problem [19–22]. Additionally, differential SAR tomographic methods [23–25] allow a joint spatio-temporal inversion of the coherent scatterers in layover, i.e., the position along the elevation axis as well as the deformation velocity of the interfering scatterers are simultaneously estimated. Therefore, differential SAR tomography has been proposed as an add-on to PSI techniques to improve deformation sampling by resolving the scatterers in layover that are rejected in the PSI processing [26–29]. Inevitably, a detection strategy is again required to classify whether the detection of one or more scatterers in the same resolution cell is true or false. In this context, it is pertinent to carry forward the same quality criteria as used in the prior PSI analysis so that the combined use of PSI and tomography holds compatibility.

The prevailing detection mechanisms for SAR tomography, such as the generalized likelihood ratio tests in [13,24,30], consider an additive noise model for the received complex signal vector. The source of the noise is attributed to the clutter in the resolution cell. However, the instabilities in the observed interferometric phases, albeit considered additive in the phase-only model, naturally represent themselves as multiplicative noise in the tomographic signal model. Therefore, in order to carry the impact of the phase instabilities from an interferometric to tomographic analysis, the detection strategy employed for hypothesis testing needs to account for the phase instabilities as a multiplicative disturbance in tomographic inversion.

Keeping in view the aforementioned concerns, this paper describes a strategy for the detection of single and double scatterers with SAR tomography whereby the hypothesis testing is directly linked to the MICC-based test statistic for PS detection in the prior PSI processing. As a whole, this paper is a follow-up to the earlier works in [12,27,31]. Section 2 presents the mathematical models typically used for SAR interferometry and tomography, as well as the associated detection mechanisms. Section 3 presents the processing methodology adopted in the paper. The data stack for empirical analysis is introduced in Section 4. The results obtained are presented in Section 5, followed by a discussion in Section 6.

2. Models

We consider the availability of a coregistered, single-reference interferometric SAR data stack comprising M layers of repeat-pass interferograms. For a given range-azimuth resolution cell in an interferometric layer, we denote the received single-look complex (SLC) signal as $y_m = z_m \exp\left(-j\varphi_m\right)$, where $z_m = |y_m| \in \mathbb{R}$ is the amplitude of the received signal, and φ_m is the observed interferometric phase. The subscript m, where $m \in \{0, 1, \ldots, M-1\}$, is used to indicate a specific layer in the interferometric stack. In the following text, an underlined symbol represents a quantity that has been modeled as stochastic, or when the distinction between *observables* versus *observations* is emphasized. Bold symbols represent vectors, or matrices when capitalized.

2.1. Interferometric Phase Model

The interferometric phase observable, $\underline{\varphi}_m$, is generally modeled as a sum of several phase contributions [32,33]:

$$\underline{\varphi}_m = \varphi_m^{\text{disp}} + \varphi_m^{\text{geo}} + \varphi_m^{\text{th.exp}} + \underline{\varphi}_m^{\text{atm}} + \underline{\varphi}_m^{\text{decor}} + 2\pi p, \tag{1}$$

where φ^{disp} is the phase change due to the linear displacement of the target as a function of time within the resolution cell:

$$\varphi_m^{\text{disp}} = \frac{4\pi}{\lambda} v t_m. \tag{2}$$

λ is the wavelength, v is the deformation velocity in the line of sight (LOS), and t_m is the temporal baseline for the mth interferogram. φ^{geo} is the phase variation due to sensor-to-target geometry. Neglecting higher order terms [16,34],

$$\varphi_m^{\text{geo}} \approx -\frac{4\pi b_m^\perp s}{\lambda\left(\rho_0 - b_m^\parallel\right)}, \tag{3}$$

where $b_m^\perp$ and $b_m^\parallel$ are the orthogonal and parallel components of the spatial baseline for the mth interferogram, respectively. ρ_0 is the range distance from the sensor to the target location for the reference acquisition. s represents the elevation, i.e., the position of the target in the axis perpendicular to the LOS. In case of thermal expansion, the additional phase variations are linearly modeled as follows [27,35]:

$$\varphi_m^{\text{th.exp}} = \eta T_m, \tag{4}$$

where T_m is the temperature change (with respect to the temperature for the reference layer), and η is the phase-to-temperature sensitivity. The term $2\pi p$, where $p \in \mathbb{Z}$, is added to account for phase wrapping. The phase variations $\underline{\varphi}_m^{\text{atm}}$ are due to the optical path length variations while propagation through the atmosphere. They are modeled as stochastic variables due to the temporally varying nature of atmospheric refractivity [36–39]. The phase decorrelation term, $\underline{\varphi}_m^{\text{decor}}$ is, by definition, a random quantity, which is typically modeled as an additive phase noise. The parameters s, v and η are treated as deterministic unknowns in this work.

The interferometric phase model in Equation (1) is implicitly assuming the presence of a single coherent scatterer in the resolution cell. In case of multiple coherent scatterers in the same resolution cell, it is not possible to write the interferometric phase, $\underline{\varphi}_m$ as a sum of the aforementioned sources of phase variations, independently of the reflectivity of the individual scatterers.

2.2. PSI: Model of Observation Equations

While several approaches to parameter estimation with PSI have been proposed over time, as in [1–6], the functional model of interferometric phase observation equations common to these approaches is as follows [33]:

$$\boldsymbol{\varphi} = A\mathbf{p} + \underline{\mathbf{w}}, \tag{5}$$

where $\boldsymbol{\varphi}$ is the $M \times 1$ vector of interferometric phase observables, A is the design matrix, and $\mathbf{p}$ is the vector of the aforementioned unknown parameters. $\underline{\mathbf{w}}$ is the $M \times 1$ vector of phase residuals which collectively represent the phase instabilities owing to decorrelation, uncompensated atmospheric phases and model imperfections. The residuals in each layer are assumed to be zero-mean and independent random variables: $E\{\underline{\mathbf{w}}\} = 0$; and $D\{\underline{\mathbf{w}}\} = E\{\mathbf{w}\mathbf{w}^H\} = Q_{ww}$ is the covariance matrix for the residuals. If it can be assumed there are no phase unwrapping issues, and the data stack can be phase calibrated by compensating for the atmospheric phase with external data—although both assumptions are simplistic—then the remaining unknowns are s, v and η. The design matrix is then constituted by the coefficients of these parameters (from Equations (2–4)) [1,33]. Under Gauss-Markov conditions, the best linear unbiased estimate of the parameter vector using weighted least squares is given as [33]:

$$\underline{\hat{\mathbf{p}}} = \left(A^T Q_{ww}^{-1} A\right)^{-1} A^T Q_{ww}^{-1} \boldsymbol{\varphi}. \tag{6}$$

The covariance matrix of the estimated parameter vector, $Q_{\hat{p}\hat{p}} = D\{\underline{\hat{\mathbf{p}}}\}$ is as follows:

$$Q_{\hat{p}\hat{p}} = \left(A^T Q_{ww}^{-1} A\right)^{-1}. \tag{7}$$

The quality of the estimates is, therefore, dependent on the dispersion of the residuals. The vector of the estimated phase residuals is as shown below:

$$\underline{\hat{\mathbf{w}}} = \boldsymbol{\varphi} - A\underline{\hat{\mathbf{p}}}. \tag{8}$$

2.3. PSI: Statistics for PS Detection

For each PS candidate, we distinguish between the following two hypotheses:

$\mathcal{H}^0$ –the null hypothesis. The range-azimuth resolution cell does not contain any coherent scatterer and comprises merely clutter;

$\mathcal{H}^1$ –the alternative hypothesis. The cell contains a phase coherent single scatterer, i.e., a PS.

In the presence of a coherent scatterer whose phase response is well-matched to the model in Equation (1), the phase residuals are expected to have a low dispersion around the expected value of zero. Contrarily, in the absence of a coherent scatterer, the observed phase and the residuals are expected to have a wider dispersion. With these considerations, we assume that the phase residuals generally follow a von Mises (*circular* normal) distribution. The probability density function (PDF) is given by [40]:

$$g(w; \mu, \kappa) = \frac{1}{2\pi I_o(\kappa)} e^{\kappa \cos(w-\mu)}, \tag{9}$$

where the support of the distribution is any 2π interval. The parameter $\mu = E\{\underline{w}\}$ represents the 'preferred direction', which we consider to be zero under both $\mathcal{H}^0$ and $\mathcal{H}^1$. The support is then the interval $[-\pi, \pi)$ and the distribution is symmetric about zero. The parameter $\kappa \geq 0$ is a measure of 'concentration' of the distribution around the mean value, i.e., κ^{-1} behaves analogously to the dispersion of a linear random variable. $I_o(\kappa)$ is the modified Bessel function of the first kind and order zero. Under $\mathcal{H}^1$, we consider the residuals to exhibit a higher concentration around μ.

NB: The term *circular distribution* as used in this paper refers to a directional distribution with support on the circumference of unit circle [40].

2.3.1. Test Statistic

A commonly used statistic to test among the two hypotheses is the *ensemble* coherence, as defined below [5,32]:

$$\gamma \triangleq E\left\{\exp\left(j\underline{w}\right)\right\}. \tag{10}$$

An unbiased estimator of the coherence, given M interferometric layers, is the multi-interferogram complex coherence (MICC) [12,32]:

$$\hat{\underline{\gamma}} = \frac{1}{M}\sum_m \exp\left(j\underline{\hat{w}}_m\right) \tag{11}$$

$$= \frac{1}{M}\left(\sum_m x_m + j\sum_m y_m\right) \tag{12}$$

$$= \frac{1}{M}\left(X + jY\right) \tag{13}$$

$$= \bar{X} + \bar{Y} = |\hat{\underline{\gamma}}|e^{j\bar{\mu}}, \tag{14}$$

where X and Y are the sum of cosine and sine terms in the expression, respectively, and the length of the resultant, $R = \sqrt{X^2 + Y^2}$. The overscore indicates sample mean. Hereafter, we refer to MICC simply as the sample coherence. The phase residuals $\underline{\hat{w}}_m$ are assumed to be independent and identically distributed (i.i.d.) random variables.

In the context of interferometry, we typically use the coherence values normalized between 0 and 1, i.e., $|\hat{\underline{\gamma}}|$, instead of the resultant $R = M|\hat{\underline{\gamma}}|$. However, in the directional statistics literature, the use of the term R is more common. Here, we state both to facilitate cross-referencing with the literature. The sample mean direction $\bar{\mu}$, computed with sample coherence for any random sample $(w_1, w_2, \ldots, w_m)$ from a von Mises population, is the maximum likelihood estimator of the preferred direction μ when $\bar{R}$ is well-defined [41,42]. This property is characteristic of von Mises populations on a circle, analogous to a similar property holding for Gaussian distribution on a real line whose location parameter is estimated with maximum likelihood by the sample mean [40,43].

2.3.2. Statistics under $\mathcal{H}^0$

The statistics of the sample coherence depend on the distribution of the phase residuals. With reference to Equation (11), the phase residuals can be considered as angles subtended by phasors of unit length. Under $\mathcal{H}^0$, when the phasors have no preferred direction, we consider the limiting case of von Mises distribution when $\kappa \to 0$ [40]:

$$\lim_{\kappa \to 0} g\left(w; \mu = 0, \mathcal{H}^0\right) = U\left(w\right) = \frac{1}{2\pi}, \quad -\pi \le w < \pi, \tag{15}$$

where $U\left(w\right)$ is the circular uniform distribution. In this case, $E\left\{x\right\} = E\left\{y\right\} = 0$; therefore, $E\left\{\hat{\underline{\gamma}}\right\} = 0$. The second order moments are $E\left\{x^2; \mathcal{H}^0\right\} = E\left\{y^2; \mathcal{H}^0\right\} = \frac{1}{2}$. The terms x and y are not independent (as $x^2 + y^2 \equiv 1$), but they are uncorrelated as $E\left\{x \cdot y\right\} = 0$ [12,40]. The variance of the addends in the Equation (12) is finite. Therefore, under the assumption of a large sample size, multivariate central limit theorem holds, and we consider the joint distribution of $(\bar{X}, \bar{Y})$ to be converging to a Gaussian distribution, $\mathcal{N}_2\left(0, \Sigma_{\hat{\gamma}}\right)$ where

$$\Sigma_{\hat{\gamma}} = \begin{bmatrix} \dfrac{1}{2M} & 0 \\ 0 & \dfrac{1}{2M} \end{bmatrix}.$$

R is then approximately Rayleigh-distributed, and its PDF is as follows [40]:

$$f_{\hat{\gamma}}\left(r; \mathcal{H}^0\right) \simeq \frac{2r}{M} \exp\left(-\frac{r^2}{M}\right),$$ (16)

where $0 \leq r \leq M$. Referring to [12,40], the probability of false alarm can be computed as the upper tail of the Rayleigh distribution, as follows:

$$\Pr\{R > r_{\text{th}} | \mathcal{H}^0\} = \exp\left(-\frac{r_{\text{th}}^2}{M}\right).$$ (17)

It can be equivalently expressed as

$$\Pr\{|\hat{\gamma}| > T_\gamma | \mathcal{H}^0\} = \exp\left(-MT_\gamma^2\right),$$ (18)

where T_γ is the detection threshold such that $0 \leq T_\gamma \leq 1$.

2.3.3. Statistics under $\mathcal{H}^1$

In case of $\mathcal{H}^1$, the probability distribution of R is given by [40],

$$f(r) = \frac{I_0(\kappa r)}{I_0^M(\kappa)} r \int_0^\infty J_0(rt) J_0^M(t) \, t \, dt.$$ (19)

J_0 is the Bessel function of the first kind and zero-order. A closed form expression for the PDF is not available. We again assume a large sample size and invoke the multivariate central limit theorem. It allows us to consider the joint distribution of $\bar{X}$ and $\bar{Y}$ to be asymptotically normal, and expectation and the variance of the sample coherence can be approximated as follows [44]:

$$E\{|\hat{\gamma}|\} \simeq v_1,$$ (20)

$$\text{var}\{|\hat{\gamma}|\} \simeq \frac{1 - 2v_1^2 + v_2}{2M},$$ (21)

where $v_j = E\{\cos(jw)\}$:

$$v_1 = \frac{I_1(\kappa)}{I_0(\kappa)},$$ (22)

$$v_2 = 1 - 2\left(\frac{I_1(\kappa)}{\kappa \cdot I_0(\kappa)}\right).$$ (23)

For sufficiently large κ, the von Mises distribution for the phase residuals can be approximated by a linear normal distribution with $\sigma_w^2 = \kappa^{-1}$ [40]. The coherence in this case is given by [12,31]:

$$E\{|\hat{\gamma}|\} = E\{\hat{\gamma}\} = \exp\left(-\frac{\sigma_w^2}{2}\right).$$ (24)

For a discussion on the details about the corresponding probability of detection, interested readers are referred to earlier works in the literature [12,13].

Since exact closed-form expressions for the PDF of $|\hat{\gamma}|$ are not available, we resort to numerical methods to compare the estimate of the coherence magnitude for the general case of $\kappa > 0$ against the estimate in case of the aforementioned linear normal approximation. For selected values of κ between $[1, 10]$, we perform 10^5 Monte Carlo simulations of the residual phase vector, **w** (comprising M instances of von Mises distributed random variables), and compute the coherence magnitude.

The results are shown in Figure 1 for three different values of M. The estimate under the normal approximation (Equation (24)) is also shown. It can be seen that the normal approximation for the limiting case tends to overestimate the coherence magnitude. The overestimation decreases for increasing values of κ. For $\kappa > 3$, the difference between the coherence estimate under the assumption of von Mises distribution and the normal approximation is less than 5% on average. With increasing number of acquisitions, the variance in the estimation of the coherence magnitude decreases (in agreement with Equation (21)).

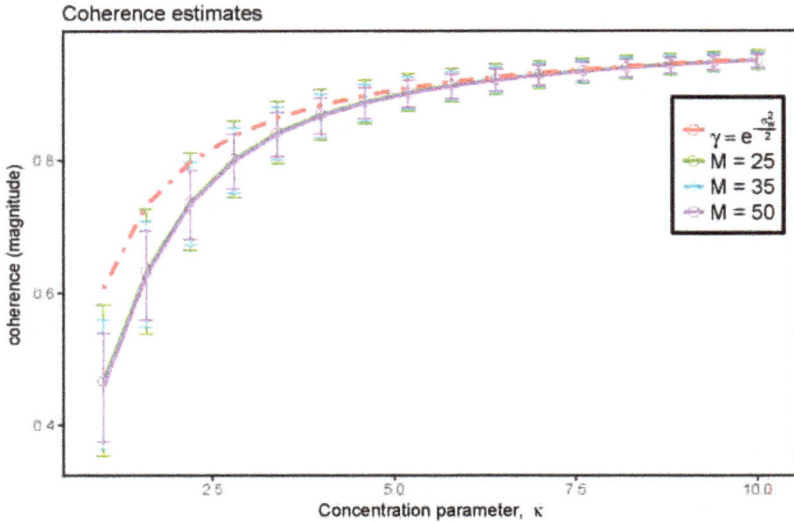

Figure 1. Estimates of the coherence magnitude obtained with 10^5 Monte Carlo iterations assuming the residual phases have a von Mises distribution with concentration parameter, κ. Each solid line indicates the estimates for a specific number of acquisitions, M in the data stack. The vertical bars represent ± 1-σ from the mean. The dashed line shows the coherence magnitude under the assumption that the residual phases follow a linear normal distribution, cf. Equation (24) (assuming $\sigma_w^2 = \kappa^{-1}$).

2.4. SAR Tomography: Mathematical Model

In the absence of noise, for a given range-azimuth resolution cell, the mathematical model for SAR tomography (3-D SAR) can be written as [16,19,21,26,45]:

$$y_m = \int_{\mathcal{I}_s} \alpha(s) \exp\left[-j\varphi_m^{geo}(s)\right] ds, \tag{25}$$

where α is the complex reflectivity and $\mathcal{I}_s$ is the support of s. This model assumes there has been no displacement in the line of sight during the observation time period. Differential SAR tomography [23,25] with extended phases models [25,27,46] allows modeling linear displacement as well as seasonal or temperature-induced motion:

$$y_m = \iiint_{\mathcal{I}_s \mathcal{I}_v \mathcal{I}_\eta} \alpha(s,v,\eta) \exp\left\{-j\left[\psi_m(s,v,\eta)\right]\right\} dsdvd\eta, \tag{26}$$

where ψ_m is the sum of the deterministically modeled phase components as a function of the unknown parameters, i.e.,

$$\psi_m\left(s, v, \eta\right) = \varphi_m^{\text{geo}}\left(s\right) + \varphi_m^{\text{disp}}\left(v\right) + \varphi_m^{\text{th.exp}}\left(\eta\right). \tag{27}$$

It is assumed that the phase terms (and hence the spatial and temporal baselines, and temperature changes) are mutually independent of each other. A general mathematical model for SAR tomography can be defined as follows [27,47]:

$$y_m = \int_{\mathbb{P}} \alpha\left(\mathbf{p}\right) \exp\left[-j\psi_m\left(\mathbf{p}\right)\right] d\mathbf{p}, \tag{28}$$

where $\mathbb{P}$ represents the support of the parameter vector (i.e., the parameter space), and $\mathbf{p} \in \mathbb{P}$. It is analogous to a multi-dimensional Fourier transform [48]. In case the resolution cell contains a single point source with dirac delta response, $\alpha(\mathbf{p}) = \tau_1 \delta(\mathbf{p} - \mathbf{p}_1)$, with $\tau_1 \in \mathbb{C}$, Equation (28) reduces to the following:

$$y_m = \tau_1 \exp\left[-j\psi_m\left(\mathbf{p}_1\right)\right]. \tag{29}$$

For the general case of Q point sources in the presence of clutter, the tomographic model is further extended as follows:

$$\underline{y}_m = \sum_{q=1}^{Q} \tau_q \exp\left[-j\psi_m(\mathbf{p}_q)\right] + \underline{n}_m \tag{30}$$

$$= d_m + \underline{n}_m, \tag{31}$$

where $\underline{n}_m$ represents additive noise which is typically modeled as zero-mean complex Gaussian (with symmetric variances for the real and imaginary parts). We assume the noise samples are i.i.d. across the stack, i.e., $D\left\{\underline{n}\right\} = \sigma_n^2 I_M$, with $\sigma_n^2 > 0$. d_m represents the coherent sum of the deterministic components in the signal vector. τ_q is the reflectivity, and $\psi_m(\mathbf{p}_q)$ is the modeled phase for the qth scatterer.

2.5. SAR Tomography: Model Inversion and Parameter Estimation

We use single-look beamforming for the inversion of the general tomographic model to estimate the unknown scatterer reflectivity as a function of the parameter vector $\mathbf{p}$ for a given range-azimuth resolution cell as follows [13,16]:

$$\hat{\alpha}\left(\mathbf{p}\right) = \frac{1}{M} \langle y, \mathbf{a}\left(\mathbf{p}\right)\rangle, \tag{32}$$

where $\langle ., .\rangle$ represents the inner product, $\mathbf{a}\left(\mathbf{p}\right)$ is the steering vector as a function of $\mathbf{p}$, and y is the vector comprising the SLC observations:

$$y = \left[\begin{array}{cccc} y_0 & y_1 & \cdots & y_{M-1} \end{array}\right]^T. \tag{33}$$

The steering vector is structured as follows:

$$\mathbf{a}\left(\mathbf{p}\right) = \left[\begin{array}{cccc} e^{-j\psi_0(\mathbf{p})} & e^{-j\psi_1(\mathbf{p})} & \cdots & e^{-j\psi_{M-1}(\mathbf{p})} \end{array}\right]^T. \tag{34}$$

For the estimation of the unknown parameters, we use the estimated absolute reflectivity as the objective function in the following maximization:

$$\hat{\mathbf{p}}_1 = \arg\max_{\mathbf{p}\in\mathbb{P}}\left(|\hat{\alpha}\left(\mathbf{p}\right)|\right). \tag{35}$$

As more than one coherent scatterer may be present in the same resolution cell, successive maxima after the global maximum may indicate the presence of more scatterers. Assuming a maximum of two scatterers, an estimate of the parameter vector for the second scatterer is obtained as follows:

$$\hat{\mathbf{p}}_2 = \arg\max_{\mathbf{p} \in \mathbb{P} \backslash \{\hat{\mathbf{p}}_1 \pm \delta \mathbf{p}\}} (|\hat{a}(\mathbf{p})|), \tag{36}$$

where $\delta\mathbf{p}$ indicates the Rayleigh resolution for the tomographic profile along each of the unknown parameters.

Equation (32) implies that noise in the SLC vector will cause errors in the reconstructed target reflectivity. As a consequence, errors will propagate in the estimation of the parameters using the aforementioned maximizations. Therefore, a scatterer detection strategy is needed to classify whether a given resolution cell contains one or more phase coherent scatterers, or is merely clutter.

2.6. SAR Tomography: Statistics for Scatterer Detection

A commonly used test statistic for coherent scatterer detection in the context of tomography is the absolute value of the estimated reflectivity, $|\hat{a}|$. The same hypotheses are carried forward as introduced in Section 2.3, except for the change that now we consider them for multiple coherent scatterer candidates for each pixel. We consider a maximum of two candidates per pixel. In case only one of the candidates fulfills $\mathcal{H}^1$, we call the pixel a single scatterer. In case both the candidates fulfill $\mathcal{H}^1$, the pixel is called a double scatterer.

2.6.1. Statistics under $\mathcal{H}^0$

In case the received signal is merely clutter, the received signal vector $y = n$. Using Equation (32),

$$E\left\{\hat{a}(\mathbf{p}); \mathcal{H}^0\right\} = \frac{1}{M} E\left\{\langle \underline{n}, \mathbf{a}(\mathbf{p}) \rangle\right\} \tag{37}$$

$$= \frac{1}{M} E\{\hat{\underline{n}}\} \tag{38}$$

$$= \frac{1}{M} E\{\underline{n}\} = 0, \tag{39}$$

where the third equality follows from rotational invariance of the Gaussian distributed samples, and, therefore, the inconsequential difference between $\hat{\underline{n}}$ and $\underline{n}$ will be dropped. Since $\varphi_m = \angle n_m$ under $\mathcal{H}^0$, the observed interferometric phase (and the residual phase in this case) follows a uniform distribution [12]. Along similar lines as in Section 2.3, the joint distribution of the real and imaginary parts of $\hat{a}$ is a zero-mean Gaussian with the following covariance matrix:

$$\Sigma_{\alpha|z_m} = \begin{bmatrix} \frac{1}{2M^2} \sum_m z_m^2 & 0 \\ 0 & \frac{1}{2M^2} \sum_m z_m^2 \end{bmatrix}.$$

The PDF of $|\hat{a}|$ in this case is Rayleigh, and the right tail probability to compute the probability of false alarm is as follows:

$$\Pr\{|\hat{a}| > T_\alpha | \mathcal{H}^0, y\} = \exp\left(-\frac{M^2 T_\alpha^2}{\|y\|_2^2}\right), \tag{40}$$

where $\|y\|_2^2 = \sum_m z_m^2$ is the squared L2-norm of the observed signal vector.

2.6.2. Statistics under $\mathcal{H}^1$

In general, the received signal contains clutter besides the possibility of backscattering contribution from point-like sources. We assume that, under $\mathcal{H}^1$, the deterministic backscatter from the point sources is dominant over the clutter, i.e., $|d_m| \gg |n_m| \; \forall \; m$. This assumption allows us to consider that the observed phase owes primarily to the vector sum of the backscatter from point-like sources (and not the clutter). Using Equations (30) and (32), the expression for the estimated reflectivity can then be stated as follows:

$$\hat{\underline{a}}(\mathbf{p}) = \frac{1}{M} \langle \mathbf{d}, \mathbf{a}(\mathbf{p}) \rangle + \langle \underline{n}, \mathbf{a}(\mathbf{p}) \rangle \tag{41}$$

$$= \frac{1}{M} \sum_m |d_m| \exp\left\{-j[\underline{\varphi}_m - \psi_m(\mathbf{p})]\right\} + \underline{n} \tag{42}$$

$$= \frac{1}{M} \sum_m |d_m| \exp\left\{j\underline{\hat{w}}_m(\mathbf{p})\right\} + \underline{n}. \tag{43}$$

Formally, the origin of phase instability, $\underline{\hat{w}}_m$ in Equation (43) is not the clutter, rather it is phase disturbances such as uncompensated atmospheric phase delay variations or residual motion [31], or phase model imperfections. Using Equations (10) and (39),

$$E\left\{\hat{\underline{a}}(\mathbf{p}); \mathcal{H}^1\right\} = \frac{1}{M} \sum_m |d_m| \, E\left\{\exp[j\underline{\hat{w}}_m(\mathbf{p})]\right\} \tag{44}$$

$$= \frac{1}{M} \sum_m |d_m| \, \gamma. \tag{45}$$

From Equation (44), it is clear that phase instability is disturbing tomographic reconstruction in a multiplicative sense. The ensemble coherence has a direct impact on the expected value of the retrieved reflectivity profile, and thereby on the hypothesis testing. Closed-form expression s for the PDF of $|\hat{a}|$ are not available when the residuals are assumed to follow a von Mises distribution with $\kappa > 0$. A Rician approximation can be taken, as suggested in [31], when the residuals can be considered to be normally distributed (i.e., the limiting case when $\kappa \to \infty$). The probability of detection, f_D for a fixed false alarm rate can then be studied as the area under the upper tail of the Rician distribution [49].

Nonetheless, we resort to Monte Carlo simulation to study the probability of detection numerically in terms of the inverse coefficient of variation (iCV) as defined below for the text statistic $\hat{\underline{a}}$:

$$\text{iCV}^2 \triangleq \frac{\left|E\left\{\hat{\underline{a}}(\mathbf{p}); \mathcal{H}^1\right\}\right|^2}{\text{var}\left\{\hat{\underline{a}}(\mathbf{p}); \mathcal{H}^1\right\}}. \tag{46}$$

This definition has been referred to as the signal-to-noise ratio (SNR) in [31]. Although, in the field of signal processing iCV is often referred to as the SNR, we avoid referring it so. In our context, formally the denominator in Equation (46) is not representing the noise power, neither additive (σ_n^2) nor multiplicative (σ_w^2), but rather the dispersion of the test statistic.

Considering $\underline{n} \approx 0$, and dropping the dependence on $\mathbf{p}$ to simplify notation,

$$\text{var}\left\{\hat{\underline{a}}; \mathcal{H}^1\right\} = \frac{1}{M^2} \sum_{l=1}^{M} \sum_{k=1}^{M} |d_l| \, |d_k| \cdot \text{cov}\left\{e^{j\underline{\hat{w}}_l}, e^{-j\underline{\hat{w}}_k}\right\}. \tag{47}$$

Using the assumption that the residual phases are i.i.d. random variables, the covariance term in Equation (47) simplifies as follows:

$$\text{cov}\left\{e^{j\underline{\hat{w}}_l}, e^{-j\underline{\hat{w}}_k}\right\} = (1 - |\gamma|^2) \cdot \delta[l - k], \tag{48}$$

where $\delta\,[.]$ is the unit sample function. Using this result, Equation (47) reduces to the following [31]:

$$\mathrm{var}\left\{\hat{\underline{a}};\mathcal{H}^1\right\} = \frac{1-|\gamma|^2}{M^2}\sum_m |d|_m^2.\tag{49}$$

Therefore,

$$\mathrm{iCV}^2 = \frac{|(\sum_m |d_m|\,\gamma)|^2}{(1-|\gamma|^2)\sum_m |d|_m^2} = \left(\frac{|\gamma|^2}{1-|\gamma|^2}\right)\frac{\|\boldsymbol{y}\|_1^2}{\|\boldsymbol{y}\|_2^2}.\tag{50}$$

Since $\|\boldsymbol{y}\|_2 \le \|\boldsymbol{y}\|_1 \le \sqrt{M}\,\|\boldsymbol{y}\|_2$ [50], we reach the following bounds on the iCV for a given level of coherence:

$$\frac{|\gamma|^2}{1-|\gamma|^2} \le \mathrm{iCV}^2 \le M\cdot\frac{|\gamma|^2}{1-|\gamma|^2}.\tag{51}$$

iCV is a function of the ensemble coherence as well as on the ratio of the L1 to L2 norm of the signal vector. While the coherence is in turn a function of the concentration of the phase residuals (as shown in Figure 1), the L1–L2 ratio is influenced by the (1) number of acquisitions and (2) the number of point-like scatterers in the same resolution cell. Figure 2 shows the variation of the empirically estimated iCV against the concentration parameter for different numbers of scatterers, for $M = 50$ acquisitions as an example. In addition, 10^5 realizations of the phase residue are generated under a von Mises distribution for each value of κ selected between $(0, 20]$. The dashed lines in Figure 2 highlight the upper and lower bounds on iCV. The upper bound is reached theoretically when $\|\boldsymbol{y}\|_1 = \sqrt{M}\,\|\boldsymbol{y}\|_2$. Therefore, the greater the number of acquisitions, the higher is the achievable iCV. At a given concentration of phase residuals, the iCV decreases for an increasing number of scatterers. The iCV estimates for $Q = 1$ converge at the upper bound. The impact of the number of scatterers on the iCV is further discussed in Appendix A.

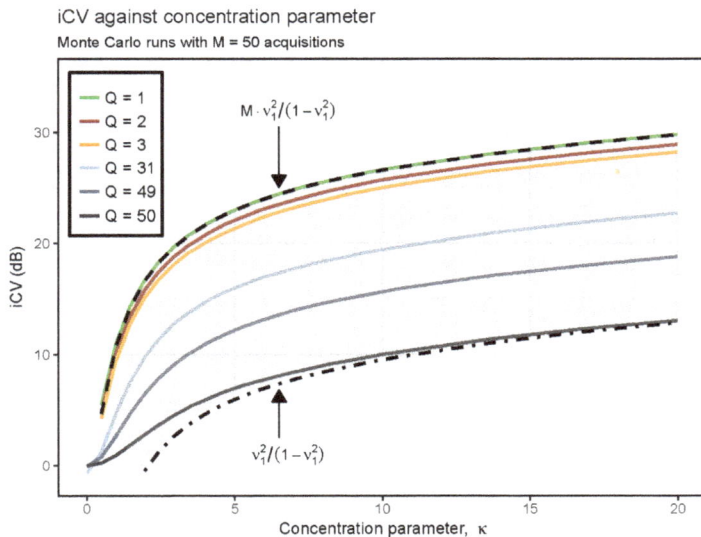

Figure 2. Empirically estimated inverse coefficient of variation (iCV) of the test statistic $\hat{\underline{a}}$ against concentration parameter for von Mises distributed phase residuals, for different number of scatterers, Q in the same resolution cell. The dashed lines enclosing the gray region indicate the theoretical bounds on the iCV (cf. Equation (51)), where $v_1 = \frac{I_1(\kappa)}{I_0(\kappa)}$ (Equation (22)).

Figure 3a is a plot of the numerically estimated f_D against the iCV. The detection thresholds are set to ensure a fixed level of probability of false alarm, $f_F \in \{10^{-2}, 10^{-3}, 10^{-4}\}$ given $M = 50$ acquisitions. Lower levels of f_F provide higher f_D, indicating the trade-off typically observed for statistical detectors [49]. At the same time, we observe a slight dependency of f_D on the number of scatterers. Even for a fixed level of iCV, the f_D is lower for a higher number of scatterers. However, this dependency diminishes as the level of the false alarm is relaxed.

Figure 3b shows f_D against iCV while fixing f_F at 10^{-3} for single and double scatterers, for different number of acquisitions in the stack, $M \in \{25, 35, 50, 75\}$. We observe slight dependency of f_D on M, though it tends to diminish as the number of acquisitions in the stack grow larger.

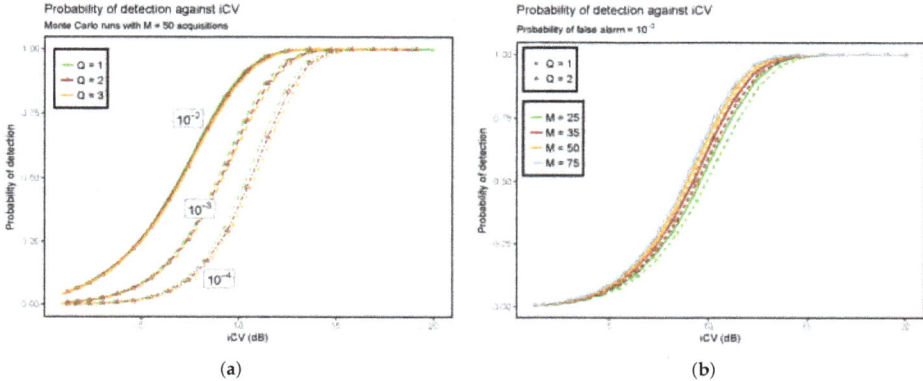

Figure 3. Numerically estimated probability of detection against inverse coefficient of variation (iCV) of the test statistic $\hat{\alpha}$, using 10^5 Monte Carlo realizations of the phase residuals, (**a**) for $M = 50$ acquisitions and fixed levels of false alarm, $f_F \in \{10^{-2}, 10^{-3}, 10^{-4}\}$, and number of scatterers, $Q \in \{1, 2, 3\}$; (**b**) and for different number of different number of acquisitions, $M \in \{25, 35, 50, 75\}$ when $f_F = 10^{-3}$.

The aforementioned simulations have been performed in the absence of clutter. We repeat them next with varying levels of clutter, expressed in terms of the signal-to-clutter ratio (SCR): $\|d\|_2^2 / \sigma_n^2$. Samples to simulate clutter are generated as instances of zero-mean Gaussian noise with variance σ_n^2. Figure 4a shows the iCV observed for the case of single and double scatterer for three different, but fixed, levels of SCR $\in \{6, 3, 0\}$ dB. As expected, the iCV decreases with decreasing SCR. The case of SCR = 0 dB, i.e., when the intensity of the deterministic backscatter from point scatterers equals that of the clutter, contradicts the assumption used in deriving Equation (43). Nonetheless, we perform the simulation as a worst-case analysis. Figure 4b shows f_D against the iCV for this case. The plots shown are nearly identical to those shown in Figure 3a. This is an auspicious finding as it implies that, for fixed levels of iCV, f_D can be characterized nearly independently of the origin (additive or multiplicative) and level of noise.

The simulation results in Figures 3 and 4 collectively imply that when the number of acquisitions are sufficiently large and the false alarm setting is not too strict, the empirically estimated iCV can be considered to fully characterize the f_D, even in the presence of clutter.

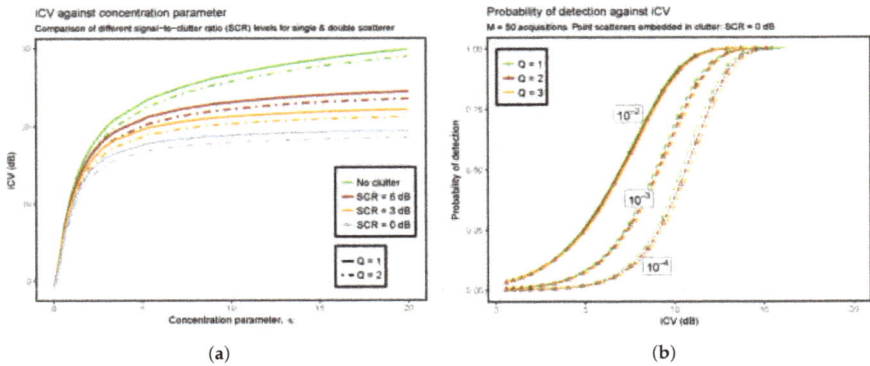

Figure 4. Numerical analysis of the inverse coefficient of variation (iCV) of the test statistic $\hat{\alpha}$ when point scatterers are embedded in different clutter levels, for $M = 50$ acquisitions. (**a**) iCV against concentration of the phase residuals for different levels of signal-to-clutter ratio (SCR) and number of scatterers, $Q \in \{1,2\}$; (**b**) probability of detection against iCV for fixed levels of false alarm, $f_F \in \{10^{-2}, 10^{-3}, 10^{-4}\}$, and $Q \in \{1,2,3\}$.

3. Methods

This section presents the overall methodology adopted for the interferometric and tomographic processing of a real interferometric data stack. The models discussed in the previous section form the basis of this methodology. The data undergoes several preprocessing steps. A reference scene is selected, and a multilooked intensity image of the reference scene is used to geocode and coregister all the acquisitions in the stack. An external digital elevation model (DEM) is used in the process [51,52]. A suitable reference point is selected to compute double-differenced interferograms.

3.1. Interferometric Processing with IPTA

We use the IPTA [3,8] framework for the PSI processing, whereby parameter estimation and phase calibration of the data stack are performed side by side using an iterative approach to least squares regression. An initial list of PS candidates is prepared on the basis of high temporal stability of the backscattering and low spectral diversity. The phase model assumed is as given in Equation (1). Point differential interferograms are obtained by subtracting the topographic phase computed using the DEM. A multiple linear regression is used for each candidate to obtain an initial estimate of s and v, as well as the phase unwrapping integer, p. The quality of the estimates is assessed in terms of the root-mean-square (RMS) phase deviation, $\hat{\sigma}_w$ of the residual phase. At the initial stage, atmospheric phases in each interferometric layer have not be corrected, and the possible temperature-induced phase variations of candidates on structures experiencing thermal expansion have also not been accounted for. Therefore, the residual phase typically exhibits a high dispersion. The PS candidates for which $\hat{\sigma}_w$ is higher than a pre-selected threshold, σ_c are masked out. The residue of the remaining candidates is analyzed further. Assuming the atmospheric phase screen (APS) to be spatially low-frequency and temporally uncorrelated, we estimate it by spatial filtering and unwrapping of the phase residue in the neighborhood of the candidates that satisfied the quality criterion. The estimated APS is subtracted and point-differential interferograms are re-computed for the full list of PS candidates, and this time the phases related to the initial estimates of residual height, linear deformation and the atmospheric phase are subtracted as well. The resulting point differential interferograms are unwrapped and the regression is iterated. It is expected that the quality of the candidates would improve since an estimate of the atmospheric phase has been subtracted prior to the regression. σ_w is computed again for all the candidates, and compared against σ_c to mask out those with relatively low quality. For the retained candidates, the newly estimated regression coefficients (residual height and deformation velocity) act as 'corrections' on the previous estimates. The new phase residue is added to the previous estimate of the atmospheric phase, re-filtered and unwrapped to give a new estimate of the atmospheric phase. The process is iterated several times. In this way, there is progressive improvement in the quality of the estimates in consecutive iterations. For more details on various time-series processing strategies using the IPTA framework, the interested readers are referred to earlier works [3,8,9,53].

For the candidates that are potentially undergoing thermal expansion, another regression-based routine is used that models it assuming that the corresponding phase variations are linearly dependent on the temperature changes [54–56]. The estimated regression coefficient is the phase-to-temperature sensitivity, $\hat{\eta}$. Further details are available in the earlier work in [35].

After several iterations, the APS is well isolated and we obtain iteratively-refined estimates of the parameter vector $\hat{\mathbf{p}}$ for the PS candidates that satisfy the quality criterion. Assuming that these PS are of sufficiently good quality that the limiting case of von Mises distribution for the phase residuals being approximated by a linear normal distribution is justified, we compute the sample coherence threshold corresponding to σ_c using Equation (24) as follows:

$$T_{\gamma_c} = \exp\left(-\frac{\sigma_c^2}{2}\right).$$

(52)

In turn, the corresponding probability of false alarm (theoretical) is computed using Equation (18). It is important to mention here that the aforementioned assumption is not mandatory to choose the threshold; in fact, a threshold can be set directly on the coherence (as typically done for interferometric processing) [2,12,57]. In our context where we perform PSI processing with the IPTA toolbox (which allows quality assessment in terms of the residual phase statistics), the relation in Equation (52) provides a means to compute the coherence threshold corresponding to the quality criteria in our PSI processing.

3.2. Single-Look Differential SAR Tomography with Extended Phase Model

Prior to tomographic inversion, the interferometric data stack requires a precise phase calibration. For the pixels containing PS, we already have an estimate of the atmospheric phases from the PSI processing. Given a sufficient distribution of the PS over the imaged scene or the region of interest, we interpolate these phases over the surrounding pixels that may or may not have been PS candidates. Single-look differential tomographic inversion is applied for each pixel. The extended phase model, given in Equation (27), is used to set up the steering vectors. The reflectivity profile, $\hat{\alpha}(\mathbf{p})$ is estimated as a function of the unknown parameters. Scatterer localization and parameter estimation for a maximum of two scatterers in each resolution pixel is performed, as stated in Equations (35) and (36). The amplitude of the estimated reflectivity is compared against a threshold for each potential scatterer to accept or reject the null hypothesis.

We propose to set the detection threshold in such a way that the desired probability of false alarm from PSI processing is carried forward for the detection of coherent scatterers at this stage. Equating the Equations (18) and (40), we set the detection threshold for tomography as follows:

$$T_{\alpha_c} = \frac{T_{\gamma_c}\,\|\mathbf{y}\|_2}{\sqrt{M}}.$$

(53)

In turn, the decision between $\mathcal{H}^1$ and $\mathcal{H}^0$ is made for each candidate as follows:

$$|\hat{\alpha}|\,(\mathbf{p}) \underset{\mathcal{H}^0}{\overset{\mathcal{H}^1}{\gtrless}} T_{\alpha_c}.$$

(54)

In this way, the same quality criterion that is used for setting the threshold T_{γ_c} in the PSI processing also determines the threshold for scatterer detection in tomographic processing. Hence, a consistency is achieved for the synergistic use of tomography as an add-on to PSI.

It is to be noted that Equation (53) is independent of how the threshold T_{γ_c} for PSI processing was selected, whether as a direct choice on the coherence, or using the standard deviation of the residual phase according to Equation (52) under the assumption of linear normal distribution of the residual phases for the PS. Therefore, this assumption is not a limiting factor for the application of the proposed detection strategy in general.

4. Data

The interferometric data stack used in the work comprises 50 TerraSAR-X stripmap acquisitions over the city of Barcelona, Spain in repeated passes. It is the same stack as used in our earlier work in [27]. The temporal span of the acquisitions extends from 2007 to 2012. The images have been oversampled by a factor of 2 to allow for more accurate coregistration. The resolution in range and azimuth is 1.2 m and 3.3 m, respectively. The orthogonal component of the total spatial baseline is 503 m, which provides resolution in elevation axis of $\sim$19 m. The distribution of the spatial and temporal baselines, as shown in Figure 5a, is highly non-uniform. The corresponding 2-D point spread function (PSF) is shown in Figure 5b. The PSF represents the impulse response of the tomographic system for the given distribution of the baselines, for an ideal point scatterer at zero elevation and with no deformation. The footprint of the acquisitions in map coordinates is shown in Figure 5c. Apart from a dense urban stretch, some part of the viewed scene extends over the Balaeric sea.

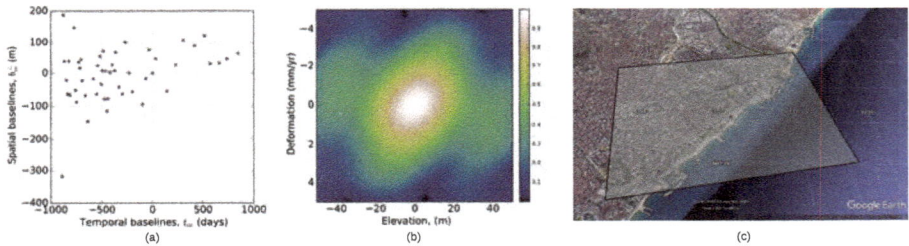

Figure 5. Data characteristics. (**a**) distribution of spatial (orthogonal component) and temporal baselines; (**b**) 2-D point spread function (PSF); (**c**) footprint of the reference acquisition over Barcelona, Spain.

5. Results on Real Data

This section presents the results obtained on the real interferometric data stack introduced in the previous section.

5.1. Interferometric Processing

An initial list of PS candidates was prepared on the basis of low spectral diversity and high stability of the backscattering amplitude that is characteristic of single dominant scatterers [3]. There was no candidate in unexpected areas, such as the water surface or radar shadows. After several iterations of the least squares regression within the IPTA framework, as outlined in Section 3.1, a subset of the initial candidates is retained such that $\sigma_w \leq \sigma_c = 1.1$ rad for each candidate. Figure 6 shows these candidates from the last iteration. These are 936,649 in number, and spread over an area of nearly 4 km^2. In sub-figure a–c, the color coding represents the estimated parameters, namely residual height, deformation velocity in the LOS and phase-to-temperature sensitivity, respectively. The sample coherence for these candidates is as shown in sub-figure d.

Corresponding to $\sigma_c = 1.1$ rad, the coherence threshold $T_{\gamma_c} = 0.55$ according to Equation (52), and the theoretical probability of false alarm according to Equation (18) is 3.3×10^{-7}. As stated in Section 3.1, the use of Equation (52) to convert a threshold in terms of residual phase standard deviation to corresponding threshold on coherence requires the assumption that the von Mises distribution can be approximated by a linear normal distribution (for the case of ideal, noise-free PS, with $\kappa \to \infty$). In order to assess the suitability of this assumption, we require estimates of the concentration parameter.

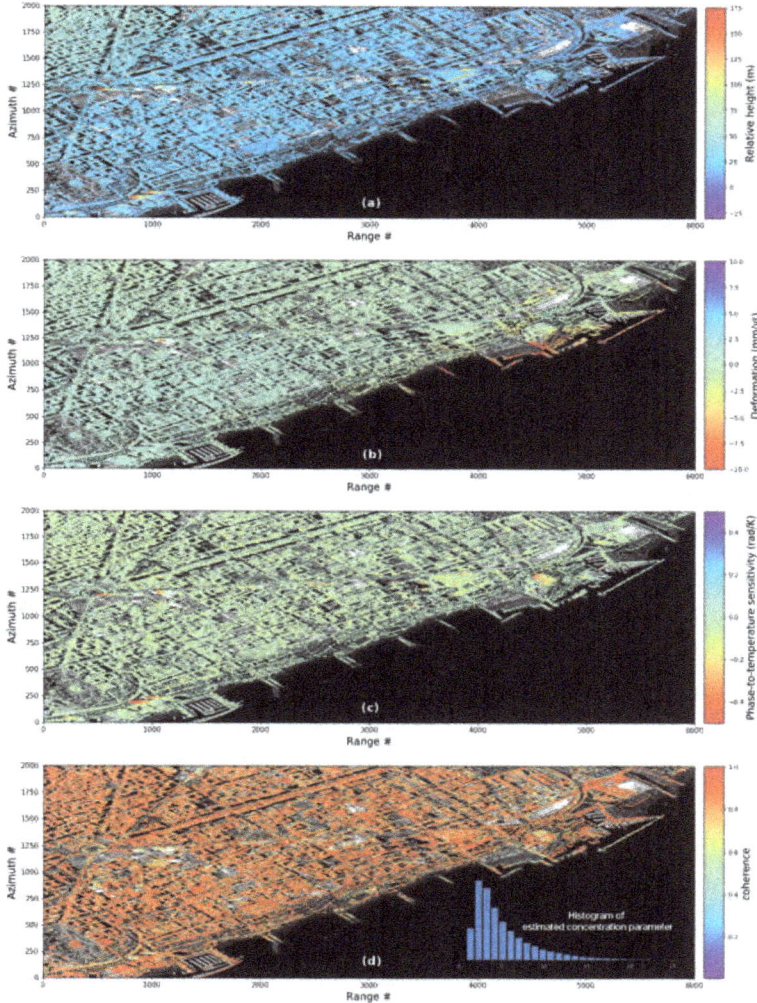

Figure 6. PSI solution obtained with iterative least-squares regression-based processing using the interferometric point target analysis (IPTA) toolbox. The colored dots are the PSs identified in the PSI processing. (**a**) estimated height, relative to the WGS-84 reference ellipsoid; (**b**) deformation velocity in the line-of-sight; (**c**) phase-to-temperature sensitivity; (**d**) sample coherence, and histogram of the estimated concentration parameter (shown as inset).

Using Equations (20) and (22), $E\{|\hat{\gamma}|\} \simeq \frac{I_1(\kappa)}{I_0(\kappa)}$; to estimate κ, this expression needs to be inverted, which involves inversion of the ratio of modified Bessel functions (first kind) of first and zero order. We do not have a closed-form expression for such an inverse relation; we use the following piece-wise defined approximation [43,58]:

$$
\hat{\kappa} \simeq \begin{cases} 2\,|\hat{\gamma}| + |\hat{\gamma}|^3 + \dfrac{5}{6}|\hat{\gamma}|^5 & |\hat{\gamma}| < 0.53, \\[2mm] -0.4 + 1.39\,|\hat{\gamma}| + \dfrac{0.43}{1-|\hat{\gamma}|} & 0.53 \le |\hat{\gamma}| < 0.85, \\[2mm] \dfrac{1}{3\,|\hat{\gamma}| - 4\,|\hat{\gamma}|^2 + |\hat{\gamma}|^3} & |\hat{\gamma}| \ge 0.85. \end{cases} \tag{55}
$$

The concentration parameter is estimated for each PS, and a histogram of the parameters is shown as an inset in Figure 6d. The mean and the median values are 5.4 and 4.1, respectively. In existing literature in the field of directional statistics, we can find precedence where concentration parameters greater than 2 are considered reasonable to approximate von Mises distribution as *wrapped* normal distribution (i.e., linear normal distribution wrapped between $-\pi$ to π rad) [58].

5.2. Tomographic Processing and Empirical Analysis of False Alarms

The APS isolated in the IPTA-based PSI processing is extrapolated over the scene and compensated for over the entire scene in each layer of the interferometric stack. In this way, each pixel is considered to be phase calibrated so that tomographic inversion can be applied next. Given that the city of Barcelona has several high-rise buildings, the elevation extent, $\mathcal{I}_s$ is set as $[-60, 300]$ m. The parameter space for the deformation parameters is as follows: $\mathcal{I}_v \in [-10, 10]$ mm/yr and $\mathcal{I}_\eta \in [-1, 1]$ rad/K. The discretization in each dimension is $1/2.5$ times the Rayleigh resolution, followed by a local refinement of the estimated reflectivity around the two candidate peaks at one-tenth the resolution. Using Equations (52) and (53), and keeping $\sigma_c = 1.1$ rad, we threshold the reflectivity of the two candidates to perform the detection process. The point cloud of single scatterers thus detected is shown in Figure 7. 1454 false alarms occur over the water surface.

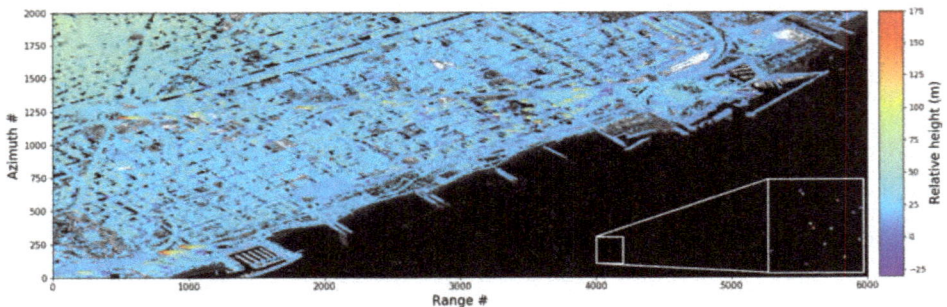

Figure 7. Point cloud of single scatterers obtained with differential SAR tomography. The detection threshold is set corresponding to $\sigma_c = 1.1$ rad under the proposed detection scheme (see Equations (52) and (53)). The color coding represents the estimated height. Some false alarms can be seen over the water surface , as highlighted in the inset.

A significant portion of the viewed sea extends over the sea, which is favorable in our context as it can be used as a test bed to conduct an empirical analysis of the false alarm rate. We perform sample coherence-based detection, as well as tomographic inversion and detection, for the range-azimuth pixels over the sea and observe the variation of the false alarm rate. These pixels constitute 1.4 million independent resolution cells. The results are shown in Figure 8. The solid lines in the figure represent different cases of tomographic inversion and detection: (1) [$\hat{a}(s, v, \eta)$; 3-D inv.]: 3-D inversion and detection on the reflectivity, α retrieved as a function of elevation (s), deformation (v) as well as thermal expansion (η) where the support in each dimension is as for the results shown in Figure 7, (2) [$\hat{a}(s, v)$; 2-D inv.]: 2-D inversion i.e., thermal expansion is not considered, (3) [$\hat{a}(s)$; 1-D inv.]: 1-D inversion, whereby the reflectivity is retrieved only along the elevation profile, (4) [$\hat{a}(s)$; reduc. supp.]: 1-D inversion with the elevation support reduced to $[-25, 50]$ m, and (5) [$\hat{a}$ (no fitting)]: 1-D inversion without the maximizations to detect peaks in the reflectivity, i.e., no fitting is performed in the parameter space to estimate the unknown elevation and deformation parameters. (6) [$\hat{\gamma}$ (no fitting)]: The dot-dashed line represents the PSI case whereby the thresholds are applied on the sample coherence without any parameter fitting. (7) [$\hat{\gamma}$ (analytical)]: The black curve with diamond symbols shows the probability of false alarm (theoretical) according to the Equation (18). The bottom x-axis in the figure shows the detection thresholds, T_γ and T_a (normalized between 0 and 1 as per Equation (53)), while the top x-axis shows the equivalent standard deviation of the residual phase according to Equation (52). The area shaded in gray indicates the region in the figure where the results may not be sufficiently accurate due to a limited number of independent range-azimuth resolution cells over the water surface. Given we have only 1.4 million of these cells, and assuming the test statistics are normally distributed over the scene, we can estimate a probability of false alarm no less than 1.1×10^{-3} with a relative absolute error of 5% for 95% of the time [49].

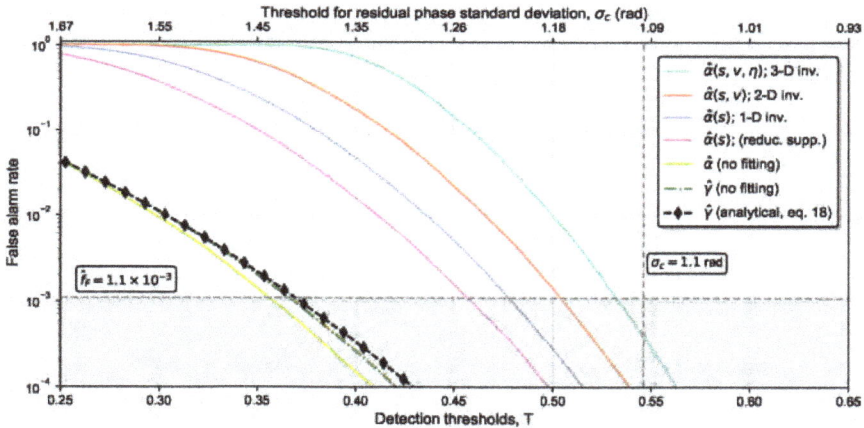

Figure 8. False alarm rate observed over the sea patch in the viewed scene at different detection thresholds. The colored solid lines represent the case of 3/2/1-D tomographic inversion. The detection is performed on the retrieved reflectivity, $|\hat{a}|$ according to Equation (53). The dot-dashed lines shows the case of PSI whereby the detection is performed on the sample coherence, $|\hat{\gamma}|$ without fitting any phase model to the observed interferometric phases.

Figure 9 shows the point cloud of single scatterers obtained with tomographic inversion and detection with $\sigma_c = 1.0$ rad. In comparison with Figure 7, we can see a reduction in the false alarms. Now, we observe only 194 false alarms. 3-D tomographic inversion has been applied with the same support in each dimension as for the results shown in Figure 7. We have estimates of height, deformation velocity as well as the phase-to-temperature sensitivity. Figure 10 shows the point cloud of double scatterers obtained with the same threshold. They are separated as lower and upper scatterers, according to the estimated height for each of the two scatterers in layover. 2.14×10^6 single scatterers and 1.01×10^4 double scatters (lower + upper) have been detected. The inset in the sub-figures in Figure 10 shows a commercial complex, namely Diagonal Mar, in focus. The red polygon encloses a high-rise building, which is partly in layover with the roof of a nearby building. These are the same test sites as in our earlier work in [27].

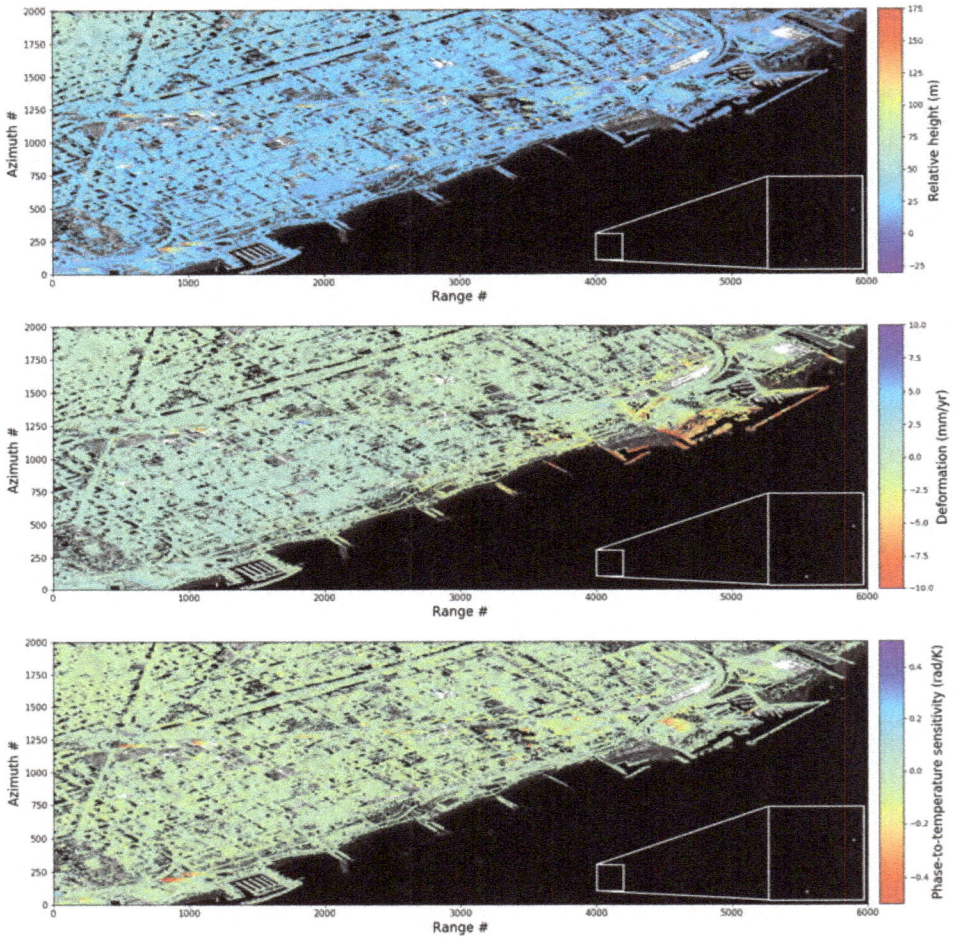

Figure 9. Point cloud of single scatterers obtained with differential SAR tomography. The detection threshold is set corresponding to $\sigma_c = 1.0$ rad under the proposed detection scheme, see Equations (52) and (53). (**Top**) Estimated height, relative to the WGS-84 reference ellipsoid. (**Middle**): Deformation velocity in the line-of-sight. (**Bottom**) Phase-to-temperature sensitivity. In comparison with Figure 7 where a more relaxed detection threshold (corresponding to $\sigma_c = 1.1$ rad) is used, fewer false alarms are observed here, as highlighted in the inset.

6. Discussion

This section provides an itemized discussion of the results presented in the previous section.

6.1. Interferometric Processing

The PSI solution, as shown in Figure 6, provides a good coverage over the viewed scene, which is typical with high resolution X-band interferometric imagery over urban areas such as Barcelona city [59,60]. The PS heights fit reasonably with actual 3-D structures, as shown for selected buildings in our earlier work in [26,27]. The PSI solution reveals deformation along the shoreline, which was partly observed in [59] as well. Several PS on high-rise buildings show temperature-dependent phase variations, which can be attributed to thermal expansion of the structures [30,35,46,61,62]. The observed coherence is high, and the estimated concentration parameters are all non-zero. With reference to Figure 1, the fact that the mean and the median value of $\hat{\kappa}$ are greater than 3 substantiates the assumption of linear normal statistics for the PS (since the approximation of von Misses as linear normal distribution is accurate to within 5% error on average).

Interestingly, we do not observe false alarms over the sea patch in the scene. This is due to the fact that we have used high stability of the backscattering amplitude and low spectral diversity as pre-classifiers to set up the initial PS candidate list. These classifiers are proxies for temporally coherent, single dominant scattering; therefore, they already preclude PS candidates from appearing on the water surface. Hence, no PSI solution has been sought (no regression fitting) on the pixels over the sea patch. In the context of tomography, these pre-classifiers cannot be used since they would tend to reject double scatterers as well.

6.2. Tomographic Processing and Empirical Analysis of False Alarms

We applied tomographic processing over the entire scene, regardless of any surface classification. The point cloud shown in Figure 7 is obtained using the same cut-off phase standard deviation, $\sigma_c = 1.1$ rad, as for the iterative least squares based PSI processing. Nevertheless, several false alarms are visible over the sea patch. A simple mask (based on SAR multi-look intensity with spatial constraints for example) could have allowed us to remove the sea patch from the processing, but we choose to show these false alarms to highlight that similar false alarms may arise (due to noise) within the urban stretch as well though they may remain unnoticed.

Figure 8, which shows the results of a false alarm analysis exclusively conducted over the sea patch, reveals that the false alarm rates can typically be higher in practice in comparison with the theoretical probability of false alarm (as the area under the upper tail of Rayleigh distribution). The maximizations (Equations (35) and (36)) allow degrees of freedom to fit the data; when the noise is fit incorrectly with the data model, it may lead to a false alarm. The false alarm rate can be seen to decrease from 3-D to 2-D inversion, as reducing the dimensionality reduces the degrees of freedom to fit the data. Similar reduction in false alarms is observed when moving from 2-D to 1-D inversion, or when we reduce the support of the elevation in case of 1-D inversion. These findings imply that in case some a priori information is available—e.g., if significant thermal expansion is not expected (as is usually the case for buildings of low height [27]), or if the support of deformation velocity can be reduced on the basis of local leveling measurements, or if the support for height corrections can be reduced given a digital surface model is available—then a reduction in false alarm rate can be achieved in practice.

Figure 8 also shows the case where no parameter fitting is performed, for both tomography as well as sample coherence based detection. The latter case, i.e., [$\hat{\gamma}$ (no fitting)], matches closely with the theoretical relationship in Equation (18), indicating that the area under the upper tail of the PDF of $|\hat{\gamma}|$ approaches that of a Rayleigh distribution. However, in the former case, i.e., [$\hat{a}$ (no fitting)], it can be observed that the estimated false alarm rate is slightly lower than the probability of false alarm according to the analytical expression for MICC-based detection, in turn implying deviation from the statistics of a Rayleighian process. It can be explained following the findings in an earlier work in [13]. In this work, a generalized likelihood ratio test (GLRT) was compared against MICC for scatterer detection in the presence of additive noise with Gaussian statistics. It is to be noted that in our case the false alarm analysis is conducted on cells over the water surface; therefore, the origin of noise in the observed SLC values lies in the backscattering characteristics (rather than phase mis-calibration). In this particular context, an additive noise model is appropriate, and, consequently, the detection for a scatterer under Equation (54) in our work becomes identical to the GLRT in [13]. It was found in [13] that the GLRT provides a lower probability of false alarm compared to MICC (as we observed). For a discussion on the performance analysis of radar detectors where the actual PDF of the amplitude of complex-valued noise/clutter deviates from Rayleigh statistics, interested readers are referred to [63–66].

Figures 9 and 10 show the single and double scatterers, respectively, detected with $\sigma_c = 1.0$ rad. As expected, we observe fewer false alarms, and at the same time fewer scatterers are detected. Double scatterers constitute <1% of the total scatterers detected over the scene. The gain in deformation sampling due to double scatterer detections [27], relative to the PSI solution, are around 2% for Diagonal Mar complex and 4% for the selected building marked in red, respectively. If the threshold is relaxed to $\sigma_c = 1.1$ rad, the gain improves to 6.4% for Diagonal Mar and 17% for the individual building.

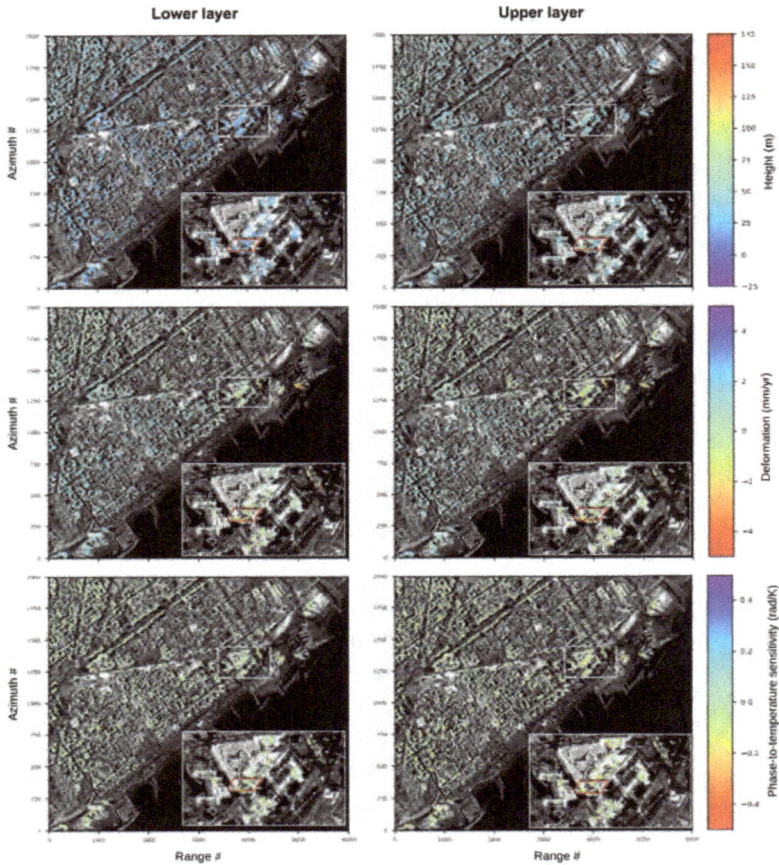

Figure 10. Point cloud of double scatterers obtained with differential SAR tomography. The detection threshold is set corresponding to $\sigma_c = 1.0$ rad under the proposed detection scheme. (**Top**) Estimated height, relative to the WGS-84 reference ellipsoid. (**Middle**) Deformation velocity in the line-of-sight. (**Bottom**) Phase-to-temperature sensitivity. The left column shows the lower layer and the right column shows the upper layer of the double scatterers, respectively. The inset focuses on a commercial complex (Diagonal Mar). The red polygon encloses a single building, part of which is in layover with a nearby building of shorter height.

The interferometric data stack and the test sites in this work are the same as in our earlier work in [27]. The detection strategies are, however, different. The sequential GLRT with cancellation (SGLRTC), as proposed in [24], was used for hypothesis testing in the earlier work. The quality of the detected scatterers was empirically evaluated only after the detection, and in turn compared with the quality of the PS (obtained independently in the prior PSI processing). In other words, the detection threshold for hypothesis testing had to be adjusted a posteriori to achieve comparable quality. The results thus obtained in [27] show a gain in deformation sampling of around 2.5% for Diagonal Mar complex and 10% for the selected building. On the other hand, the detection strategy proposed in this work allows the use of quality criterion during the hypothesis testing itself. Nonetheless, it needs to be noted that the SGLRTC and the proposed strategy are not directly comparable. SGLRTC explicitly assumes an additive noise model for SAR tomography, thus it cannot formally address multiplicative noise arising due to phase instabilities such as atmospheric disturbances. Moreover, it is a subspace method where the first scatterer is canceled out before a second scatterer is searched for [24]. Therefore, the test statistics (and the corresponding threshold settings) for double scatterer detection under the proposed detection strategy are not the same as in SGLRTC.

7. Conclusions

In the context of SAR tomography as an add-on to PSI to potentially improve deformation coverage, following the directions set in earlier works in [12,27,31], this paper reports the application of a detection strategy that allows for extending the same quality considerations to tomography as used in the prior PSI processing. In interferometric processing, the quality is typically assessed on the basis of the residual phase, either in terms of the phase dispersion (phase standard deviation) or the ensemble coherence computed using the residue of the fit. In both cases, under the proposed detection strategy, the quality parameters can be used to set up the threshold for hypothesis testing of coherent scatter candidates following tomographic inversion. Moreover, the theoretical probability of false alarm remains the same between the PSI and tomography. The paper also highlighted that while the instabilities in phase are typically modeled as additive noise, their impact on tomography is multiplicative in nature. The experiments performed in this work with simulated data consider both multiplicative noise as well as additive disturbances (clutter) in the tomographic model. It is shown that the inverse coefficient of variation is a suitable parameter to assess the probability of detection, irrespective of the origin of noise. The proposed detection strategy is also tested on real data. An assessment of the variation of the observed false alarm rates against the thresholds set according to the proposed detection strategy has been conducted. An interferometric data stack comprising 50 Terra-SAR-X acquisitions over the city of Barcelona, Spain is used. Single-look beamforming for 1/2/3-D tomographic inversion, depending on whether the phase model used considers only the scatterer height, or height plus deformation velocity, or additionally thermal expansion, is performed. The results show that higher dimensionality and larger support sizes in each dimension lead to higher false alarm rates due to larger parameter space that may incorrectly fit noise to the data model. These results also suggest that in case a priori information can reduce the dimensionality and/or support sizes, it should be adopted by the user to reduce the false alarm rate in practice. For the case of 3-D tomographic inversion, with detection thresholds set in accordance with residual phase standard deviation below 1.1 rad for the prior PSI processing, the empirically estimated false alarm rate is $<1.1 \times 10^{-3}$. The gain in deformation sampling (due to layover resolutions) is 17% for a selected high-rise building. For a commercial complex in Diagonal Mar locality, it is 6.4%. As a whole, the number of double scatterers detected in the urban scene are <1% of the total detected scatterers. These results show that, for urban areas like Barcelona, when using interferometric data stacks comprising the typical stripmap products, the application of SAR tomography as an add-on to PSI is mainly useful for a detailed analysis of selected urban zones or individual buildings in layover.

Author Contributions: M.A.S. designed and performed the experiments; developed the tomographic inversion and the proposed detection strategy, and drafted the manuscript. U.W. provided the initial PSI solution. M.A.S. and O.F. developed several routines for data preprocessing and PSI/tomography integration. All authors contributed to interpreting the results and preparing the manuscript.

Acknowledgments: This research project has been partially funded by the Swiss Space Office, State Secretariat for Education and Research of the Swiss Confederation (SER/SSO), in the frame of the "Space Technology Studies" MdP2012 project "GAMMA software module for spaceborne SAR tomography". TerraSAR-X SAR data used in this project was obtained courtesy of the German Aerospace Center DLR under proposal MTH1717. SRTM is copyrighted by USGS.

Conflicts of Interest: The authors declare no conflict of interest. The external funding sponsors had no role in the design of the study; in the collection, analyses, or interpretation of data; in the writing of the manuscript, and in the decision to publish the results.

Abbreviations

The following abbreviations are used in this manuscript:

APS	Atmospheric phase screen
DEM	Digital elevation model
GLRT	Generalized likelihood ratio test
iCV	Inverse coefficient of variation
IPTA	Interferometric point target analysis
MICC	Multi-interferogram complex coherence
PDF	Probability density function
PS	Persistent scatterer
PSF	Point spread function
PSI	Persistent scatterer interferometry
SAR	Synthetic aperture radar
SCR	Signal-to-clutter ratio
SGLRTC	Sequential generalized likelihood ratio test with cancellation
SLC	Single-look complex

Appendix A

The equality in Equation (51) holds when all elements in the vector y are identical. In our context, theoretically, it occurs for the case of a single point scatterer in the given range-azimuth resolution cell [31]. Interestingly, it is the PSI case wherein the PS is defined to be a single point-like scatterer. It can be explained by considering the SLC values as samples of the Fourier spectrum of the target situated along the elevation axis [14]. In case of point (dirac delta) scattering, the absolute value of the spectrum is a constant, and therefore, all samples (magnitude of the SLCs) are identical. Conversely, for a target that is extended continuously along the elevation axis, albeit deterministically, its spectrum is delta-like. In other words, the target can be considered as closely spaced sequence of several point-like scatterers along the elevation axis, in the same range-azimuth resolution cell. This is the case when the vector y tends to be 1-sparse, and, in turn, the ratio $\|y\|_1 / \|y\|_2$ approaches 1. A real example of such an extended scatterer can be a mountain slope with a nearly zero local incidence angle (considering it to be vegetation-free and exhibiting a stable response).

References

1. Ferretti, A.; Prati, C.; Rocca, F. Permanent scatterers in SAR interferometry. *IEEE Trans. Geosci. Remote Sens.* **2001**, *39*, 8–20. [CrossRef]
2. Berardino, P.; Fornaro, G.; Lanari, R.; Sansosti, E. A new algorithm for surface deformation monitoring based on small baseline differential SAR interferograms. *IEEE Trans. Geosci. Remote Sens.* **2002**, *40*, 2375–2383. [CrossRef]

3. Werner, C.; Wegmüller, U.; Strozzi, T.; Wiesmann, A. Interferometric point target analysis for deformation mapping. In Proceedings of the IEEE International Geoscience Remote Sensing Symposium, Toulouse, France, 21–25 July 2003; pp. 4362–4364.

4. Hooper, A. A new method for measuring deformation on volcanoes and other natural terrains using InSAR persistent scatterers. *Geophys. Res. Lett.* **2004**, *31*. [CrossRef]

5. Kampes, B. *Radar Interferometry: Persistent Scatterer Technique*; Springer: Dordrecht, The Netherlands, 2006.

6. Blanco-Sánchez, P.; Mallorquí, J.J.; Duque, S.; Monells, D. The Coherent Pixels Technique (CPT): An advanced DInSAR technique for nonlinear deformation monitoring. *Pure Appl. Geophys.* **2008**, *165*, 1167–1193. [CrossRef]

7. Crosetto, M.; Monserrat, O.; Cuevas-González, M.; Devanthéry, N.; Crippa, B. Persistent scatterer interferometry: A review. *ISPRS J. Photogramm.* **2015**, *115*, 78–89. [CrossRef]

8. Wegmuller, U.; Walter, D.; Spreckels, V.; Werner, C. Nonuniform ground motion monitoring with TerraSAR-X persistent scatterer interferometry. *IEEE Trans. Geosci. Remote Sens.* **2010**, *48*, 895–904. [CrossRef]

9. Werner, C.; Wegmuller, U.; Wiesmann, A.; Strozzi, T. Interferometric point target analysis with JERS-1 L-band SAR data. In Proceedings of the IEEE International Geoscience Remote Sensing Symposium, Toulouse, France, 21–25 July 2003; Volume 7, pp. 4359–4361.

10. Henry, C.; Souyris, J.; Marthon, P. Target detection and analysis based on spectral analysis of a SAR image: A simulation approach. In Proceedings of the IEEE International Geoscience Remote Sensing Symposium, Toulouse, France, 21–25 July 2003; Volume 3, pp. 2005–2007.

11. Souyris, J.C.; Henry, C.; Adragna, F. On the use of complex SAR image spectral analysis for target detection: assessment of polarimetry. *IEEE Trans. Geosci. Remote Sens.* **2003**, *41*, 2725–2734. [CrossRef]

12. Colesanti, C.; Ferretti, A.; Novali, F.; Prati, C.; Rocca, F. SAR monitoring of progressive and seasonal ground deformation using the permanent scatterers technique. *IEEE Trans. Geosci. Remote Sens.* **2003**, *41*, 1685–1701. [CrossRef]

13. De Maio, A.; Fornaro, G.; Pauciullo, A. Detection of single scatterers in multidimensional SAR imaging. *IEEE Trans. Geosci. Remote Sens.* **2009**, *47*, 2284–2297. [CrossRef]

14. Reigber, A.; Moreira, A. First demonstration of airborne SAR tomography using multibaseline L-Band data. *IEEE Trans. Geosci. Remote Sens.* **2000**, *38*, 2142–2152. [CrossRef]

15. Gini, F.; Lombardini, F.; Montanari, M. Layover solution in multibaseline SAR interferometry. *IEEE Trans. Aerosp. Electron. Syst.* **2002**, *38*, 1344–1356. [CrossRef]

16. Fornaro, G.; Serafino, F.; Soldovieri, F. Three-dimensional focusing with multipass SAR data. *IEEE Trans. Geosci. Remote Sens.* **2003**, *41*, 507–517. [CrossRef]

17. Frey, O.; Meier, E. 3D time-domain SAR imaging of a forest using airborne multibaseline data at L-and P-bands. *IEEE Trans. Geosci. Remote Sens.* **2011**, *49*, 3660–3664. [CrossRef]

18. Ferretti, A.; Bianchi, M.; Prati, C.; Rocca, F. Higher-order permanent scatterers analysis. *EURASIP J. Adv. Signal Process.* **2005**, 3231–3242. [CrossRef]

19. Lombardini, F.; Montanari, M.; Gini, F. Reflectivity estimation for multibaseline interferometric radar imaging of layover extended sources. *IEEE Trans. Signal Process.* **2003**, *51*, 1508–1519. [CrossRef]

20. Lombardini, F.; Cai, F.; Pasculli, D. Spaceborne 3D SAR tomography for analyzing garbled urban scenarios: Single-look superresolution advances and experiments. *IEEE J. Sel. Top. Appl. Earth Obs. Remote Sens.* **2013**, *6*, 960–968. [CrossRef]

21. Zhu, X.; Bamler, R. Very high resolution spaceborne SAR tomography in urban environment. *IEEE Trans. Geosci. Remote Sens.* **2010**, *48*, 4296–4308. [CrossRef]

22. Fornaro, G.; Lombardini, F.; Pauciullo, A.; Reale, D.; Viviani, F. Tomographic processing of interferometric SAR Data: Developments, applications, and future research perspectives. *IEEE Signal Process. Mag.* **2014**, *31*, 41–50. [CrossRef]

23. Lombardini, F. Differential tomography: A new framework for SAR interferometry. *IEEE Trans. Geosci. Remote Sens.* **2005**, *43*, 37–44. [CrossRef]

24. Pauciullo, A.; Reale, D.; De Maio, A.; Fornaro, G. Detection of double scatterers in SAR tomography. *IEEE Trans. Geosci. Remote Sens.* **2012**, *50*, 3567–3586. [CrossRef]

25. Zhu, X.; Bamler, R. Compressive sensing for high resolution differential SAR tomography–the SL1MMER algorithm. In Proceedings of the IEEE International Geoscience Remote Sensing Symposium, Honolulu, HI, USA, 25–30 July 2010; pp. 17–20.

26. Siddique, M.; Hajnsek, I.; Wegmüller, U.; Frey, O. Towards the integration of SAR tomography and PSI for improved deformation assessment in urban areas. In Proceedings of the ESA FRINGE Workshop, Frascati, Italy, 23–27 March 2015.

27. Siddique, M.; Wegmuller, U.; Hajnsek, I.; Frey, O. Single-Look SAR tomography as an add-on to PSI for improved deformation analysis in urban areas. *IEEE Trans. Geosci. Remote Sens.* **2016**, *54*, 6119–6137. [CrossRef]

28. Siddique, M.; Hajnsek, I.; Strozzi, T.; Frey, O. On the combined use of SAR tomography and PSI for deformation analysis in layover-affected rugged alpine areas. In Proceedings of the ESA FRINGE Workshop, Helsinki, Finland, 5–9 June 2017.

29. Siddique, M.; Strozzi, T.; Hajnsek, I.; Frey, O. A case study on the use of differential SAR tomography for measuring deformation in layover areas in rugged alpine terrain. In Proceedings of the IEEE International Geoscience and Remote Sensing Symposium, Fort Worth, TX, 23–28 July 2017, pp. 5850–5853.

30. Budillon, A.; Schirinzi, G. GLRT based on support estimation for multiple scatterers detection in SAR tomography. *IEEE J. Sel. Top. Appl. Earth Obs. Remote Sens.* **2016**, *9*, 1086–1094. [CrossRef]

31. Tebaldini, S.; Guarnieri, A. On the role of phase stability in SAR multibaseline applications. *IEEE Trans. Geosci. Remote Sens.* **2010**, *48*, 2953–2966. [CrossRef]

32. Ferretti, A.; Prati, C.; Rocca, F. Nonlinear subsidence rate estimation using permanent scatterers in differential SAR interferometry. *IEEE Trans. Geosci. Remote Sens.* **2000**, *38*, 2202–2212. [CrossRef]

33. Hanssen, R.F. *Radar Interferometry: Data Interpretation and Error Analysis*; Springer Science & Business Media: Berlin, Germany, 2001; Volume 2.

34. Bamler, R.; Hartl, P. Synthetic aperture radar interferometry. *Inverse Probl.* **1998**, *14*, R1–R54. [CrossRef]

35. Wegmuller, U.; Werner, C. Mitigation of thermal expansion phase in persistent scatterer interferometry in an urban environment. In Proceedings of the Joint Urban Remote Sensing Event, Lausanne, Switzerland, 30 March–1 April 2015; pp. 1–4.

36. Goldstein, R. Atmospheric limitations to repeat-track radar interferometry. *Geophys. Res. Lett.* **1995**, *22*, 2517–2520. [CrossRef]

37. Massonnet, D.; Feigl, K.L. Discrimination of geophysical phenomena in satellite radar interferograms. *Geophys. Res. Lett.* **1995**, *22*, 1537–1540. [CrossRef]

38. Tarayre, H.; Massonnet, D. Atmospheric propagation heterogeneities revealed by ERS-1 interferometry. *Geophys. Res. Lett.* **1996**, *23*, 989–992. [CrossRef]

39. Zebker, H.A.; Rosen, P.A.; Hensley, S. Atmospheric effects in interferometric synthetic aperture radar surface deformation and topographic maps. *J. Geophys. Res. B Solid Earth* **1997**, *102*, 7547–7563. [CrossRef]

40. Jammalamadaka, S.; Sengupta, A. *Topics in Circular Statistics*; World Scientific: Singapore, 2001; Volume 5.

41. Bingham, M.S.; Mardia, K.V. Maximum likelihood characterization of the von Mises distribution. In *A Modern Course on Statistical Distributions in Scientific Work*; Patil, G.P., Kotz, S., Ord, J.K., Eds.; Springer: Dordrecht, The Netherlands, 1975; pp. 387–398.

42. von Mises, R. Uber die "Ganzzahligkeit" der Atomgewicht und verwandte Fragen. *Phys. Z.* **1918**, *19*, 490–500.

43. Mardia, K.V.; Jupp, P.E. *Directional Statistics*; John Wiley & Sons: Hoboken, NJ, USA, 2009; Volume 494.

44. Mardia, K.V.; Jupp, P.E. Fundamental theorems and distribution theory. In *Directional Statistics*; John Wiley & Sons, Inc.: Hoboken, NJ, USA, 2008; pp. 57–82.

45. Frey, O.; Hajnsek, I.; Wegmuller, U. Spaceborne SAR tomography in urban areas. In Proceedings of the IEEE International Geoscience Remote Sensing Symposium, Melbourne, Australia, 21–26 July 2013 ; pp. 69–72.

46. Reale, D.; Fornaro, G.; Pauciullo, A. Extension of 4-D SAR imaging to the monitoring of thermally dilating scatterers. *IEEE Trans. Geosci. Remote Sens.* **2013**, *51*, 5296–5306. [CrossRef]

47. Fornaro, G.; Pauciullo, A.; Reale, D.; Verde, S. Multilook SAR tomography for 3D reconstruction and monitoring of single structures applied to COSMO-SKYMED data. *IEEE J. Sel. Top. Appl. Earth Obs. Remote Sens.* **2014**, *7*, 2776–2785. [CrossRef]

48. Reigber, A.; Scheiber, R. Airborne differential SAR interferometry: First results at L-band. *IEEE Trans. Geosci. Remote Sens.* **2003**, *41*, 1516–1520. [CrossRef]

49. Kay, S.M. *Fundamentals of Statistical Signal Processing. Volume II: Detection Theory*; Prentice Hall: Upper Saddle River, NJ, USA, 1998.

50. Allaire, G.; Kaber, S.M. *Numerical Linear Algebra*; Springer: Berlin, Germany, 2008; Volume 55.

51. Wegmuller, U. Automated terrain corrected SAR geocoding. In Proceedings of the IEEE International Geoscience Remote Sensing Symposium, Hamburg, Germany, 28 June–2 July 1999; Volume 3, pp. 1712–1714.

52. Frey, O.; Santoro, M.; Werner, C.L.; Wegmuller, U. DEM-based SAR pixel area estimation for enhanced geocoding refinement and radiometric normalization. *IEEE Geosci. Remote Sens. Lett.* **2013**, *10*, 48–52. [CrossRef]

53. Wegmuller, U.; Frey, O.; Werner, C. Point density reduction in persistent scatterer interferometry. In Proceedings of the Europe Conference on SAR, Nuremberg, Germany, 23–26 April 2012; pp. 673–676.

54. Crosetto, M.; Monserrat, O.; Iglesias, R.; Crippa, B. Persistent scatterer interferometry: Potential, limits and initial C- and X-band comparison. *Photogramm. Eng. Remote Sens.* **2010**, *76*, 1061–1069. [CrossRef]

55. Gernhardt, S.; Adam, N.; Eineder, M.; Bamler, R. Potential of very high resolution SAR for persistent scatterer interferometry in urban areas. *Ann. GIS* **2010**, *16*, 103–111. [CrossRef]

56. Monserrat, O.; Crosetto, M.; Cuevas, M.; Crippa, B. The thermal expansion component of persistent scatterer interferometry observations. *IEEE Geosc. Remote Sens. Lett.* **2011**, *8*, 864–868. [CrossRef]

57. Ferretti, A.; Fumagalli, A.; Novali, F.; Prati, C.; Rocca, F.; Rucci, A. A new algorithm for processing interferometric data-stacks: SqueeSAR. *IEEE Trans. Geosci. Remote Sens.* **2011**, *49*, 3460–3470. [CrossRef]

58. Fisher, N.I. *Statistical Analysis of Circular Data*; Chapter Models; Cambridge University Press: Cambridge, UK, 1993; pp. 39–58.

59. Crosetto, M.; Devanthéry, N.; Cuevas-González, M.; Monserrat, O.; Petracca, D.; Crippa, B. Systematic exploitation of the persistent scatterer interferometry potential. *Proced. Technol.* **2014**, *16*, 94–100. [CrossRef]

60. Crosetto, M.; Monserrat, O.; Cuevas-González, M.; Devanthéry, N.; Crippa, B. Analysis of X-Band very high resolution persistent scatterer interferometry data over urban areas. In Proceedings of the International Archives of the Photogrammetry, Remote Sensing and Spatial Information Sciences, Hannover, Germany, 21–24 May 2013; pp. 47–51.

61. Crosetto, M.; Monserrat, O.; Cuevas-González, M.; Devanthéry, N.; Luzi, G.; Crippa, B. Measuring thermal expansion using X-band persistent scatterer interferometry. *ISPRS J. Photogramm. Remote Sens.* **2015**, *100*, 84–91. [CrossRef]

62. Budillon, A.; Johnsy, A.C.; Schirinzi, G. Extension of a fast GLRT algorithm to 5D SAR tomography of urban areas. *Remote Sens.* **2017**, *9*, 844. [CrossRef]

63. Conte, E.; Longo, M.; Lops, M. Modelling and simulation of non-Rayleigh radar clutter. *IEE Proc. F Radar Signal Process. IET* **1991**, *138*, 121–130. [CrossRef]

64. Conte, E.; Lops, M.; Ricci, G. Asymptotically optimum radar detection in compound-Gaussian clutter. *IEEE Trans. Geosci. Remote Sens.* **1995**, *31*, 617–625. [CrossRef]

65. Eltoft, T.; Hogda, K.A. Non-Gaussian signal statistics in ocean SAR imagery. *IEEE Trans. Geosci. Remote Sens.* **1998**, *36*, 562–575. [CrossRef]

66. Ward, K.D.; Watts, S.; Tough, R.J. *Sea Clutter: Scattering, the K Distribution and Radar Performance*; IET: Stevenage, UK, 2006; Volume 20.

remote sensing

MDPI

Article

Super-Resolution Multi-Look Detection in SAR Tomography

Cosmin Dănişor [1,*], Gianfranco Fornaro [2], Antonio Pauciullo [2], Diego Reale [2] and Mihai Datcu [1,3]

[1]   Department of Applied Electronics and Information Engineering, University Politehnica of Bucharest, 060042 Bucharest, Romania; mihai.datcu@dlr.de

[2]   Institute for Electromagnetic Sensing of the Environment, National Research Council of Italy, 00185 Rome, Italy; fornaro.g@irea.cnr.it (G.F.); pauciullo.a@irea.cnr.it (A.P.); reale.d@irea.cnr.it (D.R.)

[3]   Remote Sensing Technology Institute, German Aerospace Center, D-82234 Wessling, Germany

*   Correspondence: cosmin.danisor@upb.ro; Tel.: +40-21-402-4623

Received: 29 September 2018; Accepted: 22 November 2018; Published: 27 November 2018

Abstract: Synthetic Aperture Radar (SAR) Tomography (TomoSAR) allows extending the 2-D focusing capabilities of SAR to the elevation direction, orthogonal to the azimuth and range. The multi-dimensional extension (along the time) also enables the monitoring of possible scatterer displacements. A key aspect of TomoSAR is the identification, in the presence of noise, of multiple persistent scatterers interfering within the same 2-D (azimuth range plane) pixel. To this aim, the use of multi-look has been shown to provide tangible improvements in the detection of single and double interfering persistent scatterers at the expense of a minor spatial resolution loss. Depending on the system acquisition characteristics, this operation may require also the detection of multiple scatterers interfering at distances lower than the Rayleigh resolution (super-resolution). In this work we further investigated the use of multi-look in TomoSAR for the detection of multiple scatterers located also below the Rayleigh resolution. A solution relying on the Capon filtering was first analyzed, due to its improved capabilities in the separation of the responses of multiple scatterers and sidelobe suppression. Moreover, in the framework of the Generalized Likelihood Ratio Test (GLRT), the single-look support based detection strategy recently proposed in the literature was extended to the multi-look case. Experimental results of tests carried out on two datasets acquired by TerraSAR-X and COSMO-SkyMED sensors are provided to show the performances of the proposed solution as well as the effects of the baseline span of the dataset for the detection capabilities of interfering scatterers.

Keywords: multi-look SAR tomography; multiple PS detection; Capon estimation; Generalized Likelihood Ratio Test

1. Introduction

Synthetic Aperture Radar (SAR) provides high resolution 2-D (azimuth and range) microwaves images of the illuminated scene at night and day and in all-weather conditions. This results in a systematic acquisition capability, which is an essential feature with reference to the environmental risk monitoring applications.

Advanced Differential Interferometric SAR (A-DInSAR) [1] techniques are routinely used for the accurate monitoring of slow, long-term displacements of ground targets. Among them, the class of Persistent Scatterer Interferometry (PSI) methods, typically operating at full resolution [2], relies on the assumption that the scattering response is spatially concentrated and persistent over the observation time interval, hence the name of Permanent Scatterer (PS). This scattering assumption, along with the use of a multi-acquisition model, allows accurately estimating the *scatterers parameters*, which are the residual topography (RT) and the related deformation parameters, typically given by the mean deformation velocity (MDV) and the thermal dilation (TD).

SAR Tomography (TomoSAR or 3-D imaging) [3] is a method that exploits multiple observations over different orbits typically achieved by repeated passes to synthesize a large antenna also along the direction orthogonal to the azimuth and range, referred to as elevation or slant height. The resulting fine-beam can be steered (by ground data processing) to scan the object of interest to achieve a high (meter) resolution of the scattering profile along slant height. Upon extension to the time and more dimensions (Multi-D imaging) [4], similar to PSI, TomoSAR allows the estimation of the scatterers parameters related to possible deformations (i.e., MDV and TD) as well.

A simple algorithm to reconstruct the vertical scattering profile is based on Beam-Forming (BF) [5]: its capability of separating the responses of multiple scatterers along the elevation is however limited to the so-called Rayleigh resolution [6]. Alternative methods, e.g., Compressive Sensing (CS) [7,8] Capon filtering [9] or based on proper transformation [10], allow improving such a separation and at the same time achieving better sidelobe suppression. Different from CS, Capon is typically characterized by a spatial (azimuth and range) resolution loss, although a full resolution version has been proposed [11].

Low frequency (L- or P-Band) TomoSAR applications regard the imaging of volume scattering, such as forests and ice mapping [3,12,13]. However, even in the case of limited microwaves penetration capability (e.g., X-Band), TomoSAR allows achieving improved 3-D reconstruction and monitoring in complex scenes, such as urban areas, due to the geometric distortions (layover) induced by the vertical development of the scattering [8]. In this case, TomoSAR extends PSI, improving the PSs identification and the estimation of the parameters of interest.

PSs identification, i.e., the discrimination in each image pixel of different and interfering PS mechanisms, is a key problem in PSI and TomoSAR. Such a problem can be approached in the framework of the detection theory, thus exploiting algorithms (detectors) that allow controlling the false alarm rate (FAR). Among them, the Generalized Likelihood Ratio Test (GLRT) achieves, at least asymptotically, the best Detection Rate (DR) for a given FAR. For a simple model assuming a single (dominant) PS immersed in additive white Gaussian noise, it has been shown in [14] that the GLRT statistic is simply provided by a normalization of the BF reconstruction.

The extraction of multiple (typically two) scatterers, interfering in the same image pixel, is a more complex issue and different solutions have been proposed in the literature. A simple strategy, based on sequential projections, has been proposed in [15]. Such approach, referred to as sequential GLRT with cancellation (SGLRTC), is affected by low detection performances when the interfering scatterers are located at elevation differences close to or lower than the Rayleigh resolution. It also suffers from the effects of leakage related to the influence of sidelobes.

More effective approaches for the detection of multiple scatterers are based on the joint testing of multiple directions in the data space. The support based GLRT (sup-GLRT) [16], which performs a Maximum Likelihood Estimation (MLE) of the elevation support, has been proposed as a refinement of the SGLRTC for improving the performances in the scatterers detection for elevation separation below the Rayleigh resolution. Its disadvantage is the higher computational requirement, related to the need to test the data power distribution in subspaces spanned by multiple directions. The fast sup-GLRT approach [17] provides an interesting improvement that allows retaining almost the same computation efficiency of SGLRTC while keeping, to some degree, the super-resolution capability of sup-GLRT. The rationale of this method relies on testing the signal power in higher dimensionality subspaces by sequentially adding a single direction starting from the first one provided by BF.

Inspired by the small baseline approach [18], which improves the interferometric analysis by increasing the signal to noise ratio through a spatial multi-look, recent PSI-based methods have included local averaging. The SqueeSAR approach [19] and the Component extrAction and sElectionSAR (CAESAR) [20], which is based on the Principal Component Analysis, are examples along this line.

With reference to the TomoSAR context, the use of local spatial averaging has been shown to significantly improve the detection performances as well as the accuracy in the estimation of the scatterers parameters, obviously at the expense of a spatial resolution loss. This is the case of the CAESAR based tomographic processing [21] and of the noise robust multi-look version of the single-look GLRT, hereafter referred to as M-GLRT [22]. M-GLRT is based on a proper normalization of the multi-look BF and shows (asymptotically) optimal performances on the detection of single scatterers. Following the line of the single-look SGLRTC, a M-GLRT detection scheme accounting for the presence of multiple scatterers has also been investigated [22]. Significant improvements in the detection performances have also been observed in this case. However, similar to its single-look counterpart, such a detection scheme does not allow achieving super-resolution and may suffer of the effects of leakage.

To fully address the issue of effective detection of multiple scatterers, in this work, we deepened the investigation of noise robust detection schemes with reference to their capability of separating scatterers located also below the Rayleigh resolution (super-resolution).

In the context of multi-look SAR Tomography, the Capon filtering is known to guarantee sidelobes reduction and improve separation of the responses of multiple scatterers. For this reason, the Capon based detection algorithm in [23] was first analyzed and framed in the context of the GLRT detection, to investigate the possible effects of the improved separation in terms of super-resolution detection. Limitations of such a super-resolution tomographic detection method were highlighted. Accordingly, a multi-look extension of the sup-GLRT, the M-sup-GLRT, was therefore proposed along with its fast (computationally efficient) version. It benefits from the super-resolution capabilities of the sup-GLRT as well as the improvements achieved by use of multi-look. This scheme was tested on two datasets acquired by Very High Resolution (VHR) SAR systems with different baseline spans, i.e., elevation resolutions, to appreciate on real data the different detection performances.

The paper is organized as follows. Section 2 presents a review of the multi-look TomoSAR, with a particular emphasis to the Capon and BF reconstruction algorithm. The problem of PSs detection is then addressed, summarizing the GLRT schemes already derived at full (single-look) and reduced (multi-look) spatial resolution and providing a deeper analysis of the Capon based detector proposed in [23]. The multi-look extension of the sup-GLRT detector derived in [16] is described in Section 2.5. Section 3 is devoted to the analysis of the estimation and detection results on data acquired by operative systems. Conclusions and suggestions for further developments are provided in Section 4.

2. Material and Methods

2.1. Multi-Look SAR Tomography: Problem Formulation and Filter Design

Let us consider a stack of N azimuth-range focused SAR images, co-registered with respect to a given reference (master) image. We assume that the dataset has compensated for the atmospheric phase screen (APS) as well as for possible nonlinear deformation resulting from a small scale (lower resolution) analysis. In a given image pixel, the signal is a noisy version of the integrated backscattering function over the so-called multi-dimensional parameter space.

The vector collecting the parameters of interest (i.e., RT for 3-D case; RT and MDV for 4-D case; and RT, MDV, and TD for the 5-D case) is referred to as parameter vector $\mathbf{p}$. It spans the parameter space, which is discretized in K bins, corresponding to the parameter vectors $\mathbf{p}_1, \ldots, \mathbf{p}_k$. RT can be referred to the elevation direction (orthogonal to the range and azimuth) or to the vertical direction: in the following, both definitions are used and specified according to the context.

In a given pixel, the N-length data vector, e.g., $\mathbf{g}$, is modeled as

$$\mathbf{g} = \mathbf{A}\gamma + \mathbf{w} \tag{1}$$

where the dependence on the pixel has been omitted for sake of simplicity. In Equation (1), the K-length vector

$$\gamma = [\gamma_1, \ldots, \gamma_K]^T \tag{2}$$

where $(\cdot)^T$ is the transposition operator, collects the samples of the backscattering distribution function over the bins (backscattering coefficients), whereas $\mathbf{w}$ is the additive noise contribution. Furthermore, $\mathbf{A}$ is the $N \times K$ system matrix whose columns are referred to as *steering vectors*.

With reference to the kth bin, the corresponding steering vector, e.g., $\mathbf{a}(\mathbf{p}_k)$, is a structured versor ($\|\mathbf{a}(\mathbf{p}_k)\| = 1$) whose nth component is

$$\{\mathbf{a}(\mathbf{p}_k)\}_n = \frac{1}{\sqrt{N}} \exp\left(-j2\pi \boldsymbol{\xi}_n^T \mathbf{p}_k\right) \tag{3}$$

where $\boldsymbol{\xi}_n$ is the vector collecting the Fourier mate variables of the parameter vector. Such variables, whose spans determine the Rayleigh resolution (i.e., the resolution capability of the imaging system), depend on the adopted system parameters (transmitted wavelength, spatial baseline distribution, acquisition epochs, and so on) (see references [4,24,25] for the 3-D, 4-D, and 5-D cases, respectively).

At this point, it is worth noting that Equation (1) models the signal component as a superposition over the bins of contributions (steering vectors) weighted according to the corresponding backscattering coefficients. On the one hand, along the elevation dimension, a real scenario can certainly involve the superposition of contributions associated to different bins corresponding to physically distinct scatterers. The same reasoning cannot be extended along the velocity and thermal dilation. For instance, for the 4-D case, the velocity can be interpreted as a spectral variable describing the harmonic content composing a generic, non-linear scatterer deformation. For PSI/TomoSAR it follows that, along the velocity (4-D) and thermal dilation (5-D) directions, the presence of backscattering distribution over multiple bins does not imply the presence of multiple (physical) PS, but only a spreading of the backscattering contribution associated with a deformation of a (physical) PS that cannot be described by a linear motion (4-D) or linear motion and a thermal dilation (5-D) according to the available temperatures. For the sequel, we assume that, for a given elevation, the backscattering is impulsive along the velocity and thermal dilation directions.

It is as well as understood that Equation (1) neglects the presence of decorrelation across the data-stack.

From a statistical point of view, the data vector is typically modeled as a zero-mean complex circular Gaussian random vector, with covariance matrix

$$\mathbf{R_g} = E(\mathbf{g}\mathbf{g}^H) = \sum_{k=1}^{K} \sigma_{\gamma_k}^2 \mathbf{a}(\mathbf{p}_k)\mathbf{a}^H(\mathbf{p}_k) + \sigma_w^2 \mathbf{I}_N \tag{4}$$

where $\sigma_{\gamma_k}^2$ and σ_w^2 are the variance of the backscattering coefficient corresponding to the kth bin contribution and the power spectral density (PSD) of the (white) noise contribution, respectively; $E(\cdot)$ and $\mathbf{I}_N$ are the statistical expectation operator and the $N \times N$ identity matrix, respectively; and $(\cdot)^H$ stands for hermitian operator. Notice that, according to the previous assumption on the backscattering distribution, the variance $\sigma_{\gamma_k}^2$ turns out to be concentrated in a single bin along the additional directions with respect to the elevation (4-D/5-D spaces).

Multi-look SAR tomography is aimed at reconstructing, pixel by pixel, the backscattering distribution along the bins (γ) from a set of L independent and homogeneous looks, e.g., $\mathbf{g}_1, \ldots, \mathbf{g}_L$.

A proper filter, e.g., $\mathbf{h}_k$, is exploited to carry out, look by look, an estimate $\hat{\gamma}_{k,l} = \mathbf{h}_k^H \mathbf{g}_l$ of the backscattering coefficient γ_k. To mitigate the noise effect, such an estimate is subsequently averaged over all the looks, although this leads to an unavoidable spatial (range-azimuth) resolution loss. The multi-look reconstruction is thus

$$|\hat{\gamma}_k|^2 = \mathbf{h}_k^H \hat{\mathbf{R}}_{\mathbf{g}} \mathbf{h}_k \tag{5}$$

where

$$\hat{R}_g = \frac{1}{L}\sum_{l=1}^{L} g_l g_l^H \tag{6}$$

is the sampling covariance matrix of the data. It is worth noting that, under the Gaussian assumption, the sampling covariance matrix in Equation (6) is also the Maximum Likelihood Estimate (MLE) of the statistical covariance matrix in Equation (4).

Since the data and the unknowns in Equation (1) are related by a (typically non-uniformly sampled) Fourier operator, some filter design criteria coming from the spectral estimation theory have been effectively exploited. Among them, the minimum output energy (MOE) criterion allows recovering the spectral component corresponding to the kth bin while limiting as much as possible the effects of interfering contribution. The problem is cast as:

$$h_k = \operatorname*{argmin}_{\zeta} \zeta^H R_g \zeta$$
$$\text{subject to } \zeta^H a(p_k) = 1 \tag{7}$$

R_g being the statistical covariance matrix in Equation (4).

The Capon filter [26] is the solution of the problem in Equation (7). Standard Lagrangian optimizations lead to the expression

$$h_C(p_k) = \frac{R_g^{-1} a(p_k)}{a^H(p_k) R_g^{-1} a(p_k)} \tag{8}$$

showing that the Capon is an adaptive (data dependent) filter, because of the presence of the (inverse) statistical covariance matrix. In practical situations, the matrix R_g is unknown and, therefore, its sampling (multi-look) estimation $\hat{R}_g$, defined as in Equation (6), is exploited. Accordingly, by substituting the filter expression in Equation (8) within the multi-look reconstruction in Equation (5), the Capon reconstruction can be written as:

$$|\hat{\gamma}_k|_C^2 = \frac{1}{a^H(p_k) \hat{R}_g^{-1} a(p_k)} \tag{9}$$

A consideration is now in order.

The use of the inverse sampling covariance matrix makes the Capon filter to be intrinsically a multi-look processing. In other words, the Capon reconstruction in Equation (9) cannot be straightforwardly specialized to the single-look case, although a single-look Capon based TomoSAR algorithm has been proposed in paper [11].

A much simpler nonparametric filter usually exploited for the reconstruction of the backscattering profile is Beam Forming (BF). Interestingly, BF can be considered as the solution of the problem in Equation (7) when the data are assumed to be a white process, that is, when R_g is proportional to the identity matrix. Such a condition leads in fact to the solution:

$$h_{BF}(p_k) = a(p_k) \tag{10}$$

which provides the multi-look reconstruction

$$|\hat{\gamma}_k|_{BF}^2 = a^H(p_k) \hat{R}_g a(p_k) \tag{11}$$

obtained by substituting Equation (10) into Equation (5).

Differently from the Capon, the BF in Equation (10) is a non-adaptive (data independent filter). To emphasize this aspect, the Capon is sometimes referred to as adaptive BF (ABF). Moreover, the BF reconstruction in Equation (11) can be straightforwardly specialized to the single-look case ($L = 1$), which makes it preferable to the Capon when a full spatial resolution analysis is required.

However, despite its implementation complexity compared to the BF, the ABF is expected to provide better sidelobes suppression and, as a consequence, mitigation of the leakage between interfering scatterers as well as (tomographic) super-resolution [5,23].

2.2. Multi-Look Detection in SAR Tomography: Problem Formulation and Solution Strategies

The assumed correspondence between (physical) PSs and bins allows identifying the presence of the stronger PSs interfering in the same pixel and estimating their tomographic parameters by selecting the highest peaks of the tomographic reconstruction.

However, the disturbance (noise and clutter) in the processed data along with the leakage level introduced by the exploited reconstruction technique could determine a misinterpretation of the results that makes necessary a further processing aimed at testing the reliability of the revealed PSs.

In this context is framed the TomoSAR detection problem, which in the multi-look case consists of determining, pixel by pixel, the number $m \leq M$ of present PSs and estimating the corresponding parameters vectors from a set of L independent and homogeneous looks. Such a problem can be conveniently cast in terms of the multiple (composite) hypothesis test

$$\mathcal{H}_0 : \mathbf{g}_l = \mathbf{w}_l$$
$$\mathcal{H}_m : \mathbf{g}_l = \mathbf{A}_m \gamma_{m,l} + \mathbf{w}_l \tag{12}$$
$$l = 1, \ldots, L \text{ and } m = 1, \ldots, M$$

where $\mathbf{g}_1, \ldots, \mathbf{g}_L$ are the exploited looks. The mth hypothesis in Equation (12) assumes the presence of m PSs, characterized by the (unknown) parameters vectors $\mathbf{p}_1, \ldots, \mathbf{p}_m$ which are look-independent because of the looks homogeneity. The corresponding steering vectors are collected by the system matrix

$$\mathbf{A}_m = [\mathbf{a}(\mathbf{p}_1), \ldots, \mathbf{a}(\mathbf{p}_m)] \tag{13}$$

whereas the backscattering coefficients form the look-dependent vector

$$\gamma_{m,l} = [\gamma_{1,l}, \ldots, \gamma_{m,l}]^T \tag{14}$$

Finally, each look is corrupted by an additive noise contribution $\mathbf{w}_l$, usually modelled as a white complex circular Gaussian random vectors, with (unknown) PSD σ_w^2.

In the following, Equation we deal with the problem of detecting up to two interfering scatterers (i.e., $M \leq 2$).

From a statistical point of view, the joint probability density function (pdf) of the looks can be modeled according to two different complex multivariate Gaussian distributions. The first one leads to the *zero-mean* model, which characterizes the backscattering coefficients in Equation (14) as uncorrelated zero-mean complex circular Gaussian random variables with (unknown) variances $\sigma_{\gamma_k}^2$, $k = 1, \ldots, m$ The second one corresponds to the *nonzero-mean* model, which considers, instead, the backscattering coefficients in Equation (14) as (unknown) deterministic parameters.

Different detection strategies can be followed.

A first possibility is to start from the tomographic reconstruction and exploit proper indexes to extract the information about peaks corresponding to possible persistent scatterers. In this case, the exploited reconstruction technique plays a key role in terms of achievable tomographic resolution and estimation accuracy: a particular interest is thus for the Capon based algorithms.

A second strategy is strictly framed in the detection theory context and exploits schemes based on the Generalized Likelihood Ratio Test (GLRT). In this case, according to the assumed statistical model, the estimates of the unknown parameters are based on the Least Squares (LS) criterion that, in the case of single scatterers, provides the same solution achievable by selecting the highest peak of the BF reconstruction. Moreover, this strategy also has the advantage of allowing to control the false alarm rate (FAR).

2.3. GLRT Detection

The multi-look GLRT for the binary hypotheses test $(\mathcal{H}_i, \mathcal{H}_j)$ is:

$$\frac{\max_{\theta_i} f(\mathbf{g}_1, \ldots, \mathbf{g}_L; \theta_i | \mathcal{H}_i)}{\max_{\theta_j} f(\mathbf{g}_1, \ldots, \mathbf{g}_L; \theta_j | \mathcal{H}_j)} \underset{\mathcal{H}_j}{\overset{\mathcal{H}_i}{\gtrless}} T \tag{15}$$

where, under $\mathcal{H}_k$ $(k = i, j)$, $f(\cdot|\mathcal{H}_k)$ is the joint pdf of the looks and θ_k is the vector collecting all the unknown parameters. Moreover, T is the detection threshold, set according to the desired FAR.

As for the detection of single scatterers, the single-look and, more recently, the multi-look GLRT detector for the binary hypothesis test $(\mathcal{H}_1, \mathcal{H}_0)$ have been derived in references [14,22] respectively.

By assuming the zero-mean model for the joint pdf of the looks, the test in Equation (15) leads to the multi-look GLRT detector [22]:

$$\frac{\sum_{l=1}^{L} |\mathbf{g}_l^H \mathbf{a}(\hat{\mathbf{p}})|^2}{\sum_{l=1}^{L} \|\mathbf{g}_l\|^2} = \frac{\mathbf{a}^H(\hat{\mathbf{p}})\hat{\mathbf{R}}_{\mathbf{g}}\mathbf{a}(\hat{\mathbf{p}})}{\text{tr}(\hat{\mathbf{R}}_{\mathbf{g}})} \underset{\mathcal{H}_0}{\overset{\mathcal{H}_1}{\gtrless}} T \tag{16}$$

where

$$\hat{\mathbf{p}} = \arg\max_{\zeta} \left[\mathbf{a}^H(\zeta)\hat{\mathbf{R}}_{\mathbf{g}}\mathbf{a}(\zeta) \right] \tag{17}$$

is the multi-look MLE under $\mathcal{H}_1$ of the parameter vector $\mathbf{p}$ associated with the present PS and $\hat{\mathbf{R}}_{\mathbf{g}}$ is the sample covariance matrix, defined as in Equation (6).

It is worth noting that the test in Equation (16) is a BF-based detector, since its statistic represents the highest normalized peak (belonging to the interval [0, 1]) of the BF reconstruction in Equation (11). It can be rewritten as

$$\rho_{BF} = \frac{\text{tr}[\hat{\mathbf{R}}_{\mathbf{g}}\mathbf{a}(\hat{\mathbf{p}})\mathbf{a}^H(\hat{\mathbf{p}})]}{\text{tr}(\hat{\mathbf{R}}_{\mathbf{g}})\text{tr}[\mathbf{a}(\hat{\mathbf{p}})\mathbf{a}^H(\hat{\mathbf{p}})]} \tag{18}$$

which is the correlation index (according to the Frobenius inner product) between the estimated (sample) covariance matrix $\hat{\mathbf{R}}_{\mathbf{g}}$ and the estimated, so-called *signature matrix* $\mathbf{a}(\hat{\mathbf{p}})\mathbf{a}^H(\hat{\mathbf{p}})$, associated with the ideal response *(signature)* $\mathbf{a}(\hat{\mathbf{p}})$ of a PS with parameter vector $\hat{\mathbf{p}}$ [22]. Equation (18) is also referred to as BF correlation index, because of the relation $\mathbf{h}_{BF}(\hat{\mathbf{p}}) = \mathbf{a}(\hat{\mathbf{p}})$ that defines the BF filter at the bin $\hat{\mathbf{p}}$.

It can be easily shown that, under $\mathcal{H}_1$, the test statistic ρ_{BF} in Equation (18) increases with the signal to noise ratio (SNR) of the present PS, defined as

$$\text{SNR} = \frac{\sigma_\gamma^2}{\sigma_w^2} \tag{19}$$

As final remarks, it is worth underlining that the detector in Equation (16) has the CFAR property that reflects in the possibility to use the same detection threshold for processing all pixels with the same (constant) FAR. Moreover, the reduction of the noise effect induced by the multi-look processing translates in a higher detection rate (DR). It has been demonstrated, indeed, that, for a fixed FAR, the DR increases with the number L of the exploited looks [22].

As for the double scatterers case, a GLRT-based detector for the ternary hypothesis test $(\mathcal{H}_0, \mathcal{H}_1, \mathcal{H}_2)$ has been proposed for the single-look case in [15] and subsequently extended to the multi-look case in [22]. It is a simple and efficient two-stage scheme, referred to as sequential GLRT with cancelation (SGLRTC), that sequentially tests the pairs of hypotheses $(\mathcal{H}_2, \overline{\mathcal{H}}_2)$ and $(\mathcal{H}_1, \mathcal{H}_0)$, where $\overline{\mathcal{H}}_2$ denotes the complement of $\mathcal{H}_2$ (i.e., $\mathcal{H}_1$ or $\mathcal{H}_0$). At the first stage, aimed at testing the presence of two scatterers, the decision rule from Equation (16) is applied to the vector obtained by canceling from the data the contribution of the dominant scatterer. The latter is in turn provided by the highest peak of the BF reconstruction. If hypothesis $\overline{\mathcal{H}}_2$ is selected, the final decision is demanded to the second stage that select among the null and the single scatterer hypothesis, again according to the rule in Equation (16).

The SGLRTC results to be in practice CFAR. Moreover, from the computational point of view, it is a very efficient detection scheme, since the cancelation step allows just doubling the effort required by the GLRT detector for single scatterers. Unfortunately, this is paid for by a reduction of the (tomographic) resolution capabilities, since the cancellation process does not allow locating scatterers whose distance in the parameter space is below the Rayleigh resolution. Moreover, it has limited capabilities in contrasting the leakage.

To provide super-resolution capabilities, a single-look double-stage GLRT-based detector, referred to as sup-GLRT (since it deals with a signal support estimation problem), has been proposed in [16]. Similar to the single-look SGLRTC, the sup-GLRT assumes the nonzero-mean model for the pdf of the data. However, differently from the SGLRTC that implements the binary tests starting from the higher hypothesis, the sup-GLRT sequentially tests the pairs of hypotheses $(\mathcal{H}_0, \overline{\mathcal{H}}_0)$ and $(\mathcal{H}_1, \mathcal{H}_2)$ ($\overline{\mathcal{H}}_0$ being the complement of $\mathcal{H}_0$), thus starting from the lower hypothesis. More specifically, letting $\mathbf{g}$ be the (single) exploited look, the first stage implements the rule

$$\frac{\mathbf{g}^H \mathbf{P}(\hat{\mathbf{p}}_1, \hat{\mathbf{p}}_2)\mathbf{g}}{\|\mathbf{g}\|^2} = 1 - \frac{\mathbf{g}^H \mathbf{P}^\perp(\hat{\mathbf{p}}_1, \hat{\mathbf{p}}_2)\mathbf{g}}{\|\mathbf{g}\|^2} \underset{\mathcal{H}_0}{\overset{\overline{\mathcal{H}}_0}{\gtrless}} T_1 \tag{20}$$

where

$$(\hat{\mathbf{p}}_1, \hat{\mathbf{p}}_2) = \operatorname*{argmin}_{\zeta_1, \zeta_2}\left[\mathbf{g}^H \mathbf{P}^\perp(\zeta_1, \zeta_2)\mathbf{g}\right] \tag{21}$$

is the joint MLE of the parameter vectors associated with the two present PSs under $\mathcal{H}_2$,

$$\mathbf{P}^\perp(\hat{\mathbf{p}}_1, \hat{\mathbf{p}}_2) = \mathbf{I}_N - \mathbf{P}(\hat{\mathbf{p}}_1, \hat{\mathbf{p}}_2) \tag{22}$$

is the projector in the noise subspace, which is the orthogonal complement to the signal subspace spanned by the two estimated directions $\mathbf{a}(\hat{\mathbf{p}}_1)$ and $\mathbf{a}(\hat{\mathbf{p}}_2)$, and

$$\mathbf{P}(\hat{\mathbf{p}}_1, \hat{\mathbf{p}}_2) = \mathbf{A}(\hat{\mathbf{p}}_1, \hat{\mathbf{p}}_2)[\mathbf{A}^H(\hat{\mathbf{p}}_1, \hat{\mathbf{p}}_2)\mathbf{A}(\hat{\mathbf{p}}_1, \hat{\mathbf{p}}_2)]^{-1}\mathbf{A}^H(\hat{\mathbf{p}}_1, \hat{\mathbf{p}}_2) \tag{23}$$

is the projector in the signal subspace, $\mathbf{A}(\hat{\mathbf{p}}_1, \hat{\mathbf{p}}_2)$ being the matrix collecting the two estimated directions.

The second stage acts when the first one selects $\overline{\mathcal{H}}_0$. It implements the rule

$$1 - \frac{\mathbf{g}^H \mathbf{P}^\perp(\hat{\mathbf{p}}_1, \hat{\mathbf{p}}_2)\mathbf{g}}{\mathbf{g}^H \mathbf{P}^\perp(\tilde{\mathbf{p}}_1)\mathbf{g}} \underset{\mathcal{H}_1}{\overset{\mathcal{H}_2}{\gtrless}} T_2 \tag{24}$$

where $\hat{\mathbf{p}}_1$ and $\hat{\mathbf{p}}_2$ are still given by Equation (21), and

$$\tilde{\mathbf{p}}_1 = \operatorname*{argmin}_{\zeta}\left[\mathbf{g}^H \mathbf{P}^\perp(\zeta)\mathbf{g}\right] \tag{25}$$

is the MLE of the parameter vector associated with the present PS under $\mathcal{H}_1$. Moreover,

$$\mathbf{P}^\perp(\tilde{\mathbf{p}}) = \mathbf{I}_N - \mathbf{a}(\tilde{\mathbf{p}}_1)\mathbf{a}^H(\tilde{\mathbf{p}}_1) \tag{26}$$

is the projector in the noise subspace, which is the orthogonal complement to the signal subspace spanned by the estimated direction $\mathbf{a}\left(\tilde{\mathbf{p}}_1\right)$.

The sup-GLRT is a CFAR detection scheme, and very effective compared to SGLRTC in terms of resolution capability. Indeed, by exploiting the join estimation procedure from Equation (21), it is also able to detect scatterers whose separation in the parameter space is below the Rayleigh resolution. Moreover, for separation above the Rayleigh resolution, the avoidance of the cancellation step exploited in the SGLTRC allows improving the handling of leakage associated to sidelobes. However, the super-resolution capability is paid for by a high computational effort, which turns out to be combinatorial [16]. Accordingly, the sup-GLRT is very computationally demanding, especially when the dimensionality of the parameter space increases and its discretization is carried out with a high number of bins [27].

To overcome the high computational complexity limitation of the sup-GLRT, the so-called (single-look) fast-sup-GLRT is introduced in [17]. The basic idea is on splitting the joint estimate $(\hat{\mathbf{p}}_1, \hat{\mathbf{p}}_2)$ of the two parameters vectors, performed as in Equation (21), into two decoupled estimates $\tilde{\mathbf{p}}_1$ and $\tilde{\mathbf{p}}_2$. To this aim, the estimate $\tilde{\mathbf{p}}_1$ is firstly carried out by Equation (25). Subsequently, the estimate $\tilde{\mathbf{p}}_2$ is performed as

$$\tilde{\mathbf{p}}_2 = \arg\min_{\zeta}\left[\mathbf{g}^H\mathbf{P}^\perp(\tilde{\mathbf{p}}_1, \zeta)\mathbf{g}\right] \tag{27}$$

The reduction of the computational effort is paid for by a higher leakage effect on the dominant scatterer when a secondary scatterer is present. Nevertheless, the fast sup-GLRT has been demonstrated to achieve better detection performances with respect to the SGLRTC. This improvement depends on the fact that, differently from the cancelation carried out by the SGLRTC, the estimation procedure in Equation (27) does not constrain the steering vectors associated with $\tilde{\mathbf{p}}_1$ and $\tilde{\mathbf{p}}_2$ to be orthogonal.

2.4. Capon-Based Detection

The attractive characteristic of the Capon reconstruction related to the mitigation of the leakage effect has encouraged the derivation of multi-look Capon-based detectors which should be able to achieve better performances in presence of multiple PSs interfering in the same pixel. One possibility could be the modification of the GLRT in Equation (16), which is a BF-based detector, in a Capon-based detection scheme. To this aim, the BF correlation index from Equation (18) could be substituted with the Capon correlation index

$$\rho_C = \frac{\text{tr}[\hat{\mathbf{R}}_g \mathbf{h}_C(\hat{\mathbf{p}})\mathbf{h}_C^H(\hat{\mathbf{p}})]}{\text{tr}(\hat{\mathbf{R}}_g)\text{tr}[\mathbf{h}_C(\hat{\mathbf{p}})\mathbf{h}_C^H(\hat{\mathbf{p}})]} \tag{28}$$

where $\mathbf{h}_C(\hat{\mathbf{p}})$ is the Capon filter at the bin $\hat{\mathbf{p}}$ corresponding to the highest peak of the Capon reconstruction. Unfortunately, because of the presence of the inverse sampling covariance matrix in the filter (see Equation (8)), the use of the index in Equation (28) results to be critical in terms of numerical instability related to the finite precision number representation, especially for high values of SNR.

Another possibility is given by the multi-look Capon-based iterative detector addressed in [23]. It exploits the L looks $\mathbf{g}_1, \ldots, \mathbf{g}_L$ to sequentially test the hypotheses in Equation (12), that is

$$\mathcal{H}_m : \mathbf{g}_l = \mathbf{A}_m \gamma_{m,l} + \mathbf{w}_l \tag{29}$$

$m \geq 1$, where the matrix $\mathbf{A}_m = [\mathbf{a}(\hat{\mathbf{p}}_1), \ldots, \mathbf{a}(\hat{\mathbf{p}}_m)]$ accounts for the vector parameters $\hat{\mathbf{p}}_1, \ldots, \hat{\mathbf{p}}_m$ corresponding to the m highest peaks of the Capon reconstruction.

The mth iteration verifies the presence of a further scatterer with respect to the $m-1$ already detected in the previous one, by ending the algorithm if the additional scatterer is declared to be absent. To this aim, the following joint condition is tested

$$(\varepsilon_m < T_1) \cap (\text{SNR}_m > T_2) \tag{30}$$

where ε_m is a properly defined fitting error exploited to limit the FAR,

$$\text{SNR}_m = \frac{\sigma_{\gamma_m}^2}{\sigma_w^2} \tag{31}$$

is the SNR corresponding to the mth scatterer, and T_1 and T_2 are two fixed (independent from m) thresholds.

The fitting error is a normalized index accounting for the data contribution within the noise subspace, i.e., the orthogonal complement to the signal subspace

$$\varepsilon_m = \frac{\sum_{l=1}^{L} \left\| \mathbf{g}_l - \mathbf{A}_m \hat{\gamma}_{m,l} \right\|^2}{\sum_{l=1}^{L} \left\| \mathbf{g}_l \right\|^2} \tag{32}$$

where

$$\hat{\gamma}_{m,l} = \left(\mathbf{A}_m^H \mathbf{A}_m \right)^{-1} \mathbf{A}_m^H \mathbf{g}_l \tag{33}$$

is the least squares (LS) estimate of the vector $\gamma_{m,l}$.

It is worth noting that, by increasing the iteration number m, which represents the dimensionality of the signal subspace, the fitting error from Equation (32) decreases. Accordingly, it is enough to verify the fitting error condition just for $m = 1$, whereas for $m > 1$ the detection rule in Equation (30) reduces to

$$\text{SNR}_m > T \tag{34}$$

As for the SNR in Equation (31), the power $\sigma_{\gamma_m}^2$ of the mth scatterer is evaluated as

$$\hat{\sigma}_{\gamma_m}^2 = \frac{1}{L} \sum_{l=1}^{L} \left| \hat{\gamma}_{m,l} \right|^2 \tag{35}$$

where $\hat{\gamma}_{m,l}$ is the last (mth) component of $\hat{\gamma}_{m,l}$ in Equation (33).

Regarding to the noise power σ_w^2, in [23] the authors did not provide any information on the adopted estimation strategy. However, assuming the nonzero-mean Gaussian model for the looks (see Section 2.1), Equation (33) turns out to be the MLE of the backscattering coefficients and, thus, the MLE of the noise level can also be exploited [15]

$$\hat{\sigma}_w^2 = \frac{1}{L} \sum_{l=1}^{L} \frac{\left\| \mathbf{g}_l - \mathbf{A}_m \hat{\gamma}_l \right\|^2}{N - m} \tag{36}$$

Some considerations are now in order.

It can be shown that, for $m = 1$, the fitting error condition is equivalent to the one on the SNR, thus the test turns out to always be equivalent to that in Equation (34) or, in other words, the condition on the fitting error is redundant.

Furthermore, it can be shown that the test in Equation (34) is equivalent to the multi-look GLRT for the pair $(\mathcal{H}_{m-1}, \mathcal{H}_m)$ derived by assuming the looks following the nonzero-mean Gaussian model. However, differently from the GLRT, which exploits the MLEs of all the unknown parameters, the test in Equation (34) makes use of the parameter vectors provided by the highest peaks of the Capon reconstruction. It has been empirically demonstrated, however, that simple peaks location leads to the worst results in terms of resolution even with respect to the method based on signal orthogonal projections. Additionally, the corresponding test statistic is always lower than the one of the "full" GLRT rule (which is maximized by the MLEs).

These considerations make evident the difficulty to derive, in the context of super-resolution multi-look detection, an effective Capon-based detection scheme. On the other hand, multi-look

SGLRTC does not provide super-resolution capabilities, differently from the sup-GLRT proposed in article [16], which however has been designed for the single-look case. Accordingly, in the next subsection, we derive the multi-look version of the sup-GLRT.

2.5. Proposed Multi-Look Sup-GLRT Detection Algorithm

The single-look sup-GLRT detector and its fast implementation proposed in references [16,17], respectively, are described in Section 2.3. Such schemes carry out the estimation of the parameters vector without imposing the orthogonally of the corresponding directions, thus achieving satisfactory performances also when multiple interfering PSs below the (tomographic) Rayleigh resolution are present in the same pixel. Accordingly, to profitably exploit such feature also when the processed datasets are characterized by low SNRs, in this section, we propose the extension of the sup-GLRT detector to the multi-look case. Additionally, for the fast implementation of the multi-look sup-GLRT, a Capon reconstruction is also proposed.

Let us consider a set of L independent and homogeneous looks, modeled according to the nonzero-mean statistical characterization (see Section 2.1), which extends the model exploited in [16] to the multi-look case. By supposing the presence of up to two scatterers, the first stage of the multi-look sup-GLRT implements the decision rule

$$1 - \frac{\text{trace}\left[\mathbf{P}^{\perp}(\hat{\mathbf{p}}_1, \hat{\mathbf{p}}_2)\hat{\mathbf{R}}_{\mathbf{g}}\right]}{\text{trace}\left[\hat{\mathbf{R}}_{\mathbf{g}}\right]} \underset{\mathcal{H}_0}{\overset{\mathcal{H}_0}{\gtrless}} T \tag{37}$$

where $\hat{\mathbf{R}}_{\mathbf{g}}$, defined as in Equation (6), represents the sampled correlation matrix of the exploited looks, and $\mathbf{P}^{\perp}(\hat{\mathbf{p}}_1, \hat{\mathbf{p}}_2)$ is given by Equation (22). Moreover,

$$(\hat{\mathbf{p}}_1, \hat{\mathbf{p}}_2) = \underset{\zeta_1, \zeta_2}{\arg\min} \text{trace}[\mathbf{P}^{\perp}(\zeta_1, \zeta_2)\hat{\mathbf{R}}_{\mathbf{g}}] \tag{38}$$

is the joint MLE of the parameters vector under $\mathcal{H}_2$.

The second stage, instead, operates the decision according to the rule

$$1 - \frac{\text{trace}[\mathbf{P}^{\perp}(\hat{\mathbf{p}}_1, \hat{\mathbf{p}}_2)\hat{\mathbf{R}}_{\mathbf{g}}]}{\text{trace}[\mathbf{P}^{\perp}(\tilde{\mathbf{p}}_1)\hat{\mathbf{R}}_{\mathbf{g}}]} \underset{\mathcal{H}_1}{\overset{\mathcal{H}_2}{\gtrless}} T \tag{39}$$

where $\hat{\mathbf{p}}_1$ and $\hat{\mathbf{p}}_2$ are still given by Equation (21), and

$$\tilde{\mathbf{p}}_1 = \underset{\zeta}{\arg\min} \text{trace}[\mathbf{P}^{\perp}(\zeta)\hat{\mathbf{R}}_{\mathbf{g}}] \tag{40}$$

is the MLE of the parameter vector under $\mathcal{H}_1$.

It is worth noting that, assuming the zero-mean statistical model for the exploited looks, Equations (37) and (39) can be shown to be still the decision rules of the multi-look sup-GLRT only if $\sigma_{\gamma_1}^2 = \sigma_{\gamma_2}^2$ (scatterers with the same power level).

Similar to the single-look counterpart, a fast implementation of the proposed multi-look detector can be obtained by splitting the joint estimate $(\hat{\mathbf{p}}_1, \hat{\mathbf{p}}_2)$ into two decoupled estimates $\tilde{\mathbf{p}}_1$ and $\tilde{\mathbf{p}}_2$, the latter performed as

$$\tilde{\mathbf{p}}_2 = \underset{\zeta}{\arg\min} \text{trace}[\mathbf{P}^{\perp}(\tilde{\mathbf{p}}_1, \zeta)\hat{\mathbf{R}}_{\mathbf{g}}] \tag{41}$$

The estimate $\tilde{\mathbf{p}}_1$ is given, instead, by the position corresponding to the highest peak of the multi-look tomographic reconstruction. However, differently from the (single-look) fast implementation proposed in [17,27] which exploits the Beam-Forming reconstruction, we prefer to use the Capon reconstruction carried out on the selected looks. Indeed, the capability of the Capon filter to mitigate the leakage effect should guarantee a better estimate of the parameters vector $\tilde{\mathbf{p}}_1$ when multiple interfering PSs are actually present in the same pixel, thus improving also the subsequent estimate in Equation (41).

3. Results

To test the effectiveness of the proposed detection scheme, experiments were carried out on two datasets involving highly urbanized environments. The first dataset was acquired by the TerraSAR-X sensors over the city of Bucharest in the area of the national arena; the second one by the COSMO-SkyMED sensor on the city of Rome. Sensors and datasets characteristics are summarized in Table 1.

Table 1. Detected single and double PSs for single-look (SL) and multi-look (ML) sup-GLRT.

	TSX	CSK
number of acquisitions	32	29
time span	July 2011–December 2012	April 2011–October 2012
incidence angle	37.32°	34°
range resolution	1.17 m	1.47 m
azimuth resolution	3.3 m	3 m
Rayleigh resolution	23 m	7.89 m
wavelength	3.1 cm	3.1 cm
sensor altitude	500.54 km	628.15 km
beam ID	strip_010	H4-05
acquisition mode	stripmap	stripmap
orbit direction	descending	ascending

The two datasets were analyzed with both single- and multi-look sup-GLRT. The estimate of the parameter vector associated to the dominant was used for implementing the fast version of the detector. In the proposed multi-look solution, this vector was provided by the Capon reconstruction, which is expected to reduce the leakage effect. With this regard, before starting with the analysis of the detection performances, a comparison of the elevation estimates achieved by Capon and BF was first performed with reference to the Bucharest dataset, which is characterized by a poor vertical resolution.

The TerraSAR-X dataset over the Bucharest national arena area is composed by 32 (SLC) images of 800 (azimuth) × 1200 (range) pixels, acquired between July 2011 and December 2012 with the system operating in the stripmap, single polarization mode. Orbits are descending, the spatial 2-D resolution is 2 m in azimuth and 1 m in range. Figure 1 shows the (temporal) multi-look amplitude image corresponding to the investigated area, i.e., the result of an averaging on all the available acquisitions.

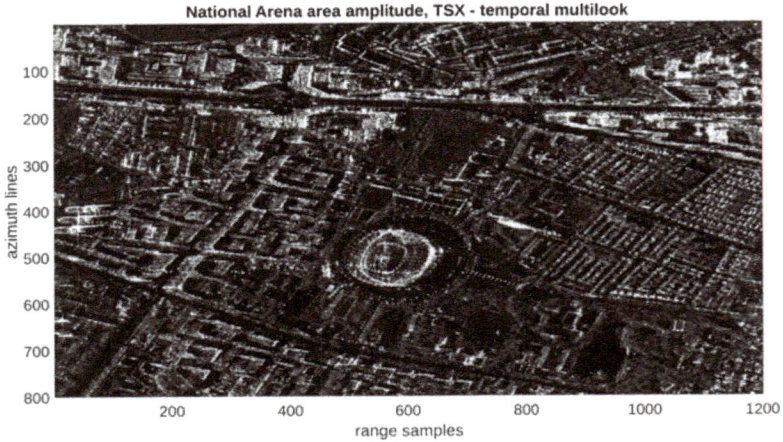

Figure 1. Amplitude of the test area from the TerraSAR-X dataset (National Arena in Bucharest, Romania), averaged across dataset's acquisitions.

The acquisitions are distributed on the so-called spatial/temporal baselines domain, as shown in Figure 2, with a span of 432 m and 528 days. The corresponding Rayleigh elevation resolution is about 23 m; this number converts vertically to about 14 m, which compares with the average building height in the area. The Rayleigh resolutions along the linear deformation rate and thermal dilation equals 1 cm/year and 0.35 mm/°C, respectively. Based on the limited spatial extension of the analyzed area, the data were calibrated for the APS by compensating for a constant phase offset.

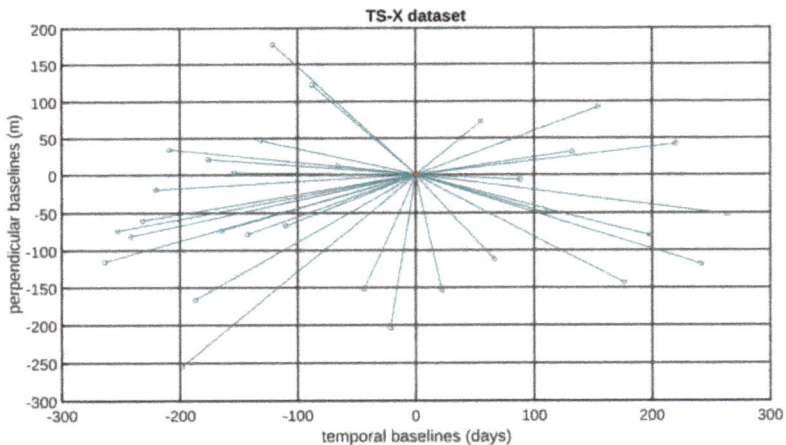

Figure 2. Distribution of the acquisitions, depicted as circles, in the temporal/spatial baseline domain. Red circle corresponds to the master acquisition.

The adaptive (spatial) multi-look operation was carried out via a Kolmogorov–Smirnov (KS) test [19,28]: The search window was set to 9 × 9 pixels and 25 pixels were selected inside each window.

The reconstructions in Equations (9) and (11) were carried out in the 4-D space, by exploiting the statistically similar looks selected as described above for the covariance matrix estimation. The RT (elevation) interval was scanned in the [−60, 60] m interval with a 3 m spacing, the MDV interval was set to [−2.5, 2.5] cm/year with a spacing of 0.25 cm/year.

The elevation map of the dominant scatterers, i.e., for each pixel, the positions of the highest peak of the tomographic reconstruction, is shown in Figure 3 for the Beam-Forming (Figure 3a) and Capon (Figure 3b). The comparison of the two maps shows that the better leakage mitigation associated with Capon leads to a generally less noisy reconstruction with respect to the plain BF. For this reason, although characterized by a higher computational cost due to the inversion of the covariance matrix, the Capon filter is considered in the implementation of the detection multi-look fast sup-GLRT scheme.

This detection scheme was then applied to select the monitored PSs. The adaptive KS test was again used for the $\hat{R}_g$ estimation. Due to the presence of metallic structures in the test area, which can be sensitive to the phenomenon of thermal dilations, the 5-D case was considered. The TD coefficient interval $[-1.6, 1.6]$ mm/$^\circ$C was scanned with a sampling step of 0.2 mm/$^\circ$C.

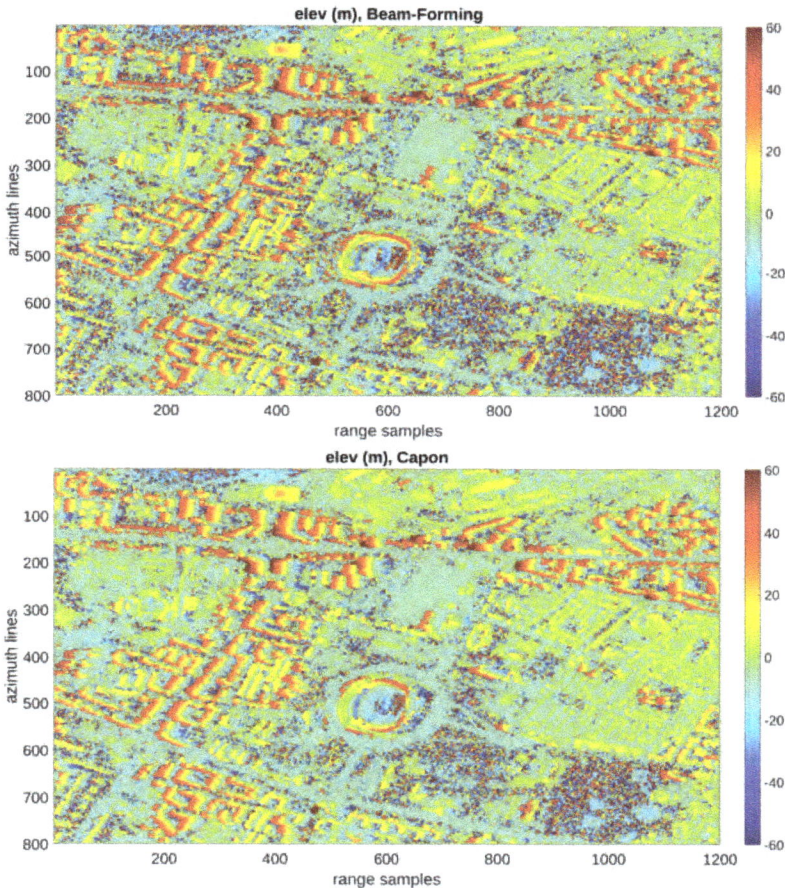

Figure 3. Maps of the estimated dominant scatterers elevation, corresponding to the peaks of the BF (**a**) and Capon (**b**) reconstructions. Colormap is in meters and set according to the estimated elevation.

Figure 4 shows the maps of the detected single (Figure 4a,b) and higher double (Figure 4c,d) PSs, when the single-look (Figure 4a,c) and multi-look (Figure 4b,d) fast sup-GLRT were used. Colormaps were set according to the estimated RT (elevation). The results show a significant increase of the detection performance achieved by the multi-look detector with respect to the single-look processing. This increase is particularly evident for the double scatterers, which were mainly identified on almost all the structures extending vertically. All pixels belonging to the top of the stadium, as well as to the roofs of the residential buildings in the neighborhood, were in fact detected as double scatterers, whereas single scatterers were mostly located on the ground. The thermal dilation map of the detected PSs, which is not reported for sake of brevity, shows that the highest values corresponded to the metallic structures, including the stadium.

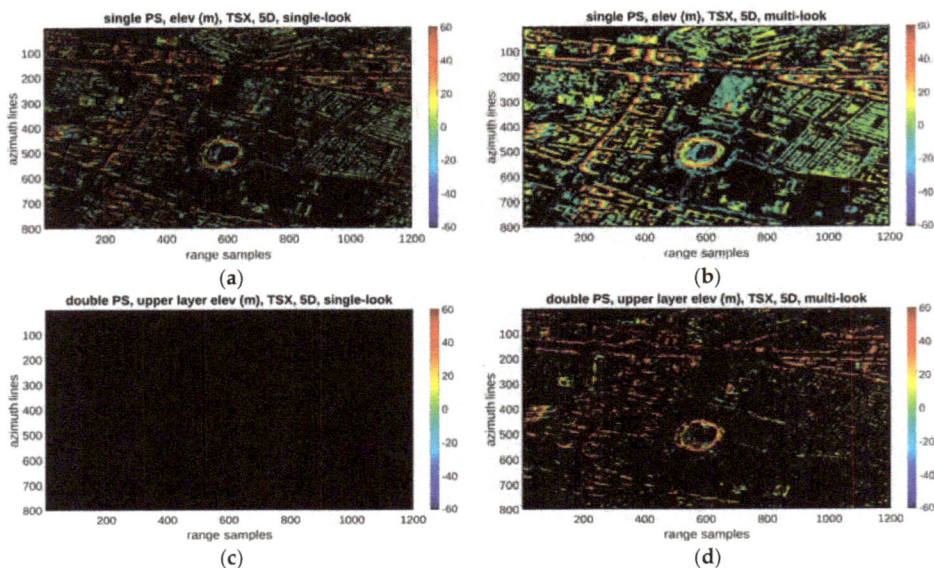

Figure 4. Distribution of the detected single scatterers (**a,b**) and higher double (**c,d**) scatterers, for single-look (**a,c**) and multi-look (**b,d**) analysis. Colormap is in meters and set according to the estimated elevation.

The COSMO-SkyMED dataset over the city of Rome is composed by 29 (SLC) images of 500 (azimuth) × 700 (range) pixels, acquired between April 2011 and October 2012 by the Cosmo-SkyMed constellation, operating in the stripmap image mode on ascending orbits, with a resolution of 3 m in azimuth and 1.45 m in range.

Figure 5 shows multi-look master image amplitude of the investigated area. The adaptive multi-look method based on the KS test was implemented also in the case of this dataset, again by setting the search window to 9 × 9 pixels and selecting 25 pixels inside each window. The acquisitions are distributed on the spatial/temporal baselines domain as in Figure 6, with a span of 1469.2 m and 556 days. The Rayleigh resolution is equal to 7.89 m in elevation and therefore much better than the Bucharest case. For linear deformation (MDV), the resolution is 1 cm/year, hence comparable to the previous case, whereas for the thermal coefficient the resolution decreases to 0.8 mm/°C.

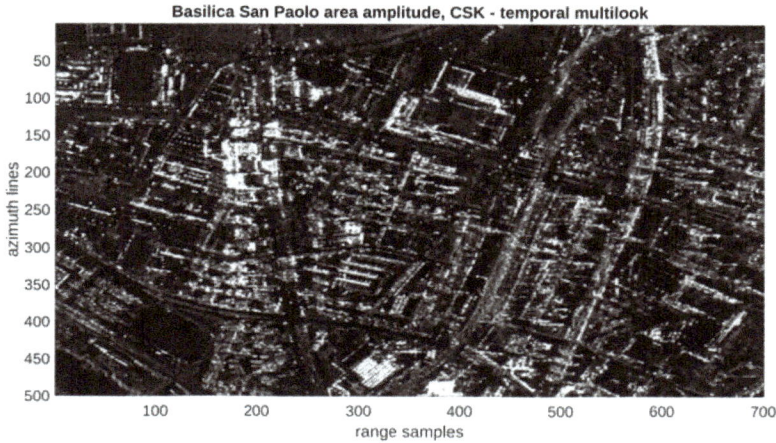

Figure 5. Amplitude of the test area from the COSMO-SkyMED dataset relevant to the Basilica S. Paolo area in Rome, Italy, averaged across dataset's acquisitions.

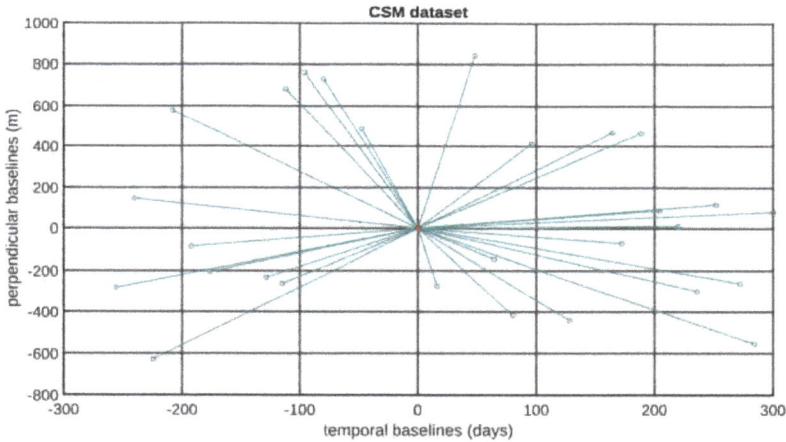

Figure 6. Distribution of the acquisitions, depicted as circles, in the temporal/spatial baseline domain. Red circle corresponds to the master acquisition.

The APS was compensated in this case by exploiting a two-scale analysis: APS was extracted at the stage of low resolution processing (see reference [5]).

The estimation and detection processes were carried in 5-D space, over a uniformly discretized grid, picked over the RT (elevation) interval [−60, 60] m with a spacing of 1 m, the MDV interval [−2.5, 2.5] cm/year with 0.25 cm/year spacing and the TD coefficient interval [−1.6, 1.6] mm/°C with a spacing of 0.4 mm/°C.

Figure 7 shows the maps of the detected single (Figure 7a,b) and higher double (Figure 7c,d) PSs, for the single-look (Figure 7a,c) and multi-look (Figure 7b,d) cases. Colormaps are set again according to the estimated RT (elevation).

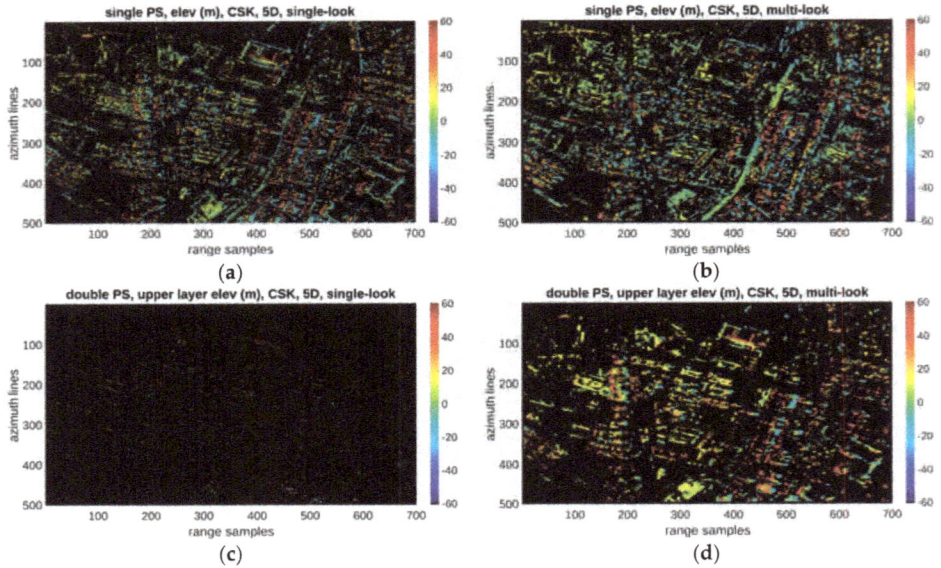

Figure 7. Distribution of the detected single scatterers (**a,b**) and higher double (**c,d**) scatterers, for single-look (**a,c**) and multi-look (**b,d**) analysis Colormap is in meters and set according to the estimated elevation.

4. Discussions

Results achieved by the processing of the two datasets shows the capabilities of multi-look sup-GLRT in providing higher densities of detected PS. Moreover, detected double scatterers well matched the building and structures developed vertically, which are subject to the effects of layover.

To quantitatively compare the detection performance of the proposed multi-look detector with that of its single-look counterpart, the number of detected single and double PSs and their percentages with respect to the total number of pixels interested by the detection of at least one stable target are synthetized in Table 2 for both processed datasets.

Table 2. Detected single and double PSs for single-look (SL) and multi-look (ML) sup-GLRT.

Dataset	PS	SL Detections	ML Detections	SL Percentage	ML Percentage
TSX	single	11,6065	25,0899	97.17%	76.7%
	double	3380	7,5934	2.83%	23.3%
CSK	single	5,3415	7,2279	91.99%	59.9%
	double	4648	4,8220	8.01%	40.1%

Table 2 shows that, with respect to the single-look, the multi-look processing provides a significant increase on the total number of the detected scatterers (singles plus twice the doubles). Moreover, the percentages highlight the better capability of the multi-look processing in detecting the double PSs, also with respect to the single-look analysis reported in the literature [29,30]. The numerical results in Table 2, along with the visual evidence in Figures 4 and 7, allow concluding that many single scatterers detected by the single-look processing were detected as double scatterers by the multi-look processing. Table 2 also shows a significant increase of the detected double scatterers in the case of Cosmo-SkyMed dataset with respect to the TerraSAR-X case: this can be ascribed to the higher elevation resolution. Lower Rayleigh resolution along the elevation calls for super-resolution detection capability but this impacts negatively the detection rate, as demonstrated in reference [25] where an analysis of the detection rates as a function of the super-resolution ratio is performed for the single-look case.

In the following, we follow lines for a validation of the results.

First, it is worth noting that a visual comparison of the estimated heights for the single- and multi-look case achieved on both datasets, shown in Figure 4 for TerraSAR-X and Figure 7 for the COSMO-SkyMED case, provides evidence of a spatial consistency of the results for the single as well as double scatterers. For both single- and multi-look processing, high estimated heights for the double scatterer case are located in areas where single scatterers also show the presence of buildings. This aspect can be analyzed on those pixels where the single-look processing declares the presence of single PSs, whereas the multi-look processing detects double PSs. It is reasonable to expect that the noise reduction induced by the multi-look enables to revealing the presence of an additional (weaker) scatterer, which the single-look (noisier) processing cannot detect. This is confirmed by the results in Figure 8, which is related to the COSMO-SkyMED dataset. It shows the histogram of the difference, on a common grid of pixels, between the heights associated with the single scatterers detected by the single-look processing and the closest (Figure 8a) and farthest (Figure 8b) double scatterers detected by the multi-look scheme.

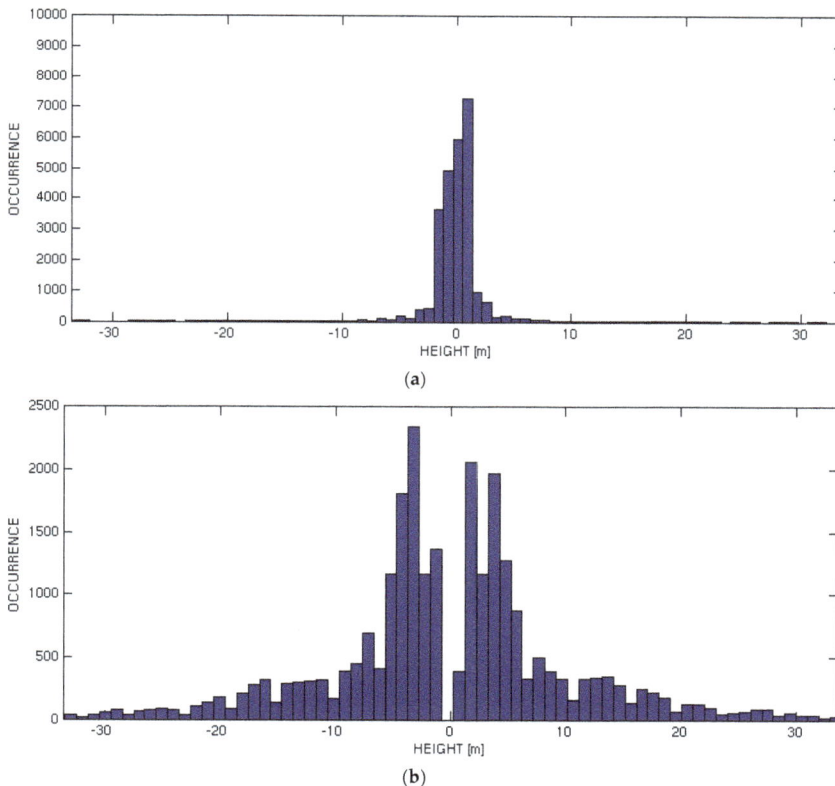

(a)

(b)

Figure 8. Histograms of the differences on a common grid of pixels between the residual topography (height) associated with the single scatterers single-look and the closest (**a**) and farthest (**b**) double scatterer resulting from the multi-look processing.

Mean and standard deviation of the differences are −0.04 m and 2.6 m for the closest case, whereas for the farthest case are −0.26 m and 10.5 m, respectively. The presence of pronounced peeks in the first histogram is an indication of the consistency of the results corresponding to the single scatterers and double scatterers. Moreover, the distribution of the RT (height) separation for the farthest scatterers in the bottom histogram in Figure 8 compares well to the distribution of the heights of the buildings in the analyzed area.

Furthermore, a quantitative validation of the results, though not exhaustive, was carried out on both the processed datasets.

For the National Arena of Bucharest (TerraSAR-X dataset), 15 buildings were selected with heights between 6 and 35 m. Heights derived by Google Earth were compared with the building heights estimated by the multi-look processing. Mean and standard deviation of the absolute height differences result to be 0.65 and 0.5 m, respectively, thus confirming the effectiveness of the estimation process.

A different validation procedure was carried out on the Rome (COSMO-SkyMED) dataset.

The detected targets over a proper selected area were geocoded and overlapped to the Google Earth layer corresponding to the same area. The results of this operation are shown in Figure 9. Single PSs are shown in Figure 9a,b, whereas all (single and double) PSs are shown in Figure 9c,d. Figure 9a,c, and Figure 9b,d, instead, shows single-look and multi-look processing, respectively.

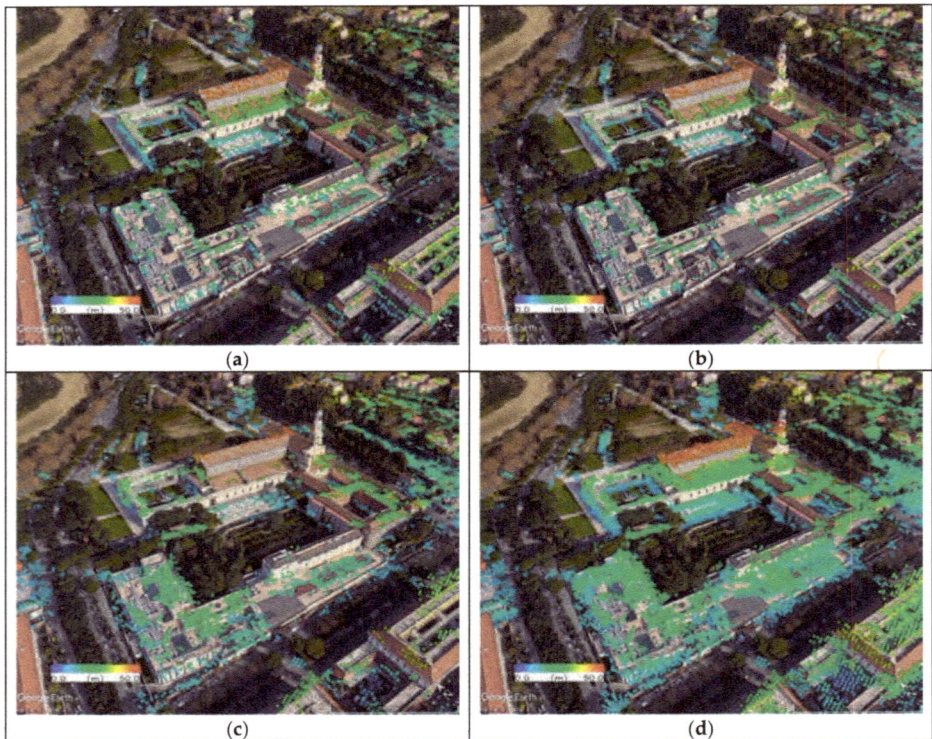

Figure 9. Geocoded scatterers detected on Rome dataset for the single-look (**a,b**) and multi-look (**c,d**) cases for the single scatterers (**a,c**) and both single and double scatterers (**b,d**).

The tomographic geocoded points fit well the shape of the buildings in the optical layer, thus providing evidence of the reliability of detected targets and excluding the presence of substantial errors in the estimated height parameters.

Remote Sens. **2018**, *10*, 1894

5. Conclusions

The use of multi-look allows providing improved detection of multiple scatterers with reference to the application of SAR tomography to the context of building reconstruction and monitoring. Depending on the system resolution along the elevation direction and the scene height distribution, the identification of multiple scatterers may be affected by the limitations in the detection of interfering scatterers below the Rayleigh resolution as well as by the effects of leakage.

This study deepened the investigation of possibilities related to the use of multi-look for improved detection of multiple persistent scatterers even below the Rayleigh resolution. A multi-look extension of the so-called sup-GLRT, which in the single-look case has been shown to have super-resolution capabilities, was derived. Following the literature for single-look, a fast version of such a detector was considered. The Capon inversion was exploited in this case for the reconstruction of the backscatter distribution in the tomographic domain to take benefit of its leakage mitigation characteristics. The detector showed good performance in urban areas, being able to achieve the discrimination of double interfering scatterers associated with the layover phenomenon, typically affecting highly urbanized areas. In particular, the double scatterers detection scheme was tested on TerraSAR-X data characterized by poor elevation resolution. Application of the method to Cosmo-SkyMED data, characterized by higher resolution along the elevation, further showed the importance of handling the layover problem in urban areas. Reported densities of double scatterers compare favorably to percentages reported in the literature for the single-look case.

Author Contributions: Conceptualization, C.D. G.F. and A.P; Methodology, C.D., G.F., A.P., and D.R.; Software, C.D.; Validation, G.F., A.P. and D.R.; Writing—Original Draft Preparation, C.D., A.P. and G.F.; and Supervision, G.F., A.P. and M.D.

Funding: This research received no external funding.

Acknowledgments: The authors wish to thank the German Aerospace Center for providing the TerraSAR-X dataset within the LAN1628 proposal and the Italian Space Agency for providing the COSMO-SkyMED dataset within the ASI AO Project ID2246. Selected algorithms presented in this work will be implemented within the frame of research project *SPERO—Space technologies used in the management of disasters and major crises, manifested at local, national and regional levels*, funded by the Romanian Minister of Research and Innovation, UEFISCDI, project reference: PN-III-P2-2.1-SOL-2016-03-0046. This work was supported by the I-AMICA project (PONa3_00363).

Conflicts of Interest: The authors declare no conflict of interest.

References

1. Fornaro, G.; Pauciullo, A. Interferometric and Tomographic SAR. In *Novel Radar Techniques and Applications*; Scitech Publishing: London, UK, 2018; pp. 361–402.
2. Ferretti, A.; Prati, C.; Rocca, F. Permanent scatterers in SAR interferometry. *IEEE Trans. Geosci. Remote Sens.* **2001**, *39*, 8–20. [CrossRef]
3. Reigber, A.; Moreira, A. First demonstration of airborne SAR tomography using multibaseline L-band data. *IEEE Trans. Geosci. Remote Sens.* **2000**, *38*, 2142–2152. [CrossRef]
4. Fornaro, G.; Reale, D.; Serafino, F. Four-Dimensional SAR Imaging for Height Estimation and Monitoring of Single and Double Scatterers. *IEEE Trans. Geosci. Remote Sens.* **2009**, *47*, 224–237. [CrossRef]
5. Fornaro, G.; Lombardini, F.; Pauciullo, A.; Reale, D.; Viviani, F. Tomographic Processing of Interferometric SAR Data: Developments, applications, and future research perspectives. *IEEE Signal Process. Mag.* **2014**, *31*, 41–50. [CrossRef]
6. Fornaro, G.; Pauciullo, A.; Reale, D.; Zhu, X.; Bamler, R. SAR Tomography: An advanced tool for spaceborne 4D radar scanning with application to imaging and monitoring of cities and single buildings. *IEEE Geosci. Remote Sens. Soc. Newslett.* **2012**, 9–17.
7. Budillon, A.; Evangelista, A.; Schirinzi, G. Three-Dimensional SAR Focusing from Multipass Signals Using Compressive Sampling. *IEEE Trans. Geosci. Remote Sens.* **2011**, *49*, 488–499. [CrossRef]
8. Zhu, X.; Bamler, R. Very High Resolution Spaceborne SAR Tomography in Urban Environment. *IEEE Trans. Geosci. Remote Sens.* **2010**, *48*, 4296–4308. [CrossRef]

9. Gini, F.; Lombardini, F.; Montanari, M. Layover solution in multibaseline SAR interferometry. *IEEE Trans. Aerosp. Electron. Syst.* **2002**, *38*, 1344–1356. [CrossRef]

10. Biondi, F. SAR tomography optimization by interior point methods via atomic decomposition—The convex optimization approach. In Proceedings of the 2014 IEEE International Geoscience and Remote Sensing Symposium (IGARSS), Quebec City, QC, Canada, 13–18 July 2014; pp. 1879–1882.

11. Lombardini, F.; Viviani, F. Single-look light-burden superresolution differential SAR tomography. *Electron. Lett.* **2016**, *52*, 557–558. [CrossRef]

12. Banda, F.; Dall, J.; Tebaldini, S. Single and Multipolarimetric P-Band SAR Tomography of Subsurface Ice Structure. *IEEE Trans. Geosci. Remote Sens.* **2016**, *54*, 2832–2845. [CrossRef]

13. Moreira, A.; Prats-Iraola, P.; Younis, M.; Krieger, G.; Hajnsek, I.; Papathanassiou, K.P. A Tutorial on Synthetic Aperture Radar. *IEEE Geosci. Remote Sens. Mag.* **2013**, *1*, 6–43. [CrossRef]

14. De Maio, A.; Fornaro, G.; Pauciullo, A. Detection of Single Scatterers in Multidimensional SAR Imaging. *IEEE Trans. Geosci. Remote Sens.* **2009**, *47*, 2284–2297. [CrossRef]

15. Pauciullo, A.; Reale, D.; de Maio, A.; Fornaro, G. Detection of Double Scatterers in SAR Tomography. *IEEE Trans. Geosci. Remote Sens.* **2012**, *50*, 3567–3586. [CrossRef]

16. Budillon, A.; Schirinzi, G. GLRT Based on Support Estimation for Multiple Scatterers Detection in SAR Tomography. *IEEE J. Sel. Top. Appl. Earth Observ. Remote Sens.* **2016**, *9*, 1086–1094. [CrossRef]

17. Budillon, A.; Johnsy, A.; Schirinzi, G. A Fast Support Detector for Superresolution Localization of Multiple Scatterers in SAR Tomography. *IEEE J. Sel. Top. Appl. Earth Observ. Remote Sens.* **2017**, *10*, 2768–2779. [CrossRef]

18. Berardino, P.; Fornaro, G.; Lanari, R.; Sansosti, E. A new algorithm for surface deformation monitoring based on small baseline differential SAR interferograms. *IEEE Trans. Geosci. Remote Sens.* **2002**, *40*, 2375–2383. [CrossRef]

19. Ferretti, A.; Fumagalli, A.; Novali, F.; Prati, C.; Rocca, F.; Rucci, A. A New Algorithm for Processing Interferometric Data-Stacks: SqueeSAR. *IEEE Trans. Geosci. Remote Sens.* **2011**, *49*, 3460–3470. [CrossRef]

20. Fornaro, G.; Verde, S.; Reale, D.; Pauciullo, A. CAESAR: An Approach Based on Covariance Matrix Decomposition to Improve Multibaseline–Multitemporal Interferometric SAR Processing. *IEEE Trans. Geosci. Remote Sens.* **2015**, *53*, 2050–2065. [CrossRef]

21. Fornaro, G.; Pauciullo, A.; Reale, D.; Verde, S. Multilook SAR Tomography for 3-D Reconstruction and Monitoring of Single Structures Applied to COSMO-SKYMED Data. *IEEE J. Sel. Top. Appl. Earth Observ. Remote Sens.* **2014**, *7*, 2776–2785. [CrossRef]

22. Pauciullo, A.; Reale, D.; Franze, W.; Fornaro, G. Multi-Look in GLRT-Based Detection of Single and Double Persistent Scatterers. *IEEE Trans. Geosci. Remote Sens.* **2018**, 1–13. [CrossRef]

23. Lombardini, F.; Pardini, M. Superresolution Differential Tomography: Experiments on Identification of Multiple Scatterers in Spaceborne SAR Data. *IEEE Trans. Geosci. Remote Sens.* **2012**, *50*, 1117–1129. [CrossRef]

24. Fornaro, G.; Serafino, F.; Soldovieri, F. Three-dimensional focusing with multipass SAR data. *IEEE Trans. Geosci. Remote Sens.* **2003**, *41*, 507–517. [CrossRef]

25. Reale, D.; Fornaro, G.; Pauciullo, A. Extension of 4-D SAR Imaging to the Monitoring of Thermally Dilating Scatterers. *IEEE Trans. Geosci. Remote Sens.* **2013**, *51*, 5296–5306. [CrossRef]

26. Stoica, P.; Moses, R. *Spectral Analysis of Signals*; Pearson Prentice Hall: Upper Saddle River, NJ, USA, 2005.

27. Budillon, A.; Johnsy, A.; Schirinzi, G. Extension of a Fast GLRT Algorithm to 5D SAR Tomography of Urban Areas. *Remote Sens.* **2017**, *9*, 844. [CrossRef]

28. Stephens, M.A. Use of the Kolmogorov-Smirnov Cramer-Von Mises and related statistics without extensive tables. *J. R. Stat. Soc. Ser. B* **1970**, *32*, 115–122.

29. Siddique, M.A.; Wegmüller, U.; Hajnsek, I.; Frey, O. Single-Look SAR Tomography as an Add-On to PSI for Improved Deformation Analysis in Urban Areas. *IEEE Trans. Geosci. Remote Sens.* **2016**, *54*, 6119–6137. [CrossRef]

30. Zhu, X.; Bamler, R. Superresolving SAR tomography for multidimensional imaging of urban areas: Compressive sensing-based TomoSAR inversion. *IEEE Signal Process. Mag.* **2014**, *31*, 51–58. [CrossRef]

remote sensing

MDPI

Letter

Comparison of Persistent Scatterer Interferometry and SAR Tomography Using Sentinel-1 in Urban Environment

Alessandra Budillon [1,*], Michele Crosetto [2], Angel Caroline Johnsy [1], Oriol Monserrat [2], Vrinda Krishnakumar [2] and Gilda Schirinzi [1]

[1]   Dipartimento di Ingegneria, Università degli Studi di Napoli "Parthenope", 80143 Napoli NA, Italy; angel.johnsy@uniparthenope.it (A.C.J.); gilda.schirinzi@uniparthenope.it (G.S.)
[2]   Centre Tecnològic de Telecomunicacions de Catalunya (CTTC/CERCA), Division of Geomatics, 08860 Castelldefels, Spain; mcrosetto@cttc.cat (M.C.); omonserrat@cttc.cat (O.M.); vkrishnakumar@cttc.cat (V.K.)
*   Correspondence: alessandra.budillon@uniparthenope.it; Tel.: +39-08-1547-6725

Received: 25 October 2018; Accepted: 5 December 2018; Published: 8 December 2018

Abstract: In this paper, persistent scatterer interferometry and Synthetic Aperture Radar (SAR) tomography have been applied to Sentinel-1 data for urban monitoring. The paper analyses the applicability of SAR tomography to Sentinel-1 data, which is not granted, due to the reduced range and azimuth resolutions and the low resolution in elevation. In a first part of the paper, two implementations of the two techniques are described. In the experimental part, the two techniques are used in parallel to process the same Sentinel-1 data over two test areas. An intercomparison of the results from persistent scatterer interferometry and SAR tomography is carried out, comparing the main parameters estimated by the two techniques. Finally, the paper addresses the complementarity of the two techniques, and in particular it assesses the increase of measurement density that can be achieved by adding the double scatterers from SAR tomography to the persistent scatterer interferometry measurements.

Keywords: synthetic aperture radar; persistent scatterers; differential interferometry; tomography; radar detection; generalized likelihood ratio test; sparse signals

1. Introduction

The European Union and the European Space Agency (ESA), with the Copernicus program [1], and, in particular, the free availability of data from the Sentinel satellite missions, have pushed the interest in developing advanced techniques for earth monitoring. This work is focused on the data of the Sentinel-1 radar mission. Sentinel-1 [2] acquires images in C-band, covering 250 by 180 km in its standard data acquisition mode (interferometric wide swath). Sentinel-1 data are characterized by high temporal resolution (revisiting cycle of 6 days) and moderate spatial resolution (pixel size of 14 by 4 m). This new sensor offers an improved data acquisition capability for deformation monitoring with respect to previous C-band sensors (ERS-1/2, Envisat, and Radarsat), considerably increasing the monitoring potential. The Sentinel-1 coverage is well suited for wide-area monitoring using differential nterferometric synthetic aperture radar (DInSAR) and persistent scatterer interferometry (PSI).

DInSAR involves the exploitation of at least a pair of complex synthetic aperture radar (SAR) images to measure surface deformation. Several DInSAR techniques have been developed in the last couple of decades. The PSI methods, which are based on large stacks of complex SAR images, have proven to be effective and are extensively applied [3]. Both the DInSAR and PSI techniques exploit the phase of the SAR images. Most of the PSI techniques assume the presence of only one dominant scatterer per resolution cell [4–12]. This assumption can be invalid when observing ground scenes with a pronounced extension in the elevation direction, for which more than one scatterer can fall in the same range–azimuth resolution cell. This, for instance, occurs in the presence of buildings of different heights, whose backscattered signals interfere in the same resolution cell, or in urban layover areas [13]. In such areas, the PSI techniques that assume one dominant scatterer usually experience a loss of deformation measurements. This potential limitation can be overcome by using the TomoSAR [14] techniques. In fact, in such techniques, the use of a stack of complex-valued interferometric images makes the separation of the scatterers interfering within the same range–azimuth resolution cell possible. This is achieved by synthesizing apertures along the elevation direction, which results in an elevation resolution, so as to provide the full scene reflectivity profile along azimuth, range, elevation, and average deformation velocity [15]. With respect to PSI techniques, TomoSAR, in addition the scatterers' position in 3D space and their average deformation velocity, also provides their intensity distribution in 3D space, which is additional information that can be conveniently used for selecting the most reliable scatterers in the reconstructed scene. Tomographic processing techniques exploit both the phase and amplitude of the backscattered signal and consist of resolving an inversion problem [16].

The aim of this paper is to show the applicability of TomoSAR to Sentinel-1 data in performing deformation monitoring. Then, an intercomparison of the results from TomoSAR and the more mature and experimented technique of persistent scatterer interferometry are carried out. For this purpose, two study areas were analyzed using Sentinel-1 data, which include a portion of the Port of Barcelona (Spain) and a part of the city center. The contributions of the paper can be summarized as follows. Firstly, the paper presents the first analysis concerning the performance of TomoSAR applied on Sentinel-1 data, which, exhibiting reduced range and azimuth resolutions and low resolution in elevation, do not seem well suited for TomoSAR. Secondly, it compares the TomoSAR and PSI estimates, i.e., height, deformation velocity, and thermal dilation. Finally, it explores how TomoSAR can complement the PSI measurements, increasing the measurement density. This paper is organized as follows. Section 2 recalls the essential characteristics of the PSI approach used in this work. In Section 3 an introduction to the TomoSAR used in this work is provided. The comparison of the results obtained in the study area is discussed in Section 4. Conclusions follow in Section 5.

2. A PSI Technique

The purpose of this section is to describe the PSI technique used in this work. The basics of DInSAR and PSI are recalled in Reference [3]. The observation equation used in this work is:

$$\Delta\varphi_{D-Int} = \Delta\varphi_{Int} - \varphi_{Topo\_simu} = \varphi_{Displ} + \varphi_{Ther} + \varphi_{RTE} + \varphi_{Atmo} + \varphi_{Noise} + 2 \cdot k \cdot \pi, \qquad (1)$$

where $\Delta\varphi_{Int}$ is the interferometric phase, $\Delta\varphi_{D-Int}$ is the so-called DInSAR phase, φ_{Topo_simu} is the simulated topographic component (using an external digital elevation model (DEM) of the scene), φ_{Displ} is the terrain deformation component, φ_{Ther} is the thermal expansion component, φ_{RTE} is the residual topographic error (RTE) component, φ_{Atmo} is the atmospheric phase component, and φ_{Noise} is the phase noise. The last term, $2 \cdot k \cdot \pi$, where k is an integer value called phase ambiguity, is a result of the wrapped nature of $\Delta\varphi_{D-Int}$.

The main goal of the PSI techniques is to estimate φ_{Displ} from the $\Delta\varphi_{D-Int}$. This implies separating φ_{Displ} from the other phase components. This can only by accomplished if the pixels to be processed are characterized by small φ_{Noise}. A common way to select such pixels is to use the amplitude dispersion criterion [4]. This criterion tends to select pixels where the response to the radar is dominated by a strong reflecting object and is constant over time. These pixels are called persistent scatterers (PSs). This is the selection method used in our PSI procedure. The main steps of the PSI procedure used in this work are briefly summarized below, see for details Reference [9].

1. The first processing step consisted of collecting a stack of N interferometric SAR images covering the area of interest. This was followed by image co-registration and the generation of a redundant

set of M interferograms, obtained combining different couples of images (then $M \leq N(N-1)/2$ with M >> N − 1).

2. The PS candidates were then selected using the amplitude dispersion criterion [4].
3. Based on the wrapped stack of interferograms and the extended model described in Reference [8], the following three parameters were estimated for each PS candidate: The linear deformation velocity (VELO), the residual topographic error (RTE), and the thermal expansion parameter (THER). The estimation was firstly performed on arcs and then the arc values were integrated over the set of PSs. The final selection of the PSs was based on the so-called temporal coherence (or ensemble phase coherence [4]), γ:

$$\gamma(PS) = \left| \frac{1}{M} \sum_{k=1}^{M} \exp(j(\Delta\varphi_{Obs}^{k}(PS) - \Delta\varphi_{Mod}^{k}(PS)) \right|, \tag{2}$$

which describes the goodness-of-fit of the three-parameter model, $\Delta\varphi_{Mod}^{k}(PS)$, see the full formula in Reference [8], and the wrapped interferometric phase, $\Delta\varphi_{Obs}^{k}(PS)$.

4. On the wrapped phases of the selected PSs, the above estimated RTE phase component was removed. This was followed by the phase unwrapping of the resulting phases. The phases were then temporally ordered using the interferogram to phase transformation [9].
5. The phase to displacement transformation was then applied, which was followed by geocoding.

The final product was given by the geocoded deformation time series, from which the deformation velocity was estimated. It is worth noting that the atmospheric phase removal is not considered in this work because the extension of the study area is limited, i.e., the atmospheric phase contribution can be assumed to be constant on the whole image and is simply compensated by phase calibration.

3. A SAR Tomography Technique

TomoSAR can be seen as a Fourier reconstruction starting from measurements that are not uniformly spaced and providing a fully 3D reconstruction of the scene reflectivity profile. For a review of the different focusing methods, refer to Reference [16].

In TomoSAR techniques, the presence of a scatterer at a given range, azimuth, and elevation position is revealed by a not negligible value of the corresponding 3D reflectivity intensity, reconstructed using tomographic coherent processing. This value is high if the signal backscattered by the target is not negligible and is coherent enough over the multiple acquisitions. Then, the intensity of the 3D focused reflectivity takes into account both the scatterer strength and its coherence. The accuracy of the elevation estimation is related to the achievable imaging resolution in the elevation direction, which is given by the Rayleigh resolution [16], and can be expressed as $\frac{\lambda R_0}{2S_T}$, where λ is the operating wavelength, R_0 the distance between the illuminated scene and the reference antenna position, and S_T is the overall perpendicular baselines extension. Resolution can be improved using super-resolution imaging techniques [17–19] that allow to find scatterers at a distance closer than Rayleigh resolution, adopting nonlinear processing.

TomoSAR techniques have also been extended to differential tomography, which integrates the TomoSAR concept with the differential interferometry concept [19]. Differential tomography allows to estimate, in addition to the elevation, the thermal dilation of the scatterers, producing a 4D reconstruction, or even the average deformation velocity, producing a 5D reconstruction [15,19–21].

TomoSAR techniques have to face the problem of irregular and sparse sampling of the aperture synthetized in the elevation direction, as it is generally sampled less densely than required by Fourier approaches. Then, ambiguities and masking problems due to anomalous side-lobes in the reconstructed reflectivity profile may arise [17]. This problem, together with the presence of clutter and noise, produces outliers in the tomographic reconstruction that can be erroneously confused with the presence of scatterers, giving rise to false alarms. To reduce the number of false alarms, a generalized likelihood ratio test (GLRT), improving the capability of discriminating real scatterers and outliers, has been introduced. On the base of a statistical model, it allows to determine the presence of scatterers fixing a probability of false alarm (P_{FA}). In References [22–25], different GLRTs, allowing to discriminate between single and double scatterers in the 3D, 4D, and 5D cases are presented. In Reference [22], a sequential GLRT is proposed, but it does not exhibit good performance when trying to super resolve close scatterers. In Reference [23], an alternative implementation of a similar GLRT approach is analyzed that exhibits super resolution capabilities but at the expense of a high computational cost. In Reference [24], an approximated version of the GLRT presented in Reference [23] is proposed that, with a slightly loss in detection performance, achieves super resolution capabilities with a low computational cost. In Reference [25], it is extended to 5D reconstruction. In Reference [26], a detection strategy for single and double coherent scatterers, based on a statistical model considering multiplicative noise, is presented. This model takes into account the statistical distribution of the scatterers signal phase variations, while additive noise (clutter) is not considered. It has been shown that the detection performance outperforms a PS approach, but the method has not been compared with other GLRT approaches using the additive noise model. We adopt the additive noise model, as done in most tomographic approaches, which allows to consider both Gaussian clutter and thermal noise, while not including the statistical distribution of additive phase noise on the scatterer response, which is, incidentally, very difficult to model.

In this paper, the method proposed in Reference [24], denoted as Fast-Sup-GLRT, was used.

The detection test can be derived starting from the discrete TomoSAR model for a fixed range and azimuth position:

$$\mathbf{u} = \mathbf{\Phi}\boldsymbol{\gamma} + \mathbf{w}, \tag{3}$$

where $\mathbf{u}$ is an $M \times 1$ (complex valued) observation column vector, $\boldsymbol{\gamma}$ is the $N \times 1$ (complex valued) column vector whose elements are the samples of the reflectivity at different elevations, $\mathbf{\Phi}$ is an $M \times N$ measurement matrix related to the acquisition geometry, and $\mathbf{w}$ is an $M \times 1$ column vector representing noise and clutter. Each m-th row $\boldsymbol{\varphi}_m$ of matrix $\mathbf{\Phi}$ is given by $vec(\mathbf{\Phi}_{m3})$, where vec is the operator transforming a three-dimensional matrix of size $N_s \times N_v \times N_k$ in a row vector of size $N = N_s N_v N_k$, by loading in the vector all the elements of the matrix scanned in a preassigned order, and $\mathbf{\Phi}_{m3}$ is the three-dimensional matrix of size $N_s \times N_v \times N_k$, whose element of entries (i,l,n) with $i = 1, \ldots, N_s, l = 1, \ldots, N_v, n = 1, \ldots, N_k$, is given by:

$$\{\mathbf{\Phi}_{m3}\}_{i\,l\,n} = \frac{1}{\sqrt{N}} e^{j\left(\frac{4\pi}{\lambda R_0} s_i s'_m + \frac{4\pi}{\lambda} v_l t_m + \frac{4\pi}{\lambda} k_n T_m\right)}, \tag{4}$$

where we have assumed that the SAR interferometric images of the same scene have been acquired at different times t_m and with different perpendicular baselines s'_m and temperatures T_m, and the triplet (s_i, v_l, k_n) represents the discretized values of elevation (RTE), deformation velocity (VELO), and thermal dilation coefficient (THER), respectively, in each range azimuth pixel (see for details Reference [25]). We note that with respect to Equation (1), the following expressions hold:

$$\phi_{RTE} = \frac{4\pi}{\lambda R_0} s_i s'_m; \ \phi_{Displ} = \frac{4\pi}{\lambda} v_l t_m; \ \phi_{Ther} \frac{4\pi}{\lambda} k_n T_m. \tag{5}$$

The detection performed applying a GLRT [27] allows to discriminate between three statistical hypotheses if we restrict our assumption to the presence of single (H_1) and double scatterers (H_2) in the same range–azimuth resolution cell, or the absence of scatterers (only noise, H_0). The Fast-Sup-GLRT detector can be cast as a binary hierarchical test [24], implemented by sequentially applying the two statistical tests:

$$
\Lambda_1(\mathbf{u}) = \frac{[\mathbf{u}^H \mathbf{u}]}{\mathbf{u}^H \left(\mathbf{I} - \boldsymbol{\Phi}_{\hat{\Omega}_2} \left(\boldsymbol{\Phi}_{\hat{\Omega}_2}^H \boldsymbol{\Phi}_{\hat{\Omega}_2} \right)^{-1} \boldsymbol{\Phi}_{\hat{\Omega}_2}^H \right) \mathbf{u}} \begin{array}{c} H_0 \\ < \\ > \\ H_1 \end{array} T_1
$$

$$
\Lambda_2(\mathbf{u}) = \frac{\mathbf{u}^H \left(\mathbf{I} - \boldsymbol{\Phi}_{\hat{\Omega}_1} \left(\boldsymbol{\Phi}_{\hat{\Omega}_1}^H \boldsymbol{\Phi}_{\hat{\Omega}_1} \right)^{-1} \boldsymbol{\Phi}_{\hat{\Omega}_1}^H \right) \mathbf{u}}{\mathbf{u}^H \left(\mathbf{I} - \boldsymbol{\Phi}_{\hat{\Omega}_2} \left(\boldsymbol{\Phi}_{\hat{\Omega}_2}^H \boldsymbol{\Phi}_{\hat{\Omega}_2} \right)^{-1} \boldsymbol{\Phi}_{\hat{\Omega}_2}^H \right) \mathbf{u}} \begin{array}{c} H_1 \\ < \\ > \\ H_2 \end{array} T_2
$$

(6)

where $\mathbf{I}$ is the $M \times M$ identity matrix, ^{H} denotes the Hermitian, $\boldsymbol{\Phi}_{\hat{\Omega}_1}$ is a column vector of size $M \times 1$ obtained by extracting from $\boldsymbol{\Phi}$ the column that minimizes the numerator of $\Lambda_2(\mathbf{u})$, and $\boldsymbol{\Phi}_{\hat{\Omega}_2}$ is a two column matrix of size $M \times 2$ obtained adding to the column $\boldsymbol{\Phi}_{\hat{\Omega}_1}$ a second column, extracted from $\boldsymbol{\Phi}$ in such a way to minimize the denominator of $\Lambda_1(\mathbf{u})$ (and of $\Lambda_2(\mathbf{u})$). The two thresholds T_1, T_2 can be derived using a Monte Carlo simulation and following a constant false alarm rate (CFAR) approach, consisting of setting the thresholds in such a way to obtain an assigned probabilities of false alarm and false detection, respectively, $P_{FA} = P(H_1|H_0)$ and $P_{FD} = P(H_2|H_1)$.

In the following, the step-by-step TomoSAR procedure used in this work is reported.

1. The stack of M SAR interferometric images of the same scene were properly registered, with a sub-pixel accuracy, and preprocessed to remove the interferometric phase corresponding to an external DEM. Atmospheric phase removal is not considered in this work because the extension of the study area is limited and its effect is simply compensated by phase calibrating the stack of images.

2. The two probabilities of false alarm and false detection were assigned, and the thresholds were derived via Monte Carlo simulations.

3. For each range azimuth cell, $\boldsymbol{\Phi}_{\hat{\Omega}_1}$ and $\boldsymbol{\Phi}_{\hat{\Omega}_2}$ were estimated minimizing the terms, $\left[\mathbf{u}^H \left(\mathbf{I} - \boldsymbol{\Phi}_{\hat{\Omega}_i} \left(\boldsymbol{\Phi}_{\hat{\Omega}_i}^H \boldsymbol{\Phi}_{\hat{\Omega}_i} \right)^{-1} \boldsymbol{\Phi}_{\hat{\Omega}_i}^H \right) \mathbf{u} \right]$, $i = 1, 2$, through i iterative minimizations.

4. For each range–azimuth cell, the two-step GLRT (6) was applied and it was determined which of the three hypotheses was verified. The estimation of the triplets (s_i, v_l, k_n) was derived from the positions of each detected scatterer. Then, one triplet in case of H_1 and two triplets in case of H_2 were determined, on the basis of the selected columns of matrix $\boldsymbol{\Phi}$.

4. Results and Discussion

In this Section, the applicability of TomoSAR to Sentinel-1 data is firstly addressed. TomoSAR techniques for urban applications have already been tested on TerraSAR-X and COSMO-SkyMed data [16,21,25,28], achieving good results. Sentinel-1 data exhibit quite different features with respect to TerraSAR-X and COSMO-SkyMed, in terms of spatial resolutions, coverage and phase stability. Moreover, the Sentinel-1 configuration has small baselines, so that the Rayleigh elevation resolution is small (about 40 m). Then, a detailed analysis of TomoSAR results on Sentinel-1 data, together with a comparison with PSI techniques, which have already been tested successfully on Sentinel-1 data, is required. We limit our analysis to the detection of single and double scatterers, as it is usually done in urban applications. The occurrence of more than two scatterers can increase when range and azimuth resolutions decrease, as with Sentinel-1 data, but it also depends on the geometry of the ground scene and on the elevation resolution, which is small in this case. Another consideration to be made regards the effect of the reduced range and azimuth resolutions with respect to the TerraSAR-X and COSMO-SkyMed systems, which produces a decrease of the signal to clutter noise ratio, thus lowering the expected spatial density of the detected scatterers with respect to the one achieved by the high-resolution systems.

As far as the number of single scatterers detected using TomoSAR and PSI is concerned, we expect to obtain more scatterers in the TomoSAR case, due to the gain of the coherent processing in the elevation direction, which increases the signal to clutter ratio in the focused 3D domain. In TomoSAR techniques, in fact, the estimation of the scatterers' amplitude, height, and deformation velocity was performed by simultaneously exploiting all the acquisitions, thus allowing an optimal smoothing of the phase noise and of the additive clutter effect, with a subsequent increase of the estimation accuracy. A drawback of TomoSAR techniques with respect to PSI techniques is that they suffer the problem of outliers that can be reduced using statistical techniques based on proper statistical models, which are not always adherent to real conditions of complex scenarios. Moreover, in TomoSAR techniques, it is more difficult to introduce contextual spatial information among adjacent range–azimuth pixels.

As far as double scatterer detection is concerned, only TomoSAR techniques allow their localization in the 3D space and their deformation parameters estimations, provided that their intensities estimated by fully 3D coherent processing are sufficiently high with respect to the clutter noise.

Their detection can be compromised not only by the low resolutions in range and azimuth, but also by the low resolution in elevation, which is limited by the low overall baseline span.

Two test areas were considered, both located in the metropolitan area of Barcelona. In the first one, the intercomparison of the PSI and TomoSAR results is discussed, while in the second one, the single and double scatterers from TomoSAR are compared with the PSs from PSI. The test areas were covered by 61 interferometric wide swath Sentinel-1 images, over the period that goes from 6 March 2015 to 30 May 2017. The perpendicular baseline range is approximately 300 m, while in the observed period, the range of the average air temperature of the scene is 26 °C. In the PSI and TomoSAR processings, the differential interferograms were derived using a 90-m shuttle radar topography mission (SRTM) DEM. Over the study areas, this DEM basically describes the ground topography. By choosing a reference point located on the ground, most of the processed pixels showed RTE values close to zero, while the pixels corresponding to structures and buildings had high RTE values.

Considering the system parameters, the achieved Rayleigh resolutions were: 41 m for RTE, 1 mm/year for VELO, and 1 mm/°C for THER. For the PSI processing, the following search steps were used: 1 m for RTE, 0.5 mm/year for VELO, and 0.0625 mm/°C for THER. For the Fast-Sup-GLRT TomoSAR processing, the following search steps were used: 3 m for RTE, 1 mm/year for VELO, and 0.2 mm/°C for THER.

The first test area covers a part of the Port of Barcelona, see the amplitude image in Figure 1. The extension of the area is approximately 1.6 by 4.9 km. The spatial density of the scatterers detected by PSI and the Fast-Sup-GLRT is shown in Figure 2. The number of detected scatterers by the PSI approach is 8700, while over the same area, the number of scatterers detected by TomoSAR is 33,350. One may notice a remarkable difference in the measurements' density: There is a factor 4 between the two solutions. Similar results are discussed in Reference [29]. This is a direct consequence of the strategies used in the two data analyses. The PSI results were generated using an amplitude dispersion threshold of 0.2, followed by a threshold on the temporal coherence of the arcs of 0.7. For Fast-Sup-GLRT, the thresholds in Equation (6) were computed by Monte Carlo simulation for the considered system parameters and acquisition configuration and fixing $P_{FD} = P_{FA} = 10^{-5}$. For instance, for evaluating T_1, the Monte Carlo approach consists of simulating a large number of realizations of clutter plus noise signals, i.e., the data under hypothesis H_0. Then, for a given P_{FA}, the threshold is evaluated such that $\Lambda_1(\mathbf{u}) \geq T_1$ with a probability equal to the fixed P_{FA}. Choosing $P_{FA} = 10^{-5}$ implies that we expect, on the first test area, a maximum number of false alarms equal to 2. A rough criterion for checking the adequacy of the chosen thresholds is the absence of detected scatterers on the sea area and the correct height positioning of the detected scatterers. Of course, when the thresholds decrease, the number of detected scatterers increases, with a possible precision degradation.

Figure 1. Mean SAR amplitude of the first test area: 401 (range) by 351 (azimuth) pixels, which cover an extension of approximately 1.6 (range) by 4.9 (azimuth) km.

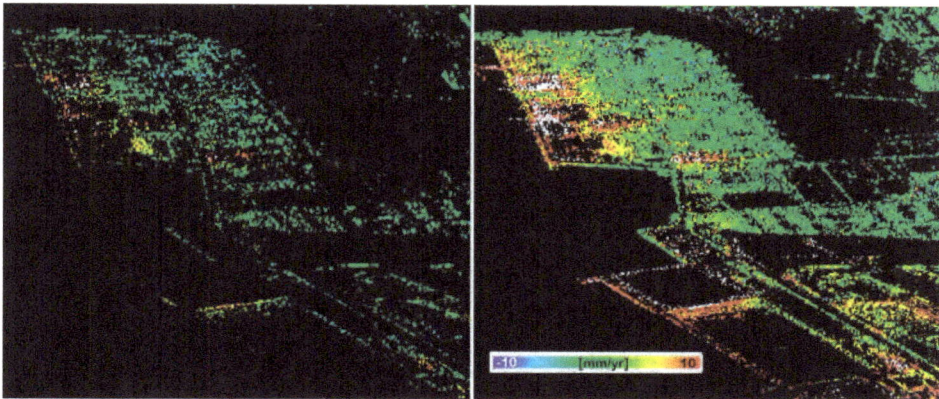

Figure 2. Deformation velocity obtained by PSI (**Left**) and by TomoSAR (**Right**).

It is worth mentioning the different computational burden of the two techniques. Considering the main processing step, i.e., the estimation of the three model parameters for each selected scatterer, to process the same dataset using equivalent computational resources and the current implementation of the two techniques, TomoSAR takes approximately 60 times the time required by PSI. For this test site, the three-parameter PSI estimation takes approximately 40 min.

The intercomparison of the results of the two techniques was computed over the set of scatterers that is common to both techniques, which includes 8684 points. The statistics of the deformation velocity intercomparison, i.e., the statistics of the differences of the deformation velocity values, are shown in Table 1, while the histogram of the velocity differences is shown in Figure 3.

Table 1. Statistics of the results of intercomparison over the first test area.

Parameter	Min.	Max.	Mean	St. Dev.
VELO [mm/year]	−4.24	5.54	0.01	1.36
RTE [m]	−18.06	32.80	0.09	3.25
THER [mm/°C]	−0.310	0.260	−0.005	0.078

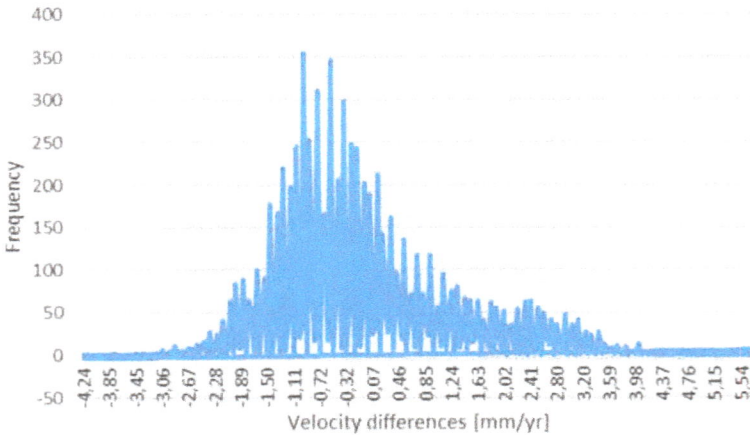

Figure 3. Histogram of the persistent scatterer interferometry (PSI) and TomoSAR velocity differences.

PSI and TomoSAR used the same reference point: The mean of the velocity differences indicates that there is basically no bias between the two experimental results. The most interesting parameter is given by the standard deviation of the velocity differences (1.36 mm/year): This indicates the dispersion of the population of the velocity differences. From this parameter, we can get valuable information on the standard deviation of the deformation velocity of each technique. For instance, by assuming the same precision for the two compared techniques and uncorrelated errors between the same techniques, it is possible to estimate the standard deviation of the deformation velocity of each technique as $1.36/\sqrt{2} = 0.96$ mm/year. The value 1 mm/year is often mentioned in the literature as the precision of the PSI deformation velocity, e.g., see Reference [30]: In this case study, this value is confirmed.

The same analysis was carried out for the RTE. Figure 4 shows the RTE maps coming from PSI and TomoSAR. The main statistics of the differences are summarized in Table 1, while the histogram of the differences is shown in Figure 5. The mean difference is close to zero: There is a negligible bias between the two datasets. An interesting experimental parameter is given by the standard deviation of the RTE differences (3.25 m). Like in the case of the velocity, we can derive from this parameter the standard deviation of the RTE of each technique (PSI and TomoSAR): 2.3 m. This represents an interesting experimental result. In similar intercomparison exercises run using European Remote Sensing (ERS) and Envisat data, the same parameter was estimated to range between 0.9 and 2 m [30]. From this case study, we can conclude that there is a relatively small impact of the reduced orbital tube of Sentinel-1 with respect to ERS and Envisat.

Finally, the analysis was focused on the THER parameter. Figure 6 shows the THER maps coming from PSI and TomoSAR; the statistics of the intercomparison are shown in Table 1, while the histogram of the THER differences is shown in Figure 7. In this case, the mean difference is −0.005 mm/°C: This is a negligible bias, thanks to the fact that the datasets use the same reference point. The standard deviation of the THER difference (0.078 mm/°C) can be used to estimate the standard deviation of the THER estimated by each technique, obtaining the value of 0.055 mm/°C. This is an interesting experimental result, which is quite close to the value (0.04 mm/°C) derived using X-band data, which is described in Reference [8].

In this first case study, very few double scatterers were found using the TomoSAR processing. This is due to the relatively low height of the buildings and structures of this area: There are very limited layover areas, where the double scatterers are usually found. For this reason, we decided to include a second study area, which includes high-rise buildings. In this study area, we used the same SAR dataset of the first case study. The experiment focused on a small area (see Figure 8), where TomoSAR found 427 single scatterers (red) and 178 double scatterers (blue), while PSI detected 172 single PSs. The common single scatterers are 166, while 6 of the 178 double scatterers coincide with single PSs. In Figure 8, the location of doubles is explained by the presence of layover due to the Agbar tower (142 m) and to the neighboring building. Many points are on the ground and on the façade of Agbar tower and of the neighboring building. The majority of the 178 double scatterers (172) correspond to pixels that were discarded during the PSI pixel selection. Regarding the 6 PSs that are detected as double scatterers by TomoSAR, most probably, one of the two scatterers dominates with respect to the other one. For this reason, such scatterers are correctly selected by PSI, but their parameter estimates are slightly different from the ones provided by TomoSAR. Figure 9 shows the range and azimuth direction and the layover area. As an example, Figure 10 shows the localization of one of the six PS detected as a double scatterer by TomoSAR: PSI identifies only one scatterer with a height of 61 m, while TomoSAR identifies two scatterers with heights of 53 m and 38 m, respectively.

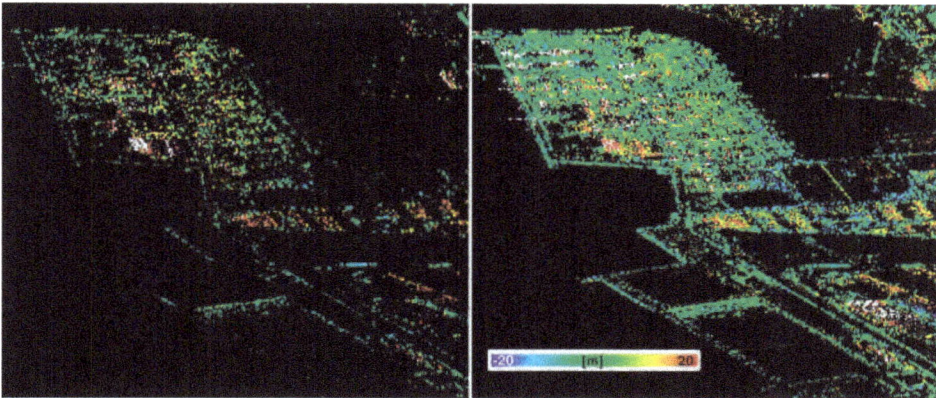

Figure 4. Residual topographic error (RTE) maps by PSI (**Left**) and by TomoSAR (**Right**).

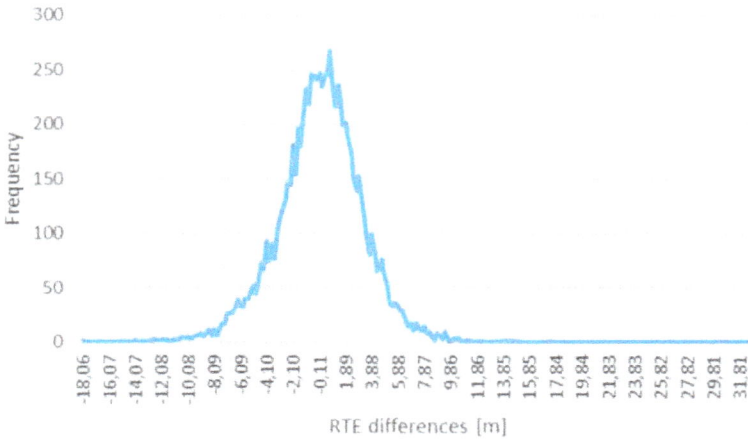

Figure 5. Histogram of the PSI and TomoSAR RTE differences.

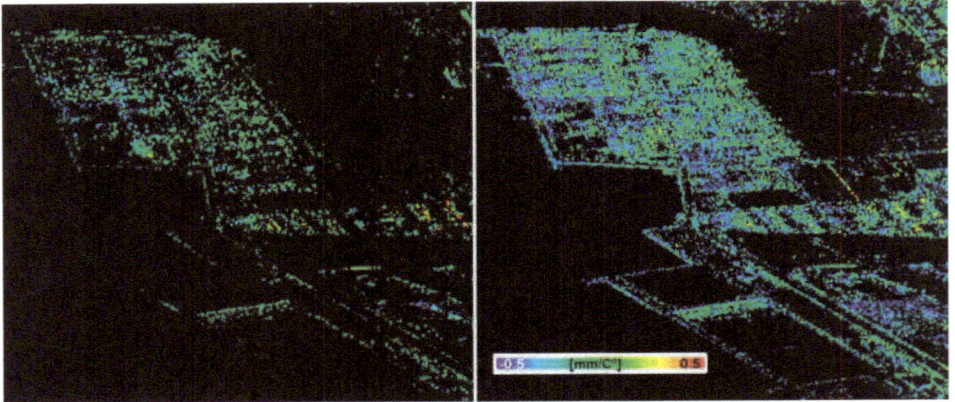

Figure 6. Thermal expansion parameter (THER) maps by PSI (**Left**) and by TomoSAR (**Right**).

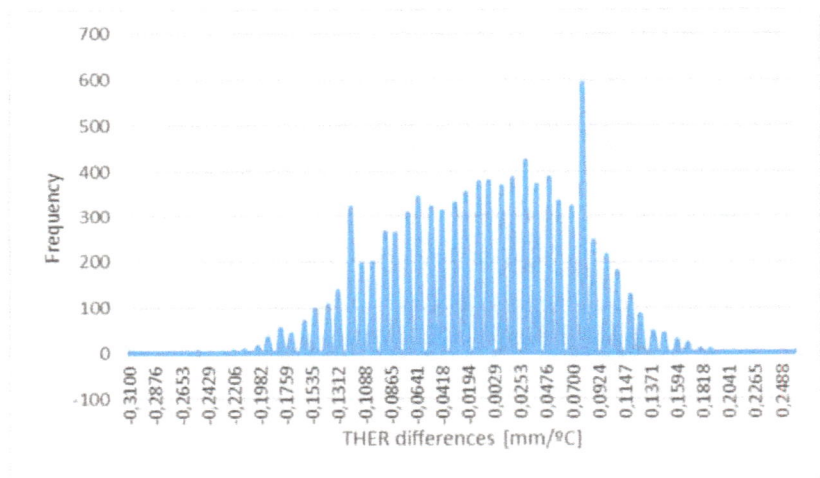

Figure 7. Histogram of the PSI and TomoSAR THER differences.

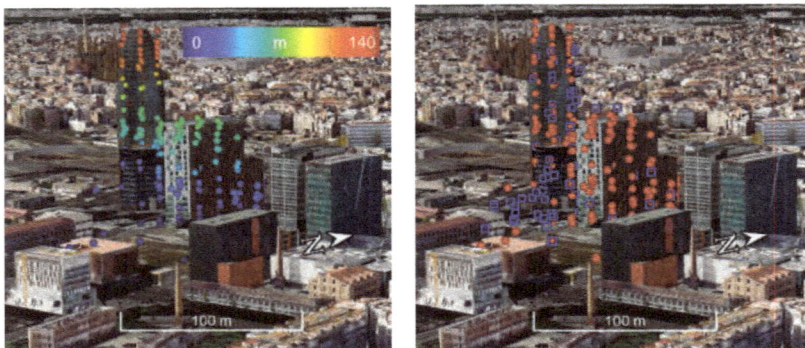

Figure 8. 3D view of the second test area with superimposed the TomoSAR height map on the (**left**) image; single (red) and double (blue) scatterers on the (**right**) image.

Figure 9. 3D view of the second test area with the SAR image superimposed.

Figure 10. Difference in height estimate between TomoSAR and PSI in a common point.

Summarizing, TomoSAR provides a larger set of single scatterers, which essentially includes all the scatterers detected by PSI, and additionally detects a significant number of double scatterers, which are mostly discarded by PSI. Then, TomoSAR allows the increase of the deformation measurement density, complementing PSI results and providing an added value.

5. Conclusions

In this paper, the TomoSAR applicability to Sentinel-1 data has been shown. An intercomparison of PSI and TomoSAR applied to Sentinel-1 data in order to monitor an urban area was carried out. The two techniques allow to estimate the following parameters over single and multiple scatterers: Height (RTE),linear deformation velocity (VELO), and the thermal expansion parameter (THER). Even if Sentinel-1 data have not been explicitly designed for tomography, in the two analyzed areas, the TomoSAR analysis has yielded good results, especially in terms of the high density of measurement points: The point density of TomoSAR is four times the density of the PSI results. The difference in the number of detected scatterers is mainly due to the coherent processing in the elevation direction of TomoSAR. By contrast, PSI processing is much less time consuming: For the same test area and similar computational resources, TomoSAR took approximately 60 times the time required by PSI.

The key experimental part of this work was devoted to the intercomparison of the PSI and TomoSAR results over two test areas. The most interesting results concern the standard deviations of the differences of the three main parameters (VELO, RTE, and THER), from which the standard deviation of each technique can be estimated. In terms of VELO, this corresponds to 0.96 mm/year for both PSI and TomoSAR. This is a value very close to 1 mm/year, which is often mentioned in the literature as the PSI precision of the deformation velocity. In terms of RTE, the standard deviation of the RTE of each technique (PSI and TomoSAR) is 2.3 m. This is an interesting experimental result, which is not so far from the values obtained in previous studies for ERS and Envisat (ranging between 0.9 and 2 m) [30]. This result indicates that there is a relatively small impact of the reduced orbital tube of Sentinel-1 with respect to ERS and Envisat. Finally, in terms of THER, the standard deviation of each technique is 0.055 mm/°C: This result is quite close to the value (0.04 mm/°C) derived using X-band data in a previous study [8].

If we consider the processing time, the PSI remains the reference technique, while TomoSAR can be used to complement the measurements done by PSI. This has been evidenced in the second case study of this work, considering the TomoSAR results in a layover area. Most of the detected double scatterers correspond to pixels that were discarded during the PSI pixel selection, which are useful to further increase the PSI deformation measurement density, providing an added value. However, if the processing time is not a major concern, TomoSAR can directly replace PSI.

Author Contributions: A.B. and G.S. developed the GLRT algorithm, M.C. and O.M. implemented the PSI algorithms; V.K. processed the PSI data; A.C.J. preprocessed real data and produced the geocoded results; all the authors contributed to performing the experiments; all the authors contributed to writing the paper.

Funding: This work has been partially funded by the Spanish Ministry of Economy and Competitiveness through the DEMOS project "Deformation monitoring using Sentinel-1 data" (Ref: CGL2017-83704-P).

Acknowledgments: We thank the European Space Agency (ESA), for providing the Sentinel 1 data and the software NEST.

Conflicts of Interest: The authors declare no conflict of interest.

References

1. European Space Agency (ESA). Copernicus Scientific Data Hub. Available online: http://www.copernicus.eu/ (accessed on 25 October 2018).
2. Rostan, F.; Riegger, S.; Pitz, W.; Torre, A.; Torres, R. The C-SAR instrument for the GMES sentinel-1 mission. In Proceedings of the 2007 IEEE International Geoscience and Remote Sensing Symposium (IGARSS), Barcelona, Spain, 23–27 July 2007; pp. 215–218.

3. Crosetto, M.; Monserrat, O.; Cuevas-González, M.; Devanthéry, N.; Crippa, B. Persistent Scatterer Interferometry: A review. *ISPRS J. Photogramm. Remote Sens.* **2016**, *115*, 78–89. [CrossRef]
4. Ferretti, A.; Prati, C.; Rocca, F. Permanent scatterers in SAR interferometry. *IEEE Trans. Geosci. Remote Sens.* **2001**, *39*, 8–20. [CrossRef]
5. Gernhardt, S.; Adam, N.; Eineder, M.; Bamler, R. Potential of very high resolution SAR for persistent scatterer interferometry in urban areas. *Ann. GIS* **2010**, *16*, 103–111. [CrossRef]
6. Crosetto, M.; Monserrat, O.; Iglesias, R.; Crippa, B. Persistent scatterer interferometry: Potential, limits and initial C- and X-band comparison. *Photogramm. Eng. Remote Sens.* **2010**, *76*, 1061–1069. [CrossRef]
7. Crosetto, M.; Monserrat, O.; Cuevas- González, M.; Devanthéry, N.; Luzi, G.; Crippa, B. Measuring thermal expansion using X-band persistent scatterer interferometry. *ISPRS J. Photogramm. Remote Sens.* **2015**, *100*, 84–91. [CrossRef]
8. Monserrat, O.; Crosetto, M.; Cuevas, M.; Crippa, B. The thermal expansion component of Persistent Scatterer Interferometry observations. *IEEE Geosci. Remote Sens. Lett.* **2011**, *8*, 864–868. [CrossRef]
9. Devanthéry, N.; Crosetto, M.; Monserrat, O.; Cuevas- González, M.; Crippa, B. An Approach to Persistent Scatterer Interferometry. *Remote Sens.* **2014**, *6*, 6662–6679. [CrossRef]
10. Ferretti, A.; Fumagalli, A.; Novali, F.; Prati, C.; Rocca, F.; Rucci, A. A new algorithm for processing interferometric data-stacks: SqueeSAR. *IEEE Trans. Geosci. Remote Sens.* **2011**, *49*, 3460–3470. [CrossRef]
11. Perissin, D.; Wang, T. Repeat-pass SAR interferometry with partially coherent targets. *IEEE Trans. Geosci. Remote Sens.* **2012**, *50*, 271–280. [CrossRef]
12. Ferretti, A.; Bianchi, M.; Prati, C.; Rocca, F. Higher-order permanent scatterers analysis. *EURASIP J. Appl. Signal Process.* **2005**, *2005*, 3231–3242. [CrossRef]
13. Gini, F.; Lombardini, F.; Montanari, M. Layover solution in multibaseline SAR interferometry. *IEEE Trans. Aerosp. Electron. Syst.* **2002**, *38*, 1344–1356. [CrossRef]
14. Reigber, A.; Moreira, A. First Demonstration of Airborne SAR Tomography Using Multibaseline L-band Data. *IEEE Trans. Geosci. Remote Sens.* **2000**, *38*, 2142–2152. [CrossRef]
15. Lombardini, F. Differential tomography: A new framework for SAR interferometry. *IEEE Trans. Geosci. Remote Sens.* **2005**, *43*, 37–44. [CrossRef]
16. Fornaro, G.; Lombardini, F.; Pauciullo, A.; Reale, D.; Viviani, F. Tomographic processing of interferometric SAR data: Developments, applications, and future research perspectives. *IEEE Signal Process. Mag.* **2014**, *31*, 41–50. [CrossRef]
17. Budillon, A.; Ferraioli, G.; Schirinzi, G. Localization Performance of Multiple Scatterers in Compressive Sampling SAR Tomography: Results on COSMO-SkyMed Data. *IEEE J. Sel. Top. Appl. Earth Obs. Remote Sens.* **2014**, *7*, 2902–2910. [CrossRef]
18. Zhu, X.X.; Bamler, R. Superresolving SAR Tomography for Multidimensional Imaging of Urban Areas: Compressive sensing-based TomoSAR inversion. *IEEE Signal Process. Mag.* **2014**, *31*, 51–58. [CrossRef]
19. Lombardini, F.; Cai, F. Temporal Decorrelation-Robust SAR Tomography. *IEEE Trans. Geosci. Remote Sens.* **2014**, *52*, 5412–5421. [CrossRef]
20. Fornaro, G.; Serafino, F.; Reale, D. 4D SAR imaging for height estimation and monitoring of single and double scatterers. *IEEE Trans. Geosci. Remote Sens.* **2009**, *47*, 224–237. [CrossRef]
21. Fornaro, G.; Reale, D.; Verde, S. Bridge thermal dilation monitoring with millimeter sensitivity via multidimensional SAR imaging. *IEEE Geosci. Remote Sens. Lett.* **2013**, *10*, 677–681. [CrossRef]
22. Pauciullo, A.; Reale, D.; De Maio, A.; Fornaro, G. Detection of Double Scatterers in SAR Tomography. *IEEE Trans. Geosci. Remote Sens.* **2012**, *50*, 3567–3586. [CrossRef]
23. Budillon, A.; Schirinzi, G. GLRT Based on Support Estimation for Multiple Scatterers Detection in SAR Tomography. *IEEE J. Sel. Top. Appl. Earth Obs. Remote Sens.* **2016**, *9*, 1086–1094. [CrossRef]
24. Budillon, A.; Johnsy, A.C.; Schirinzi, G. A Fast Support Detector for Super-Resolution Localization of Multiple Scatterers in SAR Tomography. *IEEE J. Sel. Top. Appl. Earth Obs. Remote Sens.* **2017**, *10*, 2768–2779. [CrossRef]
25. Budillon, A.; Johnsy, A.; Schirinzi, G. Extension of a fast GLRT algorithm to 5D SAR tomography of Urban areas. *Remote Sens.* **2017**, *9*, 844. [CrossRef]
26. Siddique, M.A.; Wegmüller, U.; Hajnsek, I.; Frey, O. SAR Tomography as an Add-On to PSI: Detection of Coherent Scatterers in the Presence of Phase Instabilities. *Remote Sens.* **2018**, *10*, 1014. [CrossRef]
27. Budillon, A.; Schirinzi, G. Performance Evaluation of a GLRT Moving Target Detector for TerraSAR-X along-track interferometric data. *IEEE Trans. Geosci. Remote Sens.* **2015**, *53*, 3350–3360. [CrossRef]

28. Fornaro, G.; Pauciullo, A.; Reale, D.; Verde, S. Multilook SAR Tomography for 3-D Reconstruction and Monitoring of Single Structures Applied to COSMO-SKYMED Data. *IEEE J. Sel. Top. Appl. Earth Obs. Remote Sens.* **2014**, *7*, 2776–2785. [CrossRef]
29. De Maio, A.; Fornaro, G.; Pauciullo, A. Detection of single scatterers in multidimensional SAR imaging. *IEEE Trans. Geosci. Remote Sens.* **2009**, *47*, 2284–2297. [CrossRef]
30. Crosetto, M.; Monserrat, O.; Bremmer, C.; Hanssen, R.; Capes, R.; Marsh, S. Ground motion monitoring using SAR interferometry: Quality assessment. *Eur. Geol.* **2009**, *26*, 12–15.

remote sensing

MDPI

Letter

Urban Tomographic Imaging Using Polarimetric SAR Data

Alessandra Budillon *, Angel Caroline Johnsy and Gilda Schirinzi

Dipartimento di Ingegneria, Università Degli Studi di Napoli "Parthenope", 80133 Napoli NA, Italy;
angel.johnsy@uniparthenope.it (A.C.J.); gilda.schirinzi@uniparthenope.it (G.S.)
* Correspondence: alessandra.budillon@uniparthenope.it; Tel.: +39-081-547-6725

Received: 30 November 2018; Accepted: 8 January 2019; Published: 11 January 2019

Abstract: In this paper, we investigate the potential of polarimetric Synthetic Aperture Radar (SAR) tomography (Pol-TomoSAR) in urban applications. TomoSAR exploits the amplitude and phase of the received data and offers the possibility to resolve multiple scatters lying in the same range–azimuth resolution cell. In urban environments, this issue is very important since layover causes multiple coherent scatterers to be mapped in the same range–azimuth image pixel. To achieve reliable and accurate results, TomoSAR requires a large number of multi-baseline acquisitions which, for satellite-borne SAR systems, are collected with long time intervals. Then, accurate tomographic reconstructions would require multiple scatterers to remain stable between all the acquisitions. In this paper, an extension of a generalized likelihood ratio test (GLRT)-based tomographic approach, denoted as Fast-Sup-GLRT, to the polarimetric data case is introduced, with the purpose of investigating if, in urban applications, the use of polarimetric channels allows for reduction of the number of baselines required to achieve a given scatterer's detection performance. The results presented show that the use of dual polarization data allows the proposed detector to work in an equivalent or better way than use of a double number of independent single polarization channels.

Keywords: synthetic aperture radar; tomography; polarimetry; radar detection; generalized likelihood ratio test; sparse signals

---

1. Introduction

Over the past few years, Synthetic Aperture Radar (SAR) sensors have rapidly advanced, offering multi-channel operation (polarimetry, multifrequency), improved range and azimuth resolution, and frequent revisiting of the same area (time series). A new class of SAR satellites, such as TerraSAR-X and TanDEM-X (X-band), COSMO-SkyMed (X-band), as well as Radarsat-2 (C-band), are providing images with resolution in the meter regime, and dual or full polarimetric SAR acquisition modes. In particular, the future second generation of COSMO-SkyMed will have full polarimetric SAR acquisition modes [1,2].

In polarimetric SAR (PolSAR) systems, the antennas for transmitting and receiving electromagnetic waves are configured in different polarization states. Thus, the scattering properties of the observed targets can be revealed in the alternative polarimetric combinations, providing more information compared to single polarization systems [3]. Of course, the price to pay for having enhanced polarization characteristics is a more complex sensor design and the demand for more image storage space.

PolSAR has many applications in many fields, including agricultural areas classification, oceanography (surface currents and wind field retrieval), forest monitoring and classification, disaster monitoring, and target recognition/classification [3]. PolSAR decomposition methods exploiting fully polarimetric data have been successfully applied to map vegetated areas by separating single bounce, double bounce, and volume scattering mechanisms [3]. For urban areas, the polarimetric scattering mechanisms are more complicated with respect to natural areas, due to the high variability of materials, and forms and sizes of the objects laying in the observed ground scene. Different approaches using fully polarimetric SAR data have been recently proposed [4]. In [4] the double bounce scattering form, the dihedral structure formed by the wall and the ground of a building, or the single bounce scattering from the roof or the wall, is considered, for a deterministic extraction of urban areas. However, the rotated dihedral scattering of a building, with a large orientation angle with respect to the radar look direction, results in a strong cross polarization component that can be misdetected as vegetation volume scattering [5].

An alternative approach for urban area monitoring is SAR tomography (TomoSAR). TomoSAR [6] extends the conventional two-dimensional SAR imaging principle to three dimensions by forming an additional synthetic aperture in elevation, using a stack of multi-baseline interferometric images. A fully 3-D scene reflectivity profile along azimuth, range, and elevation is provided. The use of TomoSAR techniques allows the identification of multiple scatterers in the same range–azimuth resolution cell [7]. Tomographic processing can be performed by Fourier-based techniques, beamforming, or spectral methods, such as Capon, MUSIC [8], and the more recent CS (compressive sensing) [9–11]. In CS-based approaches, TomoSAR is performed as the recovering of a sparse signal by a convex l_1 norm minimization [12] while, in [9,10], the CS approach can improve resolution in elevation and, in [11], the proposed method achieves super-resolution reconstruction, both for range and elevation.

Regardless of whichever imaging technique is adopted for the elevation reflectivity profile focusing, the discrimination between reliable scatterers and false alarms is not an easy task. Since tomographic synthetic aperture in the elevation direction is sampled sparsely, and not regularly and densely, as requested by a Fourier approach, ambiguities and masking problems from anomalous sidelobes may arise. This event, in addition to the presence of noise, produces false alarms. In [13–17], this problem has been addressed on the basis of a generalized likelihood ratio test (GLRT), that allows evaluation of the detection performance in terms of probability of detection achievable with a fixed probability of false alarm. The authors proposed a GLRT detector [14–16] that searches for the signal support (i.e., the positions of the significant samples in the unknown vector) that best matches the data. This statistical test is based on a non-linear maximization for detecting single and double scatterers with an assigned probability of false alarm. The elevation of the detected scatterers is then estimated on the basis of their position in the unknown vector.

A problem to be considered is that the performance of the GLRT detector becomes poor when the number of acquisitions or the scatterer coherence decrease [15]. A possible way for increasing the number of acquisition, while keeping the scatterer coherence high, is to exploit polarimetric systems.

Many applications of polarimetric TomoSAR (Pol-TomoSAR) are related to forest vertical structure recovering [18–21]. The vertical position of the scatterers in a forest, as well as a physical interpretation of the profile, has been shown to be more feasible by coupling polarimetric and multi-baseline information.

Recently, in polarimetric SAR tomography over urban areas [22], different spectral estimation techniques have been extended to the case of multi-pass SAR data acquired with different polarization channels, and a building layover has been studied to compare single and full polarization beamforming, Capon, and MUSIC. There are two drawbacks of these techniques: one is the use of a boxcar filter for the covariance matrix estimation, that reduces resolution in range and azimuth; the second is related to the absence of an approach for identifying multiple scatterers, which are recovered by visual inspection of the reconstructed 3-D reflectivity profile. In [23], a CS approach for polarimetric SAR tomography is proposed. In particular, it exploits the inter-signal structural correlations between neighboring pixels, as well as between polarimetric channels, applying distributed compressed sensing (DCS) theory. In this approach, the elimination of the artifacts and CS algorithm instability can be an issue.

The aim of this paper is to exploit the information given by a polarimetric SAR system, in order to detect reliable single and double scatterers in urban areas by using a reduced number of baselines with respect to the single polarization case. In particular, we will extend the signal model defined in [15], in order to take into account the different polarization channels. Following the approach presented in [22,23], we can suppose that all unknown reflectivity signals throughout polarimetric channels share, approximately, the same sparse support in the space domain, but have different nonzero coefficients. The proposed approach will be validated on dual polarimetric SAR real data.

This paper is organized as follows. Section 2 presents the signal model. In Section 3, the Pol-TomoSAR approach is presented. The results obtained on real data are discussed in Section 4. Conclusions follow in Section 5.

2. Signal Model

In Figure 1, the multi-pass SAR geometry in the range–elevation plane in a typical urban environment is shown. The three highlighted contributions of the backscattered signal are at the same distance from the platform, and will interfere in the same range–azimuth cell (azimuth axis is orthogonal to the plane). The three contributions come from the ground, the façade, and the roof of the building. In this particular case, the backscattered reflectivity elevation profile, γ, will exhibit only three samples different from zero. Moreover, γ can be assumed to be sparse with, at most, K_{max} samples different from zero, typically with $K_{max} = 2$. In order to estimate the reflectivity function γ, a stack of M range–azimuth-focused images is collected. The single channel k-th image, acquired along the orbit with the orthogonal baseline S'_k (see Figure 1), in a fixed pixel, is given by the integral superposition of the contributions of all the scatterers lying in the corresponding range–azimuth resolution cell, and located at different elevation coordinates s. A discrete estimate of γ can be found by ideally discretizing the integral operator.

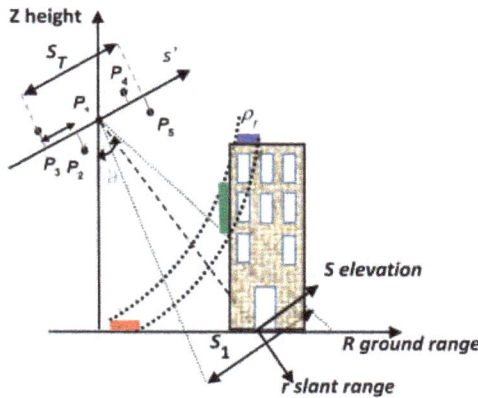

Figure 1. Multi-pass SAR geometry in the range–elevation plane (case $M = 5$).

We can denote, with γ, the $N \times 1$ column vector whose elements are the samples of the reflectivity at a fixed range and azimuth position, so the sampled received signal is related to γ by

$$u = \Phi\gamma + w,\tag{1}$$

where u is an $M \times 1$ observations column vector, w is an $M \times 1$ column vector representing noise and clutter, and Φ is an $M \times N$ measurement matrix related to the acquisition geometry, whose generic element with index kl is given by

$$\{\Phi\}_{kl} = e^{j\frac{4\pi}{\lambda R_0}S'_k s^l},\tag{2}$$

with λ the operating wavelength, and R_0 the distance between the center of the scene and a reference antenna position.

We note that in the signal model (1), we have assumed the absence of any phase miscalibration, the compensation of atmospheric delay, and the absence of temporal and thermal deformations.

When considering the fully polarimetric case, the received signal can be modeled as $3M \times 1$ observations column vector $\mathbf{u}' = \begin{bmatrix} \mathbf{u}_{HH} & \mathbf{u}_{HV} & \mathbf{u}_{VV} \end{bmatrix}^T$, and one $M \times 1$ vector for each polarimetric channel (*HH, HV, VV*). We can also assume that γ' is a $3N \times 1$ reflectivity column vector $\gamma' = \begin{bmatrix} \gamma_{HH} & \gamma_{HV} & \gamma_{VV} \end{bmatrix}^T$ and that, throughout, the three polarimetric channels share the same sparse support, as we are expecting backscatter from the same structure within a range–azimuth cell [22,23]. Under the above assumptions, we can extend model (1) to a fully polarimetric case:

$$\mathbf{u}' = \mathbf{\Phi}' \gamma' + \mathbf{w}',\tag{3}$$

where $\mathbf{w}'$ is an $3M \times 1$ column vector representing noise and clutter, and $\mathbf{\Phi}'$ is a $3M \times 3N$ block diagonal measurement matrix related to the acquisition geometry, and given by

$$\mathbf{\Phi}' = \begin{bmatrix} \mathbf{\Phi} & 0 & 0 \\ 0 & \mathbf{\Phi} & 0 \\ 0 & 0 & \mathbf{\Phi} \end{bmatrix},\tag{4}$$

where 0 is an all-zero $M \times N$ matrix.

Note that model (3) can be easily particularized to the dual polarization case by considering only the two available channels in the definition of the vectors $\mathbf{u}'$ and γ', and only two of the three blocks $\mathbf{\Phi}$ on the diagonal of $\mathbf{\Phi}'$, defined by (4).

3. Pol-TomoSAR Technique

TomoSAR techniques aims at the estimation of γ' by inverting the model (3). The inversion of (3) is ill-posed, since the M acquisitions are not uniformly spaced and usually $M < qN$, with q being the number of polarization channels. Then, false alarms can appear in the reconstructed profiles, heavily affecting the accuracy of the results. The selection of the most reliable scatterers in each range–azimuth cell can be set as a statistical detection problem assuming, as a selection criterion, the probability of false alarm (P_{FA}) is achievable using a proper statistical test.

In this paper, the polarimetric reflectivity profile γ' is estimated using a GLRT method [24], denoted as Fast-Sup-GLRT [15], which is an approximated and faster version of the GLRT proposed in [14]. Assuming that the maximum number of scatterers in each resolution cell is K_{max}, the vector γ' can be assumed to be sparse with, at most, qK_{max} significant samples, and the detection problem can be formulated as in [14,15], in the terms of the following $K_{max} + 1$ statistical hypothesis:

H_i: presence of i scatterers in each channel, with $i = 0, \ldots , K_{max}$,

assuming that urban environment K_{max} is typically set as equal to two.

The noise vector $\mathbf{w}$ can be assumed as a circularly symmetric complex Gaussian vector. Consequently, assuming deterministic scatterers, $\mathbf{u}$ is a circularly symmetric Gaussian random vector. Exploiting these statistical assumptions, the Fast-Sup-GLRT detector [15] can be extended to the polarimetric model (3). Then, at each step i, the following binary test is applied:

$$\Lambda_i'\left(\mathbf{u}'\right) = \frac{\left[\mathbf{u}'^H \mathbf{\Pi}_{\hat{\Omega}_{i-1}}^{\perp} \mathbf{u}'\right]}{\left[\mathbf{u}'^H \mathbf{\Pi}_{\hat{\Omega}_{K_{max}}}^{\perp} \mathbf{u}'\right]} \underset{H_{K \geq i}}{\overset{H_{i-1}}{\underset{>}{\lessgtr}}} T'_i,\tag{5}$$

where $\hat{\Omega}_{i-1} = \{l_1, \ldots l_{i-1}\}$ is the estimated support of cardinality $i - 1$ of each of the polarimetric vectors γ_{ab}, with $ab \in \{HH, HV, VV\}$, supposed to be $(i - 1)$-sparse, $\Pi^{\perp}_{\hat{\Omega}_{i-1}} = I - \Phi'_{\hat{\Omega}_{i-1}} \left(\Phi'^{H}_{\hat{\Omega}_{i-1}} \Phi'_{\hat{\Omega}_{i-1}} \right)^{-1} \Phi'^{H}_{\hat{\Omega}_{i-1}}$, with $\Phi'_{\hat{\Omega}_{i-1}}$ the matrix obtained by substituting in the definition (4) the matrix $\Phi_{\hat{\Omega}_{i-1}}$ in the place of Φ, where $\Phi_{\hat{\Omega}_{i-1}}$ is obtained from Φ by extracting the $i - 1$ columns of index $\hat{\Omega}_{i-1}$. Moreover, $\hat{\Omega}_{Kmax}$ is the estimated support of cardinality K_{max} of γ'_{ab}. Each support is estimated by sequentially minimizing the term $\left[u'^{H} \Pi^{\perp}_{\hat{\Omega}_k} u' \right]$ over k supports of cardinality one [15]. All thresholds can be numerically evaluated by means of Monte Carlo simulations. Assuming $K_{max} = 2$, the two thresholds will be evaluated in such a way to obtain the assigned probabilities of false alarm and false detection, respectively, $P_{FA} = P(H_1|H_0)$ and $P_{FD} = P(H_2|H_1)$.

4. Results and Discussion

In this section, results of the Fast-Sup-GLRT detector (5), that takes into account polarization channels on real data, are presented and compared with the case of only one channel. In processing real data, we limit the search to two targets ($K_{max} = 2$) per resolution cell.

We consider a total of 39 Spotlight TerraSAR-X (TSX) HH/VV images (system parameters in Table 1) and we will conduct the experiments using, in one case, all the available images in one channel (HH), and, in the other case, only a subset of twenty images in both channels (HH, VV). The resolution is in the range of about 1 m, in azimuth, 2.6 m. Considering that the overall perpendicular baseline B_p is about 800 m, the Rayleigh resolution in elevation is 14 m. The overall temporal baseline (B_t) span is about 2.3 years. The experiment consists of detecting single and double scatterers using the two datasets. The comparison between the two cases is fair, since it is done with an equal number of images.

Table 1. TSX system parameters.

System Parameters	
Wavelength	0.031 m
View angle	28°
Range distance	579 Km
Chirp bandwidth	120 MHz
Relative orbit	48
Orbit direction	descending
Look direction	right
Polarization	HH, VV
Perpendicular baselines extent	800 m
Rayleigh resolution in elevation	14 m

We aim to show that the dual polarization case can outperform the single polarization case that can count on a double number of perpendicular baselines. The diversity in polarization can compensate the loss in baseline diversity. We will compare, also, the results with the polarimetric beamforming and Capon approaches, as presented in [22], showing that, in order to detect multiple scatterers, a criterion for discriminating reliable scatterers from spurious sidelobes is needed.

The distribution of the 39 perpendicular baselines, considered in the single polarization case vs. the temporal baselines, is reported in Figure 2a, while the distribution of the 20 perpendicular baselines, considered in the dual polarization case vs. the temporal baselines, is reported in Figure 2b. The twenty baselines have been selected in such a way to refer to comparable baseline configurations.

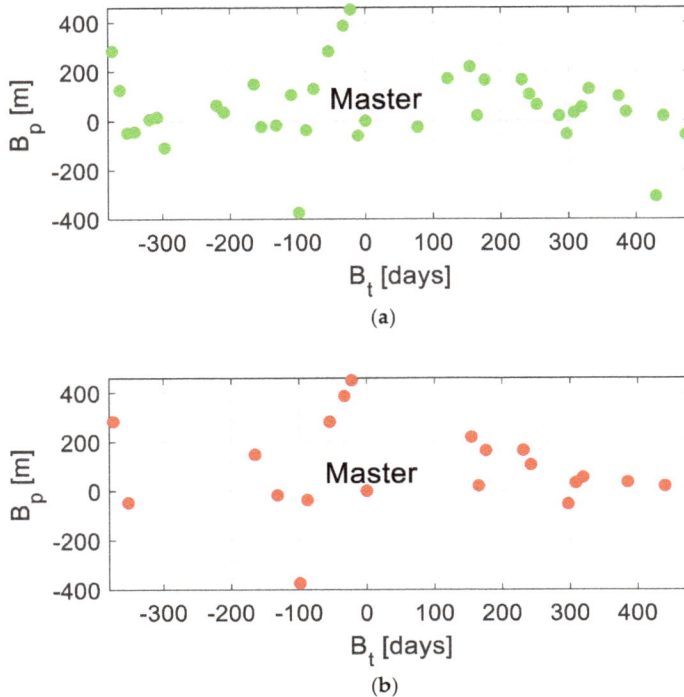

Figure 2. Distribution of the perpendicular baselines vs. the temporal baselines, in (**a**) the 39 baselines of the single polarization case, in (**b**) the 20 baselines of the dual polarization case.

A first constraint is the maintenance of the overall perpendicular baseline span, since its value determines the achievable height resolution. This constraint is easily satisfied by selecting the two acquisitions with the minimum (negative) baseline and the maximum (positive) baseline. For the selection of other baselines, in the absence of other a priori information, we selected a subset in such a way as to have a distribution of spatial and temporal baselines similar to that considered in the 39-baseline single channel case. In Figure 3, we report the normalized histogram for the spatial (a) and temporal (b) baseline distribution (in red, the 20-baseline case and, in green, the 39-baseline case). This selection criterion should guarantee having approximately the same average spatial and temporal decorrelation in the 20- and 39-baseline datasets.

In Figure 4, the intensity HH SAR image of the test area is shown. It is a small area near Toulouse, France. Two buildings are present, and both are commercial malls, with the same height of about 10–13 m. GLRT [15] has been applied to the channel HH, considering $M = 39$ baselines. The thresholds have been evaluated, setting $P_{FA} = P_{FD} = 10^{-4}$. The Rayleigh resolution in elevation is 14 m and the considered discretization step is 2 m. Since the buildings in the SAR image are not tall, we do not expect many double scatterers. There are 1328 detected single scatterers and only 2 double scatterers. The detected single and double scatterers are reported in red and blue, respectively, on the optical 3-D Google Earth image of the test area, in Figure 5a. The blue points are four, since for each double, the corresponding couple of points is reported at the estimated height in the 3-D Google Earth image. We compare these results with the case of using only 20 images but both channels HH and VV, and assuming again $P_{FA} = 10^{-4}$. In Figure 5b, the GLRT (5) has been applied considering $M = 20$ baselines, and the detected single scatterers are reported in red while the double scatterers are in blue, on the optical 3-D Google Earth image. There are 1733 singles, while the number of double scatterers is 13.

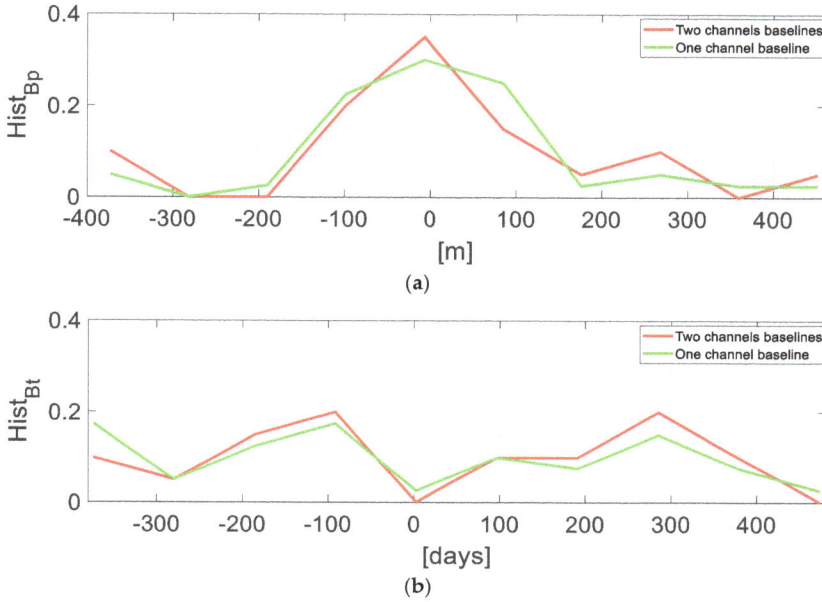

Figure 3. Normalized histogram of the perpendicular baseline distribution (**a**) and of the temporal baseline distribution (**b**); in green, the single polarization case and, in red, the dual polarization case.

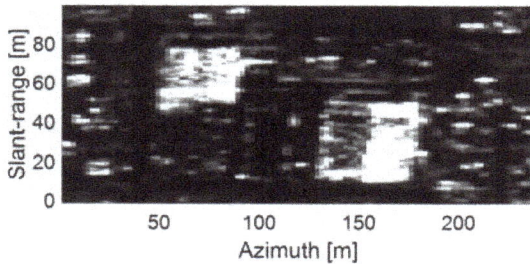

Figure 4. Intensity HH image of the test area near Toulouse, France (copyright DLR 2013–2015).

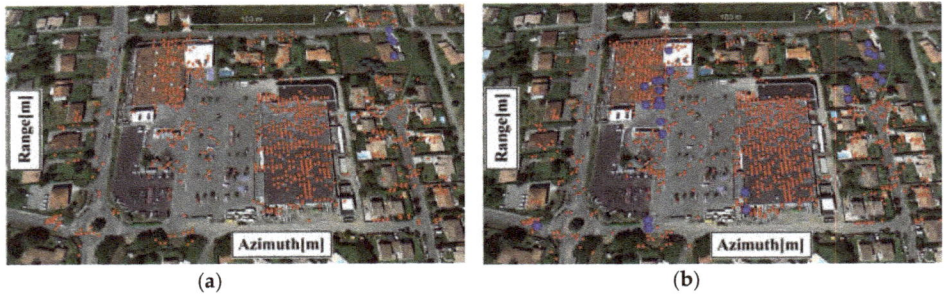

Figure 5. Targets detected by Fast-Sup-GLRT and reported on the optical image (single scatterers in red, double scatterers in blue), (**a**) single polarization, (**b**) dual polarization.

We observe a sensible increment of the total number of detected scatterers, which means that the diversity in polarization compensated for the diversity in baselines. In both cases, single and dual polarization, the GRLT approach is able to correctly locate the scatterers, which are mostly on the roof of the buildings; the double scatterers, where present, are coming from the interfering backscattering mechanism of ground and roof or façade and roof. In the single polarization case, very few double scatterers have been detected while, in the dual polarization case, we were able to identify, better, the double backscattering mechanism, even if we used fewer orthogonal baselines.

In order to find a justification as to why the single polarization approach detected less single and double scatterers, we computed the absolute values of the interferometric coherence and averaged it over all the baselines and over all the polarimetric channels. The images of these average coherence values are shown in Figure 6. We observe that, as expected, the average coherence assumes, in most of the pixels, higher values when using polarimetric data, so that a better detection performance can be achieved.

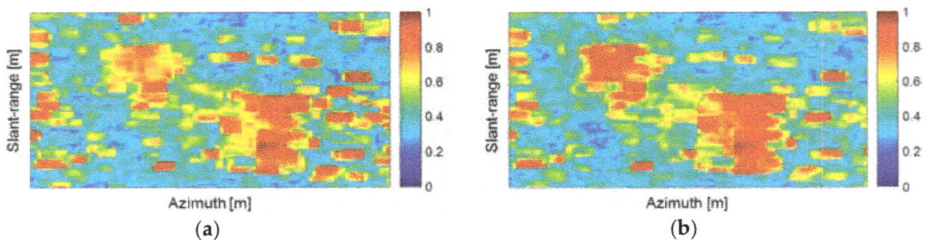

Figure 6. Average coherence (**a**) single polarization, (**b**) dual polarization.

We focus now on the discussion of the results obtained in the dual polarization case. We consider three range lines corresponding to three fixed azimuth coordinates (A, B, and C, in red in Figure 7a), crossing the left side building and localized in an area where several double scatterers (in blue) were detected.

In Figure 7b,c, the scene is reported on the optical image and described by a schematic geometrical diagram, respectively. The double scatterers are reported with the corresponding heights and for the three lines, respectively A (a), B (b), C (c), in Figure 8. In the diagram in Figure 7c, we reported the double scatterers identified in the range line C with the same color used in Figure 8c. The green markers indicate the roofs of the petrol pump (estimated height $\hat{z}$ is 4.25 m) and of the building ($\hat{z}$ = 12.13 m), while the ground has not been detected. The red markers indicate the ground ($\hat{z}$ = −0.99 m) and the roof of the building ($\hat{z}$ = 13.88), while the façade of the building has not been detected. Finally, the blue markers indicate the facade of the building ($\hat{z}$ = 6 m) and the roof ($\hat{z}$ = 13 m), while the ground has not been detected. We note that due to the presence of the petrol pump and the building layover effect, the true scatterers could be three in the three considered cells but, having fixed $K_{max} = 2$, we found the two dominant scatterers.

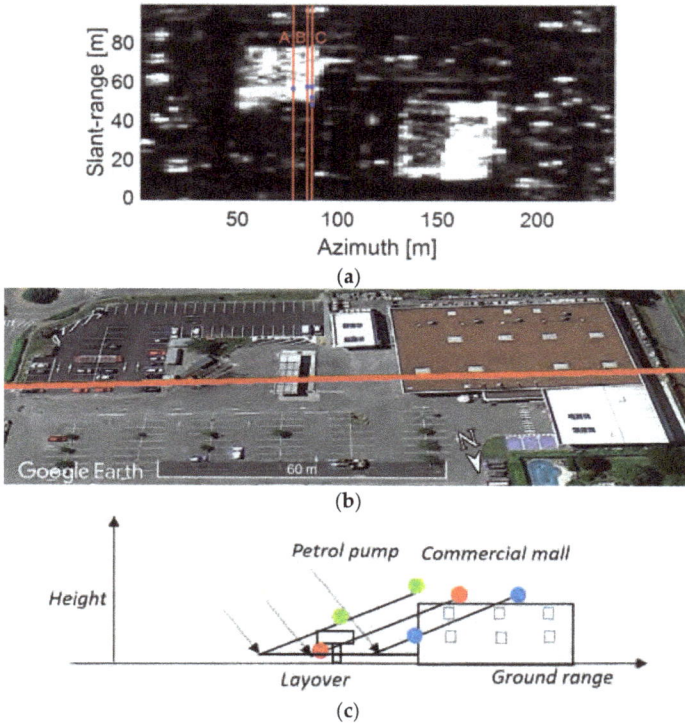

(a)

(b)

(c)

Figure 7. (**a**) Intensity HH image of the test area near Toulouse, with three range lines highlighted in red (A, B, C), (**b**) range line C reported on the optical image, (**c**) schematic geometrical diagram of the scene.

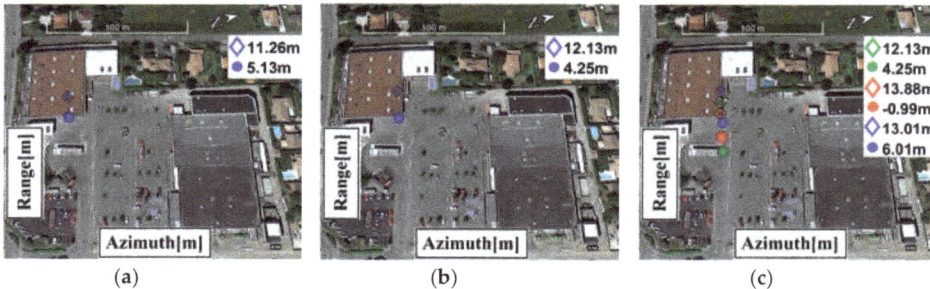

(a)   (b)   (c)

Figure 8. The double scatterers detected by Fast-Sup-GLRT in the dual polarization case, and their estimated heights, along the three range lines A (**a**), B (**b**), C (**c**).

We compare now the polarimetric Fast-Sup-GLRT with the polarimetric beamforming and Capon approaches, proposed in [22]. Firstly, we consider the reflectivity profiles obtained in the three pixels of range line C where double scatterers have been detected. In Figure 9, the beamforming and Capon reconstructed spectra are reported, respectively, in blue and red. For the Fast-Sup-GLRT, we report, in green, the two estimated samples of the reflectivity profiles assumed to be 2-sparse. We note that the positions of the scatterers identified by the Fast-Sup-GLRT (in green) are very close to the first two peaks of Capon reconstruction in all three cases. In general, the beamforming lobes are broader than the Capon lobes, as expected, and the accuracy in the localization of the scatterers is worse [22].

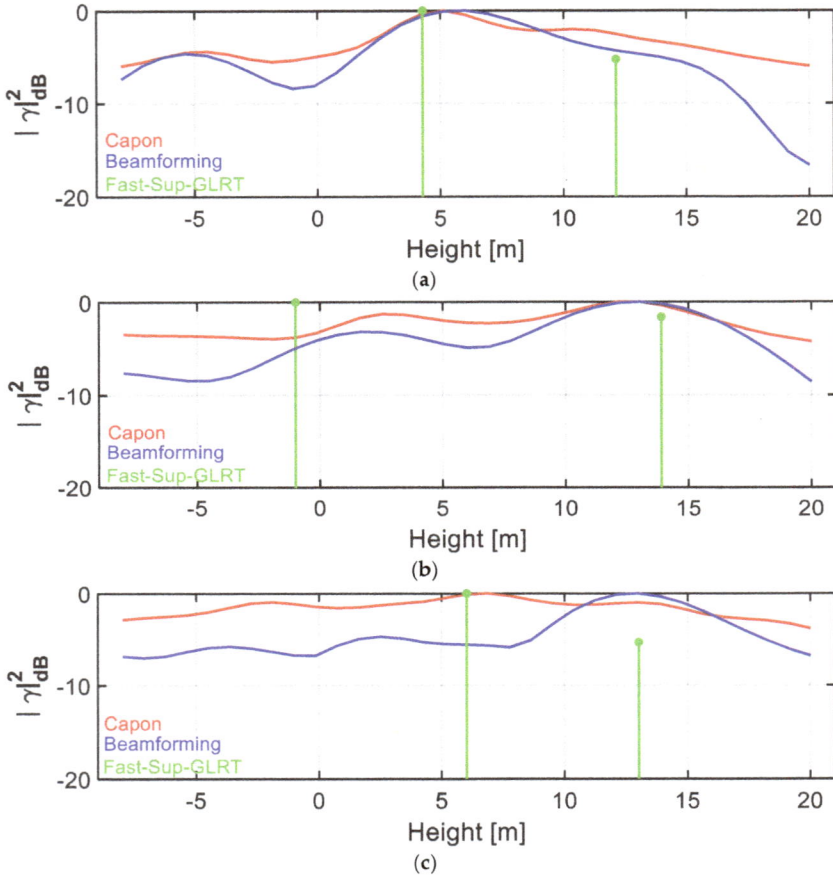

Figure 9. Reflectivity profile obtained with beamforming (blue), Capon (red), Fast-Sup-GLRT (green), in correspondence of the three double scatterers in line C, respectively in (a–c).

We compare also the tomographic slices obtained using polarimetric Fast-Sup-GLRT with the ones obtained applying the polarimetric beamforming and Capon approaches proposed in [22]. For the polarimetric beamforming and Capon approaches, in absence of a quality criterion, we report, in Figure 10, the values of the first two peaks (blue and red points) of the corresponding spectrum in all the points, without applying any threshold. For Fast-Sup-GLRT, we report the reflectivity values evaluated for the single and double scatterers when they have been detected. Double scatterers are shown with a blue circle. For each line, it is possible to see that, for Fast-Sup-GLRT (Figure 10c,f,i), the reconstruction is clear, since it is easy to identify the geometric profile of the building and the scattering contributions from ground and roof, and from façade and roof.

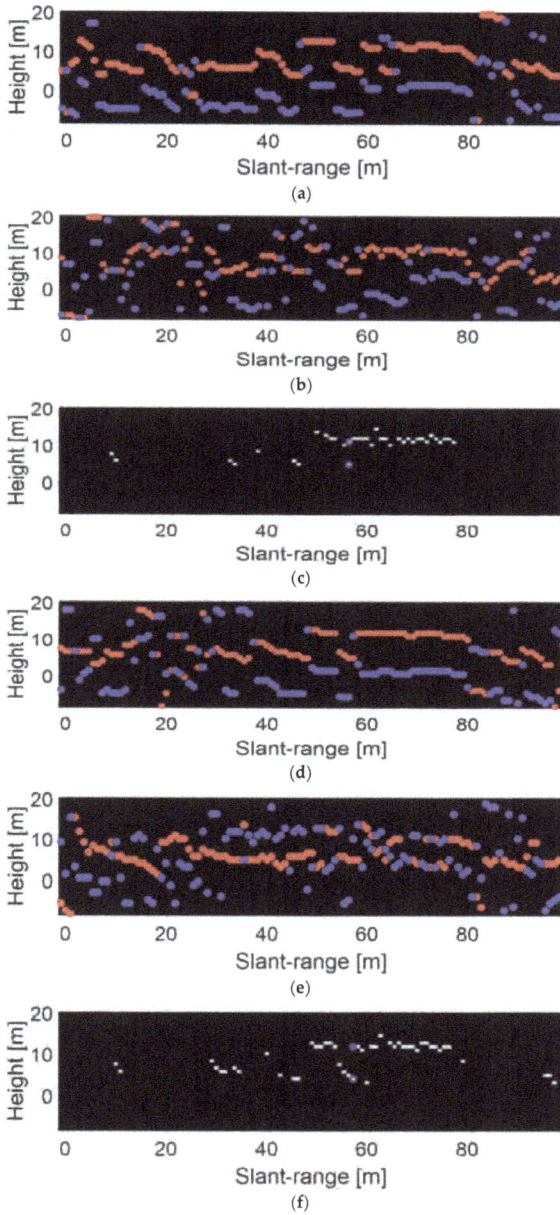

Figure 10. Tomographic slices obtained with polarimetric beamforming (**a,d,g**), Capon (**b,e,h**) and Fast-Sup-GLRT (**c,f,i**) respectively, from top to down, for line A, B, and C. Blue circles indicate the double scatterers for the Fast-Sup-GLRT, and red and blue points respectively indicate the first and second peaks of the beamforming and Capon spectra.

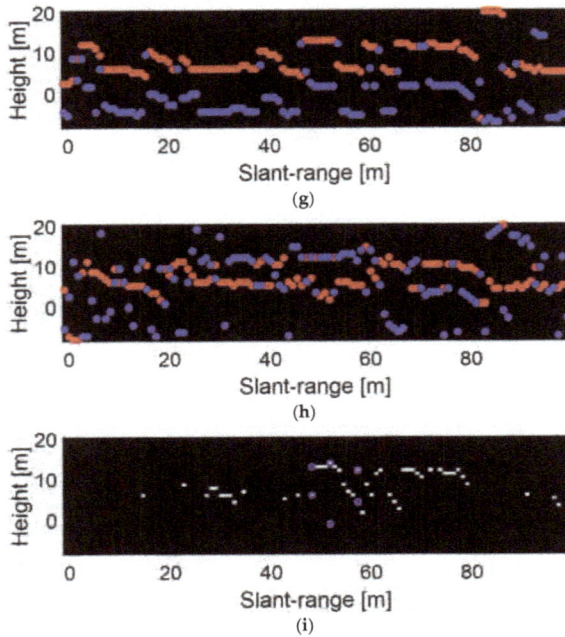

Figure 10. *Cont.*

On the contrary, the tomographic reconstructions obtained with the polarimetric beamforming (Figure 10a,d,g) and with the Capon approach (Figure 10b,e,h) are difficult to interpret and appear very noisy. Around 60 m in range, the imaging of the roof of the building is quite visible in all the reconstructions. The problem is that too many peaks are visible, and the corresponding heights cannot be associated with any structure. The presence of these peaks can be explained by the use of the boxcar filter for the covariance matrix estimation, and by the sidelobe effects, due to not uniformly spaced acquisitions. In particular, double backscattering cannot be compressed in one single range pixel, but stretches over several samples, due to spatial averaging, as also reported in [22].

5. Conclusions

In this paper, we have extended Fast-Sup-GLRT tomographic processing to the polarimetric case, and validated it on spotlight TSX real data on urban areas. In particular, we have shown that the dual polarization (HH + VV) case can outperform the single polarization case (HH), keeping the number of images constant and considering, in the polarimetric case, a lower number of baselines. The dual polarization approach gains with respect to the single polarization one, since the effect of reduced baseline diversity can be compensated by polarization diversity. For each baseline, we can count on two images acquired with two different polarizations at the same time, and then, in absence of temporal changes on the ground. The proposed approach has been compared with the polarimetric beamforming and Capon approaches, showing that, without exploiting a proper selection criterion of the spectra peaks, the tomographic slices are not clearly interpretable. Focusing on a single pixel where two dominant scatterers are identified by Fast-Sup-GLRT, we have found quite good correspondence between the localizations of the two scatterers by Fast-Sup-GLRT and Capon approaches. The polarimetric beamforming method is less accurate because the spectrum exhibits broader lobes.

Remote Sens. **2019**, *11*, 132

Author Contributions: A.B., G.S. developed the GLRT algorithm A.C.J. pre-processed real data and produced the geocoded results; all the authors contributed to performing the experiments; A.B., G.S. contributed to writing the paper.

Acknowledgments: We thank the European Space Agency (ESA), for providing software NEST. We thank for providing the TSX data the DLR (proposal MTH3182).

Conflicts of Interest: The authors declare no conflict of interest.

References

1. Caltagirone, F.; Capuzi, A.; Coletta, A.; de Luca, G.F.; Scorzafava, E.; Leonardi, R.; Rivola, S.; Fagioli, S.; Angino, G.; L'Abbate, M.; et al. The COSMOSkyMed dual use earth observation program: Development, qualification, and results of the commissioning of the overall constellation. *IEEE J. Sel. Top. Appl. Earth Obs. Remote Sens.* **2014**, *7*, 2754–2762. [CrossRef]
2. Porfilio, M.; Serva, S.; Fiorentino, C.; Calabrese, D. The acquisition modes of COSMO-Skymed di Seconda Generazione: A new combined approach based on SAR and platform agility. In Proceedings of the 2016 IEEE International Geoscience and Remote Sensing Symposium (IGARSS), Beijing, China, 10–15 July 2016; pp. 2082–2085.
3. Lee, J.-S.; Pottier, E. *Polarimetric Radar Imaging: From Basics to Applications*; CRC Press: Boca Raton, FL, USA, 2009.
4. Li, H.; Li, Q.; Wu, G.; Chen, J.; Liang, S. The impacts of building orientation on polarimetric orientation angle estimation and model-based decomposition for multilook polarimetric SAR data in Urban areas. *IEEE Trans. Geosci. Remote Sens.* **2016**, *54*, 5520–5532. [CrossRef]
5. Hong, S.H.; Wdowinski, S. Double-bounce component in cross-polarimetric SAR from a new scattering target decomposition. *IEEE Trans. Geosci. Remote Sens.* **2014**, *52*, 3039–3051. [CrossRef]
6. Gini, F.; Lombardini, F.; Montanari, M. Layover solution in multibaseline SAR interferometry. *IEEE Trans. Aerosp. Electron. Syst.* **2002**, *38*, 1344–1356. [CrossRef]
7. Reigber, A.; Moreira, A. First Demonstration of Airborne SAR Tomography Using Multibaseline L-band Data. *IEEE Trans. Geosci. Remote Sens.* **2000**, *38*, 2142–2152. [CrossRef]
8. Fornaro, G.; Lombardini, F.; Pauciullo, A.; Reale, D.; Viviani, F. Tomographic processing of interferometric SAR data: Developments, applications, and future research perspectives. *IEEE Signal Process. Mag.* **2014**, *31*, 41–50. [CrossRef]
9. Budillon, A.; Ferraioli, G.; Schirinzi, G. Localization Performance of Multiple Scatterers in Compressive Sampling SAR Tomography: Results on COSMO-SkyMed Data. *IEEE J. Sel. Top. Appl. Earth Obs. Remote Sens.* **2014**, *7*, 2902–2910. [CrossRef]
10. Zhu, X.X.; Bamler, R. Superresolving SAR Tomography for Multidimensional Imaging of Urban Areas: Compressive sensing-based TomoSAR inversion. *IEEE Signal Process. Mag.* **2014**, *31*, 51–58. [CrossRef]
11. Liang, L.; Li, X.; Ferro-Famil, L.; Guo, H.; Zhang, L.; Wu, W. Urban Area Tomography Using a Sparse Representation Based Two-Dimensional Spectral Analysis Technique. *Remote Sens.* **2018**, *10*, 109. [CrossRef]
12. Candes, E.J.; Romberg, J.; Tao, T. Stable signal recovery from incomplete and inaccurate measurements. *Commun. Pure Appl. Math.* **2006**, *59*, 1207–1223. [CrossRef]
13. Pauciullo, A.; Reale, D.; De Maio, A.; Fornaro, G. Detection of Double Scatterers in SAR Tomography. *IEEE Trans. Geosci. Remote Sens.* **2012**, *50*, 3567–3586. [CrossRef]
14. Budillon, A.; Schirinzi, G. GLRT Based on Support Estimation for Multiple Scatterers Detection in SAR Tomography. *IEEE J. Sel. Top. Appl. Earth Obs. Remote Sens.* **2016**, *9*, 1086–1094. [CrossRef]
15. Budillon, A.; Johnsy, A.C.; Schirinzi, G. A Fast Support Detector for Super-Resolution Localization of Multiple Scatterers in SAR Tomography. *IEEE J. Sel. Top. Appl. Earth Obs. Remote Sens.* **2017**, *10*, 2768–2779. [CrossRef]
16. Budillon, A.; Johnsy, A.; Schirinzi, G. Extension of a fast GLRT algorithm to 5D SAR tomography of Urban areas. *Remote Sens.* **2017**, *9*, 844. [CrossRef]
17. Siddique, M.A.; Wegmüller, U.; Hajnsek, I.; Frey, O. SAR Tomography as an Add-On to PSI: Detection of Coherent Scatterers in the Presence of Phase Instabilities. *Remote Sens.* **2018**, *10*, 1014. [CrossRef]
18. Tebaldini, S. Single and multipolarimetric SAR tomography of forested areas: A parametric approach. *IEEE Trans. Geosci. Remote Sens.* **2010**, *48*, 2375–2387. [CrossRef]
19. Frey, O.; Meier, E. Analyzing tomographic SAR data of a forest with respect to frequency, polarization, and focusing technique. *IEEE Trans. Geosci. Remote Sens.* **2011**, *49*, 3648–3659. [CrossRef]
20. Tebaldini, S.; Rocca, F. Multibaseline polarimetric SAR tomography of a boreal forest at P- and L-bands. *IEEE Trans. Geosci. Remote Sens.* **2012**, *50*, 232–246. [CrossRef]

21. Pardini, M.; Papathanassiou, K. On the estimation of ground and volume polarimetric covariances in forest scenarios with SAR tomography. *IEEE Geosci. Remote Sens. Lett.* **2017**, *14*, 1860–1864. [CrossRef]
22. Sauer, S.; Ferro-Famil, L.; Reigber, A.; Pottier, E. Three Dimensional Imaging and Scattering Mechanism Estimation over Urban Scenes Using Dual-Baseline Polarimetric InSAR Observations at L-Band. *IEEE Trans. Geosci. Remote Sens.* **2011**, *49*, 4616–4629. [CrossRef]
23. Aguilera, E.; Nannini, M.; Reigber, A. Multisignal Compressed Sensing for Polarimetric SAR Tomography. *Geosci. Remote Sens. Lett.* **2012**, *9*, 871–875. [CrossRef]
24. Budillon, A.; Schirinzi, G. Performance Evaluation of a GLRT Moving Target Detector for TerraSAR-X along-track interferometric data. *IEEE Trans. Geosci. Remote Sens.* **2015**, *53*, 3350–3360. [CrossRef]

MDPI
St. Alban-Anlage 66
4052 Basel
Switzerland
Tel. +41 61 683 77 34
Fax +41 61 302 89 18
www.mdpi.com

Remote Sensing Editorial Office
E-mail: remotesensing@mdpi.com
www.mdpi.com/journal/remotesensing